# PROJECT
## MANAGEMENT
## BEST PRACTICES

*Achieving Global Excellence*

THIRD EDITION

HAROLD KERZNER, PH.D.

WILEY

INTERNATIONAL
Institute for Learning, Inc.

Cover image: Glass Cubes © iStockphoto/Henvry
Cover design: Anne-Michele Abbott

This book is printed on acid-free paper.

Published by John Wiley & Sons, Inc., Hoboken, New Jersey
Published simultaneously in Canada

For general information about our other products and services, please contact our Customer Care Department
within the United States at (800) 762-2974, outside the United States at (317) 572-3993 or fax (317) 572-4002.

Wiley publishes in a variety of print and electronic formats and by print-on-demand. Some material included
with standard print versions of this book may not be included in e-books or in print-on-demand. If this book
refers to media such as a CD or DVD that is not included in the version you purchased, you may download this
material at http://booksupport.wiley.com. For more information about Wiley products, visit www.wiley.com.

*Library of Congress Cataloging-in-Publication Data:*

Kerzner, Harold.
    Project management best practices : achieving global excellence / Harold Kerzner. - Third edition
        pages cm
    Includes bibliographical references and index.
        ISBN 978-1-118-65701-0 (cloth); ISBN 978-1-118-83553-1 (ebk); ISBN 978-1-118-83560-9 (ebk);
    ISBN 978-1-118-83567-8 (ebk)
        1. Project management–Methodology.   2. Project management–Case studies.   I. Title.
    HD69.P75K4715 2013
    658.4'04—dc23

                                                                                    2013034142

Printed in the United States of America
10 9 8 7 6 5 4 3 2

*To*
*my wife, Jo Ellyn,*
*who showed me that excellence*
*can be achieved in*
*marriage, family, and life*
*as well as at work*

# Contents

# Preface _____

For almost 50 years, project management was viewed as a process that might be nice to have but not one that was necessary for the survival of the firm. Companies reluctantly invested in some training courses simply to provide their personnel with basic knowledge of planning and scheduling. Project management was viewed as a threat to established lines of authority, and in many companies only partial project management was used. This half-hearted implementation occurred simply to placate lower- and middle-level personnel as well as selected customers.

During this 50-year period, we did everything possible to prevent excellence in project management from occurring. We provided only lip service to empowerment, teamwork, and trust. We hoarded information because the control of information was viewed as power. We placed personal and functional interests ahead of the best interest of the company in the hierarchy of priorities, and we maintained the faulty belief that time was a luxury rather than a constraint.

By the mid-1990s, this mentality began to subside, largely due to two recessions. Companies were under severe competitive pressure to create high-quality products in a shorter period of time. The importance of developing a long-term trusting relationship with the customers had come to the forefront. Businesses were being forced by the stakeholders to change for the better. The survival of the firm was now at stake.

Today, businesses have changed for the better. Trust between the customer and contractor is at an all-time high. New products are being developed at a faster rate than ever before. Project management has become a competitive weapon during competitive bidding. Some companies are receiving sole-source contracts because of the faith that the customer has in the contractor's ability to deliver a continuous stream of successful projects using a project management methodology. All of these factors have allowed a multitude of companies to achieve some degree of excellence in project management. Business decisions are now being emphasized ahead of personal decisions.

Words that were commonplace ten years ago have taken on new meanings today. Change is no longer being viewed as being entirely bad. Today, change implies continuous improvement. Conflicts are no longer seen as detrimental. Conflicts managed well can be beneficial. Project management is no longer viewed as a system entirely internal to the organization. It is now a competitive weapon that brings higher levels of quality and increased value-added opportunities for the customer.

Companies that were considered excellent in management in the past may no longer be regarded as excellent today, especially with regard to project management. Consider the book entitled In *Search of Excellence*, written by Tom Peters and Robert Waterman in 1982 (published by Harper & Row, New York). How many of those companies identified in their book are still considered excellent today? How many of those companies have won the prestigious Malcolm Baldrige Award? How many of those companies that have won the award are excellent in project management today? Excellence in project management is a never-ending journey. Companies that are reluctant to invest in continuous improvements in project management soon find themselves with low customer satisfaction ratings.

The differentiation between the first fifty years of project management and the last ten years is in the implementation of project management on a company-wide basis. For more than three decades, we emphasized the quantitative and behavioral tools of project management. Basic knowledge and primary skills were emphasized, and education on project management was provided only to a relatively small group of people. However, within the past ten years, emphasis has been on implementation across the entire company. What was now strategically important was how to put thirty years of basic project management theory in the hands of a few into corporate-wide practice. Today it is the implementation of companywide project management applications that constitutes advanced project management. Subjects such as earned-value analysis, situational leadership, and cost and change control are part of basic project management courses today, whereas fifteen years ago they were considered advanced topics in project management. So, what constitutes applied project management today? Topics related to project management implementation, enterprise project management methodologies, project management offices, and working with stakeholders are advanced project management concepts.

This book covers the advanced project management topics necessary for implementation of and excellence in project management. The book contains numerous quotes from people in the field who have benchmarked best practices in project management and are currently implementing these processes within their own firms. Quotes in this book were provided by several CEOs, presidents, COOs, CIOs, CFOs, senior VPs, VPs, global VPs, general managers, PMO directors, and others. The quotes are invaluable because they show the thought process of these leaders and the direction in which their firms are heading. These companies have obtained some degree of excellence in project management, and what is truly remarkable is the fact that this happened in less than five or six years. Best practices in implementation will be the future of project management well into the twenty-first century. Companies have created best practices libraries for project management. Many of the libraries are used during competitive bidding for differentiation from other competitors. Best practices in project management are now viewed as intellectual property.

Excellence in project management is not achieved simply by developing a project management methodology. Instead, it is how the methodology is used again and again that creates excellence and a stream of successfully managed projects.

Project management practices and methodologies are built around the culture of companies and by determining what it takes to get people to work together, solve problems, and make decisions. Because each company most likely has its own unique culture, it is understandable that each company can have a different number of lifecycle phases, different decision points, and different success criteria. No single approach fits all companies, which is why this book discusses a variety of companies, in different industries, of different sizes, and on different continents. Hopefully, after reading this book, you will come up with ideas as to how your project management activities can improve.

Companies that are discussed in this book include:

| | |
|---|---|
| 3M | Indra |
| ABB | Johnson Controls |
| Alcatel-Lucent | Key Plastics |
| ALSTOM | Kodak |
| American Greetings | KONE |
| AT&T | maxIT-VCS |
| Aviva | MCI |
| Babcock & Wilcox | Medical Mutual |
| Bendix | Microsoft |
| Boeing | Minnesota Power & Light |
| Cassidian | Motorola |
| Chrysler | NASA |
| Chubb | Neal and Massy Holdings, Ltd. |
| Churchill Downs | Nortel |
| Comau | NXP |
| Computer Associates | Ohio Bell |
| Cooper Standard | Orange Switzerland |
| CSC | Our Lady of Lourdes Regional Medical Ctr. |
| Dell | Philips |
| Deloitte | Repsol |
| Department of Defense | Roadway Express |
| DFCU Financial | Rockwell Automation |
| Dow Chemical | SAP |
| DTE Energy | Sherwin Williams |
| EDS | Siemens |
| Eli Lilly | SigmaPM |
| Enakta | Slalom |
| Ericsson | Star Alliance |
| Fluor | Tech Mahindra Limited |
| Ford | Tecnicas Reunidas |
| General Motors | Teradyne |
| Goodyear | Thiokol |
| Harris | Tokio Marine |
| Hewlett-Packard | Visteon |
| Hitachi | Wärtsilä |
| Holcim | Westfield Group |
| IBM | World Wildlife Fund |
| ILLUMINAT | Zurich North America |

Seminars and webinar courses on project management principles and best practices in project management are available using this text and my text *Project Management: A Systems Approach to Planning, Scheduling, and Controlling*, 11th edition (Wiley, Hoboken, New Jersey, 2013). Seminars on advanced project management are also available using this text. Information on these courses, e-learning courses, and in-house and public seminars can be obtained by contacting:

Lori Milhaven, Executive Vice President, IIL:

Phone: 800-325-1533 or 212-515-5121
Fax: 212-755-0777
E-mail: lori.milhaven@iil.com

Harold Kerzner
International Institute for Learning, Inc.
2014

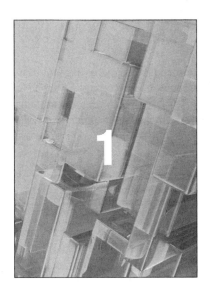

# 1 Understanding Best Practices

## 1.0 INTRODUCTION

Project management has evolved from a set of processes that were once considered "nice" to have to a structured methodology that is considered mandatory for the survival of the firm. Companies are now realizing that their entire business, including most of the routine activities, can be regarded as a series of projects. Simply stated, we are managing our business by projects.

Project management is now regarded as both a project management process and a business process. Therefore, project managers are expected to make business decisions as well as project decisions. The necessity for achieving project management excellence is now readily apparent to almost all businesses.

As the relative importance of project management permeates each facet of the business, knowledge is captured on best practices in project management. Some companies view this knowledge as intellectual property to be closely guarded in the vaults of the company. Others share this knowledge in hope of discovering other best practices. Companies are now performing strategic planning for project management because of the benefits and its contribution to sustainable business value.

One of the benefits of performing strategic planning for project management is that it usually identifies the need for capturing and retaining best practices. Unfortunately, this is easier said than done. One of the reasons for this difficulty, as will be seen later in the chapter, is that companies today are not in agreement on the definition of a best practice, nor do they understand that best practices lead to continuous improvement, which in turn leads to the capturing of more best practices. Many companies also do not recognize the value and benefits that can come from best practices.

Today, project managers are capturing best practices in both project management activities and business activities. The reason is simple: The best practices are intellectual property that encourages companies to perform at higher levels. Best practices lead to added business value, greater benefit realization, and better benefits management activities. Project management and business thinking are no longer separate activities.

## 1.1 WÄRTSILÄ[1]

**Benefits Management in Operational Development Projects in Wärtsilä**

Wärtsilä has a strong tradition in project-based businesses and project management practices. As such, a corporate-wide project management office was established in 2007 to further strengthen the focus on project management competence within the group and to develop a project management culture, processes, competences and tools.

Today the project management structures and ways of working have become a fundamental part of Wärtsilä's business thinking. The business process model has gradually shifted from being a somewhat disordered process to a harmonized model enabling the implementation of unified guidelines, targets and terminology. The company has approached this implementation of project management practices from two different but equally important aspects. Firstly, a project management tool providing, inter alia, more effective resource and schedule planning has been introduced and implemented. Secondly, the organization has been encouraged to participate actively in professional project management training and certification paths.

As the project management processes have become well defined and gained maturity, the emphasis has gradually shifted towards benefits management in operational development projects. The initiative to improve benefits management processes stems from the mission of the Wärtsilä PMO for Operational Development, which is to ensure synergies between Wärtsilä's business units that would help to enable businesses to transform their strategic ambition into daily operations. This would be achieved by providing management and expertise in terms of change management, business processes and application development.

In traditional project management, projects are often measured in terms of budget, schedule, scope or quality. Benefits management as a concept, however, focuses more on the actual value that the projects are able to deliver to the end customer. In other words, the project success is not solely measured in terms of time or money. Quite the opposite, measuring the success of a project comes from the end user: Did this solution fulfill the user's needs? As the concept of value is rather vague, it is of the utmost importance that the benefits have concrete metrics and measurements. This concerns also so-called "soft", intangible benefits. Although they could not be quantified financially, they have to be measured. Another important aspect in benefits planning is to create a valid baseline to compare the results with: instead of comparing only to a BAU (business as usual) situation, the results gained from the benefit realization measurements should be compared to other alternative scenarios ("Could this have been achieved some other way?").

---

1. Material has been provided by the Wärtsilä Project Management Office (WPMO). Copyright to Wärtsilä Corporation, ©2013. Reproduced by permission.

In operational development projects the output of the project can be, for example, an IT tool made to improve resource planning. The most crucial part of the project, however, is to make the *output* become a project *outcome*. This means that the project output (in this case an IT tool) should become a part of the end user's way of working. In order to make this happen the benefit planning must take into consideration two important aspects:

1. What does the end user want and need?
2. What has to change in order to make this happen?

With proper end user expectation management and change management the risk of the project output becoming "just another tool in the toolbox" can be avoided.

The benefits management system in a nutshell should consist of the following elements:

- **Identifying the driver for the project**: Do we really need this investment? Who else is going to benefit from it?
- **Identifying the key benefits**: What are the benefits and when will they occur? What is their proximity (How likely are they to happen?).
- **Estimating the benefits**: defining a clear baseline for the measurements allows us to define clear metrics (which apply to the entire portfolio of projects) and provides us with consistency throughout all life cycle phases, from project initiation to benefit realization. The critical question we must ask is, Do these metrics tolerate changes in the business environment?
- **Linking the benefits with change**: How does the organization have to change in order to enable the benefit realization? How can we enable this change? Plan the deployment and adjust it to (business) environmental changes (organizational changes, market situation changes etc.)
- **Who is accountable for the benefit?** Define a person/organization responsible for the benefit realization
- **Monitoring benefits:** monitor your performance with the established metrics, improve it if needed towards the defined goal and acknowledge risks in a proactive way
- **Doing a post-project evaluation:** ensure a successful deployment by communicating about the project output and honestly promoting it. Imagine yourself in the end user's position: Would I like to use this tool?
- **Learning from your mistakes:** ensure that project success points and failures are equally handled. Focus on honest communication and learning, not blaming. Examples should come all the way from the executive level.

## 1.2 PROJECT MANAGEMENT BEST PRACTICES: 1945–1960

During the 1940s, line managers functioned as project managers and used the concept of over-the-fence management to manage projects. Each line manager, wearing the hat of a project manager, would perform the work necessitated by his or her line organization and, when that was completed, would throw the "ball" over the fence in hope that some-one would catch it. Once the ball was thrown over the fence, the line managers would wash their hands of any responsibility for the project because the ball was no longer in their yard. If a project failed, blame was placed on whichever line manager had the ball at that time.

The problem with over-the-fence management was that the customer had no single contact point for questions. The filtering of information wasted precious time for both the customer and the contractor. Customers who wanted first-hand information had to seek out the manager in possession of the ball. For small projects, this was easy. However, as projects grew in size and complexity, this became more difficult.

During this time, very few best practices were identified. If there were best practices, then they would stay within a given functional area never to be shared with the remainder of the company. Suboptimal project management decision making was the norm.

Following World War II, the United States entered into the Cold War. To win a Cold War, one must compete in the arms race and rapidly build weapons of mass destruction. The victor in a Cold War is the one who can retaliate with such force as to obliterate the enemy. Development of weapons of mass destruction comprised very large projects involving potentially thousands of contractors.

The arms race made it clear that the traditional use of over-the-fence management would not be acceptable to the Department of Defense (DoD) for projects such as the B52 bomber, the Minuteman Intercontinental Ballistic Missile, and the Polaris submarine. The government wanted a single point of contact, namely, a project manager who had total accountability through all project phases. In addition, the government wanted the project manager to possess a command of technology rather than just an understanding of technology, which mandated that the project manager be an engineer preferably with an advanced degree in some branch of technology. The use of project management was then mandated for some of the smaller weapon systems such as jet fighters and tanks. The National Aeronautics and Space Administration (NASA) mandated the use of project management for all activities related to the space program.

Projects in the aerospace and defense industries were having cost overruns in excess of 200–300 percent. Blame was erroneously placed upon improper implementation of project management when, in fact, the real problem was the inability to forecast technology, resulting in numerous scope changes occurring. Forecasting technology is extremely difficult for projects that could last 10–20 years.

By the late 1950s and early 1960s, the aerospace and defense industries were using project management on virtually all projects, and they were pressuring their suppliers to use it as well. Project management was growing, but at a relatively slow rate except for aerospace and defense.

Because of the vast number of contractors and subcontractors, the government needed standardization, especially in the planning process and the reporting of information. The government established a life-cycle planning and control model and a cost-monitoring system and created a group of project management auditors to make sure that the government's money was being spent as planned. These practices were to be used on all government programs above a certain dollar value. Private industry viewed these practices as an overmanagement cost and saw no practical value in project management.

In the early years of project management, because many firms saw no practical value in project management, there were misconceptions concerning project management. Some of the misconceptions included:

- Project management is a scheduling tool such as PERT/CPM (program evaluation and review technique/critical-path method) scheduling.
- Project management applies to large projects only.
- Project management is designed for government projects only.
- Project managers must be engineers and preferably with advanced degrees.
- Project managers need a "command of technology" to be successful.
- Project success is measured in technical terms only. (Did it work?)

## 1.3 PROJECT MANAGEMENT BEST PRACTICES: 1960–1985

During this time period, with a better understanding of project management, the growth of project management had come about more through necessity than through desire, but at a very slow rate. Its slow growth can be attributed mainly to lack of acceptance of the new management techniques necessary for its successful implementation. An inherent fear of the unknown acted as a deterrent for both managers and executives.

Other than aerospace, defense, and construction, the majority of companies in the 1960s maintained an informal method for managing projects. In informal project management, just as the words imply, the projects were handled on an informal basis whereby the authority of the project manager was minimized. Most projects were handled by functional managers and stayed in one or two functional lines, and formal communications were either unnecessary or handled informally because of the good working relationships between line managers. Those individuals that were assigned as project managers soon found that they were functioning more as project leaders or project monitors than as real project managers. Many organizations today, such as low-technology manufacturing, have line managers who have been working side by side for ten or more years. In such situations, informal project management may be effective on capital equipment or facility development projects and project management is not regarded as a profession.

By 1970 and through the early 1980s, more companies departed from informal project management and restructured to formalize the project management process, mainly because the size and complexity of their activities had grown to a point where they were unmanageable within the current structure.

Not all industries need project management, and executives must determine whether there is an actual need before making a commitment. Several industries with simple tasks, whether in a static or a dynamic environment, do not need project management. Manufacturing industries with slowly changing technology do not need project management, unless of course they have a requirement for several special projects, such as capital equipment activities, that could interrupt the normal flow of work in the routine manufacturing operations. The slow growth rate and acceptance of project management were related to the fact that the limitations of project management were readily apparent yet the advantages were not completely recognizable. Project management requires organizational restructuring. The question, of course, is "How much restructuring?" Executives have avoided the subject of project management for fear that "revolutionary" changes must be made in the organization.

Project management restructuring has permitted companies to:

● Accomplish tasks that could not be effectively handled by the traditional structure
● Accomplish one-time activities with minimum disruption of routine business

The second item implies that project management is a "temporary" management structure and, therefore, causes minimum organizational disruption. The major problems identified by those managers who endeavored to adapt to the new system all revolved around conflicts in authority and resources.

Another major concern was that project management required upper-level managers to relinquish some of their authority through delegation to middle managers. In several situations, middle managers soon occupied the power positions, even more so than upper-level managers.

Project management became a necessity for many companies as they expanded into multiple product lines, many of which were dissimilar, and organizational complexities grew. This growth can be attributed to:

● Technology increasing at an astounding rate
● More money being invested in research and development (R&D)
● More information being available
● Shortening of project life cycles

To satisfy the requirements imposed by these four factors, management was "forced" into organizational restructuring; the traditional organizational form that had survived for decades was inadequate for integrating activities across functional "empires."

By 1970, the environment began to change rapidly. Companies in aerospace, defense, and construction pioneered the implementation of project management, and other industries soon followed, some with great reluctance. NASA and the DoD "forced" subcontractors into accepting project management.

Because current organizational structures are unable to accommodate the wide variety of interrelated tasks necessary for successful project completion, the need for project management has become apparent. It is usually first identified by those lower-level and middle managers who find it impossible to control their resources effectively for the diverse activities within their line organization. Quite often middle managers feel the impact of changing environment more than upper-level executives.

Once the need for change is identified, middle management must convince upper-level management that such a change is actually warranted. If top-level executives cannot recognize the problems with resource control, then project management will not be adopted, at least formally. Informal acceptance, however, is another story.

As project management developed, some essential factors in its successful implementation were recognized. The major factor was the role of the project manager, which became the focal point for integrative responsibility. The need for integrative responsibility was first identified in complex R&D projects.

The R&D technology has broken down the boundaries that used to exist between industries. Once-stable markets and distribution channels are now in a state of flux. The industrial environment is turbulent and increasingly hard to predict. Many complex facts about markets, production methods, costs, and scientific potentials are related to investment decisions in R&D.

All of these factors have combined to produce a king-sized managerial headache. There are just too many crucial decisions to have them all processed and resolved at the top of the organization through regular line hierarchy. They must be integrated in some other way. Providing the project manager with integrative responsibility resulted in:

- Total project accountability being assumed by a single person
- Project rather than functional dedication
- A requirement for coordination across functional interfaces
- Proper utilization of integrated planning and control

Without project management, these four elements have to be accomplished by executives, and it is questionable whether these activities should be part of an executive's job description. An executive in a Fortune 500 corporation stated that he was spending 70 hours each week working as both an executive and a project manager, and he did not feel that he was performing either job to the best of his abilities. During a presentation to the staff, the executive stated what he expected of the organization after project management implementation:

- Push decision making down in the organization.
- Eliminate the need for committee solutions.
- Trust the decisions of peers.

Those executives who chose to accept project management soon found the advantages of the new technique:

- Easy adaptation to an ever-changing environment
- Ability to handle a multidisciplinary activity within a specified period of time
- Horizontal as well as vertical work flow
- Better orientation toward customer problems
- Easier identification of activity responsibilities
- A multidisciplinary decision-making process
- Innovation in organizational design

As project management evolved, best practices became important. Best practices were learned from both successes and failures. In the early years of project management, private industry focused on learning best practices from successes. The government, however, focused on learning about best practices from failures. When the government finally focused on learning from successes, the knowledge of best practices came from their relationships with both their prime contractors and the subcontractors. Some of these best practices that came out of the government included:

- Use of life-cycle phases
- Standardization and consistency
- Use of templates [e.g., for statement of work (SOW), work breakdown structure (WBS), and risk management]
- Providing military personnel in project management positions with extended tours of duty at the same location
- Use of integrated project teams (IPTs)
- Control of contractor-generated scope changes
- Use of earned value measurement

## 1.4 PROJECT MANAGEMENT BEST PRACTICES: 1985–2014

By the 1990s, companies had begun to realize that implementing project management was a necessity, not a choice. By 2014, project management had spread to virtually every industry and best practices were being captured. In the author's opinion, the appearance of best practices from an industry perspective might be:

- 1960–1985: Aerospace, defense, and construction
- 1986–1993: Automotive suppliers
- 1994–1999: Telecommunications
- 2000–2003: Information technology
- 2004–2006: Health care
- 2007–2008: Marketing and sales
- 2009–Present: Government agencies

The question now was not how to implement project management, but how fast could it be done? How quickly can we become mature in project management? Can we use the best practices to accelerate the implementation of project management?

Table 1–1 shows the typical life-cycle phases that an organization goes through to implement project management. In the first phase, the embryonic phase, the organization recognizes the apparent need for project management. This recognition normally takes place at the lower and middle levels of management, where the project activities actually take place. The executives are then informed of the need and assess the situation.

There are six driving forces that lead executives to recognize the need for project management:

- Capital projects
- Customer expectations
- Competitiveness
- Executive understanding
- New project development
- Efficiency and effectiveness

**TABLE 1–1.  FIVE PHASES OF THE PROJECT MANAGEMENT LIFE CYCLE**

| Embryonic | Executive Management Acceptance | Line Management Acceptance | Growth | Maturity |
|---|---|---|---|---|
| Recognize need | Get visible executive support | Get line management support | Recognize use of life-cycle phases | Develop a management cost/schedule control system |
| Recognize benefits | Achieve executive understanding of project management | Achieve line management commitment | Develop a project management methodology | Integrate cost and schedule control |
| Recognize applications | Establish project sponsorship at executive levels | Provide line management education | Make the commitment to planning | Develop an educational program to enhance project management skills |
| Recognize what must be done | Become willing to change way of doing business | Become willing to release employees for project management training | Minimize creeping scope Select a project tracking system | |

Manufacturing companies are driven to project management because of large capital projects or a multitude of simultaneous projects. Executives soon realize the impact on cash flow and that slippages in the schedule could end up idling workers.

Companies that sell products or services, including installation, to their clients must have good project management practices. These companies are usually non–project-driven but function as though they were project-driven. These companies now sell solutions to their customers rather than products. It is almost impossible to sell complete solutions to customers without having superior project management practices because what you are actually selling is your project management expertise.

There are two situations where competitiveness becomes the driving force: internal projects and external (outside customer) projects. Internally, companies get into trouble when they realize that much of the work can be outsourced for less than it would cost to perform the work themselves. Externally, companies get into trouble when they are no longer competitive on price or quality or simply cannot increase their market share.

Executive understanding is the driving force in those organizations that have a rigid traditional structure that performs routine, repetitive activities. These organizations are quite resistant to change unless driven by the executives. This driving force can exist in conjunction with any of the other driving forces.

New product development is the driving force for those organizations that are heavily invested in R&D activities. Given that only a small percentage of R&D projects ever make it into commercialization, where the R&D costs can be recovered, project management becomes a necessity. Project management can also be used as an early-warning system that a project should be canceled.

Efficiency and effectiveness, as driving forces, can exist in conjunction with any other driving forces. Efficiency and effectiveness take on paramount importance for small companies experiencing growing pains. Project management can be used to help such companies remain competitive during periods of growth and to assist in determining capacity constraints.

Because of the interrelatedness of these driving forces, some people contend that the only true driving force is survival. This is illustrated in Figure 1–1. When the company

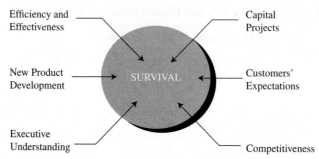

**Figure 1–1.** The components of survival. Source: Reprinted from H. Kerzner, In Search of Excellence in Project Management, Hoboken, NJ: Wiley, 1998, p. 51.

recognizes that survival of the firm is at stake, the implementation of project management becomes easier.

Enrique Sevilla Molina, PMP, formerly corporate PMO director, discusses the driving forces at Indra that necessitated the need for excellence in project management:

> The internal forces were based on our own history and business experience. We soon found out that the better the project managers, the better the project results. This realization came together with the need to demonstrate in national and international contracts, with both US and European customers, our real capabilities to handle big projects. These big projects required world class project management, and for us managing the project was a greater challenge than just being able to technically execute the project. Summarizing, these big projects set the pace to define precise procedures on how handling stakeholders, big subcontractors and becoming a reliable main point of contact for all issues related with the project.

Sandra Kumorowski discusses the driving forces at Enakta:[2]

> The company was a project-based company and it made sense to turn to project management as a tool for continuous improvement. The main issues that also drove the company to use project management were reoccurring time/cost/quality management issues, team productivity issues, and client satisfaction issues. Table 1–2 illustrates the necessity:

The speed by which companies reach some degree of maturity in project management is most often based upon how important they perceive the driving forces to be. This is illustrated generically in Figure 1–2. Non–project-driven and hybrid organizations move quickly to maturity if increased internal efficiencies and effectiveness are needed. Competitiveness is the slowest path because these types of organizations do not recognize that project management affects their competitive position directly. For project-driven organizations, the path is reversed. Competitiveness is the name of the game, and the vehicle used is project management.

---

2. Sandra Kumorowski is an assistant professor of marketing communications at Columbia College, Chicago; marketing & communications chair, Christopher & Dana Reeve Foundation, Chicago; she is also chief business advisor and owner, Enakta LLC; P: 1-224-715-5666; E: sandrakum@sbcglobal.net, skumorowski@colum.edu

**TABLE 1–2.  THE NECESSITY FOR PROJECT MANAGEMENT**

| Why? | Benefits | | | |
|---|---|---|---|---|
| **We are a project-based company.** | 1. We know how to deliver projects. | 2. PM is a <u>tool for successful delivery of actionable insights.</u> | | |
| **To build credibility, grow,** *and* ***compete:*** We should be perceived as a systematic and organized organization. We have to prevent mistakes. | 1. Earned reputation for systematic work | 2. Focused business strategy | 3. PM could be one of our competitive advantages | |
| **To control cost, time, resources:** We should establish effective project control system | 1. Decreased uncertainty | 2. Increased product quality | 3. Happy people | 4. More effective planning (project and company level) |
| **To learn as an organization and** *individuals* | 1. PM concepts as part of our <u>continuous education</u> | 2. Learning organization is always ahead of time and its competitors | | |

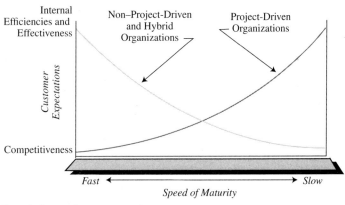

**Figure 1–2.**  Speed of maturity.

Once the organization perceives the need for project management, it enters the second life-cycle phase of Table 1–1, executive acceptance. Project management cannot be implemented rapidly in the near term without executive support. Furthermore, the support must be visible to all.

The third life-cycle phase is line management acceptance. It is highly unlikely that any line manager would actively support the implementation of project management without first recognizing the same support coming from above. Even minimal line management support will still cause project management to struggle.

The fourth life-cycle phase is the growth phase, where the organization becomes committed to the development of the corporate tools for project management. This includes the processes and project management methodology for planning, scheduling, and controlling as well as selection of the appropriate supporting software. Portions of this phase can begin during earlier phases.

The fifth life-cycle phase is maturity. In this phase, the organization begins using the tools developed in the previous phase. Here, the organization must be totally dedicated

**TABLE 1–3.   CRITICAL FACTORS IN THE PROJECT MANAGEMENT LIFE CYCLE**

| Critical Success Factors | Critical Failure Factors |
|---|---|
| **Executive Management Acceptance Phase** | |
| Consider employee recommendations | Refuse to consider ideas of associates |
| Recognize that change is necessary | Unwilling to admit that change may be necessary |
| Understand the executive role in project management | Believe that project management control belongs at executive levels |
| **Line Management Acceptance Phase** | |
| Willing to place company interest before personal interest | Reluctant to share information |
| Willing to accept accountability | Refuse to accept accountability |
| Willing to see associates advance | Not willing to see associates advance |
| **Growth Phase** | |
| Recognize the need for a corporate-wide methodology | View a standard methodology as a threat rather than as a benefit |
| Support uniform status monitoring/reporting | Fail to understand the benefits of project management |
| Recognize the importance of effective planning | Provide only lip service to planning |
| **Maturity Phase** | |
| Recognize that cost and schedule are inseparable | Believe that project status can be determined from schedule alone |
| Track actual costs | See no need to track actual costs |
| Develop project management training | Believe that growth and success in project management are the same |

to project management. The organization must develop a reasonable project management curriculum to provide the appropriate training and education in support of the tools as well as the expected organizational behavior.

By the 1990s, companies finally began to recognize the benefits of project management. Table 1–3 shows the critical success and critical failure factors that have led to changes in our view of project management. Many of these factors were identified through the discovery and implementation of best practices.

Recognizing that the organization can benefit from the implementation of project management is just the starting point. The question now becomes, "How long will it take us to achieve these benefits?" This can be partially answered from Figure 1–3. In the beginning of the implementation process, there will be added expenses to develop the project management methodology and establish the support systems for planning, scheduling, and control. Eventually, the cost will level off and become pegged. The question mark in Figure 1–3 is the point at which the benefits equal the cost of implementation. This point can be pushed to the left through training and education.

During the first decade of the twenty-first century, the understanding and acceptance of the benefits permeated all levels of senior manage rather than just those executives that interfaced with projects on a daily basis. The following three comments from senior management at American Greetings Corporation illustrate this point:

> Through project management, we've learned how to make fact-based decisions. Too often in the past we based our decisions on what we thought could happen or what we hoped would happen. Now we can look at the facts, interpret the facts honestly and make sound decisions and set realistic goals based on this information. (Zev Weiss, CEO, American Greetings)[3]

---

3. H. Kerzner, *Advanced Project Management: Best Practices on Implementation*, Hoboken, NJ: Wiley, 2004, p. 273.

**Figure 1–3.** Project management costs versus benefits.

   The program management office provides the structure and discipline to complete the work that needs to get done. From launch to completion, each project has a roadmap for meeting the objectives that were set. (Jeff Weiss, President and COO, American Greetings)[4]

   Through project management, we learned the value of defining specific projects and empowering teams to make them happen. We've embraced the program management philosophy and now we can use it again and again to reach our goals. (Jim Spira, retired president and COO, American Greetings)[5]

When all of the executives are in agreement as to the value and benefits of project management, continuous improvements in project management occurs at a rapid pace.

## 1.5  AN EXECUTIVE'S VIEW OF PROJECT MANAGEMENT

Today's executives have a much better understanding and appreciation for project management than did their predecessors. Early on, project management was seen as simply scheduling a project and then managing the project using network-based software. Today, this parochial view has changed significantly. It is now a necessity for survival.

   Although there are several drivers for this, three significant reasons seem to stand out. First, as businesses downsize because of poor economic conditions or stiffening competition, the employees remaining in the company are expected to do more with less. Executives expect the employees to become more efficient and more effective when carrying out their duties. Second, business growth today requires the acceptance of significant risks, specifically in the development of new products and services for which there may not be reasonable estimating techniques or standards. Simply stated, we are undertaking

4. Ibid.
5. Ibid.

more jobs that are neither routine nor predictable. Third, and perhaps most important, is that we believe we are managing our business as though it is a series of projects. Projects now make up a significant part of one's jobs. As such, all employees are actually project managers to some degree and are expected to make business as well as project decisions.

The new breed of executive seems to have a much broader view of the value of project management, ranging from the benefits of project management, to the selection criteria for project managers, to organizational structures that can make companies more effective. This is apparent from the four comments below, which were provided by Tom Lucas, chief information officer for the Sherwin-Williams Company:

- We have all managed projects at one time or another, but few of us are capable of being project managers.
- The difference between managing projects and professional project management is like the difference between getting across the lake in a rowboat versus a racing boat. Both will get you across the lake, but the rowboat is a long and painful process. But how do people know until you give them a ride?
- Don't be misguided into thinking professional project management is about process. It is about delivering business results.
- If you don't appreciate that implementing a PMO is a cultural transition, you are destined to fail.

The comments below from other executives clearly indicate their understanding and appreciation of project management:

Our Customers, which are multinational industrial groups, expect from COMAU Project Managers an international, multicultural and global approach. In the meantime our Shareholder is asking us for high projects governance obtained through a global Project Management effective framework. In 2006 we have adopted a world-class Project Management approach (i.e., PMI) which, together with the implementation of the best practices on the global COMAU footprint, allowed us to demonstrate that both Customers and Shareholder goals can be fulfilled. I am sure that we are on the right tracks and that this continuous improvement strategy has to be pursued in the next years with motivation and perseverance. (Riccardo Tarantini, COMAU CEO, Fiat Group)

At Dell, we are committed to delivering technology solutions that help our customers do and achieve more. A key element in our ability to accomplish this is the combination of best practice standards and flexible, next generation IT solutions that enable our customers' business outcomes. Our best practices require standardized project management across integrated, end-to-end, multiservice solutions in order to provide seamless delivery to our customers. An increasing focus on new practice adoption, coupled with our project management people, processes, and tools, allow us to improve project delivery consistency across Dell. Effective project management is crucial to producing predictable, repeatable, high-quality results for our customers. (Suresh Vaswani, president, Dell Services)

Over the past 15 years, ongoing transformation has become a defining characteristic of IBM—and a key factor in our success. Effective change in process and IT transformation doesn't just happen, it must be enabled by highly skilled Project Managers. Our Project Managers analyze processes, enabled by IT, in a way that allows us to innovate and eliminate unnecessary steps, simplify and automate. They help us become more efficient and effective by pulling together the right resources to get things done—on time and on budget. They are invaluable as we continue to make progress in our transformation journey. (Linda S. Sanford, senior vice president, Enterprise Transformation, IBM Corporation)[6]

Project managers are a critical element of our end-to-end development and business execution model. Our goal is to have sound project management practices in place to provide better predictability in support of our products and offerings. As a team, you help us see challenges before they become gating issues and ensure we meet our commitments to STG and clients. . . . We continue to focus on project management as a career path for high-potential employees and we strongly encourage our project managers to become certified, not only PMI, but ultimately IBM certified. . . . End-to-end project management must become ingrained in the fabric of our business. (Rod Adkins, senior vice president of IBM's System and Technology Group [STG])[7]

At leading IT software services providers, project management has evolved and matured from a complex process of identifying and meeting a customer's unique requirements to applying a core set of proven, second-generation best practices captured and packaged in standard offerings. The standard offerings deliver repeatable success and accelerated time to value for the customer. They also give the customer and the project manager the ability to take a phased approach to building the customer's comprehensive IT management vision. (Dave Yusuf, SVP Global PMO, Computer Associates Services)

Successful project management is mission critical to us from two points of view:

- First, as we define and implement PLM (Product Lifecycle Management) solutions, we help customers to streamline their entire product lifecycle across all functional units. This can make any large PLM project an intricate and even complex undertaking. To live up to our company mantra of "we never let a customer fail", robust and reliable project management is often the most critical component we provide aside from the PLM platform itself; the combination of the two enables our customers to achieve the business benefits they strive for by investing in PLM.
- Second, Siemens itself is one of our largest customers. This is a great opportunity and, at the same time, a great challenge. Keeping a project's objectives and scope under control with our "internal" customer is at least as challenging as with external customers; yet it is critical in order to keep our development roadmaps and deployment schedules on track. Our job is to continue to successfully develop and deploy the first and only true end-to-end industry software platform. This comprehensive platform covers the entire product lifecycle from initial requirements, through product development, manufacturing planning, controlling the shop floor and even managing the maintenance, repair and overhaul of the product in question. As a result, effective project management is vital to our success. (Dr. Helmuth Ludwig, president, Siemens PLM Software)

Project Management is vital to the success of any organization. Whether projects are focused on customer acquisition, loyalty and insight or driven by the need to increase enterprise efficiency, excellence in Project Management ensures that tangible, meaningful results are achieved on time and on budget. (Brad Jackson, CEO, Slalom Consulting)

Projects and Project Management play a vital role in our business of IT Services. While being a key enabler for delighting customers, Project Management also helps in setting the right expectations of stakeholders and more importantly, maintaining a balance

---

6. Harold Kerzner, *Project Management Best Practices: Achieving Global Excellence*, 2nd edition, John Wiley and IIL Co-publishers, 2010, p. 13
7. ibid; p. 13

between their expectations. Effective Project Management becomes a strong competitive advantage or differentiator for our delivery capabilities. Excelling in Project Management has allowed us not only to increase the quality of our services, reduce our time-to-market, decrease rework costs and increase staff motivation, but also to create a more integrated and agile organization. (A. S. Murthy, Former CEO and CTO [chief technology officer], Tech Mahindra Limited)

The foundation of our brand as a solutions company is the quality and consistency of our project managers, project management approaches and methodologies. At the same time, it is important to understand that one size does not fit all, as each project has its own requirements, challenges and personality. High-quality project management goes beyond the fundamentals and focuses on working side by side with the client to weave a solution set that meets their needs and puts them on a path to achieve their goals. (Kimberly Parrish, executive vice president solutions, maxIT-VCS)

In this age of instant communications and rapidly evolving networks, Nortel continues to maximize use of its project management discipline to ensure the successful deployment of increasingly complex projects. We foster an environment that maintains a focus on sharing best practices and leveraging lessons learned across the organization, largely driven by our project managers. We are also striving to further integrate project management capabilities with supply chain management through the introduction of SAP business management software. Project management remains an integral part of Nortel's business and strategy as it moves forward in a more services- and solutions-oriented environment. (Sue Spradley, formerly president, Global Operations, Nortel Networks)[8]

The PMO process has been essential to the success of several major IS projects within Our Lady of Lourdes Regional Medical Center. This was especially true of our recent conversion from MedCath IS support to Franciscan Missionaries of our Lady Health System (FMOLHS) IS support at our newest physician joint venture: The Heart Hospital of Lafayette. PMO built trust through transparency, accountability and a framework for real-time project assessment. Without this structure I seriously doubt we could have succeeded in bringing the conversion on time and under budget. (W. F. "Bud" Barrow, president and CEO, Our Lady of Lourdes Regional Medical Center)

In the services industry, how we deliver (i.e., the project management methodology) is as important as what we deliver (i.e., the deliverable). Customers expect to maximize their return on IT investments from our collective knowledge and experience when we deliver best-in-class solutions. The collective knowledge and experience of HP (Hewlett-Packard) Services is easily accessible in HP Global Method. This integrated set of methodologies is a first step in enabling HPS to optimize our efficiency in delivering value to our customers. The next step is to know what is available and learn how and when to apply it when delivering to your customers. HP Global Method is the first step toward a set of best-in-class methodologies to increase the credibility as a trusted partner, reflecting the collective knowledge and expertise of HP Services. This also improves our cost structures by customizing predefined proven approaches, using existing checklists to ensure all the bases are covered and share experiences and learning to improve Global Method. (Mike Rigodanzo, formerly senior vice president, HP services operations and information technology)[9]

In 1996, we began looking at our business from the viewpoint of its core processes. . . . As you might expect, project management made the short list as one of the vital, core

8. H. Kerzner, *Best Practices in Project Management: Achieving Global Excellence*, Hoboken, NJ: Wiley, 2006, p. 17
9. Ibid., p. 67.

processes to which quality principles needed to be applied. (Martin O'Sullivan, retired vice president, Motorola)[10]

The disciplines of project management constitute an essential foundation for all initiatives toward business or indeed human advancement. I can't conceive crossing the vision/ reality chasm without them. (Keith Thomas, chairman, ITC [Information Technology & Communications] Business Unit, Neal & Massy Holdings, Ltd.)

The comments by Keith Thomas clearly indicate that today's executives recognize that project management is a strategic or core competency needed for survival because it interfaces with perhaps all other business processes, including quality initiatives.

## 1.6 BEST PRACTICES PROCESS

"Why capture best practices?" The reasons or objectives for capturing best practices might include:

- Continuous improvements (efficiencies, accuracy of estimates, waste reduction, etc.)
- Enhanced reputation
- Winning new business
- Survival of the firm

Survival of the firm has become the most important reason today for capturing best practices. In the last few years, customers have put pressure on contractors in requests for proposals (RFPs) by requesting:

- A listing of the number of PMP®s in the company and how many will be assigned to this project
- A demonstration that the contractor has an enterprise project management methodology that is acceptable to the customer or else the contractor must use some other methodology approved by the customer
- Supporting documentation identifying the contractor's maturity level in project management, possibly using a project management maturity model for assessments
- A willingness to share lessons learned and best practices discovered on this project and perhaps previous projects for other customers

Recognizing the need for capturing best practices is a lot easier than actually doing it. Companies are developing processes for identifying, evaluating, storing, and disseminating information on best practices. There are nine best practices activities as shown in Figure 1–4, and most companies that recognize the value of capturing best practices accomplish all of these steps.

---

10. Ibid., p. 184.

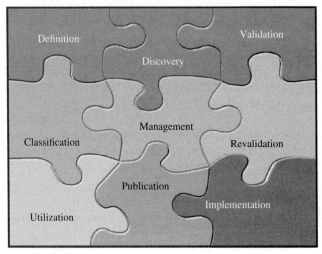

**Figure 1–4.** Best practices processes.

The processes answer the following nine questions:

- What is the definition of a best practice?
- Who is responsible for identifying the best practice, and where do we look?
- How do we validate that something is a best practice?
- Are there levels or categories of best practices?
- Who is responsible for the administration of the best practice once approved?
- How often do we reevaluate that something is still a best practice?
- How do companies use best practices once they are validated?
- How do large companies make sure that everyone knows about the existence of the best practices?
- How do we make sure that the employees are using the best practices and using them properly?

Each of these questions will be addressed in the next several sections.

## 1.7  STEP 1: DEFINITION OF A BEST PRACTICE

For more than a decade, companies have become fascinated by the expression "best practices." But now, after two decades or more of use, we are beginning to scrutinize the term, and perhaps better expressions exist.

A best practice begins with an idea that there is a technique, process, method, or activity that can be more effective at delivering an outcome than any other approach and provides us with the desired outcome with fewer problems and unforeseen complications. As a result, we supposedly end up with the most efficient and effective way of accomplishing a task based upon a repeatable process that has been proven over time for a large number of people and/or projects.

But once this idea has been proven to be effective, we normally integrate the best practice into our processes so that it becomes a standard way of doing business. Therefore, after acceptance and proven use of the idea, the better expression possibly should be a "proven practice" rather than a best practice. This is just one argument why a "best practice" may be just a buzzword and should be replaced by "proven practice."

Another argument is that the identification of a best practice may lead some to believe that we were performing some activities incorrectly in the past, and that may not have been the case. This may simply be a more efficient and effective way of achieving a deliverable. Another issue is that some people believe that best practices imply that there is one and only one way of accomplishing a task. This also may be a faulty interpretation.

Perhaps in the future the expression "best practices" will be replaced by "proven practices." However, for the remainder of this text, we will refer to the expression as "best practices," but the reader must understand that other terms may be more appropriate. This interpretation is necessary in this book because most of the companies that have contributed to this book still use the expression "best practices."

As project management evolved, so did the definitions of a best practice. Some definitions of a best practice are highly complex, while others are relatively simplistic. Yet, they both achieve the same purpose of promoting excellence in project management throughout the company. Companies must decide on the amount of depth to go into the best practice. Should it be generic and at a high level or detailed and at a low level? High-level best practices may not achieve the efficiencies desired, whereas highly detailed best practices may have limited applicability.

Every company can have its own definition of a best practice, and there might even be industry standards on the definition of a best practice. Typical definitions of a best practice might be:

- Something that works
- Something that works well
- Something that works well on a repetitive basis
- Something that leads to a competitive advantage
- Something that can be identified in a proposal to generate business
- Something the differentiates us from our competitors
- Something that keeps the company out of trouble and, if trouble occurs, the best practice will assist in getting the company out of trouble

Every company has its own definition of a best practice. There appear to be four primary reasons for capturing best practices:

- Improve efficiency
- Improve effectiveness
- Standardization
- Consistency

In each of the following definitions, you should be able to identify which of the four, or combination thereof, the company targets:

- At Orange Switzerland, a best practice is defined as an experience based, proven, and published way of proceeding to achieve an objective.[11]
- We do have best practices that are detailed in our policies/procedures and work-flows. These are guidelines and templates as well as processes that we all (members of the EPMO—enterprise project management office) have agreed to abide by as well as that they are effective and efficient methods for all parties involved. In addition, when we wrap up (conclude) a project, we conduct a formal lessons learned session (involving the project manager, sponsors, core team, and other parties impacted by the project), which is stored in a collective database and reviewed with the entire team. These lessons learned are in effect what create our best practices. We share these with other health care organizations for those vendors for which we are reference sites. All of our templates, policies/procedures, and work-flows are accessible by request and, when necessary, we set meetings to review as well as explain them in detail.(Nani Sadowski, formerly manager of the Enterprise Project Management Office at Halifax Community Health Systems).[12]
- Any tool, template or activity used by a project manager that has had a positive impact on the *PMBOK® Guide* knowledge and/or process areas and/or the triple constraint. An example of a best practice would be: Performing customer satisfaction assessments during each phase of a project allows adjustments during the project life cycle, which improves deliverables to the client, and improves overall project management. [This would be accompanied by a template for a customer satisfaction survey.] (Spokesperson for AT&T)
- Generally we view a best practice as any activity or process that improves a given situation, eliminates the need of other more cumbersome methods, or significantly enhances an existing process. Each best practice is a living entity and subject to review, amendments, or removal.[13]
- For Churchill Downs Incorporated, a best practice is any method or process that has been proven to produce the desired results through practical application. We do not accept "industry" or "professional standards" as best practices until we have validated that the method or process works in our corporate environment.

  *Examples of some of our best practices include:*
  - *Charter Signatures:* One of our best practices is requiring stakeholder signatures on project and program charters. This seems basic, but my experience is that a formal review and approval of a project's business objectives and goals is rarely documented. By documenting business objectives and their associated metrics, we have been able to proactively manage expectations and ensure alignment between various stakeholders.

---

11. H. Kerzner, *Project Management Best Practices: Achieving Global Excellence*, Hoboken, NJ: Wiley, 2006, p. 12.
12. Ibid., p. 13.
13. Ibid.

- *Process Definition:* In addition to defining the organization's project, program and portfolio management processes, the PMO has also taken an active role in mapping all of the financial processes for Churchill Downs Incorporated, from check requests and employee reimbursement requests to procedures for requesting capital expenses and purchase orders. This practice has increased corporate-wide awareness of how standardizing processes can enhance efficiency.
- *Access to Information:* The PMO developed process maps, procedures and policies for the end-to-end budgeting processes, associated workflows and templates. These have been made available company-wide via CCN, the company's intranet site.[14]

- At Indra, we consider a "best practice" in project management as a management action or activity that usually generates a positive outcome. As such, it is accepted by the management community and eventually becomes a recommended or required way of performing the task. We also consider as a "best practice," the use of predefined indicators, thresholds or metrics to make or facilitate decisions with regard to project management processes.[15]
- In the PMO, a best practice is a process, methodology or procedure that is followed in order to ensure a consistent approach and standard is utilized. Best practices within maxIT-VCS are evaluated for efficiency (internally and externally) and updated to reflect the lessons learned and the practical, real-world experience from our project management consulting base working in the field with our customers.[16] We will also be integrating industry best practices from industry and literature as well as our combined company experience and lessons learned from projects.[17]
- Sandra Kumorowski believes that a best practice is . . . either a method, tactic, or process that has been proven through implementation and tested use to add specific and measurable benefits and long-term value in terms of increased project performance outcome like decreased project cost, increased employee productivity, improved client experience (rate of retention), and an increased number of new projects. Best practices add both short-term and long-term value to the organization.

    I was responsible for providing Best Practices updates to all employees. Right after each Post Mortem session and after an idea was confirmed as a best practice, I would send an e-mail update to all employees. Typical best practices included:
- *Workload Distribution and Team Leadership:* Within a team will be equalized among the project lead (senior consultant) and junior consultant who would usually do the hard work of analysis without participating in the composition of the actual strategy. [Many times, junior people felt left out and did not feel they had contributed to the success of the project. They often felt inferior and that became a big issue that greatly decreased team productivity. The issue was part of the senior strategist job description (they were responsible for 80% of strategic thinking on a project) that could have been interpreted differently by different people.]

---

14. Comments by Chuck Millhollan, director of program management, Churchill Downs Inc.
15. Comments by Enrique Sevilla Molina, PMP, formerly corporate PMO director, Indra.
16. Comments by Marc Hirshfield, PMP, director, Project Management Office, maxIT-VCS.
17. Comments provided by Heidi Wurtz, VP, Solutions Center, maxIT-VCS.

**TABLE 1–4.  BEST PRACTICES INCENTIVES**

| Private Sector | Public Sector |
| --- | --- |
| Profit | Minimization of cost |
| Competitiveness | On-time delivery |
| Efficiency | Efficiency |
| Effectiveness | Effectiveness |
| Customer satisfaction | Stakeholder satisfaction |
| Partnerships | Sole-source procurement |

- *Project Kick-Off Meeting:* All project stakeholders must be in that meeting to discuss project scope and objectives. A specific scope statement format must be followed to clearly define objectives. During the meeting, all milestones dates must be determined and agreed on by all stakeholders.
- *Project Milestones Meetings:* Must be scheduled (some using GoToMeeting software) and included on all stakeholders' calendars right after the Kick-Off Meeting.
- *Project Post Mortem Meeting:* All team members must evaluate the project performance, fill in the Post Mortem questionnaire and discuss it with other team members. Post Mortem meetings must be scheduled no later than one week after the final client presentation to ensure the project issues and/or successes stay fresh in everyone's mind.[18]

These definitions of a best practice focus more on the private sector than on the public sector. A comparison of possible incentives for discovery and implementation of best practices in the public and private sectors is shown in Table 1–4.

## 1.8 STEP 2: SEEKING OUT BEST PRACTICES

Best practices can be captured either within your organization or external to your organization. Benchmarking is one way to capture external best practices, possibly by using the project management office as the lead for external benchmarking activities. However, there are external sources other than benchmarking for identifying best practices:

- Project Management Institute (PMI) publications
- Forms, guidelines, templates, and checklists that can affect the execution of the project
- Forms, guidelines, templates, and checklists that can affect our definition of success on a project

---

18. Comments by Sandra Kumorowski, assistant professor of marketing communications, Columbia College Chicago; marketing & communications chair, Christopher & Dana Reeve Foundation, Chicago; chief business advisor & owner, Enakta LLC P: 1–224-715-5666; E: sandrakum@sbcglobal.net, skumorowski@colum.edu

- Each of the *PMBOK® Guide* areas of knowledge or domain areas
- Within company-wide or isolated business units
- Seminars and symposiums on general project management concepts
- Seminars and symposiums specializing on project management best practices
- Relationships with other professional societies
- Graduate-level theses

With more universities offering masters- and doctorate-level work in project management, graduate-level theses can provide up-to-date research on best practices.

The problem with external benchmarking is that best practices discovered in one company may not be transferable to another company. In the author's opinion, most of the best practices are discovered internally and are specifically related to the company's use of its project management methodology and processes. Good project management methodologies allow for the identification and extraction of best practices. However, good ideas can come from benchmarking as well.

Sometimes, the identification of the drivers or metrics that affect each best practice is more readily apparent than the best practice itself. Metrics and drivers can be treated as early indicators that a best practice may have been found. It is possible to have several drivers for each best practice. It is also possible to establish a universal set of drivers for each best practice, such as:

- Reduction in risk by a certain percentage, cost, or time
- Improve estimating accuracy by a certain percentage or dollar value
- Cost savings of a certain percentage or dollar value
- Efficiency increase by a certain percentage
- Reduction in waste, paperwork, or time by a certain percentage

There are several advantages of this approach for searching for drivers. First, the drivers can change over time and new drivers can emerge rapidly. Second, the best practices process is more of a science than an art. And third, we can establish levels of best practices such as shown in Figure 1–5. In this figure, a level 4 best practice, which is the best, would satisfy 60 percent or more of the list of drivers or characteristics of the ideal best practice.

Best practices may not be transferable from company to company, nor will they always be transferable from division to division within the same company. As an example, consider the following best practice discovered by a telecommunications company:

- A company institutionalized a set of values that professed that quality was everything. The result was that employees were focusing so much on quality that there was a degradation of customer satisfaction. The company then reprioritized its values with customer satisfaction being the most important, and quality actually improved.

In this company, customer satisfaction emphasis led to improved quality. However, in another company, emphasis on quality could just as easily have led to an improvement in customer satisfaction. Care must be taken during benchmarking activities to make sure that whatever best practices are discovered are in fact directly applicable to your company.

**Figure 1–5.** Best practices levels. Each level contains a percentage of the ideal characteristics.

Best practices need not be overly complex. As an example, the following list of best practices is taken from companies discussed in this textbook, and as you can see, some of the best practices were learned from failures rather than successes:

- Changing project managers in midstream is bad even if the project is in trouble. Changing project managers inevitably elongates the project and can make it worse.
- Standardization yields excellent results. The more standardization placed in a project management methodology, usually the better the results are.
- Maximization of benefits occurs with a methodology based upon templates, forms, guidelines, and checklists rather than policies and procedures.
- Methodologies must be updated to include the results of discovering best practices. The more frequently the methodology is updated, the quicker the benefits are realized.

As stated previously, best practices need not be complex. Even though some best practices seem simplistic and based on common sense, the constant reminder and use of these best practices lead to excellence and customer satisfaction.

Another way to identify sources of best practices is from the definition of project success, critical success factors (CSFs), and key performance indicators (KPIs). Extracting best practices from the definition of success on a project may be difficult and misleading, especially if we have a poor definition of success.

Over the years, a lot of the changes that have taken place in project management have been the result of the way we define project success. As an example, consider the following chronological events that took place over the past several decades:

- *Success is measured by the triple constraints or competing constraints.* The triple constraints are time, cost, and performance (which include quality, scope, and technical performance). This was the basis for defining success during the birth of project management. Competing constraints can include safety, aesthetic value, benefits, safety, level of acceptable risk, etc.
- *Customer satisfaction must be considered as well.* Managing a project within the triple constraint is always a good idea, but the customer must be satisfied with the end result. A contractor can complete a project within the triple constraints and still find that the customer is unhappy with the end result.
- *Other (or secondary) factors must be considered as well.* These include using the customer's name as a reference, corporate reputation and image, compliance with

government regulations, strategic alignment, technical superiority, ethical conduct, and other such factors. The secondary factors may end up being more important than the primary factors of the triple constraints.

- *Success must include a business component.* Project managers are managing part of a business rather than merely a project and are expected to make sound business decisions as well as project decisions. There must be a business purpose for each project. Each project is considered as a contribution of business value to the company when completed.
- *Prioritization of constraints must occur.* Not all project constraints are equal. The prioritization of constraints is done on a project-by-project basis. Sponsorship involvement in this decision is essential.
- *The definition of success must be agreed upon between the customer and the contractor.* Each project can have a different definition of success. There must be upfront agreement between the customer and the contractor at project initiation or even at the first meeting between them on what constitutes success.
- *The definition of success must include a "value" component.* Why work on a project that does not provide the correct expected value at completion?

The problem with defining success as on time, within cost, and at the desired quality or performance level is that this is an internal definition of success only. Bad things can happen on projects when the contractor, customer, and various stakeholders are all focusing on different definitions of project success. There must be an upfront agreement on what constitutes project success. The ultimate customer or stakeholder should have some say in the definition of success, and ultimately there may be numerous best practices discovered that relate to customer/stakeholder interfacing.

Today, we recognize that the customer rather than the contractor defines quality. The same holds true for project success. There must be customer and stakeholder acceptance included in any definition of project success. You can complete a project internally within your company within time, within cost, and within quality or specification limits and yet find the project is not fully accepted by the customer or stakeholders.

At Enakta, the definition of project success is compared against the definition of project failure. According to Sandra Kumorowski,[19] the definition can appear as shown in Table 1–5.

Although companies may maintain a definition of project success (and even project failure), they may not have a clear definition of excellence in project management. This occurs when project management either is fully embedded into all of the company's work flow processes or is seen as a supportive role. Sandra Kumorowski believes:

> At Enakta, we do not have a formal definition of what is excellence in project management. However, our company wanted to view project management as a supportive role in creativity. Excellence is then achieved by complete customization to the current organizational structure and full alignment with long-term organizational goals, as shown in Figure 1–6.

19. Sandra Kumorowski is an assistant professor of marketing communications at Columbia College, Chicago, and marketing & communications chair, Christopher & Dana Reeve Foundation, Chicago; she is also chief business advisor & owner, Enakta LLC; P: 1–224-715-5666; E: sandrakum@sbcglobal.net, skumorowski@colum.edu

**TABLE 1–5.   COMPARING PROJECT SUCCESS VERSUS FAILURE: A MEASURE OF PM PROJECT SUCCESS**

| Successful | Organizational Level |
|---|---|
| ☐ More business from the client<br>☐ Clients contacting us<br>☐ Client satisfaction during/after the project<br>☐ Project team happy | • Ratio of successful to unsuccessful projects per year<br>• Increased number of successful projects per period (ROI)<br>• Everybody is on board accepting changes, no resistance<br>• Effective project portfolio management = balanced use of resources<br>• Increased client satisfaction<br>• Number of returning clients<br>• Number of recommended clients |
| **Unsuccessful** | **Project Level** |
| ☐ No more business from client<br>☐ We have to call clients.<br>☐ Client dissatisfaction during/after the project<br>☐ Project team not happy | • Reduced project costs (working on more than one project at a time effectively, etc.)<br>• Well-distributed project time, no nights/weekends per project<br>• Reduced number of unexpected events/changes throughout the project<br>• Good team dynamics, met expectations<br>• Reduced number of negative issues per project |

**Figure 1–6.**   Project management as a supportive role.

Although some definitions of project success seem quite simple, many companies have elaborated on the primary definition of project success. At Churchill Downs Incorporated (CDI), success is defined more rigorously than in most companies. According to Chuck Millhollan, director of program management:

> Project success is defined in our PMO charter as follows;
>    Based on input from CDI's executive management, the PMO considers a project to be a success when the following are true:
>
> a. Predefined business objectives and project goals were achieved or exceeded.
> b. A high-quality product is fully implemented and utilized.

    c. Project delivery met or beat schedule and budget targets.
    d. There are multiple winners:
        i. Project participants have pride of ownership and feel good about their work.
        ii. The customer's (internal and/or external) expectations are met.
        iii. Management has met its goals.
    e. Project results helped build a good reputation.
    f. Methods are in place for continual monitoring and evaluation (benefit realization).

We do not use project management "process" indicators to define project success. While schedule and budget targets are part of the criteria, sponsor acceptance, project completion, and ultimately project success, is based on meeting defined business objectives.

Enrique Sevilla Molina, PMP, formerly corporate PMO director at Indra, provides us with his company's definition of project success and program success:

- Project success is based on achieving the proposed project targets in budget, scope, performance and schedule. Many times, the economic criteria appears as the main driving factor to measure project success, but there are other factors just as important such as building a durable relationship with the customer and building strong alliances with selected partners. Another significant criteria for project success measurement is the reliability of the project data forecast. It may be the case that, when the economic results of the project are not as good as they should be, if the fact is pointed out and reported soon enough, the success of the project is equally achieved.
- Program success is based on achieving the Program's overall strategic targets defined during Program definition and, at this level, the success is measured not only by achieving the expected economic outcomes but, most of all, reaching the expected position in the market with regard to a product or a line of products, and establishing a more advantageous position with regard to our competitors. Leadership in a product line constitutes the ultimate measure of success in a Program. It is worthwhile to mention that, quite often, the success of a Program is based on the partnership concept developed with our major subcontractors at the Project level.
- Project success is defined at a business unit level by the responsible director, in accordance with the strategic goals assigned to the project.
- Program success is defined at the company level by the Chief Operations management in accordance with the program's defined mission.

AT&T defines project and program success in a similar manner. According to a spokesperson for AT&T:

Project success is defined as a Client Satisfaction rating of "Very Satisfied" and On-Time Performance of Project Delivery of 98% or greater. The Project Management Organizational Leadership Team sets the objectives, which are tracked to determine project success. Program success is defined and tracked the same way as project success.

Excellence [in project management] is defined as a consistent Project Management Methodology applied to all projects across the organization, continued recognition by our customers, and high customer satisfaction. Also our project management excellence is a key selling factor for our sales teams. This results in repeat business from our customers. In addition there is internal acknowledgement that project management is value-added and a must have.

Project success can be measured intermittently throughout the phase or gate review meetings that are part of the project management methodology. This allows a company to establish interim metrics for measuring success. An example of this will appear in Chapter 4 (Project Management Methodologies).

Another element that is becoming important in the definition of success is the word *value*. The following information has been provided by Doug Bolzman, Consultant Architect, PMP®, ITIL service manager at Hewlett-Packard:[20]

**Understanding Project Success**

At one point, customers were measuring project success as being on time and under budget. But if the project provided no real business value, what good is it being on time or under budget? Value for projects is being transformed within the planning of the project to depict the value to the user or the client of the project.

In most cases within an IT delivery organization, a project is not the end all be all. It is a means to an end, and as such, the project is viewed as an incremental gain. Projects in IT are viewed from the implementation of a new service, which can constitute a bundle of projects (or releases) down to the maintenance projects such as operating system upgrades. The success of the project attaining the objectives, producing the deliverables, and acquiring the desired outcome of the work. Value in IT is measured to how well the IT service enables the business function. Does it with lower manual labor and provide the receiver with a satisfied result?

**Defining Project Success**

A good project manager would define the success of a project from the perspective of the users or the customers of the project. This may be difficult to identify at the start of the project if the project is chartered from a different perspective. Our management team is always challenging the project managers to explain the value of the release—what is the justification of the cost and the investment of the time? We cannot afford to implement projects because someone identified a need or generated an improvement suggestion. Do we need to be at the most current release of a product or application? Does the current one meet our business needs? What is the net gain for us to upgrade? Will the cost of the project be paid for in efficiency, improved outcomes, increased revenue? If these questions cannot be answered by the project manager (or the sponsor of the project), it is not approved. The executives can determine the overall value of how the projects map to the success of a program or initiative, but the users or customers will be the entity to receive the value of the project.

**Critical Success Factors**

Typically, projects either improve something or reduce something. These improvements come in the form of capability or functionality of the company (through the employees / users). These produce addi-

20. Doug Bolzman has been with HP/EDS for more than 25 years and is currently a member of the HP Business Transformation Enablement team focusing on improving their client's IT Service Management delivery. Prior to the HP merger, EDS submitted a patent on behalf of Doug's processes titled, "System and Method for Identifying and Monitoring Best Practices of an Enterprise." Since 1995, Doug has architected and delivered an approach for clients to institute the IT Information Library (ITIL©) into their IT operations environment. Working with clients, Doug utilizes his IT Enterprise Management (ITEM) framework, along with the Project Management Body of Knowledge (*PMBOK® Guide*) and the IT Service Management (ITSM) life cycle to facilitate the client through the cultural, organizational, business, and operational changes. Doug holds an ITIL Expert Certificate, developed online ITIL Foundation Training for ITIL Version 2 along with the 2007 and 2011 editions. Doug's comments are based upon HP's relationship with its clients, especially when they are looking at the enterprise management of a business, not simply the management of the piece parts.

tional productivity either new products and services or more efficiency for existing products. Critical success factors are mapped to the overall business objectives.

## Key Performance Indicators for Success

Key performance indicators allow the customer to make a series of measurements to ensure the performance is within the stated thresholds (success factors). This is called "keeping the pulse of the company" by the executives. KPI's are determined, measured and communicated through mechanisms such as dashboards or metrics.

The comments by Doug Bolzman indicate that perhaps the single most important criterion for defining a potential best practice is that it must add value to the company and/or the client. According to a program manager at Hewlett-Packard, the following three best practices are added-value best practices:

- Project Collaboration Portals with standardized PM templates and Integrated tool kits with ability to request additional features by a Support staff.
- Project Retrospectives—very helpful for group learning and eliciting/recognizing/documenting "best practices" but indeed communication beyond the immediate team is the challenge.
- Virtual Projects—given sufficient infrastructure I feel virtual projects are more productive and effective than burning uptime and money on travel. I think HP utilizes these capabilities internally very well.

The ultimate definition of success might very well be when the customer is so pleased with the project that the customer allows you to use his or her name as a reference. This occurred in one company that bid on a project at 40 percent below its cost of doing the work. When asked why the bid was so low, company representatives responded that they knew they were losing money but what was really important was getting the customer's name on the corporate resume of clients. Therefore, the secondary factors may be more important than the primary factors.

The definition of success can also change based upon whether you are project- or non–project-driven. In a project-driven firm, the entire business of the company is projects. But in a non–project-driven firm, projects exist to support the ongoing business of production or services. In a non–project-driven firm, the definition of success also includes completion of the project *without* disturbing the ongoing business of this firm. It is possible to complete a project within time, within cost, and within quality and at the same time cause irrevocable damage to the organization. This occurs when the project manager does not realize that the project is *secondary* in importance to the ongoing business.

Some companies define success in terms of CSFs and KPIs. Critical success factors identify those factors necessary to meet the desired deliverables of the customer. CSFs and KPIs do not need to be elaborate or sophisticated metrics. Simple metrics, possibly based upon the triple constraint, can be quite effective. According to a spokesperson from AT&T:

> The critical success factors include Time, Scope, Budget and Customer Satisfaction. Key performance indicators include on-time performance for key deliverables. These include customer installation, customer satisfaction and cycle-time for common milestones.

Typical CSFs for most companies include:

- Adherence to schedules
- Adherence to budgets
- Adherence to quality
- Appropriateness and timing of signoffs
- Adherence to the change control process
- Add-ons to the contract

Critical success factors measure the end result usually as seen through the eyes of the customer. KPIs measure the quality of the process to achieve the end results. KPIs are internal measures and can be reviewed on a periodic basis throughout the life cycle of a project. Typical KPIs include:

- Use of the project management methodology
- Establish control processes
- Use of interim metrics
- Quality of resources assigned versus planned for
- Client involvement

Key performance indicators answer such questions as: Did we use the methodology correctly? Did we keep management informed and how frequently? Were the proper resources assigned, and were they used effectively? Were there lessons learned which could necessitate updating the methodology or its use? Companies that are excellent in project management measure success both internally and externally using KPIs and CSFs. As an example, consider the following remarks provided by a spokesperson from Nortel Networks[21]:

> Nortel defines project success based on schedule, cost, and quality measurements, as mutually agreed upon by the customer, the project team, and key stakeholders. Examples of key performance indicators may include completion of key project milestones, product installation/integration results, change management results, completion within budget, and so on. Project status and results are closely monitored and jointly reviewed with the customer and project team on a regular basis throughout a project to ensure consistent expectations and overall success. Project success is ultimately measured by customer satisfaction.

Here are additional definitions of CSFs and KPIs:

- CSFs:
  - Success factors are defined at the initial stages of the project or program, even before they become actual contracts, and are a direct consequence of the strategic goals allocated to the project or program. Many times these factors are

21. H. Kerzner, *Project Management Best Practices: Achieving Global Excellence*, Hoboken, NJ: Wiley, 2006, p. 26.

associated with expanding the market share in a product line or developing new markets, both technically and geographically. (Provided by Enrique Sevilla Molina, PMP, formerly corporate PMO director, Indra)

- Obviously, CSFs vary with projects and intent. Here are some that apply over a large variety of projects:
  - Early customer involvement
  - High-quality standards
  - Defined processes and formalized gate reviews
  - Cross-functional team organizational structure
  - Control of requirements, prevention of scope creep
  - Commitment to schedules—disciplined planning to appropriate level of detail and objective and frequent tracking
  - Commitment of resources—right skill level at necessary time
  - Communication among internal teams and with customer
  - Early risk identification, management, and mitigation—no surprises
  - Unequaled technical execution based on rigorous engineering. (Comments provided by a spokesperson at Motorola)[22]
- KPIs:
  - We are implementing an interim quality program to measure [the] quality of all projects and performance of project managers across the company. This program will include project reviews on a regular basis, determined by project scope, timing, and complexity, and will conclude with the customer's perspective on the quality delivered by the company and the PM. This may no longer be called an "Assignment Quality Assessment (AQA)," but will still elicit customer feedback and contribute to the PM's performance measurement plan. The QA program is a function of our Solution Center, which supports the integrated companies: maxIT-VCS. (Heidi Wurtz, VP, Solutions Center maxIT-VCS)
  - Our most common KPIs are associated to the financial projects results, for instance, project margin compliance with the allocated strategic target, new contracts figure for the business development area goals, etc. Success factors are translated into performance indicators so they are periodically checked.
  - By default, a first indication of projects health is provided by the schedule and cost performance indices (SPI and CPI) embedded into the PM tools. They are monthly provided by the project management information system and they are also available for historical analysis and review. These indicators are also calculated for each department, so they constitute an indicator of the overall cost and schedule performance of the department or business unit. (Provided by Enrique Sevilla Molina, PMP, formerly corporate PMO director, Indra)
- Postship acceptance indicators:
  - Profit and loss
  - Warranty returns

---

22. Ibid., p. 27.

- Customer reported unique defects
- Satisfaction metrics
- In-process indicators:
  - Defect trends against plan
  - Stability for each build (part count changes) against plan
  - Feature completion against plan
  - Schedule plan versus actual performance
  - Effort plan versus actual performance
  - Manufacturing costs and quality metrics
  - Conformance to quality processes and results of quality audits
  - System test completion rate and pass/fail against plan
  - Defect/issue resolution closure rate
  - Accelerated life-testing failure rates against plan
  - Prototype defects per hundred units (DPHU) during development against plan (Provided by a spokesperson at Motorola)[23]
- The SOW provides a checklist of basic indicators for the success of the project, but client satisfaction is also important. The SOW will indicate what the deliverables are and will provide information on costs and timelines that are easily tracked.

Most people seem to understand that CSFs and KPIs can be different from project to project. However, there is a common misbelief that CSFs and KPIs, once established, must not change throughout the project. As projects go through various life-cycle phases, these indicators can change. Carl Manello, PMP, solutions lead—program and project management, Slalom Consulting, believes that:

> Establishing the right Critical Success Factors and Key Performance Indicators is crucial. A project's defined success (i.e., defined success means that . . . all stakeholders are in agreement at the earliest stages of the project as to what the end-state will look like) is identified through the Critical Success Factors. A project's selection of the right Key Performance Indicators establishes the measurement metrics for tracking to determine whether the Critical Success Factors will be met. As a starting point, On Time, On Budget and with the agreed upon specifications (or within a tolerance for all three metrics) are good basic KPIs. As projects mature in their ability to deliver results, more sophisticated performance indicators may be implemented. For example, instead of using the loosely defined "agreed upon specifications," projects may choose to use the formality of a requirements volatility measure and some acceptable variation around a baseline.

In the author's experience, more than 90 percent of the best practices that companies identify come from analysis of the KPIs during the debriefing sessions at the completion of a project or at selected gate review meetings. Because of the importance of extracting these best practices, some companies are now training professional facilitators capable of debriefing project teams and capturing the best practices.

---

23. Ibid.

Before leaving this section, it is necessary to understand who discovers the best practice. Best practices are discovered by the people performing the work, namely the project manager, project team, and possibly the line manager. According to a spokesperson from Motorola[24]:

> The decision as to what is termed a best practice is made within the community that performs the practice. Process capabilities are generally known and baselined. To claim best practice status, the practice or process must quantitatively demonstrate significant improvements in quality, efficiency, cost, and/or cycle time. The management of the organization affected as well as process management must approve the new practice prior to institutionalization.

Generally, the process of identification begins with the appropriate team member. If the team member believes that he or she has discovered a best practice, they then approach their respective line manager and possibly project manager for confirmation. Once confirmation is agreed upon, the material is sent to the Project Management Office (PMO) for validation. After validation, the person that identified the best practice is given the title of "Best Practice Owner" and has the responsibility of nurturing and cultivating the best practice.

Some companies use professional facilitators to debrief project teams in order to extract best practices. These facilitators may be assigned to the PMO and are professionally trained in how to extract lessons learned and best practices from both successes and failures. Checklists and templates may be used as part of the facilitation process. As an example, consider the following statements from Sandra Kumorowski:[25]

> After each project, we conduct Post Mortem meetings in order to evaluate the project performance. I have created a standard format—list of questions—and all team members had to prepare their answers ahead of time. If there was an idea, process, or tactic that has added measurable benefit or long-term value to the organization, the top management would then add it to our Library. I was responsible for an update to all employees when a new best practice has been added to our Best Practices Library.

## 1.9 DASHBOARDS AND SCORECARDS

In our attempt to go to paperless project management, emphasis is being given to visual displays such as dashboard and scorecards utilizing and displaying CSFs and KPIs. Executives and customers desire a visual display of the most critical project performance information in the least amount of space. Simple dashboard techniques, such as traffic light reporting, can convey critical performance information. As an example:

- *Red traffic light:* A problem exists which may affect time, cost, quality, or scope. Sponsorship involvement is necessary.

24. Ibid., p. 14.
25. Sandra Kumorowski is an assistant professor of marketing communications at Columbia College Chicago, and marketing & communications chair, Christopher & Dana Reeve Foundation, Chicago; she is also chief business advisor & owner, Enakta LLC; P: 1–224-715-5666; E: sandrakum@sbcglobal.net, skumorowski@colum.edu

- *Yellow or amber light:* This is a caution. A potential problem may exist, perhaps in the future if not monitored. The sponsor is informed but no action by the sponsor is necessary at this time.
- *Green light:* Work is progressing as planned. No involvement by the sponsor is necessary.

While a traffic light dashboard with just three colors is most common, some companies use many more colors. The information technology (IT) group of a retailer had an eight-color dashboard for IT projects. An amber color meant that the targeted end date had past and the project was still not complete. A purple color meant that this work package was undergoing a scope change that could have an impact on the triple constraint.

Some people confuse dashboards with scorecards. There is a difference between dashboards and scorecards. According to Eckerson[26]:

- Dashboards are visual display mechanisms used in an *operationally* oriented performance measurement system that measure performance against targets and thresholds using right-time data.
- Scorecards are visual displays used in a *strategically* oriented performance measurement system that chart progress towards achieving strategic goals and objectives by comparing performance against targets and thresholds.

Both dashboards and scorecards are visual display mechanisms within a performance measurement system that convey critical information. The primary difference between dashboards and scorecards is that dashboards monitor operational processes such as those used in project management, whereas scorecards chart the progress of tactical goals. Table 1–6 and the description following it show how Eckerson compares the features of dashboards and scorecards.[27]

*Dashboards:* Dashboards are more like automobile dashboards. They let operational specialists and their supervisors monitor events generated by key business processes. But unlike automobiles, most business dashboards do not display events in "real time," as they occur; they display them in "right time," as users need to view them. This could be every second, minute, hour, day, week, or month depending on the business process, its volatility, and how critical it is to the business. However, most elements on a dashboard are updated on an intraday basis, with latency measured in either minutes or hours.

Dashboards often display performance visually, using charts or simple graphs, such as gauges and meters. However, dashboard graphs are often updated in place, causing the graph to "flicker" or change dynamically. Ironically, people who monitor operational processes often find the visual glitz distracting and prefer to view the data in the original form, as numbers or text, perhaps accompanied by visual graphs.

*Scorecards:* Scorecards, on the other hand, look more like performance charts used to track progress toward achieving goals. Scorecards usually display monthly snapshots of

---

26. W. W. Eckerson, *Performance Dashboards: Measuring, Monitoring and Managing Your Business*, Hoboken, NJ: Wiley, 2006, pp. 293, 295. Chapter 12 provides an excellent approach to designing dashboard screens.
27. Ibid., p. 13.

**TABLE 1–6.  COMPARING FEATURES**

| Feature | Dashboard | Scorecard |
|---------|-----------|-----------|
| Purpose | Measures performance | Charts progress |
| Users | Supervisors, specialists | Executives, managers, and staff |
| Updates | Right-time feeds | Periodic snapshots |
| Data | Events | Summaries |
| Display | Visual graphs, raw data | Visual graphs, comments |

**TABLE 1–7.  THREE TYPES OF PERFORMANCE DASHBOARDS**

| | Operational | Tactical | Strategic |
|---------|-------------|----------|-----------|
| **Purpose** | Monitor operations | Measure progress | Execute strategy |
| **Users** | Supervisors, specialists | Managers, analysts | Executives, managers, staff |
| **Scope** | Operational | Departmental | Enterprise |
| **Information** | Detailed | Detailed/summary | Detailed/summary |
| **Updates** | Intraday | Daily/weekly | Monthly/quarterly |
| **Emphasis** | Monitoring | Analysis | Management |

summarized data for business executives who track strategic and long-term objectives, or daily and weekly snapshots of data for managers who need to chart the progress of their group of project toward achieving goals. In both cases, the data are fairly summarized so users can view their performance status at a glance.

Like dashboards, scorecards also make use of charts and visual graphs to indicate performance state, trends, and variance against goals. The higher up the users are in the organization, the more they prefer to see performance encoded visually. However, most scorecards also contain (or should contain) a great deal of textual commentary that interprets performance results, describes action taken, and forecasts future results.

*Summary:* In the end, it does not really matter whether you use the term "dashboard" or "scorecard" as long as the tool helps to focus users and organizations on what really matters. Both dashboards and scorecards need to display critical performance information on a single screen so users can monitor results at a glance.

Although the terms are used interchangeably, most project managers prefer to use dashboards and/or dashboard reporting. Eckerson defines three types of dashboards, as shown in Table 1–7 and the description that follows.[28]

- *Operational dashboards* monitor core operational processes and are used primarily by front-line workers and their supervisors who deal directly with customers or manage the creation or delivery of organizational products and services. Operational dashboards primarily deliver detailed information that is only lightly summarized. For example, an online Web merchant may track transactions at the product level rather than the customer level. In addition, most metrics in an operational dashboard are updated on an intraday basis, ranging from minutes to hours depending on the application. As a result, operational dashboards emphasize monitoring more than analysis and management.

---

28. Ibid., pp. 17–18.

- *Tactical dashboards* track departmental processes and projects that are of interest to a segment of the organization or a limited group of people. Managers and business analysts use tactical dashboards to compare performance of their areas or projects, to budget plans, forecasts, or last period's results. For example, a project to reduce the number of errors in a customer database might use a tactical dashboard to display, monitor, and analyze progress during the previous 12 months toward achieving 99.9 percent defect-free customer data by 2007.
- *Strategic dashboards* monitor the execution of strategic objectives and are frequently implemented using a balanced scorecard approach, although total quality management, Six Sigma, and other methodologies are used as well. The goal of a strategic dashboard is to align the organization around strategic objectives and get every group marching in the same direction. To do this, organizations roll out customized scorecards to every group in the organization and sometimes to every individual as well. These "cascading" scorecards, which are usually updated weekly or monthly, give executives a powerful tool to communicate strategy, gain visibility into operations, and identify the key drivers of performance and business value. Strategic dashboards emphasize management more than monitoring and analysis.

There are three critical steps that must be considered when using dashboards: (1) the target audience for the dashboard, (2) the type of dashboard to be used, and (3) the frequency in which the data will be updated. Some project dashboards focus on the key performance indicators that are part of earned-value measurement. These dashboards may need to be updated daily or weekly. Dashboards related to the financial health of the company may be updated weekly or quarterly. Figures 1–7 and 1–8 show the type of information that would be tracked weekly or quarterly to view corporate financial health.[29]

## 1.10 KEY PERFORMANCE INDICATORS

Most often, the items that appear in the dashboards are elements that both customers and project managers track. These items are referred to as key performance indicators (KPIs) and were discussed previously. According to Eckerson[30]:

> A KPI is a metric measuring how well the organization or individual performs an operational, tactical or strategic activity that is critical for the current and future success of the organization.

Some people confuse KPIs with leading indicators. A leading indicator is actually a KPI that measures how the work one is doing now will affect the future.

KPIs are critical components of all earned-value measurement systems. Cost variance, schedule variance, schedule performance index, cost performance index, and time/cost at

29. J. Alexander, *Performance Dashboards and Analysis for Value Creation*, Hoboken, NJ: Wiley, 2007, pp. 87–88. Reproduced by permission of John Wiley & Sons.
30. W. Eckerson, *Performance Dashboards: Measuring, Monitoring and Managing Your Business*, Hoboken, NJ: Wiley, 2006, p. 294.

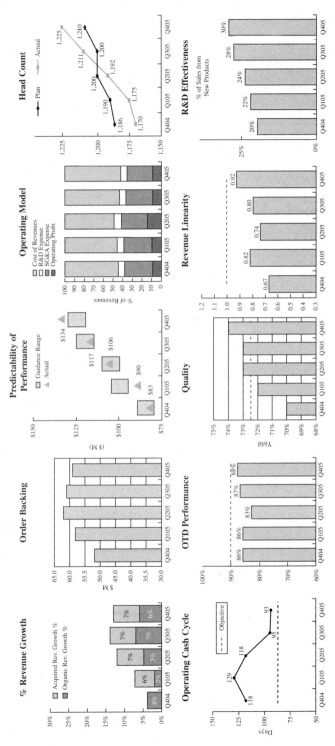

**Figure 1–7.** Typical financial health dashboards.

XYZ Company
Q4' 05 Week #7 of 13/ 54% of Q4
($ in Millions)

| Bookings | Week | Unit | QTD | Forecast | % Achieved | $ Required |
|---|---|---|---|---|---|---|
| | 0.7 | BU 1 | 15.0 | 30.0 | 50% | 15.0 |
| | — | BU 2 | 0.9 | 1.0 | 89 | 0.1 |
| | 0.5 | BU 3 | 4.0 | 6.0 | 67 | 2.0 |
| | 0.4 | BU 4 | 1.7 | 4.7 | 37 | 2.9 |
| | 0.0 | Other | 0.1 | — | | (0.1) |
| | 1.6 | Totals | 21.7 | 41.7 | 52% | $20.0 |

Bookings QTD

| Revenue | Week | Unit | QTD | Forecast | % Achieved | Backlog | Req'd Fill |
|---|---|---|---|---|---|---|---|
| | 2.0 | BU 1 | 13.0 | 28.0 | 46% | 5.0 | 10.00 |
| | 0.4 | BU 2 | 3.0 | 5.0 | 60 | 1.0 | 1.00 |
| | 0.0 | BU 3 | 3.0 | 6.0 | 50 | 2.0 | 1.00 |
| | 2.6 | BU 4 | 3.0 | 7.0 | 43 | 1.0 | 3.00 |
| | — | Other | — | — | | | |
| | 5.0 | Totals | 22.0 | 46.0 | 48% | 9.0 | 15.0 |

Revenue QTD

| Receivable Collections (Cumulative) | Week | 1 | 2 | 3 | 4 | 5 |
|---|---|---|---|---|---|---|
| | Actual | 1.0 | 5.0 | 19.0 | | |
| | Target | 4.0 | 9.0 | 17.0 | 28.0 | 35.0 |

Cumulative AR Collections

| | Day | 1 | 2 | 3 | 4 | 5 |
|---|---|---|---|---|---|---|
| Process Yield | | 77% | 80% | 81% | 68% | 82% |

Process Yield

**Figure 1–8.** Typical financial health dashboards.

completion are actually KPIs but are not referred to as such. The need for these KPIs is simple: What gets measured gets done! If the goal of a performance measurement system is to improve efficiency and effectiveness, then the KPI must reflect controllable factors. There is no point in measuring an activity if the users cannot change the outcome.

Eckerson identifies 12 characteristics of effective KPIs[31]:

- *Aligned:* KPIs are always aligned with corporate [or project] strategy and objectives.
- *Owned:* Every KPI is "owned" by an individual or group on the business [or project] side that is accountable for its outcome.
- *Predictive:* KPIs measure drivers of business [or project] value. Thus, they are "leading" indicators of performance desired by the organization.
- *Actionable:* KPIs are populated with timely, actionable data so users can intervene to improve performance before it is too late.
- *Few in number:* KPIs should focus users on a few high-value tasks, not scatter their attention and energy on too many things.

---

31. Ibid., p. 201.

- *Easy to understand:* KPIs should be straightforward and easy to understand, not based upon complex indices that users do not know how to influence directly.
- *Balanced and linked.* KPIs should balance and reinforce each other, not undermine each other and suboptimize processes.
- *Trigger changes.* The act of measuring a KPI should trigger a chain reaction of positive changes in the organization [or project], especially when it is monitored by the CEO [or customers or sponsors].
- *Standardized.* KPIs are based upon standard definitions, rules, and calculations so they can be integrated across dashboards throughout the organization.
- *Context driven.* KPIs put performance in context by applying targets and thresholds to performance so users can gauge their progress over time.
- *Reinforced with incentives.* Organizations can magnify the impact of KPIs by attaching compensation or incentives to them. However, they should do this cautiously, applying incentives only to well-understood and stable KPIs.
- *Relevant.* KPIs gradually lose their impact over time, so they must be periodically reviewed and refreshed.

There are several reasons why the use of KPIs often fails on projects, including:

- People believe that the tracking of a KPI ends at the first line manager level.
- The actions needed to regulate unfavorable indications are beyond the control of the employees doing the monitoring or tracking.
- The KPIs are not related to the actions or work of the employees doing the monitoring.
- The rate of change of the KPIs is too slow, thus making them unsuitable for managing the daily work of the employees.
- Actions needed to correct unfavorable KPIs take too long.
- Measurement of the KPIs does not provide enough meaning or data to make them useful.
- The company identifies too many KPIs to the point where confusion reigns among the people doing the measurements.

Years ago, the only metrics that some companies used were those identified as part of the earned-value measurement system. The metrics generally focused only on time and cost and neglected metrics related to business success as opposed to project success. Therefore, the measurement metrics were the same on each project and the same for each life-cycle phase. Today, metrics can change from phase to phase and from project to project. The hard part is, obviously, deciding upon which metrics to use. Care must be taken that whatever metrics are established do not end up comparing apples and oranges. Fortunately, there are several good books in the marketplace that can assist in identifying proper or meaningful metrics.[32]

---

32. Three books that provide examples of metric identification are P. F. Rad and G. Levin, *Metrics for Project Management,* Vienna, VA: Management Concepts, 2006; M. Schnapper and S. Rollins, *Value-Based Metrics for Improving Results,* Ft. lauderdale, FL: J. Ross Publishing, 2006; and D. W. Hubbard, *How To Measure Anything*; Hoboken, NJ: Wiley, 2007.

Selecting the right KPIs is critical. Since a KPI is a form of measurement, some people believe that KPIs should be assigned only to those elements that are tangible. Therefore, many intangible elements that should be tracked by KPIs never get looked at because someone believes that measurement is impossible. Anything can be measured regardless of what some people think. According to Hubbard[33]:

- Measurement is a set of observations that reduces uncertainty where the results are expressed as a quantity.
- A mere reduction, not necessarily elimination, of uncertainty will suffice for a measurement.

Therefore, KPIs can be established even for intangibles like those discussed later in this book in the chapter on value-driven project management.

Hubbard believes that five questions should be asked before we establish KPIs for measurement[34]:

- What is the decision this [KPI] is supposed to support?
- What really is the thing being measured [by the KPI]?
- Why does this thing [and the KPI] matter to the decision being asked?
- What do you know about it now?
- What is the value to measuring it further?

Hubbard also identifies four useful measurement assumptions that should be considered when selecting KPIs[35]:

- Your problem [in selecting a KPI] is not as unique as you think
- You have more data than you think
- You need less data than you think
- There is a useful measurement that is much simpler than you think

Selecting the right KPIs is essential. On most projects, only a few KPIs are needed. Sometimes we seem to select too many KPIs and end up with some KPIs that provide us with little or no information value, and the KPI ends up being unnecessary or useless in assisting us in making project decisions.

Sometimes, companies believe that the measures that they have selected are KPIs when, in fact, they are forms of performance measures but not necessarily KPIs. David Parmenter discusses three types of performance measures[36]:

- *Key results indicators* (KRIs) tell you how you have done in a perspective
- *Performance indicators* (PIs) tell you what to do
- *KPIs* tell you what to do to increase performance drastically

---

33. D. W. Hubbard, *How To Measure Anything*, Hoboken, NJ: Wiley, 2007, p. 21.
34. Ibid, p. 43.
35. Ibid, p. 31.
36. D. Parmenter, *Key Performance Indicators*, Hoboken, NJ: Wiley, 2007, p. 1.

Parmenter believes that[37]:

The ultimate success of a change strategy depends greatly on how the change is introduced and implemented, rather than on the merit of the strategy itself. Successful development and utilization of key performance indicators (KPIs) in the workplace is determined by the presence or absence of four foundation stones:

- Partnership with the staff, unions, key suppliers, and key customers
- Transfer of power to the front line
- Integration of measurement, reporting, and improvement of performance
- Linkage of performance measures to strategy

In a project environment, the performance measures can change from project to project and phase to phase. The identification of these measures is performed by the project team, including the project sponsor. Project stakeholders may have an input as well. Corporate performance measures are heavily financially oriented and may undergo very little change over time. The measurements indicate the financial health of the corporation.

Establishing corporate performance measures related to strategic initiatives or other such activities must be treated as a project in itself, and supported by the senior management team (SMT).

The SMT attitude is critical—any lack of understanding, commitment, and prioritizing of this important process will prevent success. It is common for the project team and the SMT to fit a KPI project around other competing, less important firefighting activities.

The SMT must be committed to the KPI project, to driving it down through the organization. Properly implemented, the KPI project will create a dynamic environment. Before it can do this, the SMT must be sold on the concept. This will lead to the KPI project's being treated as the top priority, which may mean the SMT's allowing some of those distracting fires to "burn themselves out."[38]

## 1.11 STEP 3: VALIDATING THE BEST PRACTICE

Previously we stated that seeking out of a best practice is done by the project manager, project team, functional manager, and/or possibly a professional facilitator trained in how to debrief a project team and extract best practices. Any or all of these people must believe that what they have discovered is, in fact, a best practice. When project managers are quite active in a project, emphasis is placed upon the project manager for the final decision on what constitutes a best practice. According to a spokesperson for AT&T, the responsibility for determining what is a best practice rests with:

The individual project manager that shows how it had a positive impact on their project.

---

37. Ibid., p. 19.
38. Ibid., p. 27. Chapter 5 of this book has excellent templates for reporting KPIs.

Although this is quite common, there are other validation methods that may involve a significant number of people. Sometimes, project managers may be removed from where the work is taking place and may not be familiar with activities that could lead to the identification of a best practice. Companies that have a PMO place a heavy reliance on the PMO for support because the approved best practices are later incorporated into the methodology, and the PMO is usually the custodian of the methodology. According to Heidi Wurtz, VP, Solutions Center maxIT-VCS:

> The methodology maintenance and update process is being defined now as a function of our Solution Center supporting the integrated maxIT-VCS. The vision is to centrally manage and maintain all methodologies, from core skills such as project management, clinician adoption, etc. through specialized methodologies such as ICD-10 and Health Analytics. Implementing new technologies and processes (e.g.ICD-10) can have a fundamental impact on an organization's entire clinical and revenue cycle processes. Project failures can result in compliance issues, delays and denials in claims payments, and significant productivity declines. Now, more than ever, it is imperative that organizations have a strong project structure in place to support this transformation. maxIT-VCS strives to fill the gap between the HIT vendors' deliverable commitments and the existing staff by developing solutions tailored to the organization's needs. Our methodology maintenance process enables subject matter expert engagement to ensure consistency in updates and methodology change management. The plan is to identify a small group of experts for each methodology, and convene them on a regular basis to review content material, feedback from users, industry best practices, internal lessons learned and contributions of new material. This body will make decisions to update, remove, replace or add to the methodology content including tools. We envision an interactive component of the web-based methodology site to facilitate feedback and contributions from users in the field.

Once the management of the organization affected initially approves the new best practice, it is forwarded to the PMO or process management for validation and then institutionalization. The PMO may have a separate set of checklists to validate the proposed best practice. The PMO must also determine whether or not the best practice is company proprietary because that will determine where the best practice is stored and whether the best practice will be shared with customers.

The best practice may be placed in the company's best practice library or, if appropriate, incorporated directly into the company's stage gate checklist. Based upon the complexity of the company's stage gate checklist process and enterprise project management methodology, the incorporation process may occur immediately or on a quarterly basis.

According to Chuck Millhollan, director of program management at Churchill Downs, Incorporated:

> We do not label our processes or methods as "best practices." We simply learn from our lessons and ensure that learning is incorporated into our methodology, processes, templates, etc.

Some organizations have committees not affiliated with the PMO that have as their primary function the evaluation of potential best practices. Anyone in the company can provide potential best practices data to the committee, and the committee in turn does the analysis. Project managers may be members of the committee. Other organizations use the PMO to perform this work. These committees and the PMO most often report to the senior levels of management.

The fourth edition of the *PMBOK® Guide* emphasizes the importance of stakeholder involvement in projects. This involvement may also include the final decision on whether or not a discovery is a best practice. According to Chuck Millhollan, director of program management, Churchill Downs, Inc.:

> Ultimately, the final decision resides with our stakeholders, both internal and external. Another way of putting this is that the PMO does not make the decision if a method or process works. We actively seek feedback from our project stakeholders and use their inputs to determine if our processes are "best practices" for Churchill Downs Incorporated. The specific best practices identified previously, among others, have even been accepted outside of the PMO as generally accepted practices.

Another example of stakeholder involvement is provided by Enrique Sevilla Molina, PMP, formerly corporate PMO director, Indra:

> The decision is taken by the corporate PMO responsible, the business unit manager, the local PMO authority, or even the cognizant authority, if it is the case. It depends on the subject and the scope of the task. Some of the management best practices have been established at corporate level, and they have been incorporated into the PM methodology. Many of them have also been incorporated into the Project Management Information Systems and the corporate PM tooling.

Evaluating whether or not something is a best practice is not time-consuming, but it is complex. Simply because someone believes that what he or she is doing is a best practice does not mean that it is in fact a best practice. Some PMOs are currently developing templates and criteria for determining that an activity may qualify as a best practice. Some items that are included in the template might be:

- Is transferable to many projects
- Enables efficient and effective performance that can be measured (i.e., can serve as a metric)
- Enables measurement of possible profitability using the best practice
- Allows an activity to be completed in less time and at a lower cost
- Adds value to both the company and the client
- Can differentiate us from everyone else

One company had two unique characteristics in its best practices template:

- Helps to avoid failure
- If a crisis exists, helps us to get out of a critical situation

Executives must realize that these best practices are, in fact, intellectual property that benefit the entire organization. If the best practice can be quantified, then it is usually easier to convince senior management of its value.

## 1.12  STEP 4: LEVELS OF BEST PRACTICES

As stated previously, best practices come from knowledge transfer and can be discovered anywhere within or outside of your organization. This is shown in Figure 1–9.

Companies that maintain best practices libraries that contain a large number of best practices may create levels of best practices. Figure 1–10 shows various levels of best practices. Each level can have categories within the level. The bottom level is the professional standards level, which would include professional standards as defined by PMI. The professional standards level contains the greatest number of best practices, but they are more of a general nature than specific and have a low level of complexity.

The industry standards level would identify best practices related to performance within the industry. The automotive industry has established standards and best practices specific to the auto industry.

As we progress to the individual best practices in Figure 1–10, the complexity of the best practices goes from general to very specific applications and, as expected, the quantity of best practices is less. An example of a best practice at each level might be (from general to specific):

- *Professional standards:* Preparation and use of a risk management plan, including templates, guidelines, forms, and checklists for risk management.

**Figure 1–9.**  Knowledge transfer.

**Figure 1–10.**  Levels of best practices.

- *Industry specific:* The risk management plan includes industry best practices such as the best way to transition from engineering to manufacturing.
- *Company specific:* The risk management plan identifies the roles and interactions of engineering, manufacturing, and quality assurance groups during transition.
- *Project specific:* The risk management plan identifies the roles and interactions of affected groups as they relate to a specific product/service for a customer.
- *Individual:* The risk management plan identifies the roles and interactions of affected groups based upon their personal tolerance for risk, possibly through the use of a responsibility assignment matrix prepared by the project manager.

Best practices can be extremely useful during strategic planning activities. As shown in Figure 1–11, the bottom two levels may be more useful for project management strategy formulation whereas the top three levels are more appropriate for the execution or implementation of a strategy.

Not all companies maintain a formal best practices library. In some companies, when a best practice is identified and validated, it is immediately placed into the stage gate process or the project management methodology. In such a case, the methodology itself becomes the best practice. Enrique Sevilla Molina, PMP, formerly corporate PMO director at Indra, states:

> In fact, our Project Management methodology constitutes our established library of best practices applicable to every project in the company. There are additional best practices libraries in different business units. There are, for instance, detailed instructions for proposal preparation or for cost and schedule estimation purposes, which are appropriate for the specific business or operations area.

When asked how many best practices they maintain at Indra, Enrique commented:

- It is hard to say because of the subject itself and the multiplicity of business areas in the company. If we consider our PM methodology as a set of "best practices", it would be difficult to count every best practice included.
- Besides our internally published Indra Project Management Methodology Manual, we have for instance specific guides at corporate level for WBS elaboration, project

**Figure 1–11.** Usefulness of best practices.

risk management, and the project's performance measurement based on earned value techniques. We have also specific instructions published for proposal preparation, costs estimation, and even detailed WBS preparation rules and formats for different business unit levels.

## 1.13 STEP 5: MANAGEMENT OF BEST PRACTICES

There are three players involved in the management of the best practices:

- The best practice's owner
- The PMO
- The best practices' library administrator who may reside in the PMO

The best practice's owner, who usually resides in the functional area, has the responsibility of maintaining the integrity of the best practice. Being a best practice owner is usually an uncompensated, unofficial title but is a symbol of prestige. Therefore, the owner of the best practice tries to enhance it and keep the best practice alive as long as possible.

The PMO usually has the final authority over best practices and makes the final decision on where to place the best practice, who should be allowed to see it, how often it should be reviewed or revalidated, and when it should be removed from service.

The library administrator is merely the caretaker of the best practice and may keep track of how often people review the best practice assuming it is readily accessible in the best practices library. The library administrator may not have a good understanding of each of the best practices and may not have any voting rights on when to terminate a best practice.

## 1.14 STEP 6: REVALIDATING BEST PRACTICES

Best practices do not remain best practices forever. Because best practices are directly related to the company's definition of project success, the definition of a best practice can change and age as the definition of success changes. Therefore, best practices must be periodically reviewed. The critical question is, "How often should they be reviewed?" The answer to this question is based upon how many best practices are in the library. Some companies maintain just a few best practices, whereas large, multinational companies may have thousands of clients and maintain hundreds of best practices in their libraries. If the company sells products as well as services, then there can be both product-related and process-related best practices in the library.

The following two examples illustrate the need for reviewing best practices.

- Once a practice has been nominated and approved to be a best practice, it is only sanctioned until the next yearly review cycle. Over time, best practices have the tendency to lose value and become ineffective if they are allowed to age. (EDS)

- Best practices are reviewed every four months. Input into the review process includes:
  - Lessons learned documents from project completed within the past four months
  - Feedback from project managers, architects, and consultants
  - Knowledge that subject matter experts (i.e., best practices owners) bring to the table; this includes information gathered externally as well as internally
  - Best practices library reporting and activity data (Computer Associates)

There are usually three types of decisions that can be made during the review process:

- Keep the best practice as is until the next review process.
- Update the best practice and continue using it until the next review process.
- Retire the best practice from service.

## 1.15 STEP 7: WHAT TO DO WITH A BEST PRACTICE

Given the definition that a best practice is an activity that leads to a sustained competitive advantage, it is no wonder that some companies have been reluctant to make their best practices known to the general public. Therefore, what should a company do with its best practices if not publicize them? The most common options available include:

- *Sharing knowledge internally only:* This is accomplished using the company's intranet to share information with employees. There may be a separate group within the company responsible for control of the information, perhaps even the PMO. Not all best practices are available to every employee. Some best practices may be password protected, as discussed below.
- *Hidden from all but a select few:* Some companies spend vast amounts of money on the preparation of forms, guidelines, templates, and checklists for project management. These documents are viewed as both company-proprietary information and best practices and are provided to only a select few on a need-to-know basis. An example of a "restricted" best practice might be specialized forms and templates for project approval where information contained within may be company-sensitive financial data or the company's position on profitability and market share.
- *Advertise to the company's customers:* In this approach, companies may develop a best practices brochure to market their achievements and may also maintain an extensive best practices library that is shared with their customers after contract award. In this case, best practices are viewed as competitive weapons.

Most companies today utilize some form of best practices library. According to a spokesperson from AT&T:

> The best practices library is Sharepoint based and very easy to use both from a submission and a search perspective. Any Project Manager can submit a best practice at any time and can search for best practices submitted by others.

Even though companies collect best practices, not all best practices are shared outside of the company even during benchmarking studies where all parties are expected to share information. Students often ask why textbooks do not include more information on detailed best practices such as forms and templates. One company commented to the author:

> We must have spent at least $1 million over the last several years developing an extensive template on how to evaluate the risks associated with transitioning a project from engineering to manufacturing. Our company would not be happy giving this template to everyone who wants to purchase a book for $85. Some best practices templates are common knowledge and we would certainly share this information. But we view the transitioning risk template as proprietary knowledge not to be shared.

## 1.16  STEP 8: COMMUNICATING BEST PRACTICES ACROSS THE COMPANY

Knowledge transfer is one of the greatest challenges facing corporations. The larger the corporation, the greater the challenge of knowledge transfer. The situation is further complicated when corporate locations are dispersed over several continents. Without a structured approach for knowledge transfer, corporations can repeat mistakes as well as missing valuable opportunities. Corporate collaboration methods must be developed. NXP has found a way to overcome several of these barriers. Mark Gray, MBA, PMP, Ph.D., formerly senior project manager at NXP Semiconductor and now CEO of SigmaPM, discusses this approach. Mark calls this: To grow oaks, you need to start with nuts . . .:

> One of the biggest problems facing project managers and their organizations today is how best to get knowledge transferred from the experts (or at least experienced) project managers to the rest of the organization. Much has been written on lessons *ignored* and equally as much has been written on having lessons learned as a value-added component in the toolbox of project management. What seems to be missing is a sound approach for getting the knowledge transmitted (the identified, captured, stored parts are all pretty obvious).
>
> At NXP we noted that there are many lessons that are captured during projects or during their review but very little evidence that these lessons were even being seen by other project managers. One solution we put in place was to use the model of a Community of Practice (CoP) as described by Wenger[39] in his work on the subject. Using the same basic idea as used by Shell in their off-shore drilling platforms, we established local forums of "experts" with the specific mandate to create an arena in which project managers would feel comfortable sharing their findings and learning's from their projects.
>
> The process itself is very simple; Lessons are identified by the project managers either from project debriefs or from peer reviews and these are then presented to the forum as a type of "war story". It's important to note here that we look for both good and "less good" incidents to learn from. In general this leads to a good (sometimes spirited) debate on the topic from which the participants can take away a genuine learning experience. The

39. E. Wenger, R. McDermott, and W. M. Snyder, *Cultivating Communities of Practice: A Guide to Managing Knowledge*, Cambridge, MA: Harvard Business School Press, Cambridge, MA, 2002.

results are of course somewhat qualitative in nature so to say we have clear measurable improvements as a direct result of these sessions would be difficult, however we have seen a general improvement in the overall performance of projects since we started this initiative.

Now for the hard part—building, and maintaining the CoP. Wenger gives some very good basics of how to construct CoP's, including the need to have a good core team, involvement of executive sponsors, clear outcomes, general house rules etc. He also talks about the need for a high level of energy input from the core team to create and maintain the momentum of the CoP. We have seen this in real life and add our own particular elements to the recipe:

- Start with a nut—You need to have at least one very extroverted and charismatic lead figure. We seek out people that are not just experts in the field but are almost fanatical in their dedication to developing and maturing project management.
- Plant it in the right place—You need to ensure the CoP does not interfere too much with either normal work or personal time. We generally hold our sessions around the lunch hour with food provided.
- You need occasionally to prune the CoP branches to promote strong growth. Groups can develop in directions which do not really promote the central theme of the CoP, in which case it's a good idea to spin them off to avoid dilution.
- When the members leave, don't panic, this is just seasonal. People come and go as their jobs change, their project pressure changes, etc. They'll be back . . .
- The CoP needs to produce more nuts which can be planted in other sites and thus grow an interlinked community of communities. In NXP's case we created a solid core team over the last few years with active communities sharing and learning in 10 different sites around the world.

There is no point in capturing best practices unless the workers know about it. The problem, as identified above, is how to communicate this information to the workers, especially in large, multinational companies. Some of the techniques include:

- Websites
- Best practices libraries
- Community of practice
- Newsletters
- E-mailings
- Internal seminars
- Transferring people
- Case studies
- Other techniques

Nortel Networks strives to ensure timely and consistent communications to all project managers worldwide to help drive continued success in the application of the global project management process. Examples of the various communication methods used by Nortel include:

- The *PM Newsflash* is published on a monthly basis to facilitate communications across the project management organization and related functions.

- Project management communications sessions are held regularly, with a strong focus on providing training, metrics reviews, process and template updates, and so on.
- Broadcast bulletins are utilized to communicate time-sensitive information.
- A centralized repository has been established for project managers to facilitate easy access to and sharing of project management–related information.[40]

The comments by Nortel make it clear that best practices in project management can now permeate all business units of a company, especially those companies that are multinational.

One of the reasons for this is that we now view all activities in a company as a series of projects. Therefore, we are managing our business by projects. Given this fact, best practices in project management are now appearing throughout the company.

Publishing best practices in some form seems to be the preferred method of communications. At Indra, Enrique Sevilla Molina, PMP, formerly corporate PMO director, states:

> They are published at corporate level and at the corresponding level inside the affected business unit. Regular courses and training is also provided for newly appointed Project Managers, and their use is periodically reviewed and verified by the internal audit teams. Moreover, the PM corporate tools automate the applications of best practices in projects, as PM best practices become requirements to the PM information systems.

According to a spokesperson from AT&T:

> We have defined a best practice as any tool, template or activity that has had a positive impact on the triple constraint and/or any of the *PMBOK® Guide* Process or Knowledge areas. We allow the individual project manager to determine if it is a best practice based on these criteria. We communicate this through a monthly project management newsletter and highlight a best practice of the month for our project management community.

Another strategic importance of best practices in project management can be seen from the comments below by Suzanne Zale, Hewlett-Packard operations director and formerly global program manager at EDS[41]:

> Driven by the world economy, there is a tendency toward an increasing number of large-scale global or international projects. Project managers who do not have global experience tend to treat these global projects as large national projects. However, they are completely different. A more robust project management framework will become more important for such projects. Planning up front with a global perspective becomes extremely important. As an example, establishing a team that has knowledge about geographic regions relevant to the project will be critical to the success of the projects. Project managers must also know how to operate in those geographic areas. It is also essential that all project team members are trained and understand the same overall project management methodology.
>
> Globalization and technology will make sound project management practice even more important.

---

40. H. Kerzner, *Project Management Best Practices: Achieving Global Excellence*, Hoboken, NJ: Wiley, 2006, p. 18.
41. Harold Kerzner, *Advanced Project Management: Best Practices on Implementation*, 2nd edition, Hoboken, NJ: Wiley, 2004; p.21

Suzanne Zale's comments illustrate the importance of extracting best practices from global projects. This could very well be the future of best practices by the end of this decade.

## 1.17 STEP 9: ENSURING USAGE OF THE BEST PRACTICES

Why go through the complex process of capturing best practices if people are not going to use them? When companies advertise to their clients that they have best practices, it is understood that tracking of the best practices and how they are used must be done. This is normally part of the responsibility of the PMO. The PMO may have the authority to regularly audit projects to ensure the usage of a best practice but may not have the authority to enforce the usage. The PMO may need to seek out assistance from the head of the PMO, the project sponsor, or various stakeholders for enforcement.

When best practices are used as competitive weapons and advertised to potential customers as part of competitive bidding, the marketing and sales force must understand the best practices and explain this usage to the customers. Unlike ten years ago, the marketing and sales force today has a good understanding of project management and the accompanying best practices.

## 1.18 COMMON BELIEFS

There are several common beliefs concerning best practices that companies have found to be valid. A partial list is:

- Because best practices can be interrelated, the identification of one best practice can lead to the discovery of another best practice, especially in the same category or level of best practices. Best practices may be self-perpetuating.
- Because of the dependencies that can exist between best practices, it is often easier to identify categories for best practices rather than individual best practices.
- Best practices may not be transferable. What works well for one company may not work for another company.
- Even though some best practices seem simplistic and based on common sense in most companies, the constant reminder and use of these best practices lead to excellence and customer satisfaction.
- Best practices are not limited exclusively to companies in good financial health. Companies that are cash rich can make a $10 million mistake and write it off. But companies that are cash poor are very careful in how they approve projects, monitor performance, and evaluate whether or not to cancel the project.

Care must be taken that the implementation of a best practice does not lead to detrimental results. One company decided that the organization had to recognize project management as a profession in order to maximize performance and retain qualified people. A project management career path was created and integrated into the corporate reward system.

TABLE 1–8.  IMPROPER APPLICATION OF BEST PRACTICES

| Type of Best Practice | Expected Advantage | Potential Disadvantage |
| --- | --- | --- |
| Use of traffic light reporting | Speed and simplicity | Poor accuracy of information |
| Use of a risk management template/form | Forward looking and accurate | Inability to see all possible risks |
| Highly detailed WBS | Control, accuracy, and completeness | More control and cost of reporting |
| Using enterprise project management on | Standardization and consistency | Too expensive on certain projects all projects |
| Using specialized software | Better decision making | Too much reliance on tools |

Unfortunately, the company made a severe mistake. Project managers were given significantly larger salary increases than line managers and workers. People became jealous of the project managers and applied for transfer into project management, thinking that the "grass was greener." The company's technical prowess diminished and some people resigned when not given the opportunity to become a project manager.

Sometimes, the implementation of a best practice is done with the best of intentions, but the final result either does not meet management's expectations or may even produce an undesirable effect. The undesirable effect may not be apparent for some time. As an example, consider the first best practice in Table 1–8. Several companies are now using traffic light reporting for their projects. One company streamlined its intranet project management methodology to include "traffic light" status reporting. Beside every work package in the work breakdown was a traffic light capable of turning red, yellow, or green. Status reporting was simplified and easy for management to follow. The time spent by executives in status review meetings was significantly reduced and significant cost savings were realized.

Initially, this best practice appeared to be beneficial for the company. However, after a few months, it became apparent that the status of a work package, as seen by a traffic light, was not as accurate as the more expensive written reports. There was also some concern as to who would make the decision on the color of the traffic light. Eventually, the traffic light system was enlarged to include eight colors, and guidelines were established for the decision on the color of the lights. In this case, the company was fortunate enough to identify the disadvantage of the best practice and correct it. Not all disadvantages are easily identified, and those that are may not always be correctable.

There are other reasons why best practices can fail or provide unsatisfactory results. These include:

- Lack of stability, clarity, or understanding of the best practice
- Failure to use best practices correctly
- Identifying a best practice that lacks rigor
- Identifying a best practice based upon erroneous judgment
- Failing to provide value

## 1.19 BEST PRACTICES LIBRARY

With the premise that project management knowledge and best practices are intellectual properties, how does a company retain this information? The solution is usually the

**Figure 1–12.** Levels of best practices.

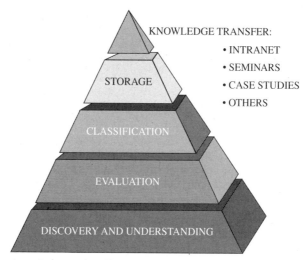

**Figure 1–13.** Creating a best practices library.

creation of a best practices library. Figure 1–12 shows the three levels of best practices that seem most appropriate for storage in a best practices library.

Figure 1–13 shows the process of creating a best practices library. The bottom level is the discovery and understanding of what is or is not a "potential" best practice. The sources for potential best practices can originate anywhere within the organization.

The next level is the evaluation level to confirm that it is a best practice. The evaluation process can be done by the PMO or a committee but should have involvement by the senior levels of management. The evaluation process is very difficult because a one-time positive occurrence may not reflect a best practice that will be repetitive. There must exist established criteria for the evaluation of a best practice.

Once a best practice is established, most companies provide a more detailed explanation of the best practice as well as providing a means for answering questions concerning its use. However, each company may have a different approach to disseminating this critical intellectual property. Most companies prefer to make maximum utilization out of the company's intranet websites. However, some companies simply consider their current forms and templates as the ongoing best practices library.

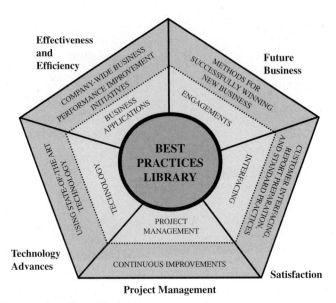

**Figure 1–14.** Best practices library.

Figure 1–12 showed the levels of best practices, but the classification system for storage purposes can be significantly different. Figure 1–14 shows a typical classification system for a best practices library.

The purpose for creating a best practices library is to transfer knowledge to the employees. The knowledge can be transferred through the company intranet, seminars on best practices, and case studies. Some companies require that the project team prepare case studies on lessons learned and best practices before the team is disbanded. These companies then use the case studies in company-sponsored seminars. Best practices and lessons learned must be communicated to the entire organization. The problem is determining how to do it effectively.

Another critical problem is best practices overload. One company started up a best practices library and, after a few years, had amassed what it considered to be hundreds of best practices. Nobody bothered to reevaluate whether or not all of these were still best practices. After reevaluation had taken place, it was determined that less than one-third of these were still regarded as best practices. Some were no longer best practices, others needed to be updated, and others had to be replaced with newer best practices.

## 1.20 HEWLETT-PACKARD: BEST PRACTICES IN ACTION[42]

**Identifying Specific Activities as Best Practices**

The focus of our organization within HP is the management of "IT," or information technology. IT consist of all of the hardware, software, net-

---

42. Material on HP has been provided by Doug Bolzman, consultant architect, PMP®, ITIL Expert at HP.

works, facilities that provide information, at the right time, to the right people, to enable the people to perform their jobs or fulfill their business obligations/responsibilities. IT departments have matured over the decades to resemble the other business units of an organization especially in the management of IT in the form of services. A best practice for the industry called ITIL®, which is the short name for IT Infrastructure Library®, was introduced in the mid-eighties and revolutionized the IT industry by promoting a single set of best practices that have been built by the global IT community. In the 90's we subscribed to the initial release of ITIL which was a collection of individual publications. At the turn of the century, we subscribed to version 2 which evolved the individual books for a grouping of practices, centered on Service Support and Service Delivery. Then we (along with the rest of the IT community) upgraded to the 2007 and 2011 editions that provided a lifecycle management approach for the management of IT Services. The reason for the history lesson is that we have subscribed to and have matured our practices with the maturity of this best practice.

IT Services are developed to support the functionality of a business unit. So a "billing system" is the IT Service to the Financial Team's function for revenue collection. The billing system is broken into IT components such as business applications, file servers, network connectivity, facilities etc. ITIL enables companies to manage those components with best practices such as Availability Management, Capacity Management, Problem Management and Change Management. Today, we have identified 28 best practices from ITIL and we developed an additional 6 that enable the implementation and management of an IT Service Lifecycle approach.

**Definition of a Best Practice**    Our brief definition of best practice is a leveragable capability. ITIL defines a capability as the ability of an organization or IT service to carry out an activity. A capability is comprised of three designs, a process design (the activities to be followed to produce the result), a people design (the structure and training of the role to equip the person to fulfill their responsibilities), and a tool design (the equipment or application used to automate the work).

To illustrate this, let's look at something we all understand, a hospital emergency room. The emergency room staff does not know what will come through the emergency doors; they have to be prepared for everything! Once a patient comes through the doors, either on their own or on a gurney the situation is identified and the proper room, tools and personnel are engaged to properly react. If someone is having a heart attack, the emergency room personnel have a crash cart prepared with all of the equipment and required drugs in addition to the proper procedures to follow, and doctors with the appropriate skills. If someone comes in with a sprained ankle, they will be sitting in the waiting room for quite a while as higher-priority patients are managed. But the sprained ankle does not wait forever, they are scheduled in.

Much like a hospital, the IT Service Desk does not know who will call with what problem, but they have to be prepared for everything. Once a call comes into the Service Desk, either a phone call or a triggered event, the situation is identified and the proper personnel and tools properly react. If someone has a question or a simple request, which can be handled at the initial call, if a vital business function is down due to an IT error, the appropriate personnel are assembled in a conference call and all emergency actions and escalations are taken until the error is corrected and the business functionality is restored.

A functional Service Desk with urgent incident response is based on the best practice of a hospital emergency room. The specific capabilities for the IT Delivery Teams or Operation Staff to carry out an activity is based on people, process and tool designs.

In order to manage each of the Best Practices consistently, each capability is documented following the Best Practice Profile template. Some of the information contained within the profile includes the description of the practice, the type, the value to the company, list of practitioners to use the practice. Each of the practices have documentation of the assets, asset status and the business drivers that have been used to develop the practice.

### Final Decision on a Best Practice

In order to understand who has the final decision of what is a best practice, we need to explore how we manage best practices. This is shown in Figure 1–15. The management of practices starts with establishing the directorate of people that will own and manage the practice. For this example, the best practice is Incident Management. All aspects of the practice will be managed; from the budgets and direction to how it fits into the overall service model to ownership and the buy-in of the stakeholders and finally to the management of improvements or design revisions (and de-commissions).

Secondly, is the understanding of how the best practice will be used, the value to be obtained and the drivers that will dictate the required design. This is where many "best practices" fail, when there is a "fits all sizes" mentality. We have 3 or 4 Incident Management designs that are the best practice for the 3 or 4 clients they support. Each best practice is designed around a specific set of attributes (business drivers) which denotes

**Figure 1–15.** Management of a best practice.

the characteristics of the practice that make it the best fit for that client. Unless each client shares the same business drivers and standards, there will never be a single best practice that can be applied to meet their needs. Some clients do request a leveraged design and are flexible to adopt the single set of attributes that they will be restricted to, allowing them a lower-cost, leveraged design that is implemented in many clients.

Thirdly, the designs for the people, process, and tools are built and integrated to support the agreed-upon attributes. This model will support a single best practice for incident, or it can be broken down further to the 25 specific capabilities such as logging an incident or determining the incident priority rating. Understanding and managing at the capability level provides us the flexibility to mix and match capability designs to meet the client's needs. If the client has governmental standards, such as Sarbanes-Oxley, or industry standards, such [as ones for the] pharmaceutical [industry], we can then identify the designs that were built to comply with those standards.

**A Library of Best Practices** The library of our IT Service Management practices are stored in a shared workgroup tool with a meta-data type of sorting. Each practice is under version control and mapped to the company that it was applied to. This way, if we have two companies in the same industry, we can always start with an existing design, then adjust based on the definition of their attributes.

### The Number of Best Practices

We have hundreds of best practices in the library since they can be at a component level (Incident Management) or a capability level (25 capabilities of Incident Management). In addition, since each capability is composed of a People, Process and tool designs so the library is best managed in a Configuration Management database.

**Getting Support for Best Practices** The best practices used by our organization are managed by ITSM Consultants and we maintain our service collateral of best practices. Each ITSM consultant service has an owner assigned to oversee the best practice designs and to apply the correct design as the starting point of a new engagement. This is a value add for a client to hire our consultants since they can be assured that we will start with the best applicable practice without having to understand the entire system. They can be the customers, provide us their attributes and we demonstrate the design based on compliance to the attributes.

## 1.21 DTE ENERGY[43]

In 2002, the Information Technology Services (ITS) Organization at DTE Energy initiated an effort to collect and document best practices for project management. Our intent was to publish, communicate, and ensure these best practices were adopted across the culture.

---

43. Material on DTE Energy provided by Joseph C. Thomas, PMP, senior project manager, and Steve Baker, PMP, CCP, principal analyst, Process and Skills Organization.

We believed this approach would lead to continuous improvement opportunities, resulting in higher quality, more timely, and less expensive software-based business solutions.

Rather than describe the ideal state of project management as we could envision it, we decided to describe the current state as it was being practiced by project managers in the organization. The goal was to establish a baseline set of standards that we knew project managers could meet because we were already doing them.

We formed a small team of our most experienced project managers and IT leaders. This team drew upon their recent project experiences to identify a set of best practices. While not every project manager uniformly followed them, the best practices represented the highest common denominator rather than the lowest. We knew that these were feasible since they represented practical experiences of our most successful project managers, and they also characterized the practices we wanted applied consistently across all projects. In this way, the bar was low enough to be achievable but high enough to be a meaningful improvement for most of the projects.

The team agreed to describe the best practices in terms of "what" rather than "how." We wanted to avoid the difficult and time-consuming task of defining detailed procedures with formal documentation. Rather, we described what project managers needed to do and the artifacts that project managers needed to produce. This provided the practicing project managers with some degree of flexibility in the methods (the "how") they employed to produce the results (the "what").

We published the best practices in a hundred-page "Project Management Standards and Guidelines" manual (Table of Contents, Table 1.9) and also posted them on our Solution Delivery Process Center (SDPC) intranet website (Screen Shot, Figure 1–16). We included

**TABLE 1–9.   PROJECT MANAGEMENT STANDARDS AND GUIDELINES**

| Software Engineering, Methods, and Staffing Project Management Standards and Guidelines | | |
|---|---|---|
| **Introduction** | . . . | 1 |
| Target Audience | . . . | 1 |
| Purpose | . . . | 1 |
| Content | . . . | 2 |
| Solution Delivery Process Center | . . . | 3 |
| Certify Your Application Checklist | . . . | 3 |
| Document Revisions | . . . | 3 |
| Document Conventions | . . . | 4 |
| **1. Project Initiating** | . . . | 7 |
| 1.1 Customer Relationship | . . . | 8 |
| 1.2 Funding | . . . | 9 |
| 1.3 Registration | . . . | 10 |
| 1.4 Charter | . . . | 12 |
| **2. Project Planning** | . . . | 17 |
| 2.1 Scope | . . . | 18 |
| 2.2 Feature/Requirement List | . . . | 20 |
| 2.3 Work Item List | . . . | 22 |

**Software Engineering, Methods, and Staffing Project Management Standards and Guidelines**

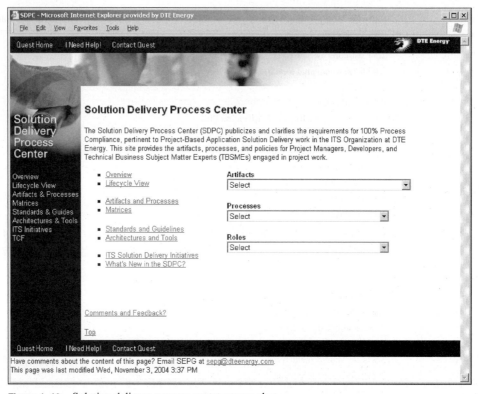

**Figure 1–16.** Solution delivery process center screen shot.

references to other resources such as standard forms, templates, and procedures that already existed.

The SDPC is our process asset library containing scores of high-level role and process descriptions, easy-to-access templates and examples, and links to various resources from other departments across ITS. We designed our digital library to be usable for a variety of perspectives, including role-based ("I am a . . ."), milestone-based ("We are at . . ."), artifact-based ("I need a . . ."), and so on.

We launched the Standards and Guidelines materials and the SDPC library with a diverse communication strategy, targeting the right message to the right audience at the right time. We included a feedback process to ensure the evolution and applicability of the standards over time. This continuous improvement process includes (a) an inbound email account for receiving comments, suggestions, and ideas; (b) a review and update process including roles and milestones; and (c) timely updates to the SDPC and targeted printings of the Standards and Guidelines manual.

As a way to capture our lessons learned from each project, we adopted the After Action Review (AAR) process. Every project conducts an AAR upon completion, and we seek improvements to, and innovations beyond, our existing processes and templates. We institutionalize these improvements within our evolving Standards and Guidelines and the SDPC library. We solved our lessons learned dilemma (best practices were dutifully archived but rarely referenced) by incorporating each discovery within our Standards and Guidelines.

Standards and Guidelines, in and of themselves, are a means to an end. We found that simply publishing and communicating them is not enough to meaningfully impact our culture. To that end, we instituted a Quality Management Group (QMG) staffed with a small, diverse team of experts. The QMG both *enables* our projects with consulting services and education, and *ensures* our projects are in compliance with published Standards and Guidelines. With five inspection milestones, the QMG demonstrates our organizational commitment to best practices and continuous improvement.

## 1.22 A CONSULTANT'S VIEW OF PROJECT MANAGEMENT AND BEST PRACTICES

When companies begin their quest for excellence in project management, emphasis is usually placed upon external benchmarking. While benchmarking studies have merit, the information obtained may have limited or no real value for the company conducting the benchmarking even though they started out with the greatest of intentions. Perhaps the better choice is to hire consultants that can bring to the table expertise from a variety of companies, both large and small. The remainder of this section has been provided by Sandra Kumorowski.[44]

**Project Management in the Small Business Environment**

Generally, in the small business environment, small business owners are more inclined to use project management practices—sometimes not intentionally and mostly through common sense—than anywhere else. I have seen quite sophisticated project estimating, budgeting, and scheduling practices in the small business environment as small business owners consider it the number one skill to master because their livelihood relies on that skill. Scope and objectives are very well defined and expectations set at the beginning of the project. That can, however, vary by industry. Let me start with two examples from construction and health care.

**Project Management in Construction**

My husband has owned his painting (residential and commercial painting) business for over a decade now. Over so many years, there was not a single project on which he had lost money. His precise estimating skills, communication style, and managerial experience enable him to create quite seamless and timely project experiences. When he works with a general contractor, where the success of the construction project lies in the precise alignment of all subcontracting services, accurate project scheduling is critical. When he works with direct clients, precision and time are even more important as clients' lives can be directly impacted by the painting project.

The number one success factor in his business is a very intensive and continuous *training of his employees*. When he hires someone new, the employee goes through an intensive training. All employees are continuously updated and trained on new innovations and techniques in painting.

---

44. Sandra Kumorowski is an assistant professor of marketing communications at Columbia College Chicago and marketing & communications chair, Christopher & Dana Reeve Foundation, Chicago; she is also chief business advisor & owner, Enakta LLC; P: 1–224-715-5666; E: sandrakum@sbcglobal.net, skumorowski@colum.edu

The number two success factor is *managed team autonomy*. Each team has a team leader who is fully responsible for the project performance and is required to update my husband every day on project progress. My husband progressed to this managed team autonomy system from the previous system where he personally almost fully controlled each project (which took up too much of his time and was not necessarily more profitable) and found it very effective. The new system was met with very positive response from the employees who felt more empowered and actually started performing better, especially in terms of time. Now, they finish projects in shorter periods of time keeping up the same if not better quality standards.

My husband's natural project management skills are well refined and he has never used any formal project management concepts or project management software. He has developed his own project management system that has been working well for him. However, there are many project management challenges in the construction industry. Therefore, I believe there is a great opportunity for these small business owners to be introduced to and educated about project management methodology from which they could benefit to a great extent. When surveyed, most of them considered it too time-consuming. But this opinion comes mainly from limited knowledge about project management and often limited computer skills.

## Project Management in Health Care/Dental Category

Based on my 7-year experience in a periodontal specialty practice, I have to admit that the project and patient management practices were quite advanced. The practice was run by one doctor and his incredibly time-efficient and advanced surgical skills allowed for a large number of patients to be seen every day. In one day, the practice could see up to 60 patients divided between the doctor, the hygienist, and the postoperative assistant. That required quite sophisticated scheduling and time-monitoring practices. We used dental practice software Eagelsoft, which, although somewhat limited in reports capability, provided effective tracking and monitoring tools. On average, a patient would see us four to six times a year. To avoid cancellations and disruptions in the schedule, we attempted to schedule all appointments in advance for each patient and established a strict cancellation policy—something quite unusual in the dental category.

Each patient was like a miniproject and without the software and methodical scheduling, we would not be able to manage them successfully. The practice, thanks to its meticulous approach to operations management and scheduling in particular, has enjoyed continuous growth.

## Project Management in Higher Education

Project work could be quite challenging for many college students. In higher education, most of the time, projects represent 50 percent of the final grade, but there is not enough attention paid to project management directions.

I did a survey in my class and, out of 44 students, only 2 were familiar with project management. About 70 percent of students perceived project work as a big challenge mainly because of the project management and team management issues.

So I have taken a proactive approach and dedicated about 10 percent of the total semester class time to project management practices and methods to make their project work more effective and successful.

**Project Management in the Marketing and Advertising Category**

Working in the marketing and advertising category for years now [has] enabled me to study and analyze how project management was understood in the agency environment. Based on my experience, first of all, there is not enough awareness about project management practices and proven benefits. People do not believe in it and, most of the time, project management is perceived as a deterrent to creativity. But it is important to state that the way client accounts and projects are managed is quite consistent with a project management methodology. The problem, however, is that account planners do not reach out to the project management discipline to improve upon processes and make projects perform better. There is no formal relationship between project management and account management in marketing/advertising category.

When I introduced project management to my company of advertising executives as a way to improve our business, it took me some time to make them believe in it.

The story of how I did it follows.

**Project Management Initiative for a Marketing Strategy Consultancy**

About the Company My company was a small (22 employees) but high-profile successful marketing strategy consultancy. Founded in 2003 by an advertising executive, it focuses on delivery of research insights and related actionable strategy based on the analysis of online conversations. It caters to Fortune 500 companies and earned its reputation on coming up with the "big idea" in each project and helping the brands in building their long-term brand equity.

Need for a Project Management Initiative? Due to the project-based nature of the company, I was approached by the top management to evaluate current project management practices and establish a more formal project management system. As the company grew, the need for a consistent system for our project planning and execution became very apparent.

Our projects ranged from two-week turnaround projects to eight-week research studies and there was a need to create a project portfolio management system to better control our future planning.

In addition, some more serious human issues began to surface. Most of the problems were in the initiation and planning phases where communication got lost mainly due to too many people being involved. Scopes were sometimes out of control and the team would find out at the end of the project that they were working on something different than what the client wanted. Main message (scope/objectives) would get lost.

Later in the project, the scoping mistakes would exponentially amplify into sometimes barely controllable situations where team members would have to stay late into the night and on weekends trying to bring the project to the satisfactory completion stage. The stress levels were sometimes very high.

Another issue that kept disturbing our projects was team dynamics. Each team consisted of two members—a senior strategist who was the project lead and a junior strategist. The role of the senior strategist on a project was defined as 80 percent strategy development and 20 percent data gathering and analysis. The junior strategist had to do 80 percent data gathering and was contributing only 20 percent to the strategy development. That led to some strong feelings of unfairness, and some team members refused to work together. That needed to be changed.

There were seven main areas where the use of project management as an instrument for continuous improvement was apparent:

1. Client relationship management (CRM)
2. Team health/dynamics
3. Project quality/creativity
4. Project costing/budgeting
5. Project scope/timing
6. Project portfolio management
7. Company culture

The needed change can be seen in Figure 1–17.

## Project Management Initiative Chronology

1. Need Recognition: Need for project management recognition based on scope creeps, unhappy team, lack of project control, and so on.
2. Initial Buy-In Meeting with Top Management: Establish immediate and relevant need
3. Interviews: Thorough interviews with each employee including top management (I created a questionnaire for each employee and then met with each of them individually to discuss it.)

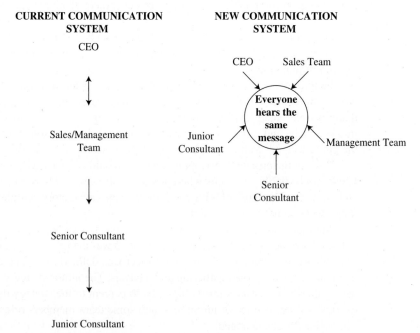

**Figure 1–17.**  The before and after communication system.

**TABLE 1–10. PROJECT MANAGEMENT INITIATIVE OVERVIEW**

| Priority Rank | Area | Action | Outcome/Benefit | Measure |
|---|---|---|---|---|
| 1. | **Client Relationship Mgmt** REASON: RETURNING CLIENT/SALES | 1. **Establish personal contact w/client** (add travel cost to contract, on longer, complex, opportunistic projects 2. **Improve planning phase** 3. **Scope/contract template** 4. **Post mortem template** 5. Establish meetings with standardized agendas & let everyone work on scope statement | • Long-term client relationships • Gaining credibility • More projects • Growing company Reputation | • Client survey • Client retention/return rate • Recommendation rate • Number of new/old client projects per month/year |
| 2. | **Team Health/Dynamics** REASON: RELATIONSHIPS | 1. **Assign 3 people in each project** (3rd for balance/control) 2. **40/60 workload distribution** (instead of 20/80 - Sr. Consultants/Jr. Consultants) 3. **New PM responsibilities** 4. Redefinition of current roles | • Better team collaboration • Less stress and issues • Jr. Consultants feel more valued • Sr. Consultants empowered with more responsibility | • Postmortem evaluation results (Jr. Consultant openly evaluates Sr. Consultant and vice versa?) • WESS (Weighted Employee Satisfaction Survey) ??? |
| 3. | **Creativity** REASON: LIMITED BY 2,4,5 | 1. **Standardize project planning** processes through project structure standardization 2. **Stick to well-defined roles** (everyone knows what to do) | • More time to create • Less time to worry | • Big idea in every project • Actionable insights in every project |
| 4. | **Project Costing/Budgeting** REASON: CONTROL | 1. Devise & implement best **customized project costing system (MS project)** 2. **Cost estimates at the beginning of the** project (MS project) | • Cost control throughout the project • Ability to price projects to avoid losses | • Earned value method • Cost variance (CV = EV – AC), *see Appendix* • CPI (Cost Performance Index) |
| 5. | **Project Scope/Timing** REASON: CONTROL | 1. Implement customized project **timing/planning system (MS project)** 2. Focus on **planning phase to avoid scope/time creep** | • Scope/time control throughout the project | • Earned value method • Schedule variance (SV = EV – PV), *see Appendix* • SPI (Schedule Perf. Index) |
| 6. | **Project Portfolio Mgmt** REASON: BIG PICTURE, STRATEGY | 1. Devise & implement all projects **tracking tool (MS project)** 2. **Analyze projects** based on cost, price, time, time-sensitivity, opportunity | • Effective project portfolio Management to make strategic business decisions • Decreased uncertainty & better future planning | • Profit margin • Effective planning & scheduling of future projects |
| 7. | **Culture** | 1. **Apply transparency** through open communication & active communication throughout each project | • Less fear & resentment • Innovation/creativity • Learning culture • Transparency | • Post-mortem evaluation results generalized & Learned from: best practices library |

4. Research and Analysis: Research of project management best practices, current processes analysis, interview analysis
5. Project Management Best Practices Research Period
6. Second Buy-In Meeting with Top Management: To reconfirm the need for project management, to review cost, timeline, and objectives of the project management initiative, to prioritize areas of focus for project management initiative
7. Project Management Initiative Goal Definition
8. Official Announcement: To all employees
9. Goal and Tactics Implementation: and testing
10. Monthly Progress Reports

Example of Project Management Initiative Initial Goals Since our company was new at project management, the changes took place in stages. Below (see Table 1–10) are some of the goals that were reached and actions taken as soon as the project management initiative was implemented.

**Goal 1 CRM:** To build long-term client relationship, increase word of mouth and reputation, increase sales
**Action:** List of all client meetings to be scheduled.
**Goal 2 Team Health/Dynamics:** To establish effective team collaboration and work load balance to achieve the best-quality product
**Action:** 1. Schedule lecture on new project management processes, effective teamwork and leadership for the whole team FRIDAY. 2. Schedule meeting on leadership with project leaders.
**Goal 3 Project Planning:** Scope management/meeting effectiveness
**Action:** 1. Finalize and use internal kick-off meeting agenda template. 2. Finalize and use internal postmortem meeting agenda template.
**Goal 4 PM Project Monitoring:** To track projects and the success of the project management project
**Action:** 1. Develop ALL projects tracking format (white board). 2. Report on project management goals presented monthly.

# From Best Practice
# to Migraine Headache

**2**

## 2.0 INTRODUCTION

For almost 40 years, project management resided in relatively few industries such as aerospace, defense, and heavy construction. These industries were project-driven and implemented project management mainly to placate customer requests. Project management was considered as something nice to have but not a necessity. As a result, best practices in project management were never really considered to be important.

Within the last two decades, project management has evolved into a management process that is mandatory for the long-term survival of the firm. Project management is now a necessity rather than a luxury. Project management permeates all aspects of a business. Companies are now managing their business by projects. Project management has become a competitive weapon. The knowledge learned from project management is treated as intellectual property and PMOs have been established as the guardians of the project management intellectual property, reporting to the senior levels of management and being given the task of capturing best practices in project management.

As with any new project management activity, benefits are accompanied by disadvantages and potential problems. Some of the problems are small and easy to correct, while others are colossal migraine headaches and keep executives awake at night. The majority of the migraine headaches emanate from either a poor understanding of the benefits of project management or having expectations that are set too high. Other potential problems occur when an activity really is not a best practice and detrimental results occur.

## 2.1 GOOD INTENTIONS BECOMING MIGRAINES

Sometimes, the best intentions can turn into migraine headaches. As an example, one company quickly realized the importance of project management and made it a career path position. This was certainly the right thing to do. Internally, people believed that the company considered project management to be a strategic competency and professionalism in project management evolved. Externally, their customers were quite pleased seeing project management as a career path discipline and the business improved.

These good intentions soon turned into problems. To show their support for excellence in project management, the project managers were provided with 14 percent salary increases, whereas project team members and line managers received 3–4 percent. Within two years after implementing a project management career path, everyone was trying to become a project manager and climb the project management career path ladder of success, including critical line managers with specialized expertise. Everyone thought that "the grass was greener" in the project manager's yard than in his or her yard. Line managers with critical skills were threatening to resign from the company if they were not given the chance to become project managers. The company eventually corrected the problem by telling everyone that every career path ladder in the company had the same career path opportunities for advancement. The large differential in salary increases disappeared and was replaced by a more equitable plan. However, the damage was done. Team members and line managers felt that the project managers exploited them, and the working relationship suffered. Executives were now faced with the headache of trying to repair the damage.

Figure 2–1 illustrates why many other headaches occur. As project management grows and evolves into a competitive weapon, pressure is placed upon the organization to implement best practices, many of which necessitate the implementation of costly internal control systems for the management of resources, costs, schedules, and quality. The project management systems must be able to handle several projects running concurrently. Likewise, obtaining customer satisfaction is also regarded as a best practice and can come at a price. As the importance of both increases, so do the risks and the headaches. Maintaining parity between customer satisfaction and internal controls is not easy. Spending too much time and money on customer satisfaction could lead to financial disaster on a given project. Spending too much time on internal controls could lead to noncompetitiveness.

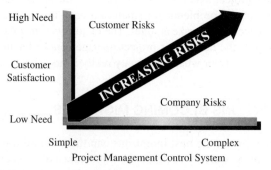

**Figure 2–1.** Risk growth.

## 2.2 ENTERPRISE PROJECT MANAGEMENT METHODOLOGY MIGRAINE _____

As the importance of project management became apparent, companies recognized the need to develop project management methodologies. Good methodologies are best practices and can lead to sole-source contracting based upon the ability of the methodology to continuously deliver quality results and the faith that the customer has in the methodology. Unfortunately, marketing, manufacturing, information systems, R&D, and engineering may have their own methodology for project management. In one company, this suboptimization was acceptable to management as long as these individual functional areas did not have to work together continuously. Each methodology had its own terminology, life-cycle phases, and gate review processes.

When customers began demanding complete solutions to their business needs rather than products from various functional units, the need to minimize the number of methodologies became apparent. Complete solutions required that several functional units work together. This was regarded by senior management as a necessity, and senior management believed that this would eventually turn into a best practice as well as lead to the discovery of other best practices.

One company had three strategic business units (SBUs), which, because of changing customer demands, now were required to work together because of specific customer solution requirements. Senior management instructed one of the SBUs to take the lead role in condensing all of their functional processes into one enterprise project management (EPM) methodology. After some degree of success became apparent, senior management tried unsuccessfully to get the other two SBUs to implement this EPM methodology that was believed to be a best practice. The arguments provided were "We don't need it," "It doesn't apply to us," and "It wasn't invented here." Reluctantly, the president of the company made it clear to his staff that there was now no choice. Everyone would use the same methodology. The president is now facing the same challenge with globalization of acceptance of the methodology. Now cultural issues become important.

## 2.3 CUSTOMER SATISFACTION MIGRAINE _____

Companies have traditionally viewed each customer as a one-time opportunity, and after this customer's needs were met, emphasis was placed upon finding other customers. This is acceptable as long as there exists a large potential customer base. Today, project-driven organizations, namely those that survive on the income from a continuous stream of customer-funded projects, are implementing the "engagement project management" approach. With engagement project management, each potential new customer is approached in a way that is similar to an engagement in marriage where the contractor is soliciting a long-term relationship with the customer rather than a one-shot opportunity. With this approach, contractors are selling not only deliverables and complete solutions but also a willingness to make their EPM methodology compatible with the customer's methodology. To maintain customer satisfaction and hopefully a long-term relationship, customers are requested to provide input on how the contractor's EPM methodology can be extended into their organization. The last life-cycle phase in the EPM methodology used by ABB (Asea, Brown, and Boveri) is called "customer satisfaction management" and is specifically designed to solicit feedback from the customer for long-term customer satisfaction.

This best practice of implementing engagement project management is a powerful best practice because it allows the company to capitalize on its view of project management, namely that project management has evolved into a strategic competency for the firm leading to a sustained competitive advantage. While this approach has merit, it opened a Pandora's box. Customers were now expecting to have a say in the design of the contractor's EPM methodology. One automotive supplier decided to solicit input from one of the Big Three in Detroit when developing its EPM approach. Although this created goodwill and customer satisfaction with one client, it created a severe problem with other clients that had different requirements and different views of project management. How much freedom should a client be given in making recommendations for changes to a contractor's EPM system? Is it a good idea to run the risk of opening Pandora's box for the benefit of customer satisfaction? How much say should a customer have in how a contractor manages projects? What happens if this allows customers to begin telling contractors how to do their job?

## 2.4 MIGRAINE RESULTING FROM RESPONDING TO CHANGING CUSTOMER REQUIREMENTS

When project management becomes a competitive weapon and eventually leads to a strategic competitive advantage, changes resulting from customer requests must be done quickly. The EPM system must have a process for configuration management for the control of changes. The change control process portion of the EPM system must maintain flexibility. But what happens when customer requirements change to such a degree that corresponding changes to the EPM system must be made and these changes could lead to detrimental results rather than best practices?

One automotive tier 1 supplier spent years developing an EPM system that was highly regarded by the customers for the development of new products or components. The EPM system was viewed by both the customers and the company as a best practice. But this was about to change. Customers were now trying to save money by working with fewer suppliers. Certain suppliers would be selected to become "solution providers" responsible for major sections or chunks of the car rather than individual components. Several tier 1 suppliers acquired other companies through mergers and acquisitions in order to become component suppliers. The entire EPM system had to be changed and, in many cases, cultural shock occurred. Some of the acquired companies had strong project management cultures and their own best practices, even stronger than the acquirer, while others were clueless about project management. To make matters even worse, all of these companies were multinational and globalization issues would take center stage. We now had competing best practices.

After years of struggling, success was now at hand for many component suppliers. The mergers and acquisitions were successful and new common sets of best practices were implemented. But, once again, customer requirements were about to change. Customers were now contemplating returning to component procurement rather than "solution provider" procurement, believing that costs would be lowered. Should this occur across the industry, colossal migraines will appear due to massive restructuring, divestitures, cultural changes, and major changes to the EPM systems. How do contractors convince customers that their actions may be detrimental to the entire industry? Furthermore, some companies that were previously financially successful as chunk or section manufacturers might no longer have the same degree of success as component manufacturers.

## 2.5 REPORTING LEVEL OF THE PMO MIGRAINE

Companies have established a PMO as the guardian of project management intellectual property. Included in the responsibilities of a PMO are strategic planning for project management, development of and enhancement to the EPM, maintenance of project management templates, forms and guidelines, portfolio management of projects, mentorship of inexperienced project managers, a hot line for project problem solving, and maintaining a project management best practices library. The PMO becomes the guardian of all of the project management best practices.

While the creation of a PMO is regarded as a best practice for most companies, it places a great deal of intellectual property in the hands of a few, and information is power. With all of this intellectual property in the hands of three or four people in the PMO, the person to whom the PMO reports could possibly become more powerful than his or her counterparts. What is unfortunate is that the PMO must report to the executive levels of management and there appears to be severe infighting at the executive levels for control of the PMO.

To allay the fears of one executive becoming more powerful than another, companies have created multiple PMOs, which are supposedly networked together and sharing information freely. Hewlett-Packard has multiple PMOs all networked together. Comau has PMOs in North America, South America, Europe, and Asia, all networked together. Star Alliance has a membership of twenty-seven airlines, each with a PMO and all networked together with a PMO in Germany as the lead. These PMOs are successful because information and project management intellectual property are shared freely.

Allowing multiple PMOs to exist may seem like the right thing to do to appease each executive, but in some cases it has created the headaches of project management intellectual property that is no longer centralized. And to make matters worse, what happens if every executive, including multinational executives, each demand their own PMO? This might eventually be viewed as an overmanagement expense, and unless the company can see a return on investment on each PMO, the concept of the PMO might disappear, thus destroying an important best practice because of internal politics.

## 2.6 CASH FLOW DILEMMA MIGRAINE

For many companies that survive on competitive bidding, the cost of preparing a bid can range from a few thousand dollars to hundreds of thousands. In most cases, project management may not appear until after the contract is awarded. The results can be catastrophic if

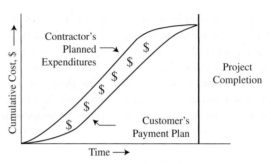

**Figure 2–2.**   Spending curve.

benefit realization at the end of the project does not match the vision or profit margin expected during proposal preparation or at project initiation. When companies develop an EPM system and the system works well, most companies believe that they can now take on more work. They begin bidding on every possible contract believing that with the EPM system they can accomplish more work in less time and with fewer resources without any sacrifice of quality.

In the summer of 2002, a large, multinational company set up a project management training program in Europe for fifty multinational project managers. The executive vice president spoke for the first ten minutes of the class and said, "The company is now going to begin turning away work." The project managers were upset over hearing this and needed an explanation. The executive vice president put Figure 2–2 on the screen and made it clear that the company would no longer accept projects where profit margins would eventually be less than 4–6 percent because they were financing the projects for their customers. The company was functioning as a banker for its clients. Benefit realization was not being achieved. To reduce the costs of competitive bidding, the company was responding to proposal requests using estimating databases rather than time-phased labor. The cash flow issue was not being identified until after go-ahead.

While project financing has become an acceptable practice, it does squeeze profits in already highly competitive markets. To maintain profit margins, companies are often forced to disregard what was told to the customer in the proposal and to assign project resources according to the customer's payment plan rather than the original project schedule provided in the proposal. While this may lead to short-term profitability, it often results in elongated schedules, potential lawsuits, and customer dissatisfaction. The balance between customer satisfaction, long-term client relationships, and profitability is creating a huge headache. The best practice of creating a world-class EPM system can lead to detrimental results if profitability cannot be maintained.

## 2.7  SCOPE CHANGE DILEMMA MIGRAINE

For companies that require successful competitive bidding for survival, the pot of gold is often the amount of scope changes that occur after go-ahead. The original contract may be underbid in the hope that lucrative customer or contractor-generated scope changes will occur. For maximization of profit, a best practices scope change control process must be part of the EPM system.

Over the years, project managers have been encouraged by their superiors to seek out any and all value-added scope changes to be funded by the customers. But these scope changes are now playing havoc with capacity-planning activities and the assigning of critical resources needed for the scope changes and other projects. As companies mature in project management, the EPM systems become Web based. All individual project schedules are rolled up into a master schedule such that senior management can get a realistic picture of resources committed for the next 90 or 180 days. This allows a company to determine how much additional work it can undertake without overtaxing the existing labor base. Furthermore, if a resource bottleneck is detected, it should be relatively clear how many additional resources should be hired and in which functional groups.

As capacity planning converts from an art to a science, the problems with obtaining qualified resources for unplanned scope changes grow. Maximization of profits on a particular project may not be in the best interest of the company, especially if the resources can be used more effectively elsewhere in the organization. Organizations today are understaffed, believing that it is better to have more work than people rather than more people than work. Executives must find a way to balance the need for added project resources, scope changes, portfolio selection of projects, and the strain on the working relationship between project and line managers. How do executives now convince project managers that scope changes are unnecessary and to forget profit maximization?

## 2.8 OUTSOURCE OR NOT MIGRAINE

One of the responsibilities of a PMO is debriefing the project team at the completion of the project. This includes capturing lessons learned, identifying opportunities for improving the EPM system, and updating the estimating database. As the estimating database improves, companies realize that they can outsource some project work at a significantly lower cost than performing the same work internally.

While this function can become an important best practice and can save the company some money, there may be detrimental results. A bank received significant negative publicity in local newspapers when it was discovered that the information systems division would be downsized in conjunction with cost-effective outsourcing. Another organization also outsourced its information systems work to such an extent that it had to begin providing its suppliers and contractors with company-proprietary data. Headaches occur when executives must balance short-term profitability with the long-term health of the corporation and community stakeholder needs and expectations.

Best practices are designed to benefit both the company and the workers. When the implementation of best practices leads to loss of employment, the relative importance of best practices can diminish in the eyes of the employees.

## 2.9 DETERMINING WHEN TO CANCEL A PROJECT MIGRAINE

Virtually every EPM system is based upon life-cycle phases. Each life-cycle phase terminates with an end-of-phase gate review meeting designed to function as a go/no-go decision point for proceeding to the next phase. Very few projects seem to be terminated at the early gate review meetings. One reason for this is that project managers do not necessarily provide all of the critical information necessary to make a viable decision. Project managers provide information in forecast reports on the estimated cost at completion and time at completion. What are missing are the expected benefits at completion and this value may be more important than time and cost. While it is understandable that this value may be difficult to obtain during early life-cycle phases, every attempt should be made to present reasonable benefits-at-completion estimates.

If a project comes in late or is over budget, the expected benefits may still be achievable. Likewise, if a project is under budget or ahead of schedule, there may be no reason to believe that the vision at the project's initiation will be met at completion. One company has initiated a concept called "map days" where periodically the team maps out its performance to date. The maps are reviewed with senior management to make sure that the project should continue. This concept can be expanded to include possible benefits at completion.

While good project management methodologies are best practices and provide valuable information for the management of projects, the system must also be capable of providing the necessary information to senior management for critical decision-making. All too often, EPM systems are developed for the benefit of project managers alone rather than for the best interest of the entire company.

## 2.10 PROVIDING PROJECT AWARDS MIGRAINE

Perhaps the biggest headache facing senior management is the establishment of an equitable project award/recognition system that is part of the Wage and Salary Administration Program. Companies have recognized that project management is a team effort and that rewarding project teams may be more beneficial than rewarding individuals. The headache is how to do it effectively.

There are several questions that need to be addressed:

- Who determines the magnitude of each person's contribution to the project's success?
- Should the amount of time spent on the project impact the size of the award?
- Who determines the size of the award?
- Will the award system impact future estimating, especially if the awards are based upon underruns in cost?
- Will the size of the awards impact future personnel selection for projects?
- Will employees migrate to project managers that have a previous history of success where large awards are provided?
- Will people migrate away from high-risk projects where rewards may not be forthcoming?
- Will employees avoid assignments to long-term projects?
- Can union employees participate in the project award system?

Providing monetary and nonmonetary recognition is a best practice as long as it is accomplished in an equitable manner. Failure to do so can destroy even the best EPM systems as well as a corporate culture that has taken years to develop.

## 2.11 MIGRAINE FROM HAVING THE WRONG CULTURE IN PLACE

Creating the right corporate culture for project management is not easy. However, when a strong corporate culture is in place and it actively supports project management such that other best practices are forthcoming, the culture is very difficult to duplicate in other companies. Some corporate cultures lack cooperation among the players and support well-protected silos. Other cultures are based upon mistrust while yet others foster an atmosphere where it is acceptable to persistently withhold information from management.

A telecommunications company funded more than twenty new product development projects, which all had to be completed within a specific quarter to appease Wall Street and provide cash flow to support the dividend. Management persistently overreacted to bad news, and information flow to senior management became filtered. The project management methodology was used sparingly for fear that management would recognize early on the seriousness of problems with some of the projects.

Not hearing any bad news, senior management became convinced that the projects were progressing as planned. When it was discovered that more than one project was in serious trouble, management conducted intensive project reviews on all projects. In one day, eight project managers were either relieved of their responsibilities or fired. But the damage was done and the problem was really the culture that had been created. Beheading the bearer of bad news can destroy potentially good project management systems and lower morale.

In another telecommunications company, senior management encouraged creativity and provided the workforce with the freedom to be creative. The workforce was heavily loaded with technical employees with advanced degrees. Employees were expected to spend up to 20 percent of their time coming up with ideas for new products. Unfortunately, this time was being charged back to whatever projects the employees were working on at the time, thus making the cost and schedule portion of the EPM system ineffective.

While management appeared to have good intentions, the results were not what management expected. New products were being developed but the payback period was getting longer and longer, while operating costs were increasing. Budgets established during the portfolio selection of the projects process were meaningless. To make matters worse, the technical community defined project success as exceeding specifications rather than meeting them. Management, on the other hand, defined success as commercialization of a product. Given the fact that as many as 50–60 new ideas and projects must be undertaken to have one commercially acceptable success, the cost of new product development was bleeding the company of cash and project management was initially blamed as the culprit. Even the best EPM systems are unable to detect when the work has been completed other than by looking at money consumed and time spent.

It may take years to build up a good culture for project management, but it can be destroyed rapidly through the personal whims of management. A company undertook two high-risk R&D projects concurrently. A time frame of twelve months was established for each in hope of making a technology breakthrough and, even if it could happen, both products would have a shelf life of about one year before obsolescence would occur.

Each project had a project sponsor assigned from the executive levels. At the first gate review meeting, both project managers recommended that their projects be terminated. The executive sponsors, in order to save face, ordered the projects to continue to the next gate review rather than terminate the projects while the losses were small. The executives forced the projects to continue on to fruition. The technical breakthroughs occurred six months late, and virtually no sales occurred with either product. There was only one way the executive sponsors could save face—promote both project managers for successfully developing two new products and then blame marketing and sales for their inability to find customers.

Pulling the plug on projects is never easy. People often view bad news as a personal failure, a sign of weakness, and a blemish on their career path. There must be an understanding that exposing a failure is not a sign of weakness. The lesson is clear: Any executive who always makes the right decision is certainly not making enough decisions, and any company where all of the projects are being completed successfully is not working on enough projects and not accepting reasonable risk.

## 2.12  MIGRAINES DUE TO POLITICS

The completion of a project requires people. But simply because people are assigned to the project doesn't necessarily mean that they will always make decisions for what is in the best interest of the project. When people are first assigned to a new project, they ask themselves, "What's in it for me? How will my career benefit from this assignment?"

This type of thinking creates severe migraines and can permeate all level of management on a project, including those responsible for the governance of the project. People tend to play politics to get what they want and this gamesmanship creates barriers that the project manager must overcome. People are motivated by the rewards they can receive from the formal structure of the company and also from the informal political power structure that exists. Barriers are created when an individual's rewards from either structure are threatened. The barriers lead to conflicts and can involve how the project will be planned, who will be assigned to specific activities especially those activities that may receive high-level visibility, which approach to take to solve a problem, and other such items that are often hidden agenda items. Some people may even want to see the project fail if it benefits them.

Political savvy is an essential skill for today's project manager. One can no longer rely solely upon technical or managerial competence when managing a project. You must understand the political nature of the people and organizations you must deal with. You must understand that politics and conflicts are inevitable and are a way of life in project management. Project managers of the future must become politically astute. Unfortunately, even though there are some books published on politics in project management, there has been limited research conducted on project management politics compared to other areas of the *PMBOK® Guide*.

**Political Risks**                    On large and complex projects, politics are often treated as a political risk especially when the project is being conducted in the host's country and subjected to government interference or political violence. The factors often considered as part of political risks include:

- Political change such as a new party elected into power
- Changes in the host country's fiscal policy, procurement policy, and labor policy
- Nationalization or unlawful seizure of project assets and/or intellectual property

- Civil unrest resulting from a coup, acts of terrorism, kidnapping, ransom, assassinations, civil war, and insurrection
- Significant inflation rate changes resulting in unfavorable monetary conversion policies
- Contract failure such as license cancellation and payment failure

We tend to include many of these risks within the scope of enterprise environmental factors that are the responsibility of the project sponsor or the governance committee. But when the project is being conducted within the host's country, it is usually the project manager that has to deal with the political risks.

The larger and more complex the project, the larger the cost overrun,, and the larger the cost overrun, the greater the likelihood of political intervention. In some countries, such as in the United States, escalating problems upward usually implies that the problem ends up in the hands of the project's sponsor. But in other countries, especially emerging market nations, problems may rise beyond the governance committee and involve high-level government officials. This is particularly true for megaprojects that are susceptible to large cost overruns.

**Reasons for Playing Politics**   There are numerous reasons why people play political games. Some of the common reasons include:

- Wanting to maintain control over scarce resources
- Seeking rewards, power, or recognition
- Maintaining one's image and personal values
- Having hidden agendas
- Fear of the unknown
- Control over who get to travel to exotic locations
- Control over important information since information is a source of power
- Getting others to do one's work
- Seeing only what one wants to see
- Refusing to accept or admit defeat or failure
- Viewing bad news as a personal failure
- Fear of exposing mistakes to others
- Viewing failure as a sign of weakness
- Viewing failure as damage to one's reputation
- Viewing failure as damage to one's career

All of these are reasons that may benefit you personally. There are also negative politics where political games are played with the intent of hurting others, which in turn may end up benefiting you personally. Some examples are:

- Wanting to see the project fail
- Fear of change if the project succeeds
- Wanting to damage someone else's image or reputation especially if they stand in the way of your career advancement
- Berating the ideas of others to strengthen your position

**Situations Where Political Games Will Occur**

While politics can exist on any project and during any life-cycle phase, there are some specific situations where history has shown us that politics are most likely to occur:

- Trying to achieve project management maturity within a conservative culture
- During mergers and acquisitions where the "landlord" and the "tenant" are at different levels of project management maturity
- Trying to get an entire organization to accept a project management methodology that was created by one functional area rather than a committee composed of members from all functional areas (i.e. the "not invented here" syndrome)
- Not believing that the project can be completed successfully and wanting to protect oneself
- Having to change one's work habits and do things differently if the project is a success
- When problems occur, not knowing where they will end up for resolution
- Believing that virtual teams are insulated from project politics
- The larger and more complex the project, the greater the chances of political interference
- The larger the size of the governance committee, the greater the chance for disagreements and political issues to appear
- Failing to understand stakeholder effective relations management practices
- The more powerful the people are on the project, the greater the chance that they will be involved in project politics
- Employees who are recognized as prima donnas are more prone to play political games that the average worker.

**The Governance Committee**

Project politics usually ends up pushing the project in a direction different from the original statement of work (SOW). The push can originate within your own senior management, some of your project team members, the customer, and even some of the stakeholders. Each will want a slightly different outcome, and your job is to try to find a way to appease everyone.

On the surface, the simplest solution appears to be the creation of a governance committee composed of senior managers from your company, representation from the customer's company, and representatives from various stakeholder groups. Now, it seems that you can let the governance committee resolve all of the conflicts among themselves and give you a unified direction for the project. Gaining support from a higher power certainly seems like the right thing to do. Unfortunately, there is still the possibility that the committee cannot come to an agreement, and even if they appear to be in agreement, certain members of the committee may still try to play politics behind the scenes. The existence of the governance committee does not eliminate the existence of project politics. People that serve on a governance committee often play the political game in order to enhance their power base.

Most companies have limited funds available for projects. The result is an executive-level competition for project funding that may serve the best interest of one functional area but not necessarily the best interest of the entire company. Executives may play political games to get their projects approved ahead of everyone else, viewing this as an increase to their power base. But the governance committee may include executives from those functional areas that lost out in the battle for project funding, and these executives may try to exert negative political influence on the project, even going so far as to hope that the project will fail. The result is a project manager who is assigned to such a project and brought on board after the project is approved never fully understanding, until well into the project, the politics that were played during project approval and initiation.

**Friends and Foes**

It is often difficult to identify quickly which people are friends and which are foes. Not all people who have political agendas are enemies. Some people may be playing the political game for your best interest. It is, therefore, beneficial to identify, if possible, from the personal agendas that people have whether they are friends or foes. This implies that you must communicate with them, perhaps more informally than formally, to understand their agendas. Reading body language is often a good way to make a first guess if someone is a friend or foe.

One possible way to classify people might be:

- *True supporters:* These are people who openly demonstrate their willingness to support you and your position on the project.
- *Fence-sitters:* These are people that you believe will support you down the road as long as you prove to them that you are deserving of their trust and support. You may need to spend extra time with them to show them your position and to gain their support.
- *True unknowns:* Unlike fence-sitters, who may be won over to your way of thinking, these people are true unknowns. They may have hidden agendas that are not in your best interest, but they are relatively quiet and may have not yet expressed their concerns. These people could pose a serious threat if they are adamantly opposed to the direction in which the project is proceeding.
- *True enemies:* These are people who have made it quite clear that they are unlikely to support your views. You understand their position and probably are quite sure how they will respond to you and the direction the project is taking.

**Attack or Retreat**

When people play political games on projects, there are two facts that we seem to take for granted. First, these people are most likely experienced in playing such games, and second they expect to win. Based upon whom the conflict is with, you must then decide whether to aggressively attack them or retreat. Simply taking no action is a form of withdrawal and you are sure to lose the battle.

The first rule in battle is to gather as much intelligence as you can about your enemy. As an example, as part of stakeholder relations management, we can map project stakeholders according to Figure 2–3. Stakeholder mapping is most frequently displayed on a grid comparing their power and their level of interest in the project.

- *Manage closely:* These are high-power, interested people who can make or break your project. You must put forth the greatest effort to satisfy them. Be aware that there are factors that can cause them to change quadrants rapidly.

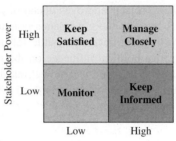

**Figure 2–3.**  Stakeholder mapping.

- *Keep satisfied:* These are high-power, less interested people who can also make or break your project. You must put forth some effort to satisfy them but not with excessive detail that can lead to boredom and total disinterest. They may not get involved until the end of the project approaches.
- *Keep informed:* These are people with limited power but keen interest in the project. They can function as an early warning system of approaching problems and may be technically astute to assist with some technical issues. These are the stakeholders that often provide hidden opportunities.
- *Monitor only:* These are people with limited power and may not be interested in the project unless a disaster occurs. Provide them with some information but not with too much detail such that they will become disinterested or bored.

When you go on offense and attack the people playing politics, you must have not only ammunition but also backup support if necessary. You must be prepared to show how the political decision might affect the constraints on the project as well as the accompanying baselines. Based upon the power and influence level of your opponent according to Figure 2–3, you may need other stakeholders to help you plead your case. It is highly beneficial to have supporters at the same level of position power or higher than the people playing the political game.

Not all political battles need to be won. People who play politics and possess a great deal of power may also have the authority to cancel the project. In such cases, retreating may be the only viable option. If you truly alienate the people playing power games, the situation can deteriorate even further. There is always the chance that you may have to work with the same people in the future. In any case, the best approach is to try to understand the people playing politics, the reason why they are playing politics, and how much power and influence they have over the final decision.

**The Need for Effective Communication**

While it is not always possible to tell when someone is playing or intends to play the political game on your project, there are some telltale signs that this may be happening. Some of the signs include:

- People do not care about your feelings.
- People avoid discussing critical issues.
- People never ask you about your feeling on the matter.

- People procrastinate about making decisions.
- People have excuses for not completing action items.
- People discuss only those items that may benefit them personally.

While project managers may not have any control over these tell-tale signs, project managers can make the situation worse through ineffective communication. To minimize the political impact on a project, the project manager should consider using the following practices:

- Listen carefully before speaking and do not jump at conclusions.
- Make sure you understand what others are saying and try to see the issue from their point of view.
- All informal communications should be followed up by a memo outlining what was discussed and to make sure that there were no misunderstandings.
- Before stating your point of view, make sure that you have gathered all of the necessary supporting information.
- Make sure that you have a clear understanding of how culture affects the way that people are communicating with you.
- If you must provide criticism, make sure that it is constructive rather than personal criticism.
- When resolving political issues, there will be winners and losers. It is not a matter of just picking a winner. You must also explain to everyone why you selected this approach and likewise why the other approaches were not considered. This must be done tactfully.
- If the situation cannot be managed effectively, do not be embarrassed to ask senior management for advice and assistance.
- Ineffective communication encourages lying which, in turn, generates additional political games to be played accompanied by a great deal of mistrust.

Project managers must be careful when discussing politics with team members, the client and stakeholders. The information could be misunderstood or filtered especially if people hear what they want to hear. The result could be additional politics that were unexpected, and friends could easily turn into foes.

## Power and Influence

Effective communication skills alone cannot resolve all political situations. To understand why, we must look at how project management generally works. If all projects stayed within the traditional hierarchy, someone would have the ultimate authority to resolve political issues. But since most projects are managed outside the traditional hierarchy, the burden for the resolution of conflicts and political issues usually falls upon the shoulders of the project manager even if a governance committee is in place. The governance committee may very well be the cause of the conflict.

On the surface, it seems like the simplest solution would be to give the project manager sufficient authority to resolve political issues. But projects are usually executed outside of the traditional hierarchy, thus limiting the authority that the project manager will possess. This lack of formal authority makes the project manager's job difficult. While project charters do give project managers some degree of authority for a given project, most project managers still have limitations because:

- The project managers must negotiate with functional managers for qualified resources.
- The project managers may not be able to remove employees from a project without the functional manager's concurrence.
- The project managers generally have no direct responsibility for wage and salary administration.
- The project managers may possess virtually no reward or punishment power.
- If employees are assigned to multiple projects, the project managers may not be able to force the employees to work on their projects in a timely manner.

With a lack of position power which comes from the traditional hierarchy, and without the ability to reward or punish, the project manager must rely upon other forms of power and the ability to influence people. Behavioral skills such as effective communication, motivation techniques, conflict management, bargaining and negotiations are essential to resolve political disputes. Unfortunately, most project managers lack political savvy and have poor conflict resolution skills.

**Managing Project Politics**

While project politics are inevitable, there are actions the project manager can take to minimize or control political issues. Some of these actions include:

- Gather as much information as you can about the political situation.
- Make sure that everyone fully understands the impact of the political situation upon the project's baselines.
- Try to see the picture through the eyes of the person playing politics.
- Try to form a coalition with the people playing politics.
- See if your sponsor or the governance committee can insulate you from the political games.
- Having a structured decision-making process as part of your project management methodology can reduce some of the political games.
- Try to determine people's political position by reading their body language.
- If the political situation cannot be resolved quickly, demonstrate a willingness to compromise as long as the integrity of the project is not sacrificed.

Power breeds politics, and politics in turn breeds power. Expecting to manage a project without any political interference is wishful thinking rather than reality. We cannot predict customer and stakeholder behavior. Sometimes the political situation occurs without any early warning signs.

Nobody can agree on a definition of organizational or project politics. Politics can appear in many shapes, forms, and sizes. Therefore, the project manager must develop superior behavioral skills to deal with political situations. The danger in not being able to manage political situations correctly is redirection or misdirection of the project.

## 2.13 MIGRAINES CAUSED BY THE SEVEN DEADLY SINS _____

For more than forty years, the project management landscape has seen textbooks, journal articles, and presented papers discussing the causes of project failures. Unfortunately, many of the failure analyses seem to look at failure superficially rather than in depth. When trying to discover the root cause of a failure, we usually look first in the contractor's company for someone to blame rather than in our own company. If that doesn't work, then we begin climbing the organizational hierarchy in our own company by focusing on the project team, followed by the project manager. Once we find someone to blame, the search seems to end and we feel comfortable that we have discovered the cause of the failure.

It is human nature to begin finger-pointing at the bottom of the organizational hierarchy first, rather than at the top. Yet, more often than not, the real cause of failure is the result of actions (or inactions) and decisions made at the top of the organizational chart rather than at the bottom. It is also human nature to make decisions based upon how we are affected by the Seven Deadly Sins, namely: envy, anger, pride, greed, sloth, lust, and gluttony. Decisions based upon the Seven Deadly Sins, whether they are made at the top or bottom of the organization, can have dire consequences for projects. Sometimes the sins are hidden and not easily recognized by us or others. We simply do not see or feel that were are committing a sin.

The Seven Deadly Sins affect all of us sooner or later, even though we refuse to admit it. Some of us may be affected by just one or two of the sins, whereas others may succumb to all seven. What is unfortunate is that the greatest damage can occur on projects when the sins influence the way that those at senior levels of management interface with projects, whether as a project sponsor or as a member of a governance group. Bad decisions at the top, especially if based upon emotions rather than practicality, can place the project on a destructive path even before the day the project is kicked off.

**The Seven Deadly Sins[1]**   The term "Seven Deadly Sins," also referred to as the "Capital Vices" or "Cardinal Sins," is a classification of objectionable vices. They were originally part of Christian ethics and have been used since early Christian times to educate and instruct Christians concerning fallen humanity's tendency to sin. The currently recognized version of the sins are usually given as wrath, greed, sloth, pride, lust, envy, and gluttony. Part or all of the sins have been discussed over the past four centuries from different perspectives in the religious writings of Christianity, Hinduism, Islam, Buddhism, and Judaism. Over the years, the sins have been modified and discussed by the clergy, philosophers, psychologists, authors, poets, and educators.

A brief description of each of the Seven Deadly Sins appears in Table 2–1. Each of the sins can be related to an animal, a specific color, and even a punishment in hell for committing the sin.

In a project environment, any or all of these sins can cause rational people to make irrational decisions, and this can occur at any level within the organizational hierarchy. At some levels, the existence of the sins may have a greater impact on project performance

---

1. Adapted from Wikipedia, the free encyclopedia.

**TABLE 2–1.  THE SEVEN DEADLY SINS**

| Sin | Traits | Animal | Color | Punishment in Hell |
|---|---|---|---|---|
| Envy | The desire to possess what others have | Snake | Green | Placed in freezing water |
| Anger/Wrath | A strong feeling of displeasure | Lion | Red | Dismembered alive |
| Pride | The need for inward emotional satisfaction | Peacock | Violet | Broken on the wheel |
| Greed | The desire for material wealth or gain | Toad | Yellow | Put in cauldrons of boiling oil |
| Sloth | The avoidance of work | Snail | Light blue | Thrown into a snake pit |
| Lust | A craving, but not necessarily sexual | Goat | Blue | Smothered in fire and brimstone |
| Gluttony | The desire to consume more than needed | Pig | Orange | Forced to eat rats, toads, and snakes |

than at other levels. If a sin is apparent in the beginning of a project, then poor decisions in the initiation phase can have detrimental consequences in all of the downstream phases.

### Envy

- "Envy is the art of counting the other fellow's blessing instead of your own." (Harold Coffin)
- "Envy is ignorance. Imitation is suicide." (Ralph Waldo Emerson)
- "When men are full of envy, they disparage everything, whether it be good or bad." (Publius Cornelius Tacitus)

Envy is the desire to have what others have. Resentful emotions occur when one lacks another's superior qualities such as status, wealth, good fortune, physical possessions, traits, abilities, or position. Envy may encourage someone to inflict misfortune on another person and try to undo someone else's advantage or deprive them of obtaining the advantage. Envy can also affect the relationship between people, as when one ignores a person of whom they are envious. Envy is often synonymous with jealousy, bitterness, greed, spite, and resentment.

Envy can be malicious or benign. Malicious envy has all of the characteristics just mentioned. Benign envy can be a positive motivational force if it encourages people to act in a more favorable manner such that the desires are attainable. Benign envy usually exists at the bottom of the organizational hierarchy, whereas malicious envy appears most frequently at the top.

The following four situations illustrate how envy can lead to project disasters:

**Situation 1: Reorganizational Failure.** A company had four divisions, each headed up by a senior vice president. In the past, most of the projects had stayed entirely within one division. Each division had its own project management methodology, and the number of project successes significantly outnumbered the failures. As the marketplace began to change, the company began working on projects that required that more than one division work together on the same project. Using multiple methodologies on the same project proved to be an impossible task.

The president decreed that there must be one and only one methodology, and that all of the divisions must use the same methodology for managing projects. The company created a Project Management Office (PMO), and the president assigned one of the vice presidents with control over the PMO. Employees from each of the other three divisions were then assigned on a "dotted line" relationship to the PMO for the development of the singular methodology.

The people in the PMO seemed to work well together, but the four vice presidents demanded that they have final signature authority over the adoption of the singular methodology. Each vice president believed that the project management approach used in their division should be the driving force for the creation of a singular methodology. Regardless of what design the PMO came up with, each vice president demonstrated envy and resentment, finding fault with the others' ideas, playing out the "not invented here" syndrome. While this was happening, the number of project failures began to increase, because of the lack of structure for project execution.

It also became obvious to each of the four vice presidents that whichever vice president had control of the PMO would become more powerful than the other three vice presidents because of the control over all of the project management intellectual property. Information is power, and envy for control of the information had taken its toll on the ability to manage and control projects effectively. Eventually, the president stepped in and allowed each vice president to have a PMO. However, the PMOs had to be networked together. This helped a little, but even after they agreed on a common methodology, each PMO tried to seduce the other PMOs into their way of thinking and, as expected, continuous changes were being introduced to the methodology, with the projects still suffering from a lack of direction. Envy prevented decisions from being made that would have been in the best interest of the entire company.

**Situation 2: Reward Failure.** Believing that an effective reward/ bonus system would motivate project teams, senior management announced that bonuses would be given to each project team based upon the profitability of their projects. The company survived on competitive bidding to win contracts and most of the projects were in the millions of dollars. Project managers quickly learned that large bonuses could be awarded if the project's cost estimates were grossly inflated during the bidding process and the contracts could be won. This way, the actual profits on some contracts could exceed the targeted profits.

Although the company lost a few contracts they expected to win because of the inflated costs, some of the bonuses given to the project managers at the end of the contracts were similar in size to the bonuses given to some of the executives. Many of the executives were now envious of the people below them who were receiving such large bonuses. Due to envy, the executives then changed the bonus policy, whereby part of the bonus fund would be distributed among the executives, even though the executives were not functioning as project sponsors. The bonuses given to the workers and the project managers were then reduced significantly. Some of the workers then sabotaged some of the projects rather than see the executives receive bonuses that were awarded at the expense of the workers.

**Situation 3: Failure due to Inflicting Misfortune.** Paul was the director of operations for a medium-sized company. Paul's company was in the process of establishing a PMO that would report directly to the CEO. Paul desperately wanted the new position of director of the PMO, believing it would be a steppingstone to becoming a vice president. Paul's major competitor for the position of Director of the PMO was Brenda, a twenty-year veteran of the company and considered the company's best project manager. Because of Brenda's decision-making skills, Brenda was almost always empowered with complete decision-making authority on her projects.

When Brenda was assigned to her latest project, Paul requested and was granted the position of project sponsor for Brenda's project. Paul was envious of Brenda's abilities and good fortune and believed that, if he could somehow inflict misfortune on Brenda's project without hurting himself in the process, he could easily be assigned director of the PMO. Paul placed limits on Brenda's empowerment and demanded that, as the sponsor, he approve any and all critical decisions. Paul continuously forced Brenda to select non-optimal alternatives when some decisions had to be made. Brenda's project was nearly a disaster, and Paul was later assigned as director of the PMO.

Envy can force us to inflict pain on others to get what we desire. Paul received his promotion, but the workers and Brenda knew what he had done. Paul's working relationship with the functional subject matter experts deteriorated.

**Situation 4: The Relationship Failure.** Jerry and two of his friends lived near each other and joined the company at exactly the same time; Jerry worked in project management, and the other two worked in engineering. They formed a carpool and traveled to and from work together every day. They also socialized when not at work.

Two years after joining the company, Jerry had received his second promotion, whereas the other two workers had not received any promotions. The other two workers were envious of Jerry's success to the point where they stopped socializing and carpooling with him. The jealousy became so strong that the two workers even refused to work on Jerry's projects. The workers never visibly displayed their jealousy of Jerry, but their actions spoke louder than words and made it clear how they really felt.

### Anger (Or Wrath)

- "For every minute you are angry, you lose sixty seconds of happiness." (Ralph Waldo Emerson)
- "Speak when you are angry—and you'll make the best speech you'll ever regret." (Dr. Lawrence J. Peters)
- "Anger is never without reason, but seldom a good one." (Benjamin Franklin)
- "Anger is one letter short of danger." (Anonymous)
- "Anger, if not restrained, is frequently more hurtful to us than the injury that provokes it." (Seneca)

Anger or wrath is a strong feeling of displeasure. Sometimes we become angry because the actions of others on the project have offended us. Other times, we demonstrate unnecessary anger to make someone extremely annoyed in order to stop a threatening behavior, such as continuous schedule slippages or cost overruns. Anger is often synonymous with ire, annoyance, irritation, rage, and resentment.

When we get angry, we often lose our objectivity. The anger we feel and demonstrate can appear suddenly, or it can be deliberate. There are ranges of anger. On the soft end of the spectrum, anger can be just a mild irritation, whereas on the hard end, anger can result in fury and rage. Not all anger is readily visible. For example, passive anger can be seen as a phony smile, giving someone the cold shoulder, overreacting to something, or constantly checking things. Aggressive anger can appear as bullying, expressing mistrust, talking too fast, or destructiveness.

Here are some examples of how anger can affect projects:

**Situation 5: Failure due to Unjust Anger.** While selecting the portfolio of 20 projects for the upcoming year, senior management established the budgets and schedules without any supporting data on what might or might not be realistic. To make matters worse, the executive sponsors on each project emphatically stated that they would not tolerate schedule slippages or cost overruns. The project teams developed the detailed project plan and, on eight of the twenty projects, the teams determined that the budgets and schedules provided by senior management were unrealistic. Rather than inform senior management immediately that their budget and schedule perceptions might be wrong, the teams began executing the projects and hoping for a miracle. The teams felt that this was a better approach than incurring the wrath of senior management when they were apprised of the situation.

The eight teams were unsuccessful in their quest for a miracle. After a few months, senior management performed a health check on one of the eight projects and discovered the truth: The project was in bad shape. A health check was then performed on all twenty projects, and it became apparent that eight of the projects were in trouble both financially and technically. Senior management became enraged that they had not been informed of this previously, canceled the eight troubled projects, and fired all eight project managers in one day.

Part of the blame certainly falls upon the shoulders of the project teams for not informing senior management early on. However, a lot of the blame must rest with senior management, especially when they have a history of demonstrating irate behavior that may have been unjustified. When project teams believe that they will encounter anger rather than support for problems from the top of the organizational hierarchy, project management may not succeed, and projects will fail. Bad news is often filtered to prevent the occurrence of anger.

**Situation 6: Failure due to a Hidden Agenda.** The chief information officer (CIO) became the project sponsor for a $25 million IT project scheduled to last about one year. The CIO established October 1st as the "go live" date for the project. During a July review of the status of the project, the CIO was informed by the project manager that the go live date was unrealistic. The CIO became furious and asked, "How much of the software would be operational by October 1st?" The project manager responded, "Perhaps 10 percent.

The CIO stormed out of the meeting after demonstrating anger, calling the project team "incompetent fools." The CIO then authorized significant overtime and awarded the prime contractor almost $5 million in additional costs if they could get at least 50 percent of the software operational by October 1st and 70 percent or more by November 1st. The CIO knew that his year-end corporate bonus was partially aligned with implementation of this project, and with 70 percent of the software operational, his bonus would be significant. When the project was finally completed in February, the executive committee viewed the project as a partial failure because of the $5 million cost overrun, and the project manager was reprimanded. However, the CIO received his bonus.

**Situation 7: Failure due to Information Filtering.** Senior management in a government agency established a culture that bad news would be filtered as the news proceeded up the organizational hierarchy. Allowing bad news to reach the top would be an invitation for fury and rage coming from the top back down to the projects. Therefore, by the time that the information reached the top, much of the bad news had disappeared, and the risks associated with the project were buried. The result on one project was just as the technical risk experts predicted: seven astronauts were killed when the Space Shuttle Challenger exploded during liftoff.[2] There were other factors, as well, that led to this disaster. During a congressional committee meeting reviewing the cause of the fatality, one subject matter expert was asked by the committee, "Why didn't you explain to senior management what the risks were?" The subject matter expert asserted, "I didn't report administratively to senior management. My responsibility was to report this to my boss and he, in turn, should have reported it higher up."

**Situation 8: Failure due to a Collective Belief.** A collective belief is a fervent, and often blind, desire to achieve—regardless of the cost and consequences. When a collective belief exists, especially at the senior levels of management, rational organizations begin making irrational decisions, and any deviation from the collective belief is met with anger. People who question the collective belief or challenge progress are removed from the project or severely reprimanded. In order to work on these projects, one must suppress one's anger and go with the flow, regardless of the outcome. These projects can be technical successes but financial failures, never totally fulfilling the corporate business strategy.

A good example of this is the Iridium Project.[3] The Iridium Project was an eleven-year project that missed the service launch date by one month. The service was a network of sixty-six satellites circling the earth, allowing one could talk to anyone anywhere. The project management activities performed by Motorola and Iridium LLP were outstanding, especially when we consider that the project resulted in more than 1,000 patents and 25 million lines of software code. Technically, it was a success, but financially it was a disaster, invoking anger when it became apparent that they could not get the number of subscribers they needed to break even. Throughout the project, the threat of severe anger from above, as well as the existence of the collective belief, made it almost impossible for people to challenge the projections on the number of subscribers.

Anger need not be demonstrated to inflict pain upon a project. Just the implied threat or fear of anger can limit a team's performance significantly.

## Pride

- "A proud man is always looking down on things and people; and, of course, as long as you are looking down, you can't see anything above you." (C.S. Lewis)
- "The blind cannot see—the proud will not." (Russian proverb)
- "Vanity and pride are different things, though the words are often used synonymously. A person may be proud without being vain. Pride relates more to our opinion of ourselves; vanity, to what we would have others think of us." (Jane Austen)
- "We are rarely proud when we are alone." (Voltaire)

---

2. For additional information on this case study see Harold Kerzner, *Project Management Case Studies*, 4th Edition, Hoboken, NJ: Wiley, 2013; "Case Study: The Space Shuttle Challenger Disaster," pp. 447–496.

3. For additional information see Harold Kerzner, *Project Management Case Studies*, 4th Edition, Hoboken, NJ: Wiley, 2013;" Case Study: The Rise, Fall and Resurrection of Iridium; A Project Management Perspective," pp.327–366.

Pride is an inward emotion that leads to personal satisfaction or meeting personal goals. Pride can be a virtue or simply love of oneself or an inflated sense of one's accomplishments, which leads to exhilarating emotions. Pride can have both negative and positive connotations. In a negative sense, pride can cause us to grossly inflate what we have accomplished. In a positive sense, it can be an attachment to the actions of others or a fulfilled feeling of belonging, such as national or ethnic pride, or being a member of the team on an important project.

Pride is often seen as a virtue. Overinflated pride can result in a disagreement with the truth, which sometimes comes with self-gratification. The antonyms to pride are humility and guilt.

Here are some examples of how pride can affect a project:

**Situation 9: Failure due to Too Much Expertise.** Peter was one of the most experienced engineers in the company. His technical expertise was second to none. Peter was asked to solve a problem on a project. Even though there were several possible options, Peter chose the option that was the most costly and resulted in the addition of unnecessary features, which we refer to as "bells and whistles." Peter asserted that his solution was the only practical one, and the project manager reluctantly agreed. Peter saw this project as a way of increasing his reputation in the company regardless of the impact on the project. The bells and whistles increased the final cost of the deliverable significantly. It also inflated Peter's self-esteem.

**Situation 10: Failure due to the Wrong Sponsor.** Nancy was the director of marketing. Her superior, the vice president for marketing, had requested the development of a rather sophisticated IT project for the Marketing Division. It was customary for the IT department to act as the project sponsor on all IT projects once the business case for the project was approved. Nancy knew that this project would get the attention of the senior-most levels of management. Nancy had never served in the capacity of a project sponsor but believed that, if she could be the sponsor for this project, her identification with this project could result in a promotion.

Nancy's campaign to become the sponsor was a success. Unfortunately, there were numerous IT issues that had to be resolved at the sponsor level and, because of Nancy's lack of expertise in IT, she made several wrong decisions. The project ended up being late and over budget because many of Nancy's decisions had to be changed later in the project. Nancy's quest for pride ended up having detrimental results.

### Greed (Avarice)

- "Ambition is but avarice on stilts, and masked." (Walter Savage Landor)
- "Avarice has ruined more souls than extravagance." (Charles Caleb Colton)
- "Avarice is the vice of declining years." (George Bancroft)
- "Avarice is generally the last passion of those lives of which the first part has been squandered in pleasure, and the second devoted to ambition." (Samuel Johnson)
- "Poverty wants much, but avarice, everything." (Publilius Syrus)
- "Poverty wants some things, luxury many things, avarice all things." (Benjamin Franklin)
- "Love is always a stranger in the house of avarice." (Andreas Capellanus)
- "To hazard much to get much has more avarice than wisdom." (William Penn)

Greed is a strong desire for wealth, goods, and objects of value for oneself. Greed goes beyond the basic levels of comfort and survival. Greed asks for more than we actually need or deserve. Greed can also appear as the desire for power, information, or control of resources. Synonyms for greed are "avarice" and "covetousness."

The following are several examples of how greed can affect projects:

**Situation 11: The Failure of Too Many Resources.** Karl was placed in charge of a two-year project that required 118 people, many of whom were needed on just a part-time basis. Karl convinced senior management that this project required a co-located team, with everyone assigned full time and that the team should be housed in a building away from the employees' functional managers. Senior management knew this was a bad idea but reluctantly agreed to it, knowing full well that the project was now overstaffed and overmanaged.

At the end of the first year, it became obvious that none of the employees on Karl's project had received promotions or merit increases in their salary. The functional managers were rewarding only those employees who sat near them and made them look good on a daily basis. The employees on Karl's project now felt that assignment to this project was a nonpromotable assignment. Several employees tried to sabotage the project just to get off it. Later, Karl discovered that several of the other project managers now had a strong dislike for him because his greed for resources had affected their projects.

**Situation 12: The Failure of Power.** Carol was a department manager. She was proud of the fact that she finally had become a department manager. Word had spread throughout the company that senior management was considering a downsizing of the company. Carol was afraid that her department might be eliminated and that she would lose her position as a department manager.

To protect her power position, Carol began giving conflicting instructions to the people in her department. The workers kept coming back to Carol for clarification of the conflicting instructions. Carol then told her superiors that the people in her department needed daily supervision or else the department's performance would suffer. While this technique seemed to prevent the downsizing of Carol's department, it did have a detrimental effect on the work the employees were doing on the projects. Carol's greed for power and resources proved detrimental to the company but beneficial to Carol's personal needs.

**Situation 13: The Failure of Greed for the Bonus.** The vice president for engineering was assigned as the project sponsor for a multimillion dollar Department of Defense (DoD) contract, and Ben was the project manager. A large portion of the vice president's bonus was based upon the profitability of the projects directly under his control and of which he was the sponsor. This large project, headed up by Ben, was scheduled to be completed in November; the follow-on contract, which was also quite large, was scheduled to begin in February.

The vice president and Ben agreed that a large management reserve should be established to support the project team between contracts. If the team were to be disbanded in November, there would be no guarantee that the same people would be available for the follow-on contract that would begin in February. When the project came to fruition in October, the remaining management reserve was large enough to support the resources with critical skills between October and February. These people would be working on some activities that would be needed for the follow-on contract, such as preliminary planning activities and procurement planning.

When the contract finally ended in October, the vice president told the financial people to book the management reserve as additional profit on the project. This increased the vice president's bonus significantly. However, without the management reserve, the critically skilled resources were reassigned back to their functional departments, and many were not available to work on the follow-on contract. The follow-on contract suffered from cost overruns and schedule slippages, because it had different resources and a new learning curve. The damage from the vice president's greed was now apparent.

## Sloth

- "Nothing irritates me more than chronic laziness in others. Mind you, it's only mental sloth I object to. Physical sloth can be heavenly." (Elizabeth Huxley)
- "We excuse our sloth under the pretext of difficulty." (Marcus Fabius Quintilian)
- "Diligence overcomes difficulties, sloth makes them." (Benjamin Franklin)
- "Sloth and silence are a fool's virtues." (Benjamin Franklin)
- "All things are easy to industry; all things are difficult to sloth." (Benjamin Franklin)

Sloth is the act of being physically, mentally, and/or emotionally inactive and often is characterized as laziness. Sloth can result in extreme waste in the effective use of people, things, skills, information, and even time. Sloth often forces us to overestimate the difficulty of the job.

The following are examples of how sloth can affect projects:

**Situation 14: The Failure of Laziness.** Becky was placed in charge of a one-year project that was relatively easy to accomplish and low risk. When negotiating with the functional managers for project staff, Becky overestimated the complexity and risk of the project so that she could request the more experienced people. That would certainly make Becky's job easier. The functional managers were not sure if Becky's estimates of risk and complexity were valid, but they decided that it would be better to grant her request than to provide mediocre resources and find out later that she was correct.

There wasn't much for Becky to do on the project. The subject matter experts did it all. Eventually, the experienced people on Becky's project reported back to their respective functional managers that lower pay grade resources should have been assigned. While Becky's project was considered as a success and there wasn't much for Becky to do, the other projects that really could have used the more experienced resources suffered. Sloth usually benefits a single individual and at the expense of the greater good.

**Situation 15: Failure due to the Union Standard.** A company had a powerful union that discouraged new employees who were eager to show what they could do by producing more units than the standard agreed to by the union. The new workers were told to slow down and enjoy life.

The company soon became uncompetitive in the marketplace, and their business base began to deteriorate. Senior management then told the union that either the standards must be updated, or people might lose their jobs. The union maintained their complacency and refused to budge on the standards. When management threatened to outsource much of the work and lay people off, the union workers went on strike.

Management personnel and nonunion workers began doing the work that was previously done by the union workers. They turned out 70 percent of the work using 10 percent as many nonunionized employees. human resources personnel were running drill presses and lathes, and salespeople worked on the assembly line. Management now had a clear picture of what the sin of sloth had been doing to the company for years. Management had no intention of negotiating an end to the strike. Eventually the union conceded and returned to work. However, more than 160 of the union workers were laid off after the new standards were adopted. The company was now competitive again.

## Lust

- "Lust is to other passions what the nervous fluid is to life; it supports them all, ambition, cruelty, avarice, revenge are all found on lust." (Marquis de Sade)
- "Of all the worldly possessions, lust is the most intense. All other worldly passions seem to follow its train." (Buddha)
- "Society drives people crazy with lust and calls it advertising." (John Lahr)
- "Their insatiable lust for power is only equaled by their incurable impotence in exercising it." (Winston Churchill)
- "Hell has three gates: lust, anger, and greed." (*Bhagavad Gita*)
- "It is not power itself, but the legitimation of the lust for power, which corrupts absolutely." (Richard Howard Stafford Crossman)
- "The lust of avarice is so totally seized upon mankind that their wealth seems to rather possess them than they possess their wealth." (Pliny the Elder)

Lust is the emotion or feeling of intense desire in the body. Although lust is usually described in a sexual content, it can also appear as a strong desire for power, knowledge, or control. It can lead to great eagerness or enthusiasm, which may be good, especially if it fulfills the need to gratify the senses.

Two examples of how lust can affect projects are given here:

**Situation 16: Failure due to the Lust for Power.** Ralph was elated to be assigned as the project manager for a new project that was won through competitive bidding. The chance for significant follow-on work from this client was highly likely. This would be Ralph's chance to become more powerful than the other project managers and possibly be promoted and be given a corner office. Corner offices with large windows were signs of power and prestige. For this to happen, Ralph had to slowly build his project into an empire of resources, regardless of the consequences.

By the end of the initial contract, Ralph had more resources assigned full time than planned for during project initiation. The project was significantly overstaffed, and this had an adverse effect on profits. But Ralph explained to his superiors that this would lead to increased profits in the future.

When the follow-on contract appeared, Ralph argued that he needed even more resources, and that a projectized organizational structure was needed with Ralph as it head. The company agreed. The projectized structure allowed Ralph to have all remaining part-time workers assigned to his project full time. Part way through the project, the company was notified that there would additional follow-on contracts, but these would all be awarded through competitive bidding. Ralph's power was now at an all-time high.

Unfortunately, because of the need to support his empire, the profitability of the follow-on contract that Ralph was finishing up had all of the profits going to worker salaries. Once again, Ralph argued that significant profits would be forthcoming. During competitive bidding for new follow-on work, Ralph's superiors significantly increased the price of the bid. Unfortunately, the company was now uncompetitive. Ralph and part of the empire he had built up were laid off. The lust for power resulted in that power, which had taken two years to develop, vanishing in one day.

**Situation 17: Revisiting the Failure due to the Lust for Power.** This project would be Kathy's first chance to function as a project sponsor. Kathy believed that her lust for power would thrive if she micromanaged the project team and demanded to make any and all decisions. Senior management would certainly notice this. At least that's what she thought. . . .

Kathy was correct in that senior management saw that she was making all of the decisions. Unfortunately, the subject matter experts assigned to the project, as well as the project manager, knew that Kathy had very limited knowledge with regard to some of the technical decisions that needed to be made on the project. They were also quite unhappy with being micromanaged. Many of Kathy's decisions were wrong, and the team knew it, but they went along anyway with the bad decisions without questioning them. Management also saw the bad decisions that Kathy had made and eventually Kathy was removed as the project's sponsor.

## Gluttony

- "Glutton: one who digs his grave with his teeth." (French proverb)
- "Gluttony is the source of all our infirmities, and the fountain of all our diseases. As a lamp is choked by a superabundance of oil, a fire extinguished by excess of fuel, so is the natural health of the body destroyed by intemperate diet." (Robert Burton)
- "The miser and the glutton are two facetious buzzards; one hides his store and the other stores his hide." (Josh Billings)
- "Gluttony is an emotional escape, a sign something is eating us." (Peter De Vries)
- "Gluttony kills more than the sword." (George Herbert)

Gluttony is usually defined in terms of food with terms such as "gulp down" or "swallow." We see it as an overconsumption of food. In a business environment, gluttony is the desire to consume more than what is required. It is extravagance or waste.

The example that follows shows how gluttony can lead to both success and failure.

**Situation 18: The Success of Gluttony of Resources.** Jerry was one of the directors of manufacturing reporting to the vice president for manufacturing. As technology began to change, manufacturing personnel recognized the need to create several new departments to take advantage of new technologies. Jerry had a thirst for resources. He convinced the vice president for manufacturing that these new departments belonged under his control. Within the next two years, all new departments were under Jerry's supervision. Jerry now controlled more than 75 percent of the resources in the Manufacturing Division.

When the vice president for manufacturing retired, Jerry was promoted to vice president. Jerry's first action was to break up the empire he created so that nobody could ever become as powerful as he had been. In Jerry's eyes, he now had control over all resources, regardless where they resided in the Manufacturing Division.

**TABLE 2–2.   VICES AND VIRTUES**

| Vice | Virtue |
|------|--------|
| Envy | Kindness |
| Wrath | Patience |
| Pride | Humility |
| Greed | Charity |
| Sloth | Diligence |
| Lust | Chastity |
| Gluttony | Temperance |

We have painted a bleak picture here of how the Seven Deadly Sins can have a negative impact on projects. From a project perspective, some of the sins are closely related and cannot be as easily separated and discussed as psychologists and philosophers would have us believe. This can be seen from some of the situations presented previously, for example, where the desire for control of vast resources could be considered as some form of lust, gluttony, or avarice.

It is true that, in some situations, the sins can produce positive results. They can force us to become more aggressive, take risks, accept new challenges, and add value to the company. Our fascination with pride and lust can help us turn around a distressed project and make it into a success so that we can get corp+orate-wide recognition. The greed for wanting a large bonus can likewise encourage us to make our project successful. The downside risk of the vices is that they most certainly can have a negative effect on our ability to establish on our interpersonal skills and our relationships with the project teams and functional departments.

So, should we train project managers and team members on how to identify and control the sins? Perhaps not as long as beneficial results are forthcoming. Once again, we all succumb to some or all of these sins, but in varying degrees.

The Roman Catholic Church recognizes seven virtues, which correspond inversely to each of the Seven Deadly Sins. This is shown in Table 2–2.

From a project management perspective, perhaps the best solution would be to teach the virtues in project management training courses. It is even possible that in future editions of the *PMBOK® Guide,* the Human Resources Management chapter may even discuss vices and virtues. Time will tell.

## 2.14 SOURCES OF SMALLER MIGRAINES

Not all project management headaches lead to migraines. The following list identifies some of the smaller headaches that occurred in various companies but do not necessarily lead to major migraines:

- *Maintaining original constraints:* As the project team began working on the project, work began to expand. Some people believed that within every project there was a larger project just waiting to be recognized. Having multiple project sponsors all of whom had their own agendas for the project created this problem.

- *Revisions to original mission statement:* At the gate review meetings, project redirection occurred as management rethought its original mission statement. While these types of changes were inevitable, the magnitude of redirections had a devastating effect on the EPM system, portfolio management efforts, and capacity planning.
- *Lack of metrics:* An IT organization maintained a staff of over 500 employees. At any given time, senior management was unable to establish metrics on whether or not the IT group was overstaffed, understaffed, or just right. Prioritization of resources was being done poorly, and resource management became reactive rather than proactive.
- *More metrics:* In another example, the IT management team, to help identify whether or not projects were being delivered on schedule, had recently implemented an IT balanced scorecard for projects. After the first six months of metric gathering, the conclusion was that 85 percent of all projects were delivered on time. From executive management's perspective, this appeared to be misleading, but there was no way to accurately determine whether or not this number was accurate. For example, one executive personally knew that none of his top five projects and all ten of an IT manager's projects were behind schedule. Executive management believed the true challenge would be determining appropriate metrics for measuring a project's schedule, quality, and budget data.
- *Portfolio management of projects:* When reviewing project portfolios or individual projects, all of the plans were at different levels of detail and accuracy. For example, some plans included only milestones with key dates, while other plans had too much detail. The key issue became "what is the correct balance of information that should be included in a plan and how can all plans provide a consistent level of accuracy across all projects?" Even the term accuracy was not consistent across the organization.
- *Prioritization of projects and resources:* In one company, there were no mechanisms in place to prioritize projects throughout the organization, and this further complicated resource assignment issues in the organization. For example, the CIO had his top five projects, one executive had his top ten projects, and an IT manager had his top ten projects. Besides having to share project managers and project resources across all of these projects, there was no objective way to determine that the CIO's #3 project was more/less important than an executive's #6 project or an IT manager's #1 project. Therefore, when competing interests developed, subjective decisions were made, and it was challenging to determine whether or not the right decision had been made.
- *Shared accountability for success and failure:* The organization's projects traditionally were characterized as single-resource, single-process, and single-platform projects. Now, almost every project was cross-team, cross-platform, and cross-process. This new model had not only increased the complexity and risk for many projects but also required increased accountability by the project team for the success/failure of the project. Unfortunately, the organization's culture and people still embraced the old model. For example, if one team was successful on its part of a project and another was not, the attitude would be "I am glad I was not the one who caused the project to

fail" and "Even though the project failed, I succeeded because I did my part." While there was some merit to this, overall, the culture needed to be changed to support an environment where "If the project succeeds, we all succeed" and vice versa.

- *Measuring project results:* Many of the projects that were completed were approved based on process improvements and enhanced efficiency. However, after a process improvement project was completed, there were no programs in place to determine whether or not the improvements were achieved. In fact, because the company was experiencing double-digit growth annually, the executive team questioned whether or not approved process improvements were truly scalable in the long term.

- *Integrating multiple methodologies:* Application development teams had adopted the software development methodology (SDM) and agile methodology for software development. Both of these methodologies had excellent approaches for delivering software components that met quality, budget, and schedule objectives. The challenge the organization faced was whether or not components from both of these methodologies could be adapted to projects that were not software development related and, if so, how can this be accomplished? This debate had elevated to upper management for resolution and upper management had been reluctant to make a decision one way or the other. This difference in views on how projects should be managed, regardless of whether or not the project was software development related or not, had led to several different groups lobbying for others to join their efforts to support SDM and Agile for all projects. Overall, the lobbying efforts were not adding value to the organization and were wasted effort by key resources.

- *Organizational communications:* Although there was a lot of communication about projects throughout the organization, many shortcomings existed with the existing process. For example, one executive stated that when he had his monthly status meeting with his direct reports, he was amazed when a manager was not aware of another manager's project, especially if the project was getting ready to migrate into production. The existing process led many managers to react to projects instead of proactively planning for projects. Additionally, the existing communication process did not facilitate knowledge sharing and coordination across projects or throughout the organization. Instead, the existing communication process facilitated individual silos of communication.

- *Meaning of words:* A project was initiated from the staff level. The SOW contained numerous open-ended phrases with vague language such as "Develop a world-class control platform with exceptional ergonomics and visual appeal." The project manager and his team interpreted this SOW using their own creativity. There were mostly engineers on the team with no marketing members, and the solution ended up being technically strong but a sales/marketing disaster. Months were lost in time to market.

- *Problem with success:* A project was approved with a team charter that loosely defined the boundaries of the project. During the course of the project, some early successes were realized and word quickly spread throughout the organization. As the project moved forward, certain department managers began "sliding" issues into the project scope, using their own interpretation of the SOW, hoping to advance their own agendas with this talented group. The project eventually bogged down and the team became demoralized. Senior management disbanded the group. After this, management had real trouble getting people to participate on project teams.

- *Authority challenges:* A new cross-functional project team was assembled involving technical experts from numerous departments. The project manager was a consultant from an outside contractor. During the course of this large project, resource conflicts with production schedules began to arise. Inevitably, the line managers began to draw resources away from the project. The consultant promptly reported pending delays due to this action and the staff reiterated the consultant's concerns and the need for the organization to support the project. The struggles continued through the entire length of the project, creating stressful situations for team members as they tried to balance their workloads. The project finished late with significant cost overruns and indirectly caused a great deal of animosity among many of the participants.

- *Open-ended deliverables:* A project was launched to redesign and deploy the engineering change management system. The team received strong support throughout its duration. At a project closure meeting with the executive staff, the project manager presented the team's interpretation of the deliverables. Much to his surprise, the staff determined that the deliverables were not complete. In the end, this particular team worked on "spider webs" spawning off of their original SOW for over three years (the original closing presentation took place after nine months). The team was frustrated working on a project that never seemed to have an end, and the staff grew impatient with a team they felt was "milking" a job. The project management process at the company came under fire, threatening future efforts and staff support.

- *Cost overruns:* Soon after a major product renovation project was commissioned, the project manager reported that the cost of completion was grossly understated. Unfortunately, the marketing department, in anticipation of a timely completion, had already gone to the marketplace with a promotion "blitz," and customer expectations were high in anticipation of the product's release. The senior staff was faced with a decision to have a runaway cost issue to complete the project on time or endure loss of face and sales in the marketplace to delay the project's completion.

Despite all of these headaches, project management does work and works well. But is project management falling short of expectations? Some people argue "yes" because project management is not some magical charm that can produce deliverables under all circumstances. Others argue that project management works well and nothing is wrong except that expectations of the executives are overinflated. Project management can succeed or fail, but the intent, commitment, and understanding at the executive levels must be there.

## 2.15 TEN UGLIES OF PROJECTS[4]

**Introduction**

Project management methodologies, classes, and books are adequate at explaining the mechanics of running projects and the tools used to do so. Understanding these mechanics is essential, but it is experience that distinguishes successful project managers. More specifically, it is the sum of all of the negative experi-

4. This section was provided by Kerry R. Wills, PMP®, director of portfolio management, Infrastructure Solutions Division. The Hartford. ©2005 by Kerry R. Wills. Reproduced by permission of Kerry R. Wills.

ences that project managers have in their careers that teaches them what not to do. As Vernon Law explains, "Experience is a hard teacher because she gives the test first, the lesson afterwards."

In my many years of project management experience, I have come across several areas that consistently cause projects to experience difficulties. I call these the "uglies" of projects, since these are the things that make projects turn ugly. These are also usually the things that, once recognized, are hard to fix easily.

This section will discuss the ten project uglies and propose some solutions. There are definitely other uglies out there, but these ten are the ones that seem to be the most common and have the biggest impact based on my experience.

**The Ten Uglies**

The following are the ten uglies with a description of each and some symptoms that indicate that these uglies may be happening.

1. **Lack of maintained documentation:** Often when projects are in a crunch, the first thing that gets eliminated is documentation. Sometimes documentation is not created even when projects do have the time. When documentation is created properly, as projects continue to progress, it is a rarity to see the documentation maintained.

*Symptoms*
- Requirement documents that do not match what was produced
- Technical documents that cannot be used to maintain the technology because they are outdated
- No documentation on what decisions were made and why they were made
- No audit trail of changes made

This is a problem since documentation provides the stewardship of the project. By this I mean that future projects and the people maintaining the project once it has been completed need the documentation to understand *what* was created, *why* it was created, and *how* it was created. Otherwise, they wind up falling into the same traps that happened before—in this case "he who ignores history in documentation is doomed to repeat it."

2. **Pile phenomenon:** "What is that under the rug?" is a question often asked toward the end of a project. The mainstream work always gets the primary focus on a project but it is those tangential things that get forgotten or pushed off until "later," at which point there are several piles (swept under the rug) that need to be handled. I call this the "pile phenomenon" because team members think of it as a phenomenon that all this "extra" work has suddenly appeared at the end.

*Symptoms*
- Any work that gets identified as "we will do this later" but is not on a plan somewhere
- Growing logs (issues, defects, etc.)
- Documentation assumed to be done at the end

There is no "later" accounted for in most project plans, and therefore these items either get dropped or there is a mad rush at the end to finish the work.

3. **No quality at source:** Project team members do not always take on the mantra of "quality at the source." There is sometimes a mentality that "someone else will find the mistakes" rather than a mentality of ownership of quality. Project managers do not always have the ability to review all work, so they must rely on their team members. Therefore, the team members must have the onus to ensure that whatever they put their name on represents their best work.

*Symptoms*
- Handing off work with errors before reviewing it
- Developing code without testing it
- Not caring about the presentation of work

There are several studies that show that quality issues not found at the source have an exponential cost when found later in the project.

4. **Wrong people on the job:** Project roles require the right match of skills and responsibilities. Sometimes a person's skill set does not fit well with the role that he or she has been given. I also find that work ethic is just as important as skills.

*Symptoms*
- Team members being shown the same things repeatedly
- Consistent missing of dates
- Consistent poor quality

As project managers, all we have are our resources. Not having the right fit for team members will result in working harder than necessary and impacts everyone else on the team who has to pick up the slack. There is also a motivational issue here: When team members are in the wrong roles, they may not feel challenged or feel that they are working to their potential. This has the impact of those persons not giving their best effort, not embodying a solid work ethic when they normally would, feeling underutilized, and so on.

5. **Not involving the right people:** The people who know how to make the project successful are the team members working on the project. Not involving the right team members at the right time can set the project up for failure before it begins.

*Symptoms*
- Having to make changes to work already completed
- Constant scope changes from the customer
- Lack of team buy-in to estimates
- Lack of ownership of decisions

Not involving the right people up front in a project always results in changes to work. Not involving team members in decisions and estimates causes them to feel like they have no control over their work or the outcomes of the project.

6. **Not having proper sponsorship:** Projects need internal and customer executive sponsorship to be successful. Sponsors act as tiebreakers and eliminate organizational politics/roadblocks that are holding up the project.

*Symptoms*
- Inadequate support from different areas of the organization and from customer stakeholders
- Issues taking very long before being resolved
- Decisions not being made efficiently

Not having proper sponsorship can result in projects "spinning their wheels." Also, when a change effort is involved, not having proper sponsorship can keep impacted employees from buying in to a project (i.e., not cascading the messages from the top down to the "masses").

7. **No rigor around process:** Almost every company uses a methodology for implementing projects. The success of these methodologies depends on the amount of rigor used on the project. Often, processes are not adhered to and projects run awry.

*Symptoms*
- Incomplete/nonexistent deliverables
- Inconsistencies within the project
- Lack of understanding of the project's big picture
- Lack of repeatable processes ("reinventing the wheel" unnecessarily)

Processes are only as valuable as the rigidity placed on them. In some companies, there are too many project management methodologies used. Some are necessary due to the varying nature of work, but basic project management practices and principles (and even tools, i.e., using Project vs. Excel) could easily be standardized but are not. When one manager has to transfer a project to another, this creates an extra layer of complexity, because a common language is not being used between the two people (it is like trying to interpret someone else's code when they have not followed the standards you have been using).

8. **No community plan:** Project managers spend a significant amount of time on planning, estimating, and scheduling activities. If these results are not shared with team members, then they do not know what they are working toward and cannot manage their own schedules. This includes the communication of goals and items that are a big picture for the team.

*Symptoms*
- Lack of knowledge about what is due and when it is due
- Missed dates
- Lack of ownership of deliverables
- Deliverables get forgotten

Not having a community plan will result in not having an informed community. Having a shared plan and goals helps to build a cohesiveness and a greater understanding of how the work of each individual fits overall.

9. **Not planning for rework:** Estimation techniques often focus on the time that it takes to create units of work. What usually gets left out is the time spent on rework. By this I mean work that was done incorrectly and needs to be revisited as opposed to scope management. When rework is required, it either takes the place of other work, which now comes in late or is pushed off until later (see ugly 2).

*Symptoms*
- Missed dates
- Poor quality

Never assume that anything is going to be done right the first time.

10. **Dates are just numbers:** Schedule is a major driver of project success. I am amazed at the number of people who think of dates as "suggestions" rather than deadlines. Because of interdependencies on projects, a missed date early on could ripple through the schedule for the remainder of the project.

*Symptoms*
- Consistently missed dates
- Items left open for long periods of time
- Incomplete/nonexistent deliverables
- Lack of a sense of urgency on the project team

Without structure around the management of dates, success requires a lot more effort. One other issue here is that of communication—these dates need to be communicated clearly and people must agree that this is their target. Also, they must understand what is on the critical path and what has slack, so if they slip on a critical path item, they know there is an impact on the project or on another project within the same program.

**Possible Remedies**       Upon analyzing the uglies I observed that they are all interrelated. For example, not having rigor around processes (#7) can result in not having a shared plan (#8), which can result in people not caring about dates (#10), and so on. (See Figure 2–4.) I also realized that a few remedies could mitigate these uglies. The trick here is to proactively resolve them rather than react to them since by the time you realize that there is an ugly, **Your project is already ugly.**

**Proactive Management** Proactive management means spending the appropriate amount of time up front to minimize the number of "fires" that need to get put out later. Proactive management includes the following actions:

- Creation of a detailed plan.
- Always looking at the plan to see what is coming up and preparing for it:

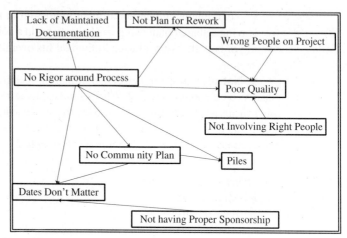

**Figure 2–4.** Observed interrelationships.

- Thinking about the upcoming work and running down any issues that may be coming. I think of the team as running a marathon and it is my job to "clear the road" in front of them so they can keep on running.
- Setting up logistics. Something as trivial as not having a conference room booked in advance can cause a schedule delay.
- Lining up the appropriate people to be ready when the work comes their way.
- Know people's vacation schedules.
- Constant replanning as information becomes more available.
- Understanding what is going on with the project. I see so many project managers in the "ivory tower" mode where they find out about things about THEIR project for the first time on a status report. By this time, as much as a week has gone by before the project manager is aware of issues.

There will always be unexpected issues that arise, but proactive management can help to mitigate those things that are controllable. This can be treated as an investment of time, in that you will spend far more time (and money) reacting to problems than you will focusing on ensuring that the process be followed properly. This is difficult for some project managers because it requires the ability to always look ahead of the current state of the project rather than just focusing on the problem of the day. A key element of proactive management is having the ability to make decisions efficiently.

**"Do It While You Do It"** Now that you are not reacting to fires, you can focus team members on maintaining their work as they go. This means staying focused on all aspects of the current work and thinking of implications. Characteristics of this include:

- Documenting as work is being done and not at the end. I am sure that this will get the knee-jerk "we don't have time" reaction but I really believe (and have proved) that documenting as you go takes far less time than doing it at the end.

- Thinking of implications as things change on the project. For example, if a document changes, the owner of that document should think about any other deliverables that may be affected by the change and communicate it to the appropriate person.
- Check all work before passing it on to others.
- Use the process/plan as a guideline for what work has to be done. I have heard this referred to as "living the plan."

The result of this technique will be an even distribution of work across the project and minimal spikes at the end. Rather than the notorious "death march," the worst case could be considered an "uncomfortable marathon."

**Empower the Team** Project managers must realize that project structures resemble an inverse pyramid where the project manager works FOR the team. It is the team members who do the work on the project, so the project manager's primary role is to support them and address obstacles that may keep them from completing their work. This includes:

- Involving team members in project planning, so they cannot say that they were just given a deadline by management.
- Ask team members how things are doing, and then act on their concerns. Asking for feedback and then doing nothing about it is worse than not asking at all because it suggests an expectation that concerns will be addressed.
- Celebrate the successes of the team with the team members.
- Be honest with the team members.

I am a big fan of W. Edwards Deming, who revolutionized the manufacturing industry. His fourteen points of management revolve around empowerment of the team and apply very much to projects. Excerpts are noted in Table 2–3 with my opinion of how they relate to project management.

Empowering the team will enable the project manager to share information with the team members and will also enable the team members to feel like they have control over their own work. The result is that each team member becomes accountable for the project.

**TABLE 2–3. DEMING POINTS OF MANAGEMENT**

| Deming Point | Observation |
|---|---|
| 8. Drive out fear, so that everyone may work effectively for the company. | This means that the "iron fist" technique of project management is not such a great idea. People will be averse to giving their opinions and doing a quality job. |
| 10. Eliminate slogans, exhortations, and targets for the workforce asking for zero defects and new levels of productivity. Such exhortations only create adversarial relationships, as the bulk of the causes of low quality and low productivity belong to the system and thus lie beyond the power of the workforce. | I take this to mean that project managers should not just throw out targets, but rather involve the team members in decisions. It also means that project managers should look at the process for failure and not the team members. |
| 12. Remove barriers that stand between the hourly worker and his (or her) right to pride of workmanship. | This is my marathon metaphor—where project managers need to remove obstacles and let the team members do their work. |
| 13. Institute a vigorous program of education and self-improvement. | Allow the team members to constantly build their skill sets. |

**TABLE 2-4. ATTRACTIVE STATE CHARACTERISTICS**

| Ugly Number | Ugly Name | Ugly State Characteristics | Proactive Management | Do It While You Do It | Empower |
|---|---|---|---|---|---|
| 1 | Maintained documentation | • No idea what decisions were made <br> • Do not know why decisions were made <br> • Cannot rely on accuracy of documents <br> • Cannot use on future projects | • Updated documentation <br> • Documentation will be planned for <br> • Anyone can understand decisions | • Done during the project <br> • No extra work at the end of the project | Team members will own documentation |
| 2 | Piles | Put off until later <br> May never get done | • Manageable work <br> • If piles do exist, they will be scheduled in the plan | Will be worked on as people go, so they should never grow out of control | Minimized because people will take ownership of work |
| 3 | Quality at the source | • No ownership of work <br> • Poor quality <br> • Expensive fixes | Better quality because you have spent appropriate time upfront | Quality will be focused on as people do their work rather than assumed at a later time. | Quality will be upheld as people take ownership of work. |
| 4 | People fit | Bad project fit | The ability to recognize resource issues and resolve them before they seriously impact the projectProper resource fit from the start | Manage work so resource issues are identified early. | Other team members may take on work for failing colleagues. |
| 5 | People involvement | Changes after work has been done <br> No ownership of work <br> No accountability for results | Involving the right people up front to avoid rework later | Involve people during work rather than have them react to it later. | Empowered team members take ownership for work. |
| 6 | Sponsorship | Cannot resolve problems <br> Caught up in organizational politics | Engaging stakeholders early will enable you to access their support when really needed. | Rapid and effective decisions as needed | May be improved due to better understanding of issues |
| 7 | Process rigor | • No rigor <br> • Poor quality <br> • Inconsistent work | • Proper rigor is the essence of proactive management. <br> • Repeatable processes <br> • Looking ahead will ensure proper attention to process. | • Rigor occur as team members follow the process. <br> • Ensures that process steps are not missed | Ownership of work will enable better rigor around process. |
| 8 | Community plan | No idea what is due and when team members do not take accountability for work—the plan is for the project manager | Have the ability to share a plan and goals with the team | Everyone is working to the same plan and knows where they are going. | • Everyone is informed—shared goals <br> • People can manage their own work. |
| 9 | Rework | Not planned for trade-off between doing other work or fixing issues | Anticipating areas where there may be rework or scope creep and working with key stakeholders early to address those planned for | • Rework will be accounted for as the team members work. <br> • By staying on top of the project, you will be aware of the magnitude of rework and can replan as needed. | Should be minimized due to motivation and ownership of work |
| 10 | Dates | • Dates do not matter <br> • No accountability <br> • Missing deliverables | Dates (and impacts of missing them) clearly communicated | Will matter and items will be closed when they are due | Team members take ownership of dates. |

**Results of the Remedies** The results of applying these remedies to the uglies are shown in Table 2–4. I call my vision of the new way of doing things the "attractive state" since it attracts people to success.

## Conclusion

Focusing on proactive management, keeping up with work, and empowering your teams are key to running a successful project. There is nothing in this section that has not been written of or spoken of hundreds of times before. Nothing should sound new to a project manager. And yet, we keep seeing the uglies over and over again. That leads me to a conclusion that it is the application of these concepts that is the challenge. I find that after I read a good paper or attend a management course, I have great enthusiasm to try out the new techniques but at the first signs of trouble I revert to my comfort zone. Therefore, I propose that there is a fourth remedy for the uglies—being conscious. This is nothing more than being aware of what is going on and how you are managing your project.

I come to work every morning a little earlier than the rest of the team so I can have my quiet time and think about what work needs to be done (not just for that day but in the upcoming days). I also give myself reminders that trigger my "step-back-and-think" mode. An excellent series that goes into this technique are the "emotional intelligence" books by Daniel Goleman.

There will always be uglies on your projects, but if you are conscious of them, then you can identify them when they are happening and you may be able to prevent them from throwing your projects into chaos. Best of luck.

## REFERENCES

Deming, W. E. *Out of the Crisis: Quality, Productivity and Competitive Position,* Cambridge: Cambridge University Press, 1982, *1986.*

Gleman, D. *Working with Emotional Intelligence,* New York: Bantam Books, 1998.

# 3 Journey to Excellence

## 3.0 INTRODUCTION

Every company has its own forces, or driving forces, as we discussed in Chapter 1, that force the company to embark upon a journey for excellence in project management. Some companies complete the journey in two or three years, while others may require a decade or more. In this chapter, we will discuss the approaches taken by a variety of companies. Each company took a different path, but they all achieved some degree of excellence in project management.

Some companies embark on the journey at the request of their own workers whereas other companies are forced into it by the actions of competitors and customers. In any event, there are driving forces that propagate the quest to excel in project management.

The driving forces for excellence, as discussed previously, include:

- Capital projects
- Customer expectations
- Competitiveness
- Executive understanding
- New product development
- Efficiency and effectiveness

Even the smallest manufacturing organization can conceivably spend millions of dollars each year on capital projects. Without good estimating, good cost control, and good schedule control, capital projects can strap the organization's cash flow, force the organization to lay off workers because the capital equipment either was not available or was not installed properly, and irritate customers with late shipment of goods. In non–project-driven organizations and manufacturing firms, capital projects are driving forces for maturity.

Customers' expectations can be another driving force. Today, customers expect contractors not only to deliver a quality product or quality services but also to manage this activity using sound project management practices. This includes effective periodic reporting of status, timely reporting of status, and overall effective customer communications. It should be no surprise that low bidders may not be awarded contracts because of poor project management practices on previous projects undertaken for the client.

The third common driving force behind project management is competitiveness. Companies such as IBM and Hewlett-Packard view project management as a competitive weapon. Project-driven companies that survive on contracts (i.e., income) from external companies market their project management skills through virtually every proposal sent out of house. The difference between winning and losing a contract could very well be based upon a firm's previous project management history of project management successes and failures.

The most common form of competitiveness is when two or more companies are competing for the same work. Contracts have been awarded based upon previous project management performance, assuming that all other factors are equal. It is also not uncommon today for companies to do single-source procurement because of the value placed upon the contractor's ability to perform. A subset of this type of competitiveness is when a firm discovers that outsourcing is cheaper than insourcing because of the maturity of their contractor's project management systems. This can easily result in layoffs at the customer's facility, disgruntled employees, and poor morale. This creates an environment of internal competition and can prevent an organization from successfully implementing and maturing in project management.

A fourth driving force toward excellence is executive buy-in. Visible and participative executive support can reduce the impact of many obstacles. Typical obstacles that can be overcome through executive support include:

- Line managers who do not support the project
- Employees who do not support the project
- Employees who believe that project management is just a fad
- Employees who do not understand how the business will benefit
- Employees who do not understand customers' expectations
- Employees who do not understand the executives' decision

Another driving force behind project management is new product development. The development of a new product can take months or years and may well be the main source of the company's income for years to come. The new product development process encompasses the time it takes to develop, commercialize, and introduce new products to the market. By applying the principles of project management to new product development, a company can produce more products in a shorter period of time at lower cost than usual with a potentially high level of quality and still satisfy the needs of the customer.

In certain industries, new product development is a necessity for survival because it can generate a large income stream for years to come. Virtually all companies are involved in one way or another in new product development, but the greatest impact may very well be with the aerospace and defense contractors. For them, new product development and customer satisfaction can lead to multiyear contracts, perhaps for as long as 20 or more years. With product enhancements, the duration can extend even further.

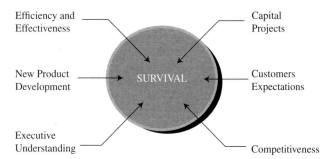

**Figure 3–1.** Components of survival.

Customers will pay only reasonable prices for new products. Therefore, any methodology for new product development must be integrated with an effective cost management and control system. Aerospace and defense contractors have become experts in earned value measurement systems. The cost overruns we often hear about on new government product development projects are attributed not necessarily to ineffective project management or improper cost control but more to scope changes and enhancements.

Improvement in the overall efficiency and effectiveness of the company is sometimes difficult, if not impossible. It often requires change in the corporate culture, and culture changes are always painful. The speed at which such changes accelerate the implementation of project management often depends on the size of the organization. The larger the organization, the slower the change.

Obviously, the most powerful force behind project management excellence is survival. It could be argued that all of the other forces are tangential to survival (see Figure 3–1). In some industries, such as aerospace and defense, poor project management can quickly lead to going out of business. Smaller companies, however, certainly are not immune.

Sometimes, there are additional driving forces:

● Increase in project size mandated by the necessity to grow
● Customers demanding faster implementation
● Customers demanding project management expertise for some degree of assurance of success completion
● Globalization of the organization mandated by the need to grow
● Consistency in execution in order to be treated as a partner rather than as a contractor

## 3.1 STRATEGIC PLANNING FOR PROJECT MANAGEMENT

For more than five decades, project management has matured from what was once considered to be just a fad that would soon disappear into a strategic competency and career path necessary for the growth and survival of the firm. Project management is now being used in virtually every industry and in all parts of the business. We have matured to the point where we believe that we are managing our business as though it is a series of projects, and where project managers are expected to make both project decisions as well as business decisions. Project managers are now considered as business people rather than just project managers.

Today, project management is recognized as a series of processes that can be used on every project, regardless of its length or complexity, the project's dollar value, or the project's exposure to risk. Yet the one part of the business where project management has been slow in being accepted, at least up to now, has been in strategic planning execution projects. One can always argue that managing strategic planning execution projects is no different from managing any other type of project. While this argument may have merit, there are several important differences that must be considered. Specifically, project managers must think strategically rather than tactically or operationally, and they may have to change from traditional project management leadership to strategic leadership based upon the complexity of the project.

**Why Strategic Plans Fail**     To understand how project management can benefit strategic planning, it is important to understand why some strategic plans fail. Some of the common reasons for failure, as seen through the eyes of project managers, include:

- Neglecting to understand how the enterprise environmental factors can influence senior management's vision of the future
- Inadequate understanding of consumer behavior or the client's actions
- Improper research prior to project approval
- Poorly defined or ill-defined scope
- Poorly documented business case resulting in the approval of the wrong project
- Failing to get executive and stakeholder buy-in right from the start
- Poor executive governance once the strategy begins to be implemented
- Constantly changing the membership of the governance team
- Overestimating resource competencies needed for project execution
- Poor capacity planning efforts resulting in understaffed projects
- Functional managers refusing to commit the proper resources for the duration of the strategic project
- Failing to get employee commitment to the project
- Failing to explain the importance of the project to the project execution team
- Failing to explain to the execution team the incentives or financial benefits of working on this long-term project
- Failing to understand the magnitude of the organizational change needed for the project to be a success
- Unable to manage change effectively
- Failing to consider the impact of changes in technology during the execution of the project
- Poor estimating of time and cost
- Having an execution team that is unable to work with ill-defined or constantly changing requirements
- Poor integration of the project across the entire organization
- Inadequate communications

There are numerous other reasons for the failure of strategic planning execution projects. These causes could occur on any project, but on strategic planning execution projects the potential damage to the firm could be quite severe.

**Project Management: An Executive Perspective**

With the ability to produce repeated successes on projects, it is no wonder that executives are now realizing the value in using project management for the execution of a strategic plan. There are several reasons why executive see value in using project management for these activities:

- Execution takes significantly more time than planning and consumes more resources. Executives do not have the time to spend possibly years coordinating and integrating work across a multitude of functional areas.
- Without a successful implementation plan, strategic planning cannot succeed.
- Project managers can successfully manage the dysfunctional separation between planning and execution.
- Long-term strategic objectives must be broken down into short-term objectives to simplify execution. This can be done easily using project management tools and a work breakdown structure.
- Project management staffing techniques, possibly with the use of a project management office (PMO), can match the proper resources to the projects.
- The organizational process assets used in project management can keep senior management updated on project status.
- Strategic planning objectives, because of the long time duration, are highly organic and subject to change. Project managers know how to manage and control change.

**Strategic Planning: A Project Management Perspective**

Strategic planning is an organization's process of defining where and how it would like to be positioned in the future. The future may be measured in a three-year, five-year or ten-year (or longer) window. The strategic plan is based upon the firm's vision, mission, social consciousness and values. Strategic planning requires an understanding of the firm and its environment. Executives, more so than project managers, have a better understanding of the enterprise environmental factors, namely products offered, markets served, present and future technologies, supplier base, labor markets, economic conditions, the political environment and regulatory requirements.

Executives establish high-level objectives for *what they want done*. Often, this is nothing more than a wish list that may or may not border on reality. The role of the project manager is to determine *if it can be done*. This requires a clear business case for each project, a scope statement, and use of the work breakdown structure (WBS) to break down the high-level objectives into sub-objectives, or lower-level objectives that are easier to understand and accomplish. If the project manager and the project team believe that is can be done, then a formalized project action plan is created. According to Wikipedia, the free online encyclopedia,

One of the core goals when drafting a strategic plan is to develop it in a way that is easily translatable into action plans. Most strategic plans address high-level initiatives

**TABLE 3–1.  THE *PMBOK* GUIDE AND THE EXECUTION OF STRATEGIC PROJECTS**

| Area of Knowledge | Strategic Planning Project Impacts |
| --- | --- |
| Integration Management | The integration of the effort may very well span the entire organization both domestically and globally. |
| Scope Management | The scope can change as technology changes. The length of the project makes it imperative that an effective scope change control process exist. The scope baseline may appear as a moving window requiring constant updates. |
| Time Management | Matching the right people and their availability to the constantly changing scope will play havoc with scheduling. Losing people due to fire fighting in the functional areas may have a serious impact. |
| Cost Management | Predicting the true cost of the project will be almost impossible. Reestimation must take place on a routine basis to make sure that the benefits and business value still exceeds the cost. |
| Quality Management | Customers' expectations of quality and competitive forces can cause major changes to the direction of the project. |
| Human Resource Management | The longer the project, the greater the likelihood that changes in resources will occur, possibly for the worse. Long-term motivation may be difficult to do. |
| Communication Management | Communication requirements can span the entire company. Changes in stakeholders will also have a serious impact on the communication plan. |
| Risk Management | The project may be required to have a dedicated risk management team. |
| Procurement Management | The length of the project may make it difficult to accurately determine procurement costs upfront. |
| Stakeholder Management | Because of the length of the project, the project manager may end up interfacing with a different set of stakeholders at the end of the project than at the beginning. |

and overarching goals, but don't get articulated (translated) into day-to-day projects and tasks that will be required to achieve the plan. Terminology or word choice, and the level at which a plan is written, are both examples of easy ways to fail at translating your strategic plan in a way that makes sense and is executable to others. Often, plans are filled with conceptual terms which don't tie into day-to-day realities for the staff [project team] expected to carry out the plan.

On the surface, it may appear that strategic planning execution projects can be treated as any other project. However, if we look at the Areas of Knowledge in the *PMBOK®* *Guide*, we can see some significant differences mostly attributed to the length of the project. A few of these differences are shown in Table 3–1.

## The Benefits of Project Management

Perhaps the primary benefit of using project management that makes it extremely attractive for strategic planning projects is to provide executives and clients with a single point of contact for status reporting. Most of today's strategic planning projects are so complex that they cannot be managed effectively by one functional manager, who may have a conflict between his/her functional duties and project duties. These projects require the coordinated effort of several functional areas such as sales, marketing, engineering, and manufacturing. Without having a single point of contact for status reporting, the executives would need to do the coordination and integration themselves, and it is highly unlikely that they would have the time to do this in addition to their other duties. Likewise, functional managers do not have sufficient time to manage their functional areas and perform integration work on various projects. The need for project management is quite clear.

**TABLE 3–2. BENEFITS OF USING PROJECT MANAGEMENT**

| Attribute | Benefit |
| --- | --- |
| Efficiency | Allows an organization to take on more work in less time without any increase in cost or degradation of quality |
| Profitability | With all other things being equal, profitability should increase |
| Scope changes | Allows for better upfront planning, which should reduce the number of scope changes downstream and prevent unwanted changes from happening |
| Organizational stability | Focuses on effective teamwork, communication, cooperation, and trust rather than organizational restructuring |
| Quality | Quality and project management are married; they both emphasize effective upfront planning. |
| Risks | Allows for better identification and mitigation of risks |
| Problem solving | The project management processes allow for informed decision making and problem solving in a timely manner. |

**TABLE 3–3. ADDITIONAL BENEFITS FOR STRATEGIC PLANNING EXECUTION PROJECTS**

| Attribute | Benefit |
| --- | --- |
| Alignment | Better alignment of projects to corporate strategic objectives |
| Underperformance | Earlier identification of underperforming investments |
| Capacity planning | Better analysis of corporate resource planning and availability of qualified resources |
| Prioritization | Combining capacity planning efforts and project management allows for better prioritization of the portfolio of projects. |
| Risk mitigation | Allows for better mitigation of business risks by using more "what if" scenarios |
| Time to market | Allows for quicker time to market |
| Decision making | More informed and timely decisions due to availability of essential information |
| Efficiency and effectiveness | Allows us to work on more projects without increasing headcount |
| Better information flow | Elimination of duplication of efforts by managers who are unaware what others are doing |
| Selection of projects | Better analysis of what is and what is not a good idea |

There are also many other benefits of using project management, some of which are shown in Table 3–2.

The benefits shown in Table 3–2 apply to just about all projects including strategic planning execution projects, complex projects, and traditional projects. But there are some additional benefits that affect strategic planning execution projects more so than other types of projects. These are illustrated in Table 3–3.

**Dispelling the Myths**   When we look at Tables 3–2 and 3–3 and see all of the advantages, we must ask ourselves, "Why is there still resistance to the acceptance of project management, especially for strategic planning execution projects?" The answer is quite clear; there are still myths about the use of project management for activities related to strategic planning.

**Myth #1: Project managers have strong technical knowledge but limited knowledge about the business.** While it is true that historically project managers came out of technical disciplines and many even possessed master's degrees and doctorates in technical disciplines, today's project manager has more of an understanding of technology than a command of technology, but has an excellent knowledge of the business. Business knowledge is essential to bridge strategy and execution effectively. Project managers who are considered "global" project managers must have a good understanding of the client's business as well as their own firm's business. This is a necessity to compete in a global marketplace. These global project managers are also being trained in stakeholder relations management, politics, culture, and religion, since all of these topics can have an impact on the client's project.

We believe today that we are managing our firm's business as though it is a series of projects, where project managers are expected to make both project decisions and business decisions. Some companies are requiring their project managers to become certified in the company's business processes or to take coursework leading to certification as a business analyst.

**Myth #2: Project managers should be assigned to a project after the project is approved and the business case is developed.** Years ago, project managers were brought on board a project at the end of the initiation, rather than at the beginning. We believed that, because project managers had limited knowledge of the business, they could not contribute anything worthwhile during the initiation process. After the projects were selected, project managers would be brought on board and told to begin execution. Today, project managers are brought on board at the beginning of the project initiation and selection process and are expected to make a valuable contribution because of their understanding of the business.

**Myth #3: If we implement project management, project managers will begin making decisions that should be made at the executive levels of management.** Strategic planning and the accompanying necessary decisions are made by executives, not by someone else for them. However, in some cases strategic planning execution decisions may be made for the executives rather than by them. Executives have always been fearful of having to empower project managers with authority and responsibility with regard to project decision making. This myth alone has been a great impediment to the successful implementation of project management.

The problem was partially resolved with the creation of the position of the executive sponsor or project sponsor. Project managers were allowed to make technical decisions, but the project sponsors reserved the right to make any and all business-related decisions. This approach worked well for reasonably short duration projects. But for strategic planning execution projects, which can be five to ten years in length, the number of decisions that must be made can be overwhelming. Therefore, to overcome this myth, it is beneficial to clearly define the empowerment of the project manager with regard to responsibilities and decision-making authority.

**Myth #4: Project managers do not know how to use the organization process assets effectively for controlled measurement systems needed for informed decision making to take place.** For the past five decades, the two primary metrics used by project managers were time and cost. This was because of the rule of inversion, which states that we often select the easiest metrics to measure and report, even though they may not provide us with a clear picture of the health of the project. Time and cost alone cannot predict the success of a project or whether the value will be there at the completion of the project. This is particularly true for strategic planning execution projects.

There are seminars in the marketplace today on measurement techniques. There are also textbooks on measurement techniques, which argue that anything can be measured if you simply understand the information at your disposal. The result has been the creation of additional metrics for project management. There is a belief that we should consider the following as core metrics for today's projects:

- Time
- Cost
- Resources
- Scope
- Quality
- Action items

These core metrics apply to all projects, but additional core metrics must be added based upon the size, nature, scope and importance of the project. Because strategic planning implementation projects can be long in duration, significant changes can take place. Therefore, we must allow metrics to change over the course of the project. Establishing a set of core metrics that can be used on every project may be difficult.

## Ways That Project Management Helps Strategic Planning

There are often special situations where project management can significantly benefit an organization. In a company that manufactures household appliances, each functional area was allowed to perform their own strategic planning. The problem occurred when functional units had to work together on the same project. In this company, new products were introduced at trade shows, and there were two trade shows each year. Missing a trade show product launch could easily result in lost revenue for six months until the next trade show.

The launching of new products was the highest priority in marketing's strategic plan. R&D, on the other hand, had more than 300 projects in the queue. On the R&D list of priorities, the new products that marketing needed for the trade shows were low on their list of priorities. Battles between marketing and R&D ensued continuously.

In another company, marketing was allowed to prioritize projects as part of their strategic planning activities. For each project, they also prioritized the attributes of the project/product that had a direct bearing on the way the product would be advertised and marketed. But when the project/product went into manufacturing, the manufacturing people often had a different set of priorities for the attributes. This time, battles over priorities ensued between marketing and manufacturing.

In both of the abovementioned examples, the issues were resolved when project management personnel requested that the organization create a single priority list for all projects in the company. The result was that R&D, engineering and manufacturing would meet once every three months and come to an agreement on the priorities of the projects. However, there were too many projects in the queue to prioritize each project. The decision was then made that only twenty projects at a time would be prioritized. This greatly benefited the project staffing process, because everyone was now working off of the same priority list.

Another effective use of project management is gap analysis and gap closure. Gap analysis is used to strengthen your company's competitive position or to reduce the competitive position of your competitors by reducing gaps. Projects are established to take advantage of best practices and lessons learned on other projects, by which gaps can be compressed. The gaps can be:

- Speed with which new products are introduced (time-to-market)
- Competitiveness on cost
- Competitiveness on quality
- Introduction of new technology or product performance

**Strategic Project Management Leadership**

We showed some of the ways that project management can benefit the execution of a strategic plan. For this to happen, the project manager may need to change his/her leadership style from traditional project management leadership, which focused heavily on situational leadership oriented towards the project team, to a strategic project management leadership style, where the end result can affect organizational change across the entire company.

"Strategic leadership" is a term usually reserved for the senior most levels of management. It entails the executive's ability to express a strategic vision for the company's future and then motivate or persuade the organization to acquire or follow that vision. Strategic leadership requires the development of action plans, and this is where project management takes on paramount importance. Visions do not help much unless plans can be developed and implemented to make the vision a reality. Managing projects that involve the implementation of a strategy are significantly different than managing the traditional projects that some of us are used to managing. Unlike action plans for traditional projects that are based upon a well-defined statement of work, the project manager has to develop action plans that can be based upon complexity, ambiguity, uncertainty, and volatile knowledge. Because of the large number of unknowns, and their ability to constantly change, project managers must understand that managing these projects requires consequential decisions that must involve the managers who have ultimate control of the resources for executing these decisions.

If the projects require certain degrees of innovation, leadership skills must be designed around getting the team to be innovative and creative. Brainstorming and problem-solving sessions could occur each week. Facilitation skills are also a necessity. The leadership skills needed for long-term innovation projects may be significantly different than the skills needed to provide a client with a simple deliverable.

To be effective in strategic project management leadership, the project manager must realize that he/she is now a manager of organizational change and, as such, may have to build prepared minds on a large scale. The project manager, and possibly the entire team, must now function as a cheerleader and enforcer at the same time in order to get people across the entire organization to agree on a common sense of purpose. For these types of projects, the project team is usually referred to as a strategy support team (SST). For the SST to function effectively, they must be willing to coach and guide the strategy process as it unfolds. The most difficult challenge for the SST will be on projects that require organizational change. There are significant roadblocks that must be overcome. The SST must also be innovators and change agents, capable of seeing the big picture and thinking strategically rather than operationally or tactically. They must give up short-term thinking and focus on the distant future.

The main objective of strategic leadership is to make the organization more strategically productive and inventive as well as efficient and effective. Workers must be encouraged to follow their own ideas when feasible and provide feedback on technical or behavioral innovations that can be captured through lessons learned and best practices. Lessons learned and best practices allow companies to focus only on the right energies that will help a company profit in the long run.

Traditionally, project managers were expected to capture project-related knowledge and send it to the PMO for analysis and storage. But with strategic leadership, more business-related knowledge must be captured and fed into a corporate knowledge repository.

**Strategic Project Management Leadership Traits**

For over four decades, we examined the skills needed to be a project manager and provide effective project leadership. The analyses were made focusing on the traditional project which may last 12–18 months or less. In addition, the statement of work is reasonably well defined, many of the people may be full time but only for a few weeks, and the outcome of the project may affect only a small number of people. The potential long time frames for strategic projects are now forcing us to revise some of these leadership skills.

It is almost impossible to create an all-inclusive list of the competencies required for a project manager to provide strategic project management leadership. However, we can show some of the possible changes in leadership that will be needed. This appears in Table 3–4. There is a valid argument that all project managers need these skills, but they may be more critical on strategic projects.

**TABLE 3–4.  DIFFERENCES BETWEEN TRADITIONAL AND STRATEGIC PROJECT MANAGEMENT LEADERSHIP STYLES**

| Traits | Differences |
|---|---|
| Authority | From leadership without authority to significant authority |
| Power | From legitimate power to judicious use of power |
| Decision making | From some decision making to having authority for significant decision making |
| Types of decisions | From project only decisions to project and business decisions |
| Willingness to delegate | The length and size of the project will force the project managers to delegate more authority and decision making than they normally would. |
| Loyalty | From project loyalty to corporate vision and business loyalty |
| Social skills | Strong social skills are needed, since we could be working with the same people for years. |
| Motivation | Learning how to motivate workers without using financial rewards and power |
| Communication skills | Communication across the entire organization rather than with a selected few |
| Status reporting | Recognizing that the status of strategic projects cannot be made from time and cost alone |
| Perspective/outlook | Having a much wider outlook, especially from a business perspective |
| Vision | Must have the same long-term vision as the executives and promote the vision throughout the company |
| Compassion | Must have a much stronger compassion for the workers, since they may be assigned for years |
| Self-control | Must not overreact to bad news or disturbances |
| Brainstorming and problem solving | Must have very strong brainstorming and problem-solving skills |
| Change management | Going from project to corporate-wide change management |
| Change management impact | Going from project to organizational change management effects |

**The Project Manager as a Manager of Change**

The role of the project is continuously evolving. As stated in Table 3–4, some strategic planning projects are designed for organizational change, and the change may affect the entire company worldwide. An example might be the implementation of a new corporate-wide security system, information system or secured e-mail system.

The new question becomes, "Who will manage the implementation of the change once the project is ready for implementation?" Historically, project managers created the deliverables and someone from the management ranks would take the lead for the implementation of the change. Today, the project manager is being asked to take the lead role in organizational change management. There may also be a project sponsor from the senior most levels of management with specialized knowledge in organizational change management.

For years, some of us have been managing strategic projects without realizing that we were doing so, and we may not have recognized the possible need for a different leadership style. But as the use of project management begins to grow in terms of its application to strategic planning execution projects, we may need to conduct more research on the specific leadership skills needed. We are in the infancy states of strategic project management applications, but we do expect this trend to take hold over the next decade or longer.

## 3.2 HITACHI LTD.

When strategic planning for project management is done correctly, the beneficial use of project management can permeate the entire company and project management may get integrated into virtually all business areas. An excellent example of this is seen in Hitachi.

**Initiatives to Strengthen Project Management Capacity at Hitachi[1]**

The business lineup at the Hitachi Group ranges widely from developing, manufacturing, sales and provision of solutions for Information & Telecommunication Systems, Power Systems, Social Infrastructure & Industrial Plant Systems, High Functional Materials, Rail Systems, Elevators & Escalators, Automotive Products and Components, Construction Machinery, Digital Media & Consumer Products, as well as related consulting and services. For each line of business, improvements to engineering technologies to support the quality of the business, and improvements to project management are essential. This is the context for the initiatives to strengthen the project management capacity for every line of business.

From the viewpoint of project management, the five perspectives indicated in Figure 3–2 are the support components for taking a project to its successful completion.

Perspective (1) refers to initiatives to provide continuous and effective training for superior project managers. Since the success or failure of a project depends to a large degree on the capabilities of the project manager, it is important to train superior personnel. To do so, it is necessary not only to build the educational systems for training

---

1. Material on Hitachi had been provided by the PM Technical Committee, Hitachi Ltd. ©2013 by Hitachi Ltd. Reproduced by permission.

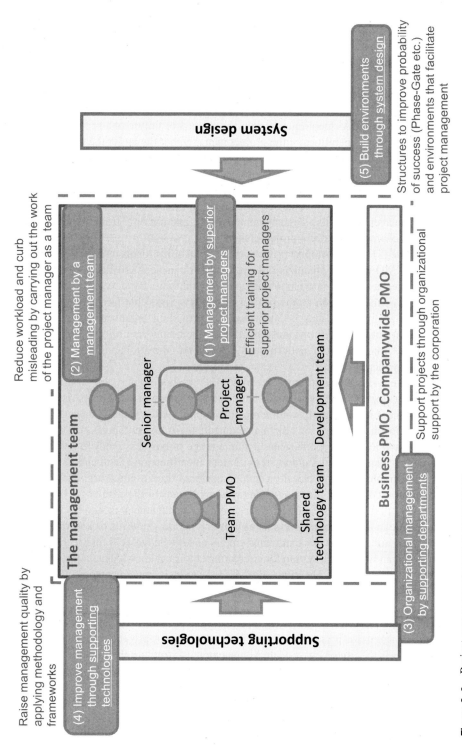

**Figure 3–2.** Project management support components.

Reduce workload and curb misleading by carrying out the work of the project manager as a team

Raise management quality by applying methodology and frameworks

System design

(5) Build environments through system design

Structures to improve probability of success (Phase-Gate etc.) and environments that facilitate project management

The management team

(2) Management by a management team

(1) Management by superior project managers

Efficient training for superior project managers

Senior manager

Project manager

Development team

Team PMO

Shared technology team

Business PMO, Companywide PMO

Support projects through organizational support by the corporation

(3) Organizational management by supporting departments

Supporting technologies

(4) Improve management through supporting technologies

personnel, but to train according to individual characteristics. In terms of project manager skills, the PMCDF (Project Manager Competency Development Framework) analyzes the relationship between the project manager's individual characteristics and performance [1, 2], and carries out initiatives to boost personnel training by leveraging the individual characteristics of project managers [3].

Perspective (2) refers to support for team building, and aims to reduce the project manager's workload and curb his/her misleading by building a team to carry out the work of the project manager. As mentioned earlier, the success or failure of a project depends to a large degree on the capabilities of the project manager, but the larger the scale, the more difficult it is for a single project manager to cover all areas. Therefore, projects need to be covered with the management skills of a project management team that includes not only the project manager, but also a senior manager, a PMO, a shared technology team and a development team. Studies are underway of evaluation models for management structures to evaluate management skills as a management team, and not only from the viewpoint of the individual characteristics of a project manager [4].

Perspective (3) aims to support projects through organizational support at the corporation with the PMO or other organizations evaluating the project situation, offering advice and carrying out assessments from a third party standpoint. By developing an external understanding of the situation as the project progresses, risks that have gone undetected by those who are involved may be identified. By providing organizational support and risk assessment not only by the management team, including the project manager for the project in question, but from a third-party standpoint, the probability of success for the project may increase [5].

Perspective (4) uses supporting technologies, methodologies, and frameworks to support project implementation activities, and the support domain includes risk management, requirement identification, communication support, knowledge management, PMO support, and so on. In terms of risk management, there are initiatives to support risk identification and to design countermeasures in a range of business fields [6–10]. In terms of requirement identification support, requirement identification for clients is supported through ethnographic surveys based on human-centered design processes, and initiatives such as building construction management systems based on the results [11]. For communication support, there are initiatives to identify management issues by visualizing communication about project progress using sensor systems. In terms of knowledge management, there are methods for extracting empirical knowledge [13] and techniques for using knowledge [14, 15] as a system for circulating empirical knowledge gained from projects across the organization. The circulation of knowledge is implemented as illustrated in Figure 3–3 by collaboration with the system design of perspective (5). There are also information system initiatives to support not only the project manager, but also the PMO [5].

Perspective (5) builds structures for increasing the probability of success as a system, certification systems for project managers, and structures for project governance. One of these structures is the use of Phase-Gate Management (a structure that divides the product process into several phases and erects gateways to review whether or not the conditions have been met before moving to the next phase) illustrated in Figure 3–4 to make decisions about continuing or stopping projects [16]. By using Phase-Gate Management, it is possible to optimize decision making to lower risk, improve design quality, and maximize management gains.

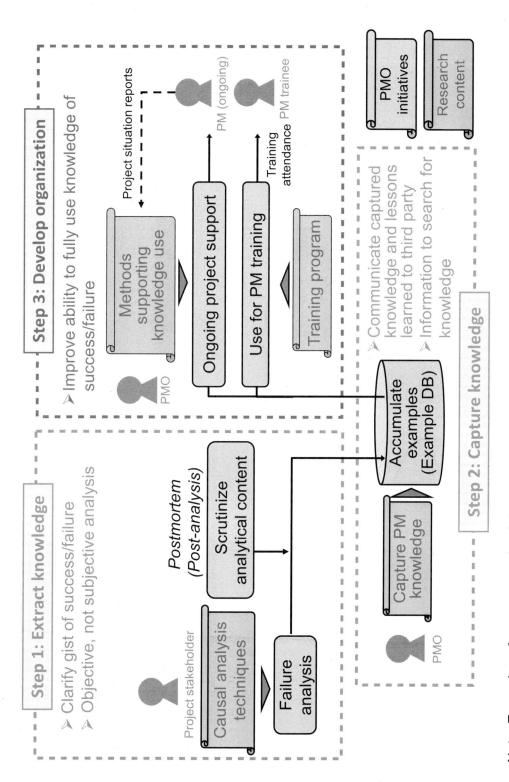

Note: Examples of success can be delivered with the same structure.
PMO: Project Management Office, PM: Project manager

**Figure 3–3.** The across-the-organization circulation process for empirical knowledge in project management.

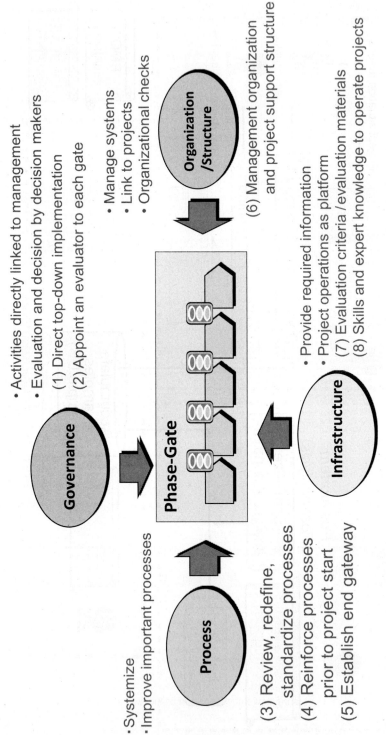

**Figure 3–4.** The Hitachi Phase-Gate Management Process.

Further, as a measure to link perspectives (1) to (5) organically, not independently, a business improvement project called D-WBS (Denryoku Work Breakdown Structure) is under way at Power Systems company [17]. The D-WBS Project has a platform (the D-WBS Platform) for drawing out business synergy related to project management as shown in Figure 3–5, and constructs business processes that unite perspectives (1) to (5) on the platform.

The abovementioned support components are required components for implementing projects without relying on business units. Even though each field of application has its own characteristics, comparing and contrasting project management techniques at Power Systems company and Information & Telecommunication Systems company finds that there are quite a few useful reference points for both sides. Considering that sharing and utilizing knowledge of initiatives to strengthen project management in the wide-ranging business areas of the Hitachi Group will become a source of competitive excellence for Hitachi, the internal Technical Committee is creating opportunities for information exchange.

This technical committee is called the Project Management Technical Committee, and it started with working-level discussions at Power Systems company and Information & Telecommunication Systems company around 2000. Then, a technical committee open to the whole company set up at a meeting of volunteers in 2005, and today, many business units are participating with the focus on the PMO. The Technical Committee is promoting stronger project management across the whole company. As well as exchange of information, it supports the research institutes by investigating solution strategies for shared issues [18]. For example, the structure for sharing project management tasks across business units [19], or activities across the whole company based on initiatives to use knowledge in Information & Telecommunication Systems company [20].

As well as organizing internal forums with the aim of educating staff about project management, or communicating Technical Committee activities for professionals to the whole Hitachi Group, the Technical Committee also conducts regular surveys of awareness of the issues for strengthening project management at each business unit.

In this way, there are initiatives under way to deliver synergy at the Hitachi Group by rolling out expertise laterally across the business units, identifying shared issues, and studying solution strategies through the activities of the Technical Committee.

## REFERENCES (ABSTRACTS PROVIDED WHERE AVAILABLE) ____

[1] Takafumi Kawasaki et al., "Practice action of project managers: The difference between highly competent PM and moderately competent PM," *Proceedings of 13th National Conference of The Society of Project Management*, 2007, 373–377, 2007-03-15. (http://ci.nii.ac.jp/naid/110007602747)

Conventionally knowledge and skills which effective project managers have been focused to explain successful performance. But highly complex situation require professional project managers to create adaptive and useful practices. This practice is known as "knowing," that is, creating new knowledge and adaptive actions. This study analyzed actions of superior project managers and less superior project managers and explained the difference in terms of promoting member's co-knowing.

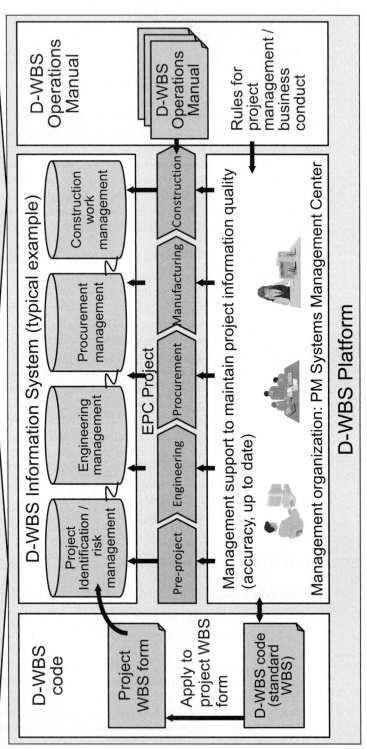

**Figure 3–5.** D-WBS Project Outline.

[2] Hitoshi Yamadera et al. "Relations between Achievement and Characteristics of Project Managers," *Proceedings of 16th National Conference of The Society of Project Management*, 2009, 209–212, 2009-03-10.
(http://ci.nii.ac.jp/naid/110007602894)

This study investigated relations between achievement of project managers and personality, work attitude, type of project activity. The results showed that extroversion, problem consciousness, and learn from others were related to achievement. Concerning to project activity five types of expertise were identified as for normal level contribution managers. Though they fully contributed to their projects, it was revealed that high-level contributors equally and consciously leveled their behavior up to promote organizational evolution.

[3] Takeshi Yokota et al., "Strengthening of Personnel Training Process of Project Managers," *Journal of the Society of Project Management* 15(2), 2013–04-15.

To improve a strike rate of construction projects, we have been developing personnel training process of project managers. We have developed the method which evaluates characteristics of project managers quantitatively. This method has a basic data which consists of replies of about 200 items of questionnaire. It evaluates project manager's characteristics in some viewpoints (project experience, behavioral trait, knowledge, etc.). This method defines the target score of project management work, and by comparing project manager's characteristics with target score, it clarifies their strength and weak points for education. Furthermore, we will support the organization of project formation by evaluating a result of comparison.

[4] Akiyuki Onaka, "Model of Project Team Assessment to Make Projects Succeed," ProMAC2010, 2010.

As for the project managers of the IT system projects in our company, their management ability is evaluated and their PM rank, divided into three levels, "small," "middle," "large," according to project size, is certified based on our own criteria "Project Manager Accreditation System." Certified project managers are appointed to a project whose size corresponds to their PM rank according to "Project Manager Appointment System."
    Some projects, especially large size ones, went wrong even though the project managers were properly qualified by above-mentioned systems. This fact caused us to infer the importance of the behaviors of project members particularly [those] expected to help project managers. The members to be considered are, for example, those who reduce manager's burdens and those who supervise and advise project managers. However, our insufficient discussion about such points of view made it difficult to assess project teams, including support members, in the same way as we assessed project managers. In order to improve such situations, we analyzed previous project data to conclude how to organize a project team. Based on the conclusion, we created an assessment model for project management teams which takes project size and development type into account. This paper describes the result of our research for the model of project management team assessment.

[5] Kenji Hatsuda et al., "PMO Information System as a Support of Project Management Office Activities," *Journal of the Society of Project Management* 5(4), 28–31, 2003–08-15.
(http://ci.nii.ac.jp/naid/110003726282)

According to the increase in recognizing the importance of project management, the project management office (PMO) has also come to play important roles as the organization for project management promotion. Project management needs to be the task to be strategically promoted, and to be systematically deployed by PMO. The roles of PMO are common base developments such as project management procedures, personnel training, and technical developments, as well as the project supports across the organizations. It is effective to build the PMO information system which supports the activities of PMO. This paper focuses on the activities based on practical PMO, and considers of development, utilization, and expected effects of the PMO information system as a support of PMO.

[6]   Minamino, Toyama, "An Application of Modern Project Management "IT" System Development Projects", ProMAC2002, 2002.

In recent years, each IT system development project has come to be diversified and complicated, and its exploitation is required in the short term. Also, changes [to] the requests during the development have also increased. IT system projects have characteristics that the whole image of the system cannot be observed as a concrete shape directly, either during or after the development. The authors have attempted the application of modern project management, especially for risk management and scope management in such system development projects. The authors provide some examples on application of modern project management to IT system development projects and describe future aspects on the application of model project management.

[7]   Takeshi Yokota et al., "Development of a Contract Risk Assessment Support System(CRARIS)," *Journal of the Society of Project Management* 7(3), 20–25, 2005-06-15. (http://ci.nii.ac.jp/naid/110003726628)

A business process necessary to support a contract risk assessment in overseas projects was examined, and ContRAct RISk assessment support system (CRARIS) that was a knowledge management system concerning the contract risk management was developed. CRARIS is based on a contract checklist made in a legal affairs section, and it characterizes in a presentation of knowhow information that relates to each check item and automatic evaluation of risks according to content of checklist inputs. Moreover, about 2000 [pieces of] knowhow information has been extracted from hearing results to specialists in an operation division and the legal affairs section, the minutes of the evaluation of actual projects, and so on. In addition, an examination of a business process and an organizational structure effective to evaluate the contract risk was executed.

[8]   Takeshi Yokota et al., "Development of a Risk Management System for Construction Projects," *Journal of the Society of Project Management* 8(5), 36–41, 2006-10-15. (http://ci.nii.ac.jp/naid/110006278350)

In order to support the risk management of a construction project, we have developed the system which uses progress simulation technology and supports evaluation of the problem of a project and the decision of countermeasures to [the] problem. This system is characterized by having progress evaluation simulation logic, which evaluates detailed progress of each work of a project serially per week. Moreover, it also has the [capability] to take into consideration situations, such as change of working efficiency, and increase the number of workers, in simulation logic. This system was evaluated using the data of an actual project, and the validity of the project evaluation result using the various functions of a system was checked.

[9] Takeshi Yokota et al., "Upgrade of Risk Management Technique for IT System Development Project," *Journal of the Society of Project Management* 14 (3), 25–30, 2012-06-15. (http://ci.nii.ac.jp/naid/110009495477)

To support the effective introduction of IT systems, we have constructed a business justification analysis support system for IT system development. It evaluates the benefit of systems, investment effects, risk factors, and the justification of development systems. By using [this] information, it clarifies the appropriateness and priority level of development investment, and supports the risk management process of development phase. To clarify characteristics of projects more accurately, we classified risk score by considering whether project managers can manage or not. By applying this technique to real projects, we [verified] that this risk classification technique is effective for project management.

[10] Yoshinobu Uchida,, "Development of the Risk Management System for Construction Projects," ProMAC2011, 2011.

To support the risk management of construction projects, we have developed a risk management system that identifies project risks and supports decisions on countermeasures to these risks. The risk management system has a project evaluation system, a risk register, and a risk management web portal. The project evaluation system provides a checklist suitable for a project and supports project evaluation. The risk register provides the worksheet and supports project risks identification and response planning development. The risk management web portal visualizes evaluation results through the project evaluation system. In this paper, we report each subsystem of the risk management system.

[11] Hisako Okada et al., "An Approach to Advance Construction Management System for Large-scale Power Plant Projects," *Journal of the Society of Project Management* 15 (1), 8–13, 2013–02-15.

Power plant construction project is large-scale and complex involving many stakeholders. To ensure "quality, schedule, cost" of this huge project, Hitachi has applied IT and system to construction area. Its focal point is (1) realization of huge project consistent and coordinated control, and (2) realization of improving construction field efficiency and quality, for reducing risk and cost. From 1990s, it was applied to actual projects and achieved an effect, but further improvement was reaching a limit in management and system centered approach. Hence, by reconsidering the basis "Construction is a Production by Human," Hitachi has conducted researching construction management system based on Human Centered Approach by focusing user/human side, and has conducted reflecting the results to projects management itself.

[12] Hideyuki Maeda et al., "Visualization of Communication using the Team Activity Measuring System and Its Application to the Project Management," *Journal of the Society of Project Management* 12(1), 5–10, 2010-02-15. (http://ci.nii.ac.jp/naid/110007573280)

By the advancement of sensor and analysis technologies, a system that can automatically measure and visualize a team activity has been developed. We have applied this system to a large-scale project and have experimentally measured and analyzed the condition of a communication within it. As a result, we have successfully quantified and visualized the condition of a project communication. A result of a quantification analysis shows that there is a strong relationship between the productivity and the time spent [in] face-to-face communication. By continually monitoring the face-to-face communication time, we have been able to help expose a project's problem in the early phase. It has also helped us offer useful information to [the] project manager for solving this exposed problem.

[13]   Yoshinobu Uchida et al., "An approach of knowledge extraction via empirical failure knowledge in project management," *Journal of the Society of Project Management* 12(4), 27–32, 2010-08-15.
       (http://ci.nii.ac.jp/naid/110007880184)

One way to make a project successful is [to have] an understanding the essence of past experiences. To develop a scheme for learning from the past experiences, it is important to accumulate [the] valuable knowledge of organization. The knowledge is extracted by analyzing and interpreting it after information obtained from the experience is arranged again. But, it is difficult to derive an objective and profitable knowledge according to the following obstruction factor. (1) The analysis based on superficial or a local situational awareness is done. (2) The consideration of buck-passing work. To solve [these] problems, we developed causal analytical method including visualization of the decision sequence to support the knowledge extraction, and defined knowledge form to understand knowledge. We evaluate the analytical method and the knowledge form, and show the effectiveness of the knowledge extraction.

[14]   Yoshinobu Uchida et al., "Proposal for Risk Management Support Method Using Failure Knowledge," *Journal of the Society of Project Management* 7(6), 3–8, 2005-12-15
       (http://ci.nii.ac.jp/naid/110006278374)

The failure of projects have a big influence on corporate performance. Many SI enterprises need to rebuild project management. We work on the extermination of the deficit project and are researching the method of using the failure knowledge in project management to aim to prevent the same failure. In this paper, we define "Process information for risk (PIR)" as information on the correspondence process on the risk, and propose the method of using the information in the project management process. Advantages of our proposal are the following. (1)PIR is extracted from the periodic report automatically. (2)A past similar case is presented as a failure case. (3)The project member deliberates [about] measures based on the failure case. It is thought that our proposal can support [preventing] the project from failing [from] the same cause.

[15]   Yoshinobu Uchida et al. "Proposal of utilization of the failure experience in project management," *Proceedings of 15h National Conference of The Society of Project Management*, 2008, 140-143, 2008-03–14.
       (http://ci.nii.ac.jp/naid/110007602790)

Understanding the essence of the failure experience and learning lessons from the failure experience are important to creating knowledge. We think that our organization can strengthen the project management by learning from failure experiences in past projects and sharing precepts in the organization. To learn from failure experiences, we should utilize the failure experience in project management.

Our approach assumes the assessment activity by Project Management Office in an ongoing project. We surveyed on current assessment activity, and found the following issues.

1. How should [assessors] understand the situation in the project? Not every assessor has the ability to clarify all the aspects in the project. Often an assessor does not have enough information to verify the effectiveness of a countermeasure.
2. How should [assessors] find the information the project manager needs in the project?

To solve [these] issues, we defined format to descriptive the situation in the project and search strategy of information the project manager needs.

[16]   Koji Okada et al., "Applying Phase-Gate Management for Diverse Business Types," *Journal of the Society of Project Management* 13(6), 29–34, 2011-12-15. (http://ci.nii.ac.jp/naid/110009425403)

Global competition has become harder and harder in every business domain. In order to wipe out unprofitable projects and to improve business profits under such situation, we started a corporate initiative to deploy phase-gate management, which has produced successful results in leading business unit, into every business unit broadly. At first, the concept of "Hitachi Phase-Gate Management," which is applicable for diverse business types, was made clear. Then common fundamental enablers, such as (1) operation guides, (2) training materials, (3) phase-gate maturity model, (4) a KPI setting guide, and (5) sharing knowledge contents, are developed/established and provided broadly Moreover, ten model business units were selected, and supported on both developing their improvement action plans of phase-gate management and performing them. According to the results, improvement of both phase-gate maturity levels and some KPIs were demonstrated in all selected model business units.

[17]   Tomoyuki Aoki et al., "The Case Study of Business Process Reengineering for EPC Project Management," *Journal of the Society of Project Management* 14(6), 5–10, 2012-12-15.

In Hitachi Power Systems Company, the BPR (Business Process Reengineering) project called "D-WBS project" was started in 2009, and we are driving forward this project's first phase for completion by FY2013. In the D-WBS project, we would like to achieve the improvement of project management capability by developing a standard management platform which [can be] used among our business segments. This platform's target is an EPC project that constructs power plant and advanced medical system.

In this paper we introduce the background and overview of the D-WBS project at first. Then we explain our project management platform that consists of four domains, a WBS code system, an IT system, an operation standard, and an operational department. We also explain the application methodology of D-WBS code for the EPC project's planning. Finally, we would like to share our BPR approach.

[18]   Kichie Matsuzaki, "Hitachi, Ltd. 100th Anniversary Series: Genealogy of the Pioneers (20) Inheriting and Reforming the Heart of Monozukuri at Hitachi – Companywide Activities toward Monozukuri," *Hitachi Review* 92(2), 136-143, 2010-02. (http://digital.hitachihyoron.com/pdf/2010/02/2010_02_pioneers.pdf)

[19] Koji Okada et al., "Challenge for Extracting Project Management Knowledge across Business Units," *Journal of the Society of Project Management* 10(3), 23–28, 2008-06-15. (http://ci.nii.ac.jp/naid/110006950594)

In order to prevent project trouble or to repeat project success, enterprises have developed their own QMS (Quality Management System) for project management. However, these improvement activities are performed in individual business units, obtained knowledge was not shared across business units. In this paper, we describe methodology for extracting and organizing project management knowledge, which is developed through real practice for extracting and organizing them. Especially, we made fundamental concepts based on commonalities, differences and specialties. Also, we developed a "knowledge extracting worksheet," "knowledge description sheet," and "knowledge map" as supporting tools as well as a procedure for extracting and organizing knowledge, through three trial cycles of real practice.

[20] Koji Okada et al., "An Analysis Method for Extracting Project Lessons Learned which are Sharable across Business Units," *Journal of the Society of Project Management* 12(6), 21–26, 2010-12-15. (http://ci.nii.ac.jp/naid/110008592927)

Improvements of project management activities are desired in every business domain. Sharing project lessons learned across business units can be an effective way to improve project management activities in an enterprise composed of various business units. In order to share project lessons learned across business units, we collected 31 failed project cases analyzed in each business unit from 8 business unit, re-analyzed them, and extracted 50 sharable project lessons learned. Moreover, we designed an analysis method to extract sharable project lessons learned that reflects analysis know-how gathered from actually performed re-analysis activities.

## 3.3 KONE: THE PROJECT MANAGEMENT CHALLENGE[2]

**Growth Challenge**

KONE's success in the High Rise and Major Projects markets since the early 2000s was based on great technical innovation and increased corporate focus on Major Projects. KONE was investing a lot in sales and sales engineering competences, utilizing our strong global supply chain for deliveries.

What naturally follows strong sales performance is a need for strong execution. These new markets and projects were more demanding, and the customers had different expectations than before. The business, and its customers, were employing professional project managers to deliver the work, and they needed the same from us. This was noticed in 2008 when analyzing our project profitability development—We were losing our margin!

This was due to:

- not having a global standard Project Management structure;
- not having a global standard for the role and responsibility of a Project Manager;
- Project Management was not seen as a desired or legitimate career;

---

2. ©2013 by KONE. Reproduced by permission. All rights reserved. Material on KONE was provided by Steve Gonzalez – Director of Major Projects and Jyri Toiviainen - Director, PM Competence Development, MP Project Management Support

- Project Management were brought into the delivery process too late;
- No system for identifying, recruiting, developing, and retaining project managers; and as a result
- management of a project being handled by an under qualified person out of convenience

It wasn't all bad news. There were a few very good project managers in the company. Each had developed their own tools and systems to manage work, contracts, and customers. Comparing the work of these people with the others, it became clear that the average result in project margin ("as-sold" vs. "delivered") was very much dependent on the competence of the project manager.

- Projects with trained and experienced PMs        +1.5 points
- Projects with untrained PMs        −4.3 points

There were two issues. The first was the lack of predictability of the business. The second was the large variation between the best and the worst. With our backlog and our ambition, we needed to move quickly. Starting over wasn't an option. Instead, we would: learn from history, secure the future.

**Project Management Competence Development Must Do's 2009**

To be able to get some measurable results in given 6 months time we set 5 "Must Do's" for our Project Management development (See Table 3–5):

1. Have Project Management skills assessment in place and working
2. Establish a training program and pilot that during 2009
3. Establish a Project Management network in KONE
4. Start auditing project regularly
5. Have Project Reviews running globally

**TABLE 3–5. PM COMPETENCE DEVELOPMENT–ACTIONS AND METRICS**

| Must Do's | Objectives | Mesurements |
|---|---|---|
| Project Management Skills Assesment | • Systematic & continuous evaluation of PM's by KONE defined competence standards. | • Self and superior assesment done as a part of annual PD/IDP program. |
| Project Management Training Program | • Establish PM training program.<br>• Have pilot traing during 2009.<br>• Implement learnings in every day operations in projects. | • Existing development plans in FL's.<br>• Customer satisfaction (NPI).<br>• Project CMII. |
| Project Management Network | • Every project manager has known network for support.<br>• Establish network for project professionals in KONE. | • Communicated contact networks.<br>• Use of contact networks. |
| Project Auditing | • Have minimum 2 projects audited in each area.<br>• Best practice sharing between project and countries (FL) | • Shared best practices.<br>• Competence gap identified.<br>• Project risks identified. |
| Project Reviews (and steering) | • Have Area/FL management commitment for the program.<br>• Have project reviews running in selected FL companies. | • Established project reviews |

### Share the existing KONE Best Practices in Project Management

Based on the project portfolio and profitability analysis, global interviews, skills assessment and the given time frame (6 months), we decided to continue with very practical approach by sharing 'The Best Practices" we have in the company already in Project Management.

By using industry-standard project management knowledge areas we collected the best known KONE practice globally in each discipline. Finally we had collection of proven, successful practices in selected focus areas in Project Management in KONE.

Successful Project Managers were used to deliver the actual training. So the approach to each topic was very practical—Project Managers discussing with colleague Project Managers successful ways of managing the projects. This brought two distinct benefits. First and foremost, it helped the Project Managers understand they were not alone in their careers. Second, it ensured that everything that was taught could be immediately applied in the individual businesses around the world. To ensure that project managers understood the available support structure as well, other corporate experts from different areas were used to share their knowledge.

### Make It Efficient and a Great Experience—Use Professional Partners

When having nonprofessional trainers sharing their experiences, the challenge is to keep the training structured with clear content and learning objectives targeting to key goals to be achieved in practice.

"What changes will each Project Manager implement in the way they currently manage their projects (goals, measures and actions)?"

To make this happen successfully, we selected a professional company (b.l.t.) to structure the training content with us. We also planned the training approach with them to include 2 face to face workshops that act as bookends for virtual sessions and "on the project coaching."

To make this happen in practice, the following key elements were defined:

- Session slides
- Session activity slides
- Session Trainer's guides
- Identified Best Practices for each topic
- Key Goals for each topic
- Individual action planning form (record of learning, plans of action)
- Clear, scalable, and repeatable model for coaching
- Coaching planning and communication framework
- Project Management Handbook

In practice the training workshops included a facilitation of the topic with lots of group activities and common discussion—everyone was encouraged to take part.

With this approach we were not only sharing the best practices from trainer to the delegates, but also collecting and sharing the best practices from each participant among the whole group.

Based on the discussion everyone has a chance to update their individual action planning for coaching and best practice implementation in their everyday work.

KONE Project Management Development Program (PMDP) was established.

## Key Success Factors

Understood challenge—Clear, common target
Passion and motivation among the team
Great team in place—Area PM's + experts—Winning together
Challenging targets and limited time—innovative approach needed
Practical approach for fast implementation
Focus on key topics—not to try to do everything in one go
Using competent partners
Having fun!

## Achievements

Higher awareness of Project Management in KONE
Significant profitability improvement
Improved customer satisfaction
Understanding what project management is in KONE business environment
Base from which to develop Project Management further
Be a professional partner in demanding and growing business for our customers

## 3.4 THE LIGHT AT THE END OF THE TUNNEL

Most people seem to believe that the light at the end of the tunnel is the creation of an enterprise project management methodology that is readily accepted across the entire organization and supports the need for survival of the firm. Actually, the goal should be to achieve excellence in project management, and the methodology is the driver for this. According to a spokesperson at AT&T, excellence can be defined as:

> A consistent Project Management Methodology applied to all projects across the organization, continued recognition by our customers, and high customer satisfaction. Also our project management excellence is a key selling factor for our sales teams. This results in repeat business by our customers. In addition there is internal acknowledgement that project management is value-added and a must have.

While there may be some merit to this belief that excellence begins with the creation of a methodology, there are other elements that must be considered, as shown in Figure 3–6. Beginning at the top of the triangle, senior management must have a clear vision of how project management will benefit the organization. The two most common visions are for the implementation of project management to provide the company with a sustained competitive advantage and for project management to be viewed internally as a strategic competency.

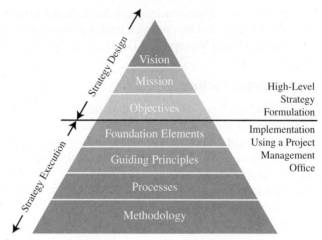

**Figure 3–6.**  Enterprise project management.

**Figure 3–7.**  Identifying the mission.

Once the vision is realized, the next step is to create a mission statement, accompanied by long- and short-term objectives that clearly articulate the necessity for project management. As an example, look at Figure 3–7. In this example, a company may wish to be recognized by its clients as a solution provider rather than as a supplier of products or services. Therefore, the mission might be to develop a customer-supported enterprise project management methodology that provides a continuous stream of successful solutions for the customers whereby the customers treat the contractor as a strategic partner rather than as just another supplier. The necessity for the enterprise project management methodology may appear in the wording of both the vision statement and the mission statement.

Mission statements can be broken down into near- and long-term objectives. For example, as seen in Figure 3–8, the objectives might begin with the establishment of metrics from which we can identify the CSFs and the KPIs. The CSFs focus on

**Figure 3–8.** Identifying the metrics.

**TABLE 3–6. FOUNDATION ELEMENTS**

| Long Term | Short Term |
| --- | --- |
| Mission | Primary and secondary processes |
| Results | Methodology |
| Logistics | Globalization rollout |
| Structure | Business case development |
| Accountability | Tools |
| Direction | Infrastructure |
| Trust | |
| Teamwork | |
| Culture | |

customer satisfaction metrics within the product, service, or solution. The KPIs are internal measurements of success in the use of the methodology. The CSFs and KPIs are the drivers for project management to become a strategic competency and a competitive advantage. Notice also in Figure 3–8 that the CSFs and KPIs can be based upon best practices.

The top three levels of the triangle in Figure 3–2 represent the design of the project management strategy. The bottom four levels involve the execution of the strategy beginning with the foundation elements. The foundation elements are the long- and short-term factors that must be considered perhaps even before beginning with the development of an enterprise project management methodology (Table 3–6). While it may be argumentative as to which factors are most important, companies seem to have accelerated to excellence in project management when cultural issues are addressed first.

To achieve excellence in project management, one must first understand the driving forces that mandate the need for excellence. Once the forces are identified, it is essential to be able to identify the potential problems and barriers that can prevent successful implementation of project management. Throughout this process, executive involvement is essential. In the following sections, these points will be discussed.

## 3.5 GOODYEAR[3]

When companies embark upon a journey towards excellence in project management, they usually establish a global project management office in order to capture and share best practices in project management. The capturing of best practices is essential if continuous improvements in project management are to occur. Goodyear shares with us three of their best practices.

**Best Practice #1: Goodyear's Global Project Management Summit**
The Goodyear Tire & Rubber Company has been hosting an annual global project management Summit since 2012 to connect project managers and team members from across all strategic business units and nearly all business functions. The Summit consists of a three-day experience focused solely on project management, leadership skills and professional networking. Participants are selected through a nomination and approval process by their respective PMO leaders and their business unit leadership. Selection criteria include the percentage of time the project manager spends on projects as well as the scope and complexity of their projects.

"Each year, more than one hundred of Goodyear's talented project managers and team members come together to learn from the experts and from each other with the focus on one thing: leveraging project management excellence to deliver expected outcomes for Goodyear's projects. This gathering illustrates Goodyear's commitment to building capabilities around project management," said Sherry Neubert, Vice President, Global Project Management Office.

Goodyear's project management leadership takes part in selecting the curriculum for the Summit — specifically focusing on topics that contribute to raising enterprise project management competency, and that have immediate application to its project managers' every day work. The schedule is a well-balanced collection of presentations and discussions led by industry experts, Goodyear's senior leaders and Goodyear's project managers.

Participant schedules are filled with two types of learning sessions: "general sessions" which are typically two–three hours long, include all attendees, and focus on subject matter that crosses all project types and experience levels; and "breakout sessions" which are hour-long interactive sessions that are focused on function and role-specific topics with each participant opting-in to attend. Sample general session topics are stakeholder management, a panel discussion about a Lessons Learned Case Study from a Mega Project, building world-class project teams, emotional intelligence and change management. Breakout sessions include topics like sponsor expectations, gaining and maintaining a PMP certification, negotiation, executive roundtable discussions and risk and scenario planning.

The first time he attended a project management Summit, the Program Director of OTR Tires Alfredo Gamboa said the energy at the Summit was contagious. "It was as if each of us took home a piece of 'Goodyear Project Management DNA.' Now we will apply these learnings to our work no matter what function we support and, more importantly,

---

3. ©2013 by Goodyear. Reproduced by permission. All rights reserved. Material on Goodyear has been provided by Sherry Neubert, VP Global Project Management Office; Andy Weimer, Sr. Manager RDE&Q Project Management Office; Jayne Cole, Enterprise Competency Manager, Global Project Management Office; Alexis Rizopulos, Knowledge Management Manager, Global Project Management Office; and John Renner, Project Manager, RDE&Q Project Management Office.

work together through the process. Collaboration is an essential part of the project management community. It enables us to work horizontally, not vertically, or in a silo."

The Project Management Summit is designed to support overall enterprise project management maturity. It is also indicative of strong middle- and senior-level support for the discipline as a means to achieve desired results. With more than 20 senior leaders participating in the Summit each year, strong support is evident.

The application of project management and leadership concepts learned at the Summit is of utmost importance to Goodyear leadership and the project management community. Participants are surveyed at the end of each day of the Summit and at the end of the entire three days for feedback on speakers, schedule and activities. Feedback is then applied to following year's Summit as part of continuous learning planning.

**Best Practice #2: The Project Management Case Study**

**Background**

In addition to managing projects, Goodyear's Project Management Office is focused on building capability, facilitating stages and gates, coaching project managers, and creating mechanisms to capture and re-use Goodyear's institutional knowledge. In 2011 Goodyear began using the traditional case study with an innovative twist as a mechanism to learn, heal and energize its project teams and projects to enable future project success.

In recognition of organizations that instill a culture of continuous improvement are able to learn from their pasts, Goodyear's PMO combined the traditional project case study with Appreciative Inquiry methodologies, as refined by Case Western Reserve's Weatherhead School of Management, to capture project lessons learned and project management best practices.

Appreciative Inquiry is a strengths-based methodology that focuses on what was done well and seeks to renew, build, and develop future successes by visualizing a future that emulates past positive experiences. Applying this twist to the traditional case study has been highly effective for Goodyear's case study process. The purpose of the case study is to document an objective review of what happened and what future projects may learn from it. Beginning the interview process on a positive note by asking "what went well" sets the tone for the interview and creates an environment where the interviewee feels comfortable enough to share key knowledge to be passed to future projects. This reflection helps bring closure and healing to the project team as well as creates a sense of satisfaction that other teams may benefit from project lessons. The case study creates an opportunity for people's voices to be heard and for projects team members to take part in documenting a valuable piece of Goodyear project management history.

**The Process**

A project is nominated for consideration for a case study by a representative in one of the company's four business units, usually by a project sponsor or regional leadership. The PMO reviews projects and decides whether to conduct a case study based on the resources and investment needed to capture the data and the organizational value of the project lessons learned to be shared.

Once the case study is approved, the collection process begins by interviewing project stakeholders and contributors, structuring the responses into key learning and best practices, and then retelling, in a narrative form, the highlights of the project's lifecycle and critical success factors. The belief is that future successes can be created by speaking about, documenting, and then focusing on emulating the critical success factors of past successes. At Goodyear, the desire is great to learn from the past and share enterprise knowledge. Emulating is a key factor in the success of this process. The "pull" from future projects is created by the desire to succeed as past projects have—and the more favorable desire to not re-learn by experience but through the learning of others.

The project manager and sponsors are consulted when preparing the interviewee list—which can range from 20–100 team members, sponsors, and stakeholders. These individuals are interviewed in a private one-on-one setting—sometimes stretching over several months and several continents. Interviewees are asked to tell a story about the most positive project experience they had ever had, regardless of project type or location. Next, they are asked to describe their role on the project being studied. And finally, they are asked to describe the factors that contributed to the project's current state.

The result is a case study that tells the project's story from its inception through current state, focusing on a main theme or key lesson learned.

It is recommended that case studies are performed directly after a project (or phase) is completed or while the project (or phase) is still in motion. This is primarily because as team members finish their work on one project, they move quickly to new projects, dispersing, and they may forget relevant details of their former project.

### The Benefits:

Goodyear has observed that the case study capture process is cathartic for persons interviewed. The process allows the team members a chance to reflect on and bring closure to the project. There is also gratification in that someone else can learn simply by reading the case study. It is so simple that knowledge transfer can be easy when we take the time to write things down. The person collecting the information is also learning and given that this person is typically a member of the PMO they increase their ability to contribute to other projects. Another benefit to the company and the people involved is the opportunity for process improvement or best practices reinforcement in future projects and project teams. The case study enables future project teams to leverage on what we already know and replicate past successes.

The Goodyear project case study is a critical success factor in accelerating Goodyear's project management maturity journey. It enables Goodyear associates from all over the world to learn from each other. In enables project teams to celebrate their successes and the project management community as a whole to be recognized and valued for its contribution to enabling the enterprise's investments. And with this appreciation for the application of a disciplined approach to project management, Goodyear has also increased the enterprise confidence in its ability to execute major projects.

**Best Practice #3: Goodyear's Approach to Coaching and Advising Project Managers**   Goodyear takes a three-pronged approach to instill Project Management principles and methods in its organization:

1. Establishment of Project Management standards (through a "Playbook")
2. Project Management training
3. Project Management coaching and advising

Early on, Goodyear found that the combination of documented standards and training alone was not enough to reach the desired level of Project Management maturity. Therefore, Goodyear introduced a third "leg of the stool," project management coaching and advising.

Projects with a certain risk profile receive coaching and advising from the PMO. Goodyear has recognized that project support provided by a project management Coach is an integral part of a successful project management lifecycle. As the title implies, a Coach assigned to a project will support the project every step of the way to enable successful project results. Coaches work directly with the Project Manager to provide a foundation of standard process work, strategic oversight for project planning, and a balanced approach to project governance within a stage gate process.

The ultimate objective of the Coach is to ensure the Project Manager has the capability to deliver, which in turn improves the likelihood that the project will deliver the expected results. The Coach relies on a give and take relationship with the Project Manager where the project plans are transparent and the Coach respectfully advises areas for opportunity and best practice application. The main focus of a Coach is to provide overall project guidance from an objective set of eyes to help the project achieve its goals. A key benefit of Coaching is the visibility that the Coach has to current and past projects—the Coach is in an excellent position to contrast and compare current and past projects and quickly transfer best practices across projects. However, the Project Manager remains the leader of the project, and is ultimately accountable for its success.

*While involved with a project, the Coach may wear several hats:*
- Mentor: one-on-one advice from past company project learnings or personal leadership experiences.
- Facilitator/Trainer: encouraging process and standard work to produce the best results.
- Mediator: advising in negotiations, risk or situations with sponsors, stakeholders, team members and cross-functional relationships.
- Seer: can help discover a project risk or provide visibility to a challenge before it's too late.
- Cheerleader: recognizing accomplishments and reinforcing good work.

While there are situations where a project manager is provided from a central pool, in the cases where a Coach is deployed the project manager typically comes from the business unit or function. As Goodyear's philosophy is to develop the competency of project management and a mindset of integration in all of its future business leaders and not to develop a company full of project managers, more than 80% of the project managers assigned have a functional or business background and not a formal project management background.

Coaching and advising is a critical success factor for Goodyear's PMO. When applied throughout the lifecycle of a project with full transparency and partnership, Goodyear believes coaching greatly increases the probability of project success.

With the addition of coaching, Goodyear has seen the project management maturity of its organization increase markedly. As Sherry Neubert, the VP Global Project Management Office puts it, "you must have a Playbook that lays out the foundation policies and procedures, and you must have training to get Project Managers up to speed on the concepts and terminology, but Coaching on the application is really the glue that holds the whole package together."

## 3.6 MANAGING ASSUMPTIONS

Whenever we discuss the journey to excellence, people expect to see a chronology of events as to how the company matured in project management. While this is certainly important, there are other activities that happen that can accelerate the maturity process. One such factor is an understanding of the assumptions that were made and a willingness to track the assumptions throughout the project. If the assumptions were wrong or have changed, then perhaps the direction of the project should change or be canceled.

Planning begins with an understanding of the assumptions. Quite often, the assumptions are made by marketing and sales personnel and then approved by senior management as part of the project selection and approval process. The expectations for the final results are based upon the assumptions made.

Why is it that, more often than not, the final results of a project do not satisfy senior management's expectations? At the beginning of a project, it is impossible to ensure that the benefits expected by senior management will be realized at project completion. While project length is a critical factor, the real culprit is changing assumptions.

Assumptions must be documented at project initiation using the project charter as a possible means. Throughout the project, the project manager must revalidate and challenge the assumptions. Changing assumptions may mandate that the project be terminated or redirected toward a different set of objectives. The journey to excellence must necessitate a way to revalidate assumptions. The longer the project, the greater the chance that the assumptions will change.

A project management plan is based upon the assumptions described in the project charter. But there are additional assumptions made by the team that are inputs to the project management plan.[4] One of the primary reasons that companies use a project charter is that project managers were most often brought on board well after the project selection process and approval process were completed. As a result, project managers needed to know what assumptions were considered.

## 3.7 MANAGING ASSUMPTIONS IN CONSERVATION PROJECTS—WWF INTERNATIONAL[5]

In 2005, in collaboration with other conservation organisations,[6] the World Wide Fund for Nature (WWF), agreed upon and began to roll out a set of *Standards for Conservation*

---

4. See *A Guide to the Project Management Body of Knowledge*®, 4th ed., Project Management Institute, Newtown Square, PA, 2008, p. 79.

5. Any reproduction in full or in part of this article must mention the title and credit WWF as the copyright owner. © text 2009 WWF–World Wide Fund For Nature (also known as World Wildlife Fund). All rights reserved. Material was provided by William Reidhead, MSc, Manager, Advisor, Design and Impact (Monitoring), Conservation Strategy and Performance Unit, WWF International.

6. The WWF Programme Standards are closely based on the Open Standards for the Practice of Conservation developed by the Conservation Measures Partnership, a partnership of 11 conservation organizations working together to seek better ways to design, manage, and measure the impacts of their conservation actions (www.conservationmeasures.org).

*Project and Programme Management ("the Programme Standards")*[7] These Standards arc rooted in a long history of project and programme planning and management within WWF, across other conservation organizations, and in other disciplines. The Programme Standards are designed to help project managers and staff describe what they intend to conserve, identify their key assumptions, develop effective strategies, measure their success, and then adapt, share, and learn over time.

## Adaptive Management and Challenges in Conservation Projects

Although there exists significant research and documentation on project management in the private sector, the principles of which apply equally to the nonprofit sector, conservation projects face additional challenges. Beyond the usual processes of project execution and control, conservation projects must operate amid significant uncertainty and complex systems influenced by biological, political, social, economic, and cultural factors.

In defining the project context, conservationists must consider uncertainty on the status of biodiversity, on the functioning of ecological systems, and on how humans bring about changes to the ecological systems and are in turn affected by them. Similarly, when designing interventions aimed at improving the status of biodiversity, conservation projects face challenges in selecting amongst a number of untested strategies, in knowing which will be the most effective, and in measuring and communicating the impact of these strategies. All of this takes place in the context of limited human and financial resources, information, and political capital, and increased calls for transparency and impact from the donors and governments supporting the projects.

As a result, the WWF Programme Standards follow an experimental approach to managing conservation projects, integrating project definition, design, management, and monitoring to systematically test assumptions in order to adapt and learn. The adaptive management process requires that project teams explicitly identify the assumptions under which they are operating and systematically test each assumption to see if it holds in their project context. This provides a method for making more informed decisions about strategies, testing the effectiveness of strategies used, and learning and adapting to improve strategies.

Described below are two tools that are recommended best practice within the WWF Programme Standards and are key for determining and managing project assumptions.

## Conceptual Models

A *conceptual model* (variously known as a "problem tree" or "map of the problematic") is a diagram representing a set of assumed causal relationships between factors that are believed to impact one or more of the biodiversity targets (species or habitats) that the project aims to conserve. A good conceptual model should explicitly link biodiversity targets to the direct threats impacting on them and the indirect threats and opportunities influencing the direct threats. It will also highlight the assumptions that have been made about causal relationships and will advise paths along which strategic activities can be used to positively influence these relationships. In summary, a conceptual model portrays the present situation at the project site and provides the basis for determining where project teams can intervene with strategic activities.

---

7. For more information on the WWF Programme Standards, please visit www.panda.org/standards.

Note that each arrow connecting two boxes in Figure 3–9 indicates causality and represents an assumption that can be tested.

**Results Chains**          Conservation project teams implement strategies that they believe will contribute to conserving the biodiversity in their site, but may not formally state their assumptions about exactly how the strategy will lead to threat reduction and conservation of biodiversity. In fact, it is likely that they have many implicit assumptions—assumptions which may even differ across team members and project partners—about how their strategies will contribute to achieving conservation. If these assumptions are not made explicit, however, they cannot be tested nor can their validity be determined over time.

A *results chain* is a tool that clarifies these assumptions, a diagram that maps out a series of causal statements that link factors in an "if . . . then" fashion. Results chains help teams to specify and model their theories of change. In some organisations, results chains are also termed "logic models" or "solutions trees." The results chains are built from the conceptual model, and as shown in Figure 3–10, are composed of a strategy (a group of activities), desired outcomes, and the ultimate impact that these results will have on the biodiversity target. A goal is a formal statement of a desired impact on a biodiversity target and an objective is a formal statement of a desired outcome, frequently the reduction of a threat.

In this manner, a well-constructed results chain will provide a project with a set of strategic activities to be executed on the ground, as well as goals and objectives, in short, a conservation action plan. Results chains also provide the basis for financial/operational plans, as well as for formulating indicators and monitoring and control plans. In addition to elucidating assumptions and developing plans for project execution, conceptual models and results chains are both useful tools for monitoring and control during project implementation, for assessing impact, and for diagnosing any bottlenecks that may arise.

It is worth nothing that the above two tools fall within the WWF Programmes Standards planning steps known as Define and Design. These steps also include other tools for assessing the viability of biodiversity targets, for ranking threats to biodiversity, for analysis of stakeholders, for assessing risks, etc. Further steps include best practices for Implementation, for Analysing results and Adapting plans, and for Sharing.

## 3.8 PROJECT GOVERNANCE

Most companies begin the journey to excellence with the development of a project management methodology. The purpose of the methodology is to provide not only a road map of how to proceed but also the project manager with the necessary and timely information for decision making. Decision making requires some form of governance and too often this is discovered late in the journey toward excellence.

A methodology is a series of processes, activities, and tools that are part of a specific discipline, such as project management, and designed to accomplish a specific objective. When the products, services, or customers have similar requirements and do not require significant customization, companies develop methodologies to provide some degree of consistency in the way that projects are managed. These types of methodologies are often based upon rigid policies and procedures.

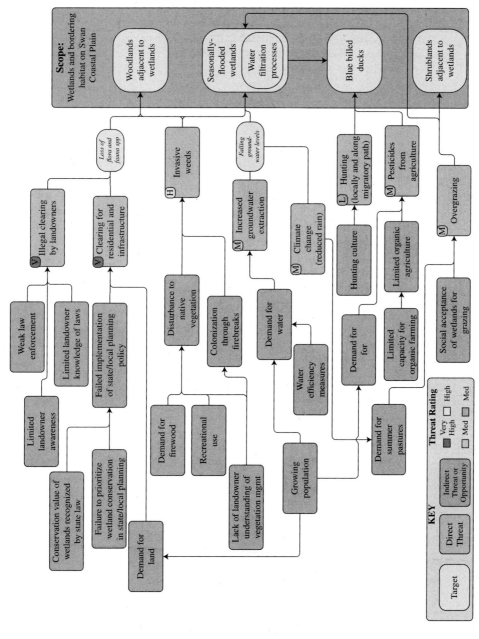

**Figure 3–9.** Example (simplified) of a conceptual model for Swan Coastal Plain in southwest Australia.

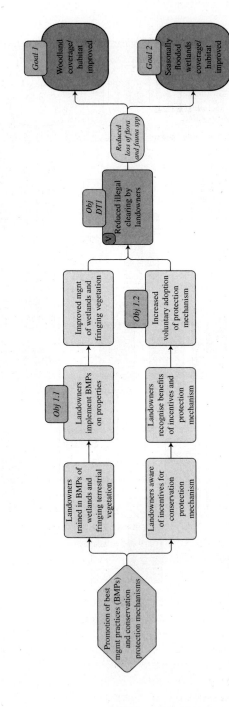

**Figure 3–10.** Example (simplified) of a results chain for Swan Coastal Plain in southwest Australia.

As companies become reasonably mature in project management, the policies and procedures are replaced by forms, guidelines, templates, and checklists. This provides the project manager more flexibility in how to apply the methodology to satisfy a specific customer's requirements. This leads to a more informal application of the project management methodology.

Today, we refer to this informal project management approach as a framework. A framework is a basic conceptual structure that is used to address an issue, such as a project. It includes a set of assumptions, concepts, values, and processes that provide the project manager with a means for viewing what is needed to satisfy a customer's requirements. A framework is a skeleton support structure for building the project's deliverables.

Frameworks work well as long as the project's requirements do not impose severe pressure upon the project manager. Unfortunately, in today's chaotic environment, this pressure appears to be increasing because:

- Customers are demanding low-volume, high-quality products with some degree of customization.
- Project life cycles and new product development times are being compressed.
- Enterprise environmental factors are having a greater impact on project execution.
- Customers and stakeholders want to be more actively involved in the execution of projects.
- Companies are developing strategic partnerships with suppliers, and each supplier can be at a different level of project management maturity.
- Global competition has forced companies to accept projects from customers that are all at a different level of project management maturity.

These pressures tend to slow down the decision-making processes at a time when stakeholders want the processes to be accelerated. This slowdown is the result of:

- The project manager being expected to make decisions in areas where he or she has limited knowledge
- The project manager hesitating to accept full accountability and ownership for the projects
- Excessive layers of management being superimposed on top of the project management organization
- Risk management is being pushed up to higher levels in the organization hierarchy
- The project manager demonstrating questionable leadership ability

These problems can be resolved using effective project governance. Project governance is actually a framework by which decisions are made. Governance relates to decisions that define expectations, accountability, responsibility, the granting of power, or verifying performance. Governance relates to consistent management, cohesive policies and processes, and decision-making rights for a given area of responsibility. Governance enables efficient and effective decision-making to take place.

Every project can have different governance even if each project uses the same enterprise project management methodology. The governance function can operate as a separate process or as part of project management leadership. Governance is designed not to replace project decision making but to prevent undesirable decisions from being made.

Historically, governance was provided by the project sponsor. Today, governance is a committee. The membership of the committee can change from project to project and industry to industry. The membership may also vary based upon the number of stakeholders and whether the project is for an internal or external client.

## 3.9 SEVEN FALLACIES THAT DELAY PROJECT MANAGEMENT MATURITY

All too often, companies embark upon a journey to implement project management only to discover that the path they thought was clear and straightforward is actually filled with obstacles and fallacies. Without sufficient understanding of the looming roadblocks and how to overcome them, an organization may never reach a high level of project management maturity. Their competitors, on the other hand, may require only a few years to implement an organization-wide strategy that predictably and consistently delivers successful projects.

One key obstacle to project management maturity is that implementation activities are often spearheaded by people in positions of authority within an organization. These people often have a poor understanding of project management, yet are unwilling to attend training programs, even short ones, to capture a basic understanding of what is required to successfully bring project management implementation to maturity. A second key obstacle is that these same people often make implementation decisions based upon personal interests or hidden agendas. Both obstacles cause project management implementation to suffer.

The fallacies affecting the maturity of a project management implementation do not necessarily prevent project management from occurring. Instead, these mistaken beliefs elongate the implementation time frame and create significant frustration in the project management ranks. The seven most common fallacies are explained here.

**Fallacy 1: Our ultimate goal is to implement project management.** Wrong goal! The ultimate goal must be the progressive development of project management systems and processes that consistently and predictably result in a continuous stream of successful projects. A successful implementation occurs in the shortest amount of time and causes no disruption to the existing work flow. Anyone can purchase a software package and implement project management piecemeal, but effective project management systems and processes do not necessarily result. Furthermore, successfully completing one or two projects does not mean that only successfully managed projects will continue.

Additionally, purchasing the greatest project management software in the world cannot and will not replace the necessity of people having to work together in a project management environment. Project management software is not:

- A panacea or quick fix to project management issues
- An alternative for the human side of project management
- A replacement for the knowledge, skills, and experiences needed to manage projects
- A substitute for human decision making
- A replacement for management attention when needed

The right goal is essential to achieving project management maturity in the shortest time possible.

**Fallacy 2: We need to establish a mandatory number of forms, templates, guidelines, and checklists by a certain point in time.** Wrong criteria! Project management maturity can be evaluated only by establishing time-based levels of maturity and by using assessment instruments for measurement. While it is true that forms, guidelines, templates, and checklists are necessities, maximizing their number or putting them in place does not equal project management maturity. Many project management practitioners—me included—believe that project management maturity can be accelerated if the focus is on the development of an organization-wide project management methodology that everyone buys into and supports.

Methodologies should be designed to streamline the way the organization handles projects. For example, when a project is completed, the team should be debriefed to capture lessons learned and best practices. The debriefing session often uncovers ways to minimize or combine processes and improve efficiency and effectiveness without increasing costs.

**Fallacy 3: We need to purchase project management software to accelerate the maturity process.** Wrong approach! Purchasing software just for the sake of having project management software is a bad idea. Too often, decision makers purchase project management software based upon the *bells and whistles* that are packaged with it, believing that a larger project management software package can accelerate maturity. Perhaps a $200,000 software package is beneficial for a company building nuclear power plants, but what percentage of projects require elaborate features? Project managers in my seminars readily admit that they use less than 20 percent of the capability of their project management software. They seem to view the software as a scheduling tool rather than as a tool to proactively manage projects.

Consider the following example that might represent an average year in a midsize organization:

- Number of meetings per project: 60
- Number of people attending each meeting: 10
- Duration of each meeting: 1.5 hours
- Cost of one fully loaded man-hour: $125
- Number of projects per year: 20

Using this information, the organization spends an average of $2.25 million (U.S.) for people to attend team meetings in one year! Now, what if we could purchase a software package that reduced the number of project meetings by 10 percent? We could save the organization $225,000 each year!

The goal of software selection must be the benefits to the project and the organization, such as cost reductions through efficiency, effectiveness, standardization, and consistency. A $500 software package can, more often than not, reduce project costs just as effectively as a $200,000 package. What is unfortunate is that the people who order the software focus more on the number of packaged features than on how much money using the software will save.

**Fallacy 4: We need to implement project management in small steps with a small breakthrough project that everyone can track.** Wrong method! This works if time is not a constraint. The best bet is to use a large project as the breakthrough project. A successfully managed large project implies that the same processes can work on small projects, whereas the reverse is not necessarily true.

On small breakthrough projects, some people will always argue against the implementation of project management and find numerous examples why it will not work. Using a large project generally comes with less resistance, especially if project execution proceeds smoothly.

There are risks with using a large project as the breakthrough project. If the project gets into trouble or fails because of poorly implemented project management, significant damage to the company can occur. There is a valid argument for starting with small projects, but the author's preference is for larger projects.

**Fallacy 5: We need to track and broadcast the results of the breakthrough project.** Wrong course of action! Expounding a project's success benefits only that project rather than the entire company. Illuminating how project management caused a project to succeed benefits the entire organization. People then understand that project management can be used on a multitude of projects.

**Fallacy 6: We need executive support.** Almost true! We need *visible* executive support. People can easily differentiate between genuine support and lip service. Executives must *walk the talk*. They must hold meetings to demonstrate their support of project management and attend various project team meetings. They must maintain an open-door policy for problems that occur during project management implementation.

**Fallacy 7: We need a project management course so our workers can become PMPs.** Once again, almost true! What we really need is lifelong education in project management. Becoming a PMP is just the starting point. There is life beyond the *PMBOK® Guide*. Continuous organization-wide project management education is the fastest way to accelerate maturity in project management.

Needless to say, significantly more fallacies than discussed here are out there, waiting to block your project management implementation and delay its maturity. What is critical is that your organization implements project management through a well-thought-out plan that receives organization-wide buy-in and support. Fallacies create unnecessary delays. Identifying and overcoming faulty thinking can help fast-track your organization's project management maturity.

## 3.10 MOTOROLA

"Motorola has been using project management for well over[BH2] 30 years in 2005," according to a spokesperson at Motorola.[8] The forces that drove the company to recognize

---

8. H. Kerzner, *Project Management Best Practices: Achieving Global Excellence*, Hoboken, NJ: Wiley, 2006, p. 88.

the need to become successful in project management were "increasing complexity of projects coupled with quality problems, and schedule and cost overruns, which drove senior management to seek an alternative management solution to what previously existed. A chronology of what Motorola did to get where it is today as well as some of the problems encountered are as follows:

- 1995: Hire a director of project management
- 1996: First hire project managers—formal role definition and shift in responsibilities for scheduling and ship acceptance
- 1998: Formal change control instituted—driven by project managers
- 1998: Stage–gates rolled out and deployed across all projects
- 2000: Deployment of time-tracking tool
- 2001: Deployment of a more formal resource tracking
- 2002: Improved resource planning and tracking
- 2004: Project cost accounting

Initially, program management was viewed as an overhead activity, with engineering managers reluctant to give up program control and status communication. It was only through senior management commitment to formal project management practices that a PMO was created and roles and responsibilities shifted. Full engineering management acceptance did not occur until after several years of project management demonstrating the value of structured program management practices which resulted in consistent on-time product delivery. These include formal, integrated, and complete project scheduling, providing independent cross-functional project oversight, communicating unbiased program status, coordinating cross-functional issue resolution, and the identification and management of program risks. Later, project management responsibilities increased to include other key areas such as customer communications, scope control and change management, cost containment, and resource planning.

Executive support was provided through sponsorship of the development of the program management function. The reporting structure of the function has been carefully kept within an appropriate area of the organization, ensuring independence from undue influences from other functional areas so that objective and independent reporting and support would be provided."

## 3.11 TEXAS INSTRUMENTS[9]

A critical question facing companies is whether the methodology should be developed prior to establishing a project management culture. Companies often make the fatal mistake of believing that the development of a project management methodology is the solution to their ailments. While this may be true in some circumstances, the excellent companies realize that people execute methodologies and that the best practices in project management might be achieved quicker if the focus is on the people rather than the tools.

9. H. Kerzner, *Advanced Project Management: Best Practices in Implementation*, Hoboken, NJ: Wiley, 2004, pp. 46–48.

**Figure 3–11.** Success pyramid.

**Figure 3–12.** Texas Instruments success pyramid.

One way to become good at project management is to develop a success pyramid as shown in Figure 3–11. Every company has their own approach as to what should be included in a success pyramid.

Texas Instruments recognized the importance of focusing on people as a way to accelerate project success. Texas Instruments developed a success pyramid for managing global projects. The success pyramid is shown in Figure 3–12. A spokesperson at Texas Instruments describes the development and use of the success pyramid for managing global projects at Texas Instruments:

By the late 1990s, the business organization for sensors and controls had migrated from localized teams to global teams. I was responsible for managing 5–6 project managers who were in turn managing global teams for NPD (new product development). These teams typically consisted of 6–12 members from North America, Europe, and Asia. Although we were operating in a global business environment, there were many new and unique difficulties that the teams faced. We developed the success pyramid to help these project managers in this task.

Although the message in the pyramid is quite simple, the use of this tool can be very powerful. It is based on the principle of building a pyramid from the bottom to the top. The bottom layer of building blocks is the *foundation* and is called "understanding and trust." The message here is that for a global team to function well, there must be a common bond. The team members must have trust in one another, and it is up to the project manager to make sure that this bond is established. Within the building blocks at this level, we provided additional details and examples to help the project managers. It is common that some team members may not have ever met prior to the beginning of a project, so this task of building trust is definitely a challenge.

The second level is called "sanctioned direction." This level includes the team charter and mission as well as the formal goals and objectives. Since these are virtual teams that often have little direct face time, the message at this level is for the project manager to secure the approval and support from all the regional managers involved in the project. This step is crucial in avoiding conflicts of priorities from team members at distant locations.

The third level of the pyramid is called "accountability." This level emphasizes the importance of including the values and beliefs from all team members. On global teams, there can be quite a lot of variation in this area. By allowing a voice from all team members, not only can project planning be more complete but also everyone can directly buy into the plan. Project managers using a method of distributed leadership in this phase usually do very well. The secret is to get people to transition from attitude of obligation to a willingness of accepting responsibility.

The next level, called "logistics," is where the team lives for the duration of the project and conducts the day-to-day work. This level includes all of the daily, weekly, and monthly communications and is based on an agreement of the type of development process that will be followed. At Texas Instruments, we have a formal process for NPD projects, and this is usually used for this type of project. The power of the pyramid is that this level of detailed work can go very smoothly, provided there is a solid foundation below it.

Following the execution of the lower levels in the pyramid, we can expect to get good "results," as shown in the fifth level. This is driven in the two areas of internal and external customers. Internal customers may include management or may include business center sites that have financial ownership of the overall project.

Finally, the top level of the pyramid shows the overall goal and is labeled "team success." Our experience has shown that a global team that is successful on a one- to two-year project is often elevated to a higher level of confidence and capability. This success breeds added enthusiasm and positions the team members for bigger and more challenging assignments. The ability of managers to tap into this higher level of capability provides competitive advantage and leverages our ability to achieve success.

At Texas Instruments, the emphasis on culture is a best practice. It is unfortunate that more companies do not realize the importance of this.

## 3.12 HEWLETT-PACKARD: RECOGNIZING THE NEED[10]

Since 1992, Hewlett-Packard's management made the decision to focus on developing maturity and excellence in project management. A new group of dedicated project resources was formed within the Services organization and given the charter to become professional project management "experts." Hewlett-Packard established an aggressive project management training program as well as an informal "mentor" program, where senior project managers would provide guidance and direction for the newly assigned people. In addition to the existing internal training courses, new project management courses were developed. When necessary, these courses were supplemented by external programs that provided comprehensive education on all aspects of project management. Efforts to achieve industry-recognized certification in project management became a critical initiative for the group.

Hewlett-Packard recognized that demonstrating superior project management skills could expand its business. In large, complex solution implementations, project management was viewed as a differentiator in the sales process. Satisfied customers were becoming loyal customers. The net result was additional support and product business for Hewlett-Packard. Hewlett-Packard recognized also that its customer's either did not have or did not want to tie up their own resources, and Hewlett-Packard was able to educate customers in the value of professional project management. Simply stated, if Hewlett-Packard has the skills, then why not let Hewlett-Packard manage the project?

According to Jim Hansler (PMP®), project manager at Hewlett-Packard, the following benefits were obtained:

> First, we are meeting the implementation needs of our customers at a lower cost than they can achieve. Second, we are able to provide our customers a consistent means of implementing and delivering a project through the use of a common set of tools, processes, and project methodologies. Third, we are leveraging additional sales using project management. Our customers now say, "Let HP do it!"

Hewlett-Packard recognized early on that it was no longer in the business of selling only products, but more in the business of providing "solutions" to its customers. HP sells solutions to its customers whereby HP takes on all of these responsibilities and many more. In the end, the customer is provided with a complete, up-and-running solution without the customer having to commit significant company resources. To do this successfully and on a repetitive basis, HP must also sell its outstanding project management capabilities. In other words, customers expect HP to have superior project management capability to deliver solutions. This is one of the requirements when customers' expectations are the driving force.

Mike Rigodanzo, former senior vice president, HP Services Operations and Information Technology, stated that:

> In the services industry, how we deliver is as important as what we deliver. Customers expect to maximize their return on IT investments from our collective knowledge and experience when we deliver best-in-class solutions.

---

10. H. Kerzner, *Project Management Best Practices: Achieving Global Excellence*, 2nd edition, Hoboken, NJ: Wiley, 2010; p. 118

The collective knowledge and experience of HP Services is easily accessible in HP Global Method. This integrated set of methodologies is a first step in enabling HPS to optimize our efficiency in delivering value to our customers. The next step is to know what is available and learn how and when to apply it when delivering to your customers.

HP Global Method is the first step toward a set of best-in-class methodologies to increase the credibility as a trusted partner, reflecting the collective knowledge and expertise of HP Services. This also improves our cost structures by customizing predefined proven approaches, using existing checklists to ensure all the bases are covered and share experiences and learning to improve Global Method.

Hewlett-Packard clearly identifies its project management capabilities in its proposals. The following material is an example of what typically is included in HP proposals.

### HP Services' Commitment to Project Management

### Why HP Services Project Management

HP Services considers strong project management a key ingredient to providing successful solutions to our customers. Our project managers are seasoned professionals with broad and deep experience in solutions, as well as managing projects. Our rigorous business processes make sure you are satisfied. A program roadmap provides an overall architecture of the project lifecycle while senior HP Services management conduct regular progress reviews to ensure quality. Our world-class project management methodology combines industry best practices with HP's experience to help keep everything on track. Our knowledge management program enables project managers and technology consultants to put our experience around the globe to work for you.

### PM Processes and Methodology

HP Services uses rigorous processes to manage our programs. The Program Roadmap provides an overall architecture for the project lifecycle. It includes the Solution and Opportunity Approval and Review (SOAR) process that approves new business as well as conducts implementation progress reviews to ensure quality and resolve problems quickly.

HP Services' project management methodology uses industry best practices with the added value of our experience implemented through web-based technology to allow quick updates and access throughout the world. It has over 20,000 web pages of information available to support our project teams. The methodology includes extensive knowledge management databases such as lessons learned and project experience from prior engagements that our project managers can use to help in managing their projects.

## 3.13 HEWLETT-PACKARD: THE JOURNEY AND THE OBSTACLES _____

When a company can recognize the driving forces for excellence in project management and understands that project management potentially could be needed for the survival of the firm, good things can happen quickly for the betterment of both the company and its clients. Doug Bolzman, Consultant Architect, PMP®, ITIL Expert at

Hewlett-Packard, describes the forces affecting project management success and some of the problems they faced and overcame. Doug's comments are addressed from the experiences and lessons learned while implementing frameworks and best practices into client environments and is not a reflection of HP directly. For Clients, our organization is involved in the consulting, strategy setting, mentoring/facilitating, and training activities to implement a framework, process, or environment at their site. All of our implementations include the basic project management principles, as professed in the *PMI PMBOK® Guide*.

Doug Bolzman discusses the significant emotional events that he has experienced in client environments:

- Loss of market share
- Not knowing the baseline timing or budgets for projects, thus not knowing if they are doing good or poorly
- Not having the ability for Speed to Market
- Understanding that there is much bureaucracy in the organization due to no design of a single project management capability

Doug Bolzman discusses three specific problems that were encountered:

*Problem 1: Management not knowing or understanding the relevance of full-time, professionally trained and certified project management staff.* This problem generated several business symptoms that were removed once the root cause of the problem was eliminated. To remove this problem, management needs to analytically understand all of the roles and responsibilities performed by project management, the deliverables produced, and the time required. Once they understand it is a significant effort, they can start to budget and plan for the role separate from the work at hand.

One manager was convinced that the engineering team was not working at the capacity they should until it was demonstrated that the amount of project management work they were required to perform was over and above their engineering responsibilities. Since the work was distributed to every engineer, their overall output was reduced. The leader of the engineering team demonstrated the roles and the time commitments that were project management related for the executive to assign the role to a full-time project manager. The output of the team was restored to expected levels.

*Problem 2: Everyone is overworked, and there is no time to implement a project management discipline.* Since this is a common problem, the way to work around this situation is to generate a tactical Project Management Governance Board to determine the standards, approaches, and templates that will be considered "best practice" from previous projects and leveraged to future projects. To not place additional scope or risk to existing projects, they are "grandfathered" from the new standards. As a project charter and team are generated, they are trained and mentored in the new discipline. The Governance Board meets when needed to approve new project management structures and measure conformance by project managers.

*Problem 3: Project managers were working at a higher level of maturity than the organization can benefit.* Project managers often use all the tools and templates at their disposal to manage a project but are incoherent of the client's level of business maturity. For such cases, the saying goes, "That manager is using 30 pounds of project management to manage a 10-pound project." If the client is in an unstable, ever-changing environment, the project manager spends most of the time formally administering change management, adjusting all of the appropriate costing tools, and does not further the project. To remedy this situation, guidelines must be given to the project managers to balance the level of project management maturity to the client's business environment. This is done while working with the client to demonstrate how the maturity of the business environment costs additional time, money, and resources.

With regard to the role of executives during project management implementation, Doug Bolzman commented:

Many executives take a mild "Management Commitment" role during the implementation since Project Management and framework implementations are foundational capabilities and are not recognized as market facing, revenue generating or exciting! Usually the executive approves a low budget plan where the majority or resources are absorbed from the organization and will be a motivational speaker at the kick off meeting. I have attended a kickoff meeting last week where the sponsor told the team that he did not expect the effort to be successful or change the culture. The team wondered if the sponsor was providing motivation by instituting a challenge.

Executives understand business language and do not tolerate or listen to Project Management techno talk. Executives are results oriented and if the Project Teams can simply translate the environment into business terms, create a business cast for incremental improvements to provide business value, the executive will be receptive to assist. If the executives are expected to generate the strategy or plan for improvement, or define project management's role within the organization, the implementation will fail.

The majority of implementation success comes from the immediate business leaders and the project managers themselves who are tired of the status quo and want to implement improvements.

A big indicator to understanding the level of management commitment prior to the start of the project is to ask the leader to provide the format for how they want the team to document the business case for the investment needed for the project. When they ask for clarification, ask them what criteria they will use to base their decision, such as:

- Business Value (improving how the overall business operates, achieving business unit objectives)
- Financial Value (decision purely based on cost and staying within a cost threshold)
- Quality Value (conformance to requirements & standards reflected in an audit)
- Integration Value (ability to support an end-to-end delivery model and maintain operational levels)
- Client Value (customer satisfaction indicators, voice of the customer surveys)
- Return on Investment (financially based where the returns are larger than the cost)
- Increased Market Share

Whatever criteria they use for justifying the release will give you the indicator as to what is important to them, and how they will support the project. Many times the criteria chosen is not a personal choice, but the criteria for how they are measured. The project criteria may mirror their objectives for acquiring their next bonus, such as team efficiency, staff size (increase or decrease), growth, and cost reductions.

**TABLE 3–7. ORGANIZATIONAL MILESTONES**

| Milestone | Activity | Value |
|---|---|---|
| 1. Establish Governance board structure | Development of all participants' roles and responsibilities for implementing the improvements | Implementing of a working "best practice" governance structure. All roles integrated and approved. |
| 2. Governance assignments | The sponsor's (executives) name who will play each role, assigning accountability and authorization | Executives establish priority through assignments. Everyone is trained in their role; expectations are set. |
| 3. Generation of attributes | The attributes describe the requirements, standards, capabilities, and metrics that will be used to define the improvements and measure the results. | Improvement is measurable, not emotional. Team can demonstrate value in business terms |
| 4. Generation of improvement plan | Incremental plans for improvement are generated based on a maturity model or business improvement objectives. | The environment is improved based upon the speed the organization can afford. The plans can be adjusted based on business changes. |
| 5. Implementation | Each implementation is a release of the environment, is measured, and demonstrates business value. | Incremental are improvements realized. Business invests incrementally, based on need. |

As for the chronology of events, Doug Bolzman continues:

When planning for the implementation of a framework, such as the implementation of a project, change, or release management (all using project management disciplines), we have learned that the organization needs to successfully progress through a series of organizational milestones. An example is shown in Table 3–7. The chronology is similar to a person who decides they need to lose weight. The person first makes the determination due to an event, such as clothes getting tight, peoples' observations, or health issues. Then the person realizes a cultural change is required, and if they are not willing to change behavior, they will not lose weight. The person has to encounter the "significant emotional event" for them to justify the discomfort of the change of behavior, such as exercising, not eating at night, or changing food types.

For clients, a basic approach is defined and reviewed. Many times the client attempts short cuts but then realizes that every step provides a foundational value for the larger journey. For one client, this approach was implemented seven years ago, is still in place, has generated nine major releases of their change management environment and has weathered five major corporate reorganizations.

Figure 3–13 reflects how the client needs to transform from a functional to a matrix directorate to establish a common framework and how the programs are then measured for how they conform to the framework.

As for project management being regarded as a profession, Doug Bolzman continues:

With the onset of PMI, most companies began to recognize project management as a profession and have developed a career path for project managers. Improvements to this career path occur with the entry point (project schedulers) coming from other parts of the business and the exit point (program managers) leading significant business units. People leaving the project management job family still utilize their training and disciplines to establish and run other parts of the business.

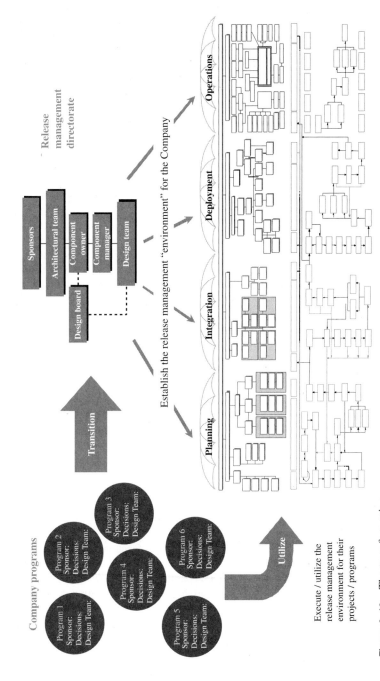

**Figure 3–13.** The transformation.

In order to enhance the opportunities for people who are currently assigned a project manager role, we have identified additional roles that the project manager can perform while meeting their commitments. We have identified various roles for IT service development, such as the component manager as illustrated in the Component Structure figure. A component manager role includes project management, but also includes a designer role, such as process design. An employee that can perform multiple roles simultaneously will provide more value to the organization than an employee that can perform only a single role. This dual role assignment will also provide the manager other opportunities or apply other skills than the routine scope management, resource management, and communication management etc. This approach also supports an environment of resource balancing. If the PMO has staff with other leveragable skills such as process design, courseware design, role design, they can be assigned other types of work, where the organization does not need to onboard new talent. This also helps when there are reductions in project management work and the staff can be re-allocated and not released.

When I was contemplating my career and opportunities for advancement were made available, such as being a team leader or a program leader, I listened to my manager and chief mentor who gave me this advice. You have one or two skills under your belt and can start being promoted. But without a broad base of skills, your advancement will be severely limited. If you are contemplating being an executive or business unit lead, think about all of the challenges that they will face and aggressively seek those types of assignments, from financial, to disaster recovery, to security so that you can fall back on your experiences when challenged with those situations. The bottom line, you do not want to be learning on the job after you are promoted and the outcome of your inexperience will impact many people and the business you are in charge of. I never forgot that advice and saw that it came true when my peers advanced rapidly, then hit a plateau in their careers as they made basic mistakes and were held back due to their experience.

Project Management provides a magnitude of experience for a person in their career path since it touches many business aspects that need to be formally managed. Project managers also have an opportunity to experience many different business situations since they can be assigned to most any project. A company would be wise to formally manage these types of skills, make opportunities available for their people's skill growth, then promote from their PMO as these people demonstrate the skills for more challenging business needs. Then as they take the reins, they will be more valuable in setting direction if they have experience in process design, data modeling, courseware development or risk management since they will be able to visualize and communicate their direction and assist in removing the roadblocks.

As for project management job descriptions, Doug Bolzman continues:

The project management role is baked into other functional roles of our resource model. The project manager role is part of our role design for release manager, component manager, master release controller, and ITSM consultant. We have won new business due the fact that our ITSM Consultants could roll up their sleeves and perform project management functions. We sell our ITSM Consulting services with the caveat that we can step in and fulfill the immediate business needs while we define the overall environment along with training and mentoring your staff to take it over. The following is an excerpt from the role design for a component manager:

**Agent Rational.** The component manager is required to lead all resources and activities for any modifications to the design or direction of an ITSM component

**Agent Description.** The component manager manages all aspect of the component for each new release including the release matrix, release plan, scope, release schedule, budget, resources, and communications. They work directly with the release manager and coordinate the activities of the design team, ensuring the release plan is followed and the design direction is completed for implementation.

**Overall Responsibilities.** Support the component owner in a project management capacity for one or more ITSM component, ensuring the design of the component is meeting the intended client needs, business drivers and business justifications

Support the release manager as the component is bundled into a larger bundle of components

Provide the planning and oversight for the management and control of release tasks, deliverables, work plans, budgets, staffing, issues, and milestones

Estimate and manage resource and financial needs, make work assignments, set priorities and establish the release schedule

Maintain the project workbook repository

## Component-Specific Responsibilities

Description: This section will detail the responsibilities that the agent encounters for each impacted component.

| | |
|---|---|
| **Service Strategy** | • Determine the specific impacts that a new IT release will have on their component in terms of process, tool, role, and training design impacts. <br> • Determine the required skillets and experience of the component design team for a specific release. <br> • Lead the component design team in defining the component test cases, plans and pass/fail conditions. |
| **Service Design** | • Lead the component design team in developing and testing all required installation, operations or training materials that will be required by the deployment or operations teams <br> • Oversee the release test team during the validation of the component designs |
| **Service Transition** | • Provide support to the transition team if the component installation is not proceeding at the customer site as designed. |

## Work Products

Description: This section will detail the deliverables or work products that are produced by this role.

| Deliverable | Description |
|---|---|
| **Release Plans** | The purpose of this document is to accurately document and communicate the scope, intent, plan, schedule for each version of the component |
| **Component Inventory (tools & processes)** | The purpose for conducting this inventory is to provide the information required to understand the current component environment. This inventory provides a single location to identify all the processes, tools, metrics, resources, and decision makers that currently make up the component |

## Required Training

Description: This section will detail the specific courses that are required for this role.

| Course ID | Name | Description | Location |
|---|---|---|---|
| SMLC-Aware | SMLC Overview Awareness | | |
| ITSM-Library | ITSM Library Structure | | |

**Tool Access**                    Description: This section will detail the tools that this role will require access in the delivery of their obligations.

| Tool Name | Access Level |
| --- | --- |
| | |

**Qualifications**                Every person filling the role of the component manager will be interviewed and will be able to demonstrate his/her qualifications to meet the needs of each version release.

- ITIL service lifecycle practitioner trained (ITIL foundation certification preferred—or obtained in first 3 months)
- 2–3 years project/program management experience or equivalent team leading experience or PMI certified

**Skills**
- Self-motivator and self-starter must be able to work without direct supervision.
- Strong communicator, both oral and written. Communicate appropriate information to all levels of the organization. Deliver executive presentations.

This list is used by multiple organizations to manage the overall resource needs, resource balancing, and resource recruiting. As every organization uses the same listing, managing resources becomes more efficient and accurate.

## 3.14 COOPER STANDARD[11]

**Best Practices at Cooper Standard**

Bill Pumphrey (Cooper Standard President NA - PMP) identified Project Management as a critical need for the success of the company. He began a project of implementation including PM Best Practices. To execute this, he brought in Dave Kandt to develop a People, Process and Structure strategy and implement Project Management in the organization. Bill's role was Project Sponsor.

Dave Kandt brought with him three decades of Project Management experience and was formerly Group Vice President of Program Management, Continuous Improvement and Quality at Johnson Controls. Dave had been responsible for Project Management in all product and customer groups globally with businesses, plant and product launches in all regions including Europe, China, South East Asia, Japan, Korea, the Americas and expansion regions including Russia and Turkey. Dave managed eight PMOs (Program Management Offices) in the various product and customer groups.

---

11. Material on Cooper Standard has been provided by Kristin L. Handley, Sr. Human Resource Manager - North America Division. ©2013 by Cooper Standard. Reproduced by permission.

Dave had been responsible for Project Management systems, organization and people, including:

- The JCI Project Management System (PLUS).
- The Project Management Academy.
- Program Management hiring and career pathing.
- The PMLDP (Program Management Leadership Development Program).
- Guidelines and procedures on the management global projects.
- Project Management Skills Assessment.
- Work Force Planning process for Simultaneous Development Teams.
- Standardized global project reporting and KPI system.
- Project Management Continuous Improvement Process (Excellence in Project Execution).

**Key Elements of the Strategy Accomplished in 2012:**

1. Establishing PM as a separate department with Senior Project Managers running Project Management for the four Business Units. This effectively clarified that the Project Manager was ultimately responsible for project success, ran the Project Teams, and had the authority to drive team decisions and commitments to customers.

2. Selection of Project Managers based on a written Skills Assessment tool that resulted in numerical ratings for each individual. This methodology added clarity to the requirements for Project Managers which included 9 basic areas: Leadership/Program Team Management /Scope Management/Launch Planning/Financial Management/Timing Management/Product Knowledge/Customer Satisfaction/Global Project Management Skills. The last area was necessary because many of the automotive projects are now managed and manufactured in multiple regions.

3. Modification of the existing PM process to simplify and clarify the Gate requirements based on industry standard Deliverables. Along with the basic gate system, other project management standards were released globally, to facilitate the global nature of the business, and allow Project Teams to function across multiple regions. A Six Gate standard was adopted as shown in Figure 3–14.

1. Company-wide training in PM with focus on CLAUS. Special focus on Engineering, Purchasing and Operations ensured alignment within all departments.

2. Implementation of a global PMO responsible for People, Process, Metrics, and Training. This allowed a global focus for the Project Managers and a clear organization.

3. Requirements for the PMs to take the IIL PMP certification material, as well as pass the PMP exam. This formalized the profession of project management in the company.

4. Collocation of the Project Teams including PMs, Account Managers, Engineering and CAD experts. Once this was done, Cost Analysts and Process Engineers were added to allow true cross functional collaboration in the Project Teams.

5. A formal Start On Time focus for future business opportunities with standardized block timing to ensure adequate development and tooling time for new products. Significant focus was added to the future business opportunities well before the timing and projects

Cooper LAUnch Standard

| Phases | Phase 1 | Phase 2 | Phase 3 | Phase 4 | Phase 5 | Phase 6 |
|---|---|---|---|---|---|---|

| Customer Events | | | Prototype Shipments to Customers | | Production Shipments to Customers | |
|---|---|---|---|---|---|---|

| | Quote | Design and Development | Prototype Tools & Equipment | DV Testing | Production Tools & Equipment | PV Testing | Product and Process Optimization | Production |
|---|---|---|---|---|---|---|---|---|

| Gates | G0 | G1 | G2 | G3 | G4 | G5 | G6 |
|---|---|---|---|---|---|---|---|

| | Quote Start | Program Start | Prototype Release | Testing Start | Production Release | Equipment & Tools In Plant | PPAP | Start of Production | Post Launch Report |
|---|---|---|---|---|---|---|---|---|---|

**Figure 3–14.** Cooper launch standard.

became critical. This also allowed better budgeting for resources, corporate expenses, capital investments and plant loading.

6. Formalization of the Project Team roles via standardized Position Descriptions. This, combined with the CLAUS gate structure and the standard deliverables allowed common understanding of Who was responsible for What Deliverables and by When across all 25 plants in NA and around the world.

7. Implementation of standard Executive Project Reviews and Launch Readiness Reviews managed by the Operations team at plant facilities for new projects. This increased the focus on project execution and put the executive team squarely in the middle of the launch process and in position to make critical decisions early in the project.

8. Implementation of a Risk Assessment process for all new business quotes. This allowed the executive team to determine clearly the skills and quantity of the project team, as well as plan for special support for challenging products or manufacturing processes.

9. Implementation of Project Charters by Gate 1 for all projects with focus on financial targets, timing, Project Team staffing and Risk Assessment for Engineering, Manufacturing, Quality, Finance and Supplier Performance. The key element of the Charter was the closing document – a standard Risk and Mitigation form with Red/Yellow/Green risk ratings and required Mitigation Strategies for each (there were no Green risks, by definition). The Project Teams at this stage of the project would request specific help from the executive team where support was needed.

10. Scope Change approval and control documents. If the scope changed based on either Cooper Standard decisions or customer requests, a revised Project Charter was required with new sign off by the executive team.

Limiting Innovation In Cycle for new business opportunities.

**Additional Improvements Targeted for 2013:**

1. Organizational development of the Project Teams at the 24 plant facilities in North America. This will be led by the PMO with greater focus on skill development and placement of project team members in the operations.
2. Training in formalized problem solving for all Project Team members. Cooper Standard has established Problem Solving as a core part of their Continuous Improvement process and will formalize the approach globally.
3. Implementation of a PM Dashboard that tracks critical Timing, Financial and Customer Satisfaction metrics. This will be provided monthly to the executive staff to allow focus on timing, financial performance, quality and customer satisfaction where needed.

**Enablers**

**Sponsor** The critical enabler of this initiative was Bill Pumphrey's role as sponsor and his extensive understanding of Project Management resulting from his PMP and personal background in Project Management. The most important Best Practice at work here is the importance of the leader having true PM background, knowledge and experience. It isn't enough to have an executive who believes in PM and supports the initiative. For true improvement with high acceleration, the leader has to have been a PM. This should be considered as a prerequisite for company presidents and leaders everywhere.

## Project Environment and Scope

As the appreciation for Project Management grew in the executive team, it became clearer that the main drivers for success were the decisions made by the executive team during the business planning process and during quoting. The key decisions that "set up the project for success" were:

1. Adequate time to execute the project.
2. Adequate staffing with team members with the right skills sets.
3. Selection of plants for production with adequate resources and capacity.
4. Accepting financial targets from the customers that are reasonable and achievable, with the appropriate focus on roadmaps and planning by the Project Manager and Project Teams.
5. Accepting reasonable product and manufacturing process scope and difficulty, with avoidance of risky "first time in the industry" innovation in a project's life cycle. This kind of innovation should be (and IS at Cooper Standard) done in a separate department, where new innovations can mature and develop in a controlled way.
6. Selection of suppliers with the appropriate level of technical and manufacturing competence, based on the product specifications and requirements.

These basic parameters are normally beyond the authority level of the project teams and are "given" to them when the project is "won." Cooper Standard's executive team became much more sensitized to the need to establish an achievable scope at the very beginning of the project, and avoid high risk projects.

**Expectations for the Future**      Cooper Standard is projected to grow significantly during the 2013 through 2018 time period. Project Management is expected to be a driver for that growth which will be fueled by improved Customer Satisfaction resulting from excellence in new product launches. Further, Cooper Standard expects a reduction in new product launch cost, a reduction in working capital resulting from better control over timing and deliverables, better use of existing resources as a result of collocation and improved integration of our global projects and Project Teams.

**Verbatims**      "My responsibility at Cooper is the North America region, currently $1.5B in annual revenue. From my perspective, I run two businesses at Cooper, both worth $1.5B+. One is already in production and one will be in production in the future (currently in development). We can work intensely to improve the business currently in production, but if we don't pay attention to the future, we will have to fix the business over and over again. It was clear when I began that we had good people, processes and structure to handle the current business, but that other one was, well, not so clear. Our future business is going to be bigger than the current business – growth is good! But we had to act fast to get ahead of the wave that was coming.

First, we needed to improve the process to become crystal clear on deliverables, timing, RASIC, etc.—accountability was needed! Then we needed a structure that supported the process and highlighted the importance of program management. It had to be a separate function, reporting directly to me. Finally, we needed people, the right people, which we could train and deploy who could identify and mitigate risk, manage complex priorities, deal effectively with people/customers and never, never, never be afraid to ask for help. We have more to do, but the organization is seeing the benefit and feeling the support that the people, process and structure can bring. As the new business starts to launch, the historical 'waste' costs (cost of poor quality, premium freight, attrition, etc.) will be noticeably improved. Customers already see the benefit and are beginning to value our competencies. We can see that already from the expanding order book. In the end, that is what it is all about—profitable growth which benefits all of us at Cooper Standard Automotive."

**Bill Pumphrey**
**Cooper Standard President, North America**

"Program Management is not simply a tool or group of individuals that reside in a functional department; rather, it is a culture that needs to be embraced from the top down, and aligned with your customers and vendors. At Cooper Standard, we have integrated the Program Management tools within our supply base to reduce risk associated with new program launches, speed up the development of new technologies and sustain long term relationships with strategic vendors."

**Ed Kiell, Director**
**Global Product Line Purchasing**

"Program Management boils down to creating an effective system within an organization which drives the right people around conducting the right conversation at the right time. It also means creating an environment that drives success around what I call the 4 P's":

1. People
2. Processes
3. Procedures
4. Purchases

**Kevin Kreiner**
**Director, Engineering Product Development**

"As we know, Cooper is now fully engaged in a Program Management system that is 'program requirement date' dependant vs. the Cooper of yesteryear that was 'sales kickoff date' dependant. Meaning . . . if our sales group does not yet have all of our full approvals from the customer lined up, but the program must go to avoid expedites, our management will approve us to proceed.

However, the initial improvements seen are:

- Programs are reviewed prior to Gate 1 at monthly staff meetings to ensure kickoffs are on time.
- The charter process enforces the need to understand the program, start to finish. Detailed program timing, gate dates, team resources, financials, noted and risks are identified.
- Management engagement is the key to remove the roadblocks to keep programs on track. This is seen weekly in Critical Issues meetings, monthly at Executive Reviews, and quarterly at the plant Launch Readiness Reviews.

These actions will assist our teams in reaching their Charter commitments and will greatly improve program success."

**Chuck Eisel**
**Senior Manager, Program Management**

## 3.15 NAVIAIR: ON TIME—ON BUDGET[12]

**How to Make Big and Complex Programs a Success**

### Recognise the setting

Air navigation service provision in Europe is one of the last market segments that has not been liberalised to any larger extent. Air navigation is—with the exception of the tower-area—still a monopoly for the 37 air navigation service providers and as Siim Kallas, Vice President of the European Commission expressed it in his opening speech at the "Single European Sky—the time for Action" conference, in Limassol on the 10th October 2012: "We are moving towards a regulatory environment which is more streamlined, coherent and based on a market economy."

In parallel this industry has been heavily regulated in the same way as the railway and the medical sectors. New demands are scoped as EU regulations, national legislation and new or updated ICAO standards are continuously rolled out with tight target dates to be met. Significant investments are made in order to meet the regulatory requirements. At the same time traffic is stagnating and even decreasing in the Danish airspace due to the fifth year of consecutive recession reported in the first quarter of 2013. In other words limited resources are available from a service provider point of view to meet the on-going complex and demanding nature of the aviation sector.

Based on the growing number of EU regulations provided by the European Commission there is an expectation that the air navigation service provision in Europe shall develop more efficient ways to perform air traffic control. In this context Naviair has formed a cooperation called COOPANS in cooperation with the Swedish, Irish, Austrian and Croatian air navigation service providers and the French supplier Thales. This cooperation shares the necessary costs and resources for the development, implementation and maintenance of a state of the art Air Traffic Management (ATM) System which is compliant with existent and future EU regulations. So far the COOPANS programme has been very successful and is today operational in four countries and in six air traffic control centrals. The seventh control central located in Zagreb will go operational next year (2014).

In this setting there is a strong need for success. Scarce resources and external pressure make this endeavour challenging. However, when we benchmarked ourselves with other similar market segments, we were proud of how successful we actually performed our programmes and projects. There is no room for failures and in Naviair we have a hit rate of nearly 100% when talking about delivering on time and—on budget.

Naviair's ability to cope with the setting and at the same time delivering on time and on budget is based on six main principles (see Figure 3–16):

- Build confidence
- Plan for success
- Manage performance and culture
- Tailor processes
- Organise and Report
- Communicate everywhere

---

12. ©2013 by Naviair. reproduced by permission. All rights reserved. Material on Naviair is provided by Mikael Ericsson, Director ATM Projects & Engineering, Steen Myhre Taschner Erichsen, Director/Manager Project Office, ATM Projects & Engineering (B.Sc.E.E.) and Michael Wibelius, Tactical Management. (M. Sc. Planning and Management)

**Figure 3–15.**  How to make big and complex programs a success.

Since the main principles are not strictly interrelated and since the success does not necessarily rely on a full roll-out of every principle, the user level of the principles can be tailored to the organisation in question as some parameters may be more useful in some organisations as opposed to others. Therefore the senior management and the project/ programme managers specifically (as they are the target audience of this essay) are free to pick and choose from the ideas contained in the description of each of the principles. However, one should bear in mind that it is recommended to maximise the use of each of the principles as described in this essay.

## Build Confidence

Change management is too often not prioritized or not taken in account when a big programme is performed. Many companies have had a negative experience from previous projects and therefore the management simply do not expect that their internal organisations to be able to run a big program smoothly. In Denmark analyses of IT projects performed by the Government revealed that as many as 75% of the projects did not deliver on time. Furthermore, a significant amount of the projects did not deliver on budget and 40% of these where heavily overspending.

When a programme is initiated in Naviair we start up by ensuring that the organisation attend to the changes that are about to come. Questions as to why the changes are necessary are welcome as well as discussions concerning alternatives. This supports a demystification of the changes in the organisation and is an important initial step towards avoiding that time has to be invested in this at a later stage where things either cannot be changed or are accompanied by great difficulties and/or expenses.

**Figure 3–16.** On time—on budget framework (Naviair).

Key for a have to be identified while at the same time recognising the fact that these may be populated by people with different backgrounds and concerns. In this context it is important to avoid reacting protectively and allowing groups with different professions to express their opinions. It is our experience that this makes the change process smoother and allows for fine-tuning of the direction in order to mitigate different risks that otherwise might turn into problems. You should make sure that you listen to all parts of the organisation and make a common view even if this will change the scope slightly. It is very easy to change the scope at this stage compared to doing it at later stages of the programme/ project. In order to ensure that all involved internal stakeholders have the same understanding of the changes, a high-level project frame forming the main benefits and measurable objectives should be agreed upon as the first thing and before any actual project preinvestigations are performed.

The governance structure must also allow the different stakeholders to discuss and to get the appropriate level of information during all phases of the programme/project. Naviair performed a very large programme containing more than 50 interrelated projects that went into operation at the end of 2007, and led to a complete new air traffic management system in Denmark. The responsibility for integrating all technical solutions from many different suppliers was put on our shoulders. Although the technological challenges where great, the change management was even greater. In fact it is a mental challenge to pull through such a programme if you expect to meet the targets spot on. Naviair managed to do so but we had to invest a lot of time and concerns in order to implement this governance structure and to secure that all stakeholders, internally as well as externally, were involved. We also had to perform regularly surveys to make sure that everybody supported the changes and sometimes certain groups had concerns that had to be addressed immediately. The mantra in this context is that such concerns are very useful in the process of making the programme successful. We never tried to defend ourselves, or to make difficult comments go away and this has become a permanent practice in Naviair today.

## Plan for Success

A golden rule in Naviair is to define a date for going operational with the new system as soon as possible. If possible we even set an exact time and in the above mentioned programme, we also had a countdown clock on the Naviair intranet front page. It is much easier if you have the courage to define a very visible target for the organisation. The pitfall is, however, that the date cannot be changed. A professional tennis player like Roger Federer does not think about a possible failure when he enters the tennis court and you must do the same: be a professional each and every day with one focus—On time and on budget.

If you succeed in getting your organisation behind such a date, which is an achievable goal, you can start to plan backwards. If you have a gate-driven approach, which is strongly recommendable, you will immediately find yourself and your teams very busy even if you have a multiyear programme. You should always remember that in the beginning the time schedule is a qualified guess. The schedule will gradually improve and be more detailed as the programme moves onwards. A programme manager who is able to follow the time schedule has a lot to gain and all work related to revision of the schedule is avoided. When the programme is ready the time schedule will be a perfect plan. However, you should never use this argument to fool yourself into postponing planning. As long as the operational date is not changed, milestones can be adjusted if necessary which is often the case with most programmes.

The expected outcome from late activities such as verification and validation, training or live tests must be addressed early. Your organisation, if not mature, will as an example most probably argue that the training cannot be planned before the system is physically in place. Such arguments should be taken seriously due to the fact that they express that the stakeholders do not know how to proceed in this early phase of the programme. Once the different parts of the organisation learn to address the topics on the right level, the work can be initiated early and the targets can be met. Inexperienced members of the programme must be supported by a PMO or similar in order to learn how to plan the activities before they enter the solution mode.

You have to communicate the plan in your governance structure repeatedly. The key to success is to obtain buy-ins from all stakeholders and some facts must be spelled out. At the same time all fora should be taken into consideration as described in the principle: "Build confidence" above. Different governance parties must be addressed at the appropriate level and some external stakeholders might be satisfied with the going-operational date if they are not affected by your tests etc.

One of the most important key messages from Naviair is never to operate with a plan B including an alternative date for operation of the system. You are allowed and advised to implement mitigate actions in relation to the risk of missing the "O-Date", e.g., by having a well-tested roll-back plan and other similar action plans. However, only one plan should be available and the management and internal stakeholders would have to agree upon this plan and communicate the following: We will make it!

## Manage Performance and Culture

A team is not automatically stronger than the individuals but with a high-performance team culture the outcome can be fantastic. A common method used by sport teams, special forces or the similar is seldom used in programme management. When a SWAT (Special Weapons And Tactics) team is gathered for the first time or when the team is changed they usually spend eleven weeks getting to know each other. At this point the task to solve as such is not even on the agenda. What is then the purpose for such a social event?

When a task is performed by a SWAT team the participants are totally depending on each other. In order to be able to trust each other 100% it takes much more than just a number of professional individuals. You also have to know the persons behind, the social factors and parts of their life histories. In a SWAT team you are about to put your life in another person's hands and that would not work with a total stranger. The same goes for a challenging programme that might affect your night sleep, your family life and leisure activities. When a programme is pulled through successfully, most of the participants would say the same: "it has been hard work but an experience for life!"

The process of building a high-performance team, as shown in Figure 3–17, should start with the social interaction in an environment that is protected from the daily interference from the office or the factory floor. In this environment the first step would be asking **Why** this change? At the same [time] the team members should get acquainted [with] each

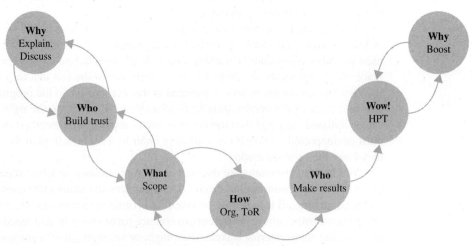

**Figure 3–17.** High-performance teams.

other. Many different methods could be used when socialising; one used in Naviair is to ask every participant to bring along a very important personal item and make a speech about this. You will find new sides of your colleagues that you never thought existed. You are now at the second step called: **Who.** At this stage you build up the trust between the team members.

Stay at step 1 and/or 2 as long as you can, at least during a seminar and at a follow-up meeting. Now you can go on to step 3 which is **What**. At this step you scope the changes and the task. This step and step 4: **How,** is straightforward for a programme organisation where you have established governance, terms of reference etc. In most organisations **What** and **How** are the starting points. Using **What** and **How** will work but will, however, only bring a mediocre performance. If you start at step 3 or 4 the process cannot be reversed due to the fact that it is very hard for most people (and at least a couple in your team) to leave the "solution mode" once it is initiated.

The next step is starting to work with your scope and your team and if you did start out the right way, it is very likely that you will experience the **WOW** step where the team is high performing. This must be maintained so that the last step will repeat the first step **WHY** again which you have to go through at least once a year or immediately after you have replaced one of your team members. If you replace one of your team members you will by definition have a new team so do not be misled to think that a high-performance culture will continue forever.

In multicultural teams, as most teams are today, you must as programme manager hold skills concerning cultural differences. Some knowledge regarding home countries, history, religion, political scenario and culture (e.g., a male- or female-dominant culture) will bring about a successful team building and support the achievement of a high performance team (HPT).

## Tailor Processes

The position as programme manager is like "being between the devil and the deep blue sea." Though you are on the top of your own governance, you have many people and instances to refer to. You have to cope with the environment as the programme moves along. Your target will be affected economically, technologically and by market fluctuations but often your programme may also be affected politically due to cooperation or alliances that your company may be a part of. These factors may add further complexity to the programme.

The Naviair project model, which contains the project processes and the templates that are used for initiating, executing, delivering and closing projects, is based on the PRINCE2 principles. However, it is tailored to the organisational setup, nature and setting of our projects. As such the Naviair project model is pragmatic in its nature with a lean paper flow and a simple-phase structure with a very clear go/no-go decision to be made by the steering group (please refer to the principle "Organise and report") between two phases. The project processes are clearly linked to the surrounding company processes such as the annual budget process, maintenance procedures etc.

The Naviair project model focuses on the initial phases in order to ascertain that the project is justified and that the right decision is made concerning the product specifications and the scope of the programme/projects before proceeding to the execution phase. The project initiation phase is based on a high-level project frame compiled by the project owner forming the main reference of the programme/project with clearly stated measurable objectives. The project manager is assigned to analyse possible solutions—if any—within the scope of the project frame. In this context and as a final step of the initial phases, the project manager makes a relatively detailed project assessment containing estimates concerning budget, resources, time, and main risks etc. to form the basis for a recommendable solution. Based on this analysis the steering group decides whether or not the project should continue into the execution phase where the progress is continuously monitored (please refer to the principle: Organise and Report). If the project is no longer justifiable, it can be terminated at any time during the project lifecycle. Once the project deliveries are complete the project handover phase is carried out before the actual closure and the lessons-learned phase is initiated. The latter phase provides for knowledge sharing and benefit realisation which in turn may lead into a new project being initiated. This approach has been very successful with a very good track record of being on time and on budget with every investment made throughout the process.

The phase for the realisation of the project is a lean one with focus on monitoring the progress and to mitigate risks and problems.

The portfolio prioritisation is performed in accordance with the Naviair prioritisation standards and these standards were implemented in order to make sure that we only realize programmes and projects that will support and strengthen our business values.

## Organise and Report

Your governance is very important and a rule of thumb is to place the sponsorship fairly high in the organisation. This person should be a member of the executive management and on the business value side, such as the COO or the CEO. If the sponsorship is held by the CFO or the CTI, it normally brings another type of focus on the programme, either a very strong focus on the financial side or on the technological side.

The sponsor should be the chairman of the programme steering group which should be manned by management representatives from each organisational key area in order to ascertain that decisions concerning prioritisation and programme/project phase shifts are holistic and aligned within the whole organisation. The programme steering group should meet regularly. The frequency of the steering-group meetings very much depends on the programme, but it can be an advantage to meet more frequently as the date of going operational approaches. If your organisational setup allows for forming one steering group for all programme and project, this would be advantageous as you can prioritise the full portfolio at once and thus benefit from one holistic view.

A complex programme should be manned with its own administrative and planning support. Richard Branson, the entrepreneur and founder of the brand Virgin, stated "I prefer a brilliant assistant." We are of the same opinion and as to Naviair's most complex programme, COOPANS; we do have such an organisation and support.

The subgroups in the programme organisation should be balanced in such a way that it reflects the different internal stakeholders in a positive way and be equipped with sufficient competencies to make the decisions related to their area of expertise in order to ensure progress. External stakeholders can be part of the organisation, but it is more likely that they are part of a group of interest or a user group. It is not important what kind of subgroups you define but how the subgroups interact with each other and with you as a programme manager.

In Naviair we prefer a pragmatic view on reporting. Our template used for status reporting is made as a simple excel tool and is based on traditional traffic light reporting. The report itself consists of six parameters where some of them represents business KPIs linked to Naviair's overall balanced scorecard. Depending on the complexity of the portfolio the reporting frequency varies from once a week to once a month. The most important thing is, however, not your present traffic lights status. That is all history and old information. The programme risks and problems, and the proactivity to mitigate for these are much more important. If you have a large portfolio with many programmes and projects, you will need more complex tools and risk management. In Naviair we use pragmatic tools for risk management, experienced and certified managers and a lot physical meetings to interact in relation to the risks and problems. If you hold the experience and have performed many projects previously, you can use your gut feeling to decide where to use your efforts. Therefore, we prioritise physical one-to-one meetings, discussions and interactions rather than extensive reporting.

In Naviair we have learnt that it is very hard for an organisation to decide whether the results of a programme are satisfactory. Often an organisation turns shaky and too detail-oriented before finally going operational with a new system—often with delays as a consequence. In Naviair we have developed an accept-criteria matrix to decide whether a programme is satisfactory to be put into operation. We have two levels: one with detailed milestones and descriptions per criteria and one steering group level which can make a quick report on a two-slide PowerPoint presentation. When all criteria are met, we are ready to go into operation. There will be no hassles or discussions as to whether we are ready or not.

**Communicate Everywhere**  Successful communication concerning a programme is not performed without a communication plan. The communication plan should be based on a stakeholder analysis, a SWOT analysis and/or similar in order to get a clear picture of the target audience and how this audience may react to certain statements. The communication plan based on the above mentioned analyses will provide a more targeted communication which in the end will see to that you achieve the result you were looking for.

You should consider any possible media as well as the frequency and timing of addressing the different internal and external stakeholders and at which level of information. The key message must be clear, consistent and easy to understand and relate to when addressing the different stakeholders. Use the producers of great beverages, cars or service companies and how they communicate their product values as prominent examples. You should address your business values and not the technological advantages which to most people are useless and only seem expensive. You should repeat the business values and key messages, until the programme has been executed.

You can target your communication towards the different key fora in many different ways and you should not confine yourself only to use well-known traditional communication tools. A successful programme manager will have to use nearly half of his working hours just to communicate and lobby in order to ensure the success of his programme. A programme manager who prioritises communication correctly will never refuse a possibility to present his programme and the related business values.

It is good practise in communication to submit articles in magazines, define a project portal to which the internal organisation will have access, publish news on both the internet and intranet, arrange kick-off meetings and open-house events and of course provide physical presentations whenever a possibility occurs. One of the keys of effective communication is to vary the means of communication and to find new creative ways of addressing the stakeholders, e.g., arrangements for merchandisers where the key message is displayed, involving the canteen, broadcast interviews with key persons of the programme/project on the news portal or display banners with key messages at often visited places, for instance by the coffee machine. The latter also provides for a more informal way of communication, since the coffee machine and/or similar places represent "safe zones" where the communication flows freely between employees. A banner with a positive message may lead the conversation into a more positive direction or simply provide for more "airtime" and visibility among the employees. Informal ways of communicating have been utilised with success in Naviair. An example is among others the weekly breakfast meetings chaired by the project manager on Fridays where employees meet and discuss the progress of the programme/project while enjoying a Danish pastry and a cup of coffee/tea. Since no high-level management is present at these meetings worries and information which would not otherwise have been discussed may occur. The same could be the case concerning rumours which the project manager will have the opportunity to spot and react upon in order to avoid that progress and key business values of the programme/project is undermined. The project manager may follow up on the breakfast meetings with an informal status e-mail to keep the people who were not present at the breakfast meeting in question up to date on the programme/project and invite people to come up with their comments if they possess any other or contradicting information.

Lack of communication and information will make people develop their own information when meeting at the coffee machine. This will lead to rumours and worries that have to be handled seriously, since rumours can take over facts. The simple cure is to attack all rumours when these are heard and they should be challenged in order to see if there are any facts behind these rumours. Often there are no facts behind occurring rumours. However, if the rumours are based on facts a solution has to be found.

You should not forget to celebrate any important milestone which has been achieved. Celebrate in an acceptable way which is compatible with the company culture. Some company cultures have no problems in providing free dinners or tickets to the opera whereas others cannot accept this for tax reasons or simply because it is not "*comme il faut.*" An inexpensive toy race car for an achieved site acceptance test could be a tool to signal gratefulness from the organisation. In Naviair we have programme managers who have nice collections of Ferraris and Lamborghinis that are proudly displayed at very visible places in their offices.

The most important and effective thing you could do in relation to projects is to praise, give credit and recognition to the staff who are involved in the project. The most powerful recognition you could give, is the one you communicate to a person from another department.

**Sum Up**                              The six main principles presented in this essay are continuously fine-tuned based on lessons learned, external inputs and as the industry for air navigation service provision in Europe develops in order to maximise

Naviair's project performance and ability to deliver on time and—on budget. The key message here is that we have to recognise the setting and adopt to it and realise that we are living in a dynamic world—no matter how good your best practices may be they may become yesterday's news if focus is not put on continuous development with a willingness to change and adapt. Finally, as the late Irish playwright, socialist, and a co-founder of the London School of Economic, George Bernard Shaw, put it: "Those who cannot change their minds cannot change anything."

## 3.16 DTE ENERGY

Several maturity models are available in the marketplace to assist companies in their quest for growth, excellence, and success in project management. Table 1-1 from Chapter is one such model. The purpose of the model is simply to provide some sort of structure to the maturity process. Tim Menke, PMP, senior continuous improvement expert, DO Performance Management, describes the growth process at DTE Energy:

> DTE Energy embraces project management to achieve timely and cost effective delivery of high quality products and services for our diversified utility customers. Our application of project management to engineering and construction efforts dates back decades. Expanding project management to areas other than major projects has been increasing in recent years. For example, our Information Technology organization started a formal drive to increase project management maturity in 2000. Individual mastery of project management skills was encouraged through training, practical experience, and formal certification. Job codes and a career progression model were created. Simultaneously, a cross-functional group of IT leaders, representing all areas of IT, was established to manage the portfolio of IT projects. Over time, other departments have migrated to a similar model including the use of formal job codes and progression paths in an effort to increase control of their projects.
>
> A focus on project management continues in our company. For example, we recently created a new department—Major Enterprise Projects—headed by a Senior Vice President, reporting directly to the CEO. This department was created to oversee major projects and their interactions. An example of one such project is the potential build of a Nuclear Power Generator.
>
> Our motivation to increase project management maturity was previously internally driven (Year 2000 IT Remediation, Merger with MichCon, Enterprise Resource Planning System Implementation, CMMI Certification, etc.). However, the recent economic shockwave being felt both nationally and internationally has resonated loudly in Southeast Michigan. We are using Lean Six Sigma continuous improvement to drive waste out of our processes and further streamline our operations. Project management is crucial to bringing our continuous improvement projects to fruition quickly and efficiently. We have revised our Continuous Improvement Project Management methodology to include a feedback mechanism to allow for placing on hold or canceling projects without losing sight of them in the future(Project Identification and Project Selection Phases). We have provided for scalability in our approach by recognizing a range of project types and corresponding requirements. All of these efforts have been championed by senior levels of DTE Energy leadership.

## 3.17 KEY PLASTICS

The journey to excellence usually begins with the creation of a project management broken down according to life-cycle phases. Methodologies and life-cycle phases need not be complex. They form the basis for standardization and repeatable success, regardless of the size of the company.

**Key Plastics Background**[13]     Key Plastics has a strong legacy as a Tier 1 automotive supplier providing complete design, engineering, manufacturing, and assembly operations for injection-molded interior trim components, exterior handles and under hood components. Key has been consistently recognized by OEMs globally for the ability to successfully deliver an excellent value, high-quality, on-time product.

Historically, Key has leveraged regional program management processes and tools. The North America process is heavily focused on tracking advanced product quality planning (APQP) requirements and with a strong design phase. The Asia Pacific team leverages the proven processes and tools of the joint venture partner. The European team follows two different processes. One process has a focus on manufacturing and a strong cost estimating and quote phase. A second European process is a balanced process containing commercial, design and manufacturing elements timed around customer build phases. See Figure 3–18.

In an effort to leverage best practices globally, Key recently kicked off the effort to create one global Key Product Realization Process (KPRP) and deploy a single global program management collaboration tool, enterProj. A steering committee was formed with the global program management leaders as well as functional team members to ensure all areas of the business were represented in the new process development.

**The Process (KPRP)**     The team agreed to use the balanced EU process as a baseline for the new KPRP (See Figure 3–19). The team shared regional things-gone-right and things-gone-wrong to determine the best practices to be captured from each of the regional processes and incorporated into the new process. Some highlights of the newly proposed KPRP include:

- "Business Planning" phase: kicks off the project into the corporate portfolio providing early global visibility into pursuits
- "Acquisition" phase: focused on detailed cost estimating to provide the best value in the product via early engagement of functional team members
- "Development and Planning" phase: centered around strong design review gates with an option for a specialized version of this phase used only for build-to-print projects
- "Industrialization" phase: containing "mini" phases managing the development of—injection molding tooling, equipment, facilities, packaging, purchased parts and resin as well as the manufacturing staffing plan
- "Serialization" phase: focused on delivery of the final product

13. Material on Key Plastics has been provided by Darlene Taylor, Director IT and Global Business Practices. Reproduced by permission

**Figure 3–18.** German process.

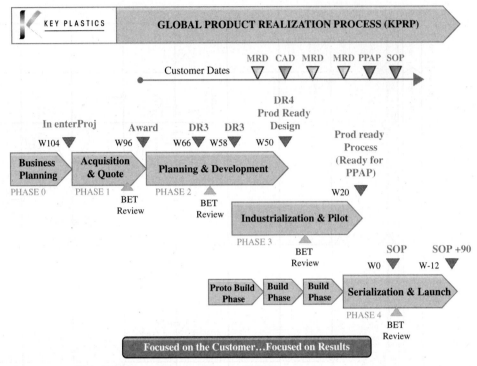

**Figure 3–19.** Global Product Realization Process (KPRP).

To ensure flexibility and to allow for a "build based" process, the customer build phases are purposely held separately. These phases are focused solely on the logical tasks required for the build. This approach allows for flexibility to deliver as few or as many prototype or production builds needed by the customer. These phases are timed separately from the core phases listed above.

"Engagement of the program management experts globally, as well as the functional team members, yielded excellent discussion and debate on the critical tasks required to deliver a high value, quality product for Key and our customers" said Darlene Taylor, Director IT and Global Business Practices. "This global input resulted in an optimized, unique process focused on all areas of the project"

**The Tool (enterProj)**            Regionally, the Key program teams leverage different localized tool-sets to effectively manage project plans and documentation. Along with the development of the global process, the team is implementing enterProj-a web-based cloud solution designed for project collaboration for both product and IT projects. See Figure 3–20.

Key is initially focused on the deployment of core enterProj including the ability to manage:

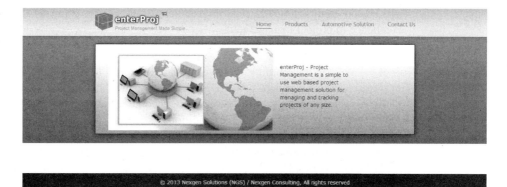

**Figure 3–20.** An example of an enterProj screenshot.

- <u>Project Summary</u>: description, key dates, measurables, product photos, summary financials
- <u>Team</u>: including all cross functional team members, custom project security for desired members, ability to add supplier team members and ability to easily reassign tasks to team members
- <u>Risk, Issue and Opportunity (RIO)</u>: collaborative approach to managing risks, issues and opportunities with the ability to add multiple actions on each RIO
- <u>Parts, Tooling and Equipment, Suppliers</u>: tracks key information on this project data as well as, the ability to create specific timing tied into the master project timing plan and forming baseline master data for the project financials
- <u>Timing</u>: creates custom project plan based on the global KPRP process and the customer build requirements. Ability to highlight specific tasks to reveal in customer plan and also export to MS Project
- <u>Change Management</u>: creates project based change request and change notice functionality, which allows the teams to program manage the change based [on] action templates as well as build custom action templates. These changes can also be linked to the financial models of the project.
- <u>Financial Management</u>: online financial model calculates the financial metrics of the project and is augmented by the ability to upload any detailed costing models.

"With the implementation of enterProj, I now have real-time visibility into the status of all of programs globally" said Terry Gohl, CEO Key Plastics "My executive team has immediate access to each project summary and detailed information from the enterProj dashboard"

## 3.18 ILLUMINAT AND THE STRATEGIC BUSINESS OF PROJECT MANAGEMENT

Throughout this chapter, we have discussed the journey to excellence. When companies are able to integrate together project management processes, business processes and strategic planning activities, project management maturity, and eventually excellence can and will be achieved. ILLUMINAT is an excellent example of a company that has achieved excellence in project management.

**ILLUMINAT Project Management Services: The Pursuit of Project Management Excellence**[14]

According to Owen Field, Regional Manager, Project Management Services at ILLUMINAT, there must be a strategic business purpose why a firm desires to become excellent in project management. Looking at the traditional strategic element, namely the products offered–markets served, excellence in project management should allow a firm the opportunity to develop more products and serve more markets than without having project management. ILLUMINAT recognized this years ago and began a journey to achieve excellence in project management.

ILLUMINAT is an industry-leading IT and telecommunications solutions provider for major organizations across the Caribbean servicing all industries, sectors and enterprises; from office equipment and supplies to enterprise applications in asset management, records management, [and] financial and human resource management; also from networking and telecommunications to technical support and project management services. Project management practices now affects all areas of our business.

The benefits of achieving excellence in project management are quite apparent. ILLUMINAT's PM Services Unit is an elite team of Professionals with expertise and certification in various methodologies to provide business, project and IT consulting solutions, enabling our clients ROI by leveraging our expertise to deliver value realization. We have created a precision set of consulting services from the project, business and information and communications technology domains providing our clients with expert delivery and guided excellence. We execute projects, manage business transformation and deliver results for our clients.

We can now return to the strategic element of products offered and markets served. ILLUMINAT provides the following services:

- Project & Program Management
- Project Management Office Consulting & Implementation
- Business Implementation & Support
- Business Solutions for IT
- Project Management & Business Solutions for IT Training

ILLUMINAT provides services:

- To Private and Public Sector
- Across Industries in Energy, Financial, Government and National Agencies, Retail and Hospitality, Utilities and Manufacturing
- Across the Caribbean from the Bahamas to Guyana
- Satisfying stakeholders from India, Europe, UK, North America and South America.

---

14. ©2013 by ILLUMINAT. Reproduced by permission. All rights reserved. Material in this section has been provided by Owen Field, Regional Manager, Project Management Services at ILLUMINAT.

We constantly:

- Monitor the Competitive Market
- Respond to Changing Market Demands
- Identify Performance Gaps in Organizations and Develop Services Solutions to Optimize Client Performance and Value Realization for Their Investments
- Satisfy Client Needs
- Commit to Professional Development

Our clients' feedback from Client Satisfaction Surveys and Impact Assessments indicate that our best practices and deliverables are being adopted within their business to optimize their business processes.

Our journey in pursuit of excellence in project management is based on the philosophy that, "We are what we repeatedly do. Excellence therefore is not an act but a habit." —Aristotle

ILLUMINAT PM Services in its current form has been operating and growing for the last 7 years.

We have striven towards excellence by being at the forefront of the profession:

- Committed to Professional Development
- Recognized at Various Award Ceremonies for Our Projects, People and Service Excellence
- Recognized by Our Clients and Stakeholders for Improving Their Business Processes through the Adoption of Our Project Management Best Practices and Deliverables
- Recognized by Various Authors for Excellence in Project Management
- Authored White Papers
- Served as Panelists in Industry Meetings
- Pioneered Research in Project Management in the Region
- Accredited as a Global Registered Education Provider by the Project Management Institute (PMI).
- Served on the Boards of Professional Societies
- Invested in National and Regional Development
- Focused towards a Global Perspective, Export Ready
- Transformed from a Cost Centre, [with] One Service Line, Internally Focused and Unknown Entity—to a Profit Centre, [with] Five Service Lines, External and Export Focused, Internationally Published and Recognized Entity for Best Practices Winning Awards for Excellence.

We achieve this by building Excellence in all that we do including;

- PMO Management Strategy and Capability
- PM Employees Competency, Certification and Business Acumen
- PM Methods, Procedures, Tools and Templates development and refinement aligned with Industry Standards, Best Practices and Regulations

- Passionate Leadership and Culture of Commitment to Service and Relationship Management
- Strategic Networks and Knowledge Repositories
- Continuous Evolution and Improvement

The PMO's contribution to the organization is an overall improvement in the following;

- Image of ILLUMINAT
- Employee Morale
- Employee Productivity
- Customer Service & Quality
- Good Business Practices increasing revenue and profitability
- Expense Management
- Receivables & Inventory Management
- Operational Efficiency

Overall impact of the PMO over a sustained period

- Customer Satisfaction – Delivering to standards and expectations with professionalism.
- Employee Satisfaction – Eliminating inter-unit conflict for resources via an enterprise PMO clearing the path for employees to deliver. The PMO is ranked as a leading department in the organization.
- Productivity – Significantly improved due to professional development and culture to achieve superior results.
- Efficiency – Developed capability through methodology and tools to do more with less.
- Utilization – Improved the standardization of professional services billable time tracking and management.
- PMO Employee Performance – Consistently above target.
- Erosion – Significantly reduced revenue leakage due to poor estimating and effort not invoiced due to poor change management. Improvements in estimating and the application of project change control enabled a 100% reduction in revenue Erosion.
- Revenue – Developed new revenue channel and revenue recovery capability. PMO moved from a cost centre to a revenue/profit generator.
- Brand reputation – Excellence in project delivery, formal and informal industry recognition.
- Organization integration – Uniform processes and standards, organizational capability advancement.
- Regional integration – Benchmarking and standards. The Regional PMO Integration is providing synergies across the Caribbean where PM expertise from one territory can be applied in another, as well as the cross-cultural innovations and learnings are shared and applied across territories.
- Knowledge assets – Employee certifications, project information, lessons learned and best practices repository continue to provide intellectual capital appreciation, yielding additional business benefits.

## 3.19 AVALON POWER AND LIGHT

Avalon Power and Light (a disguised case) is a mountain states utility company that, for decades, had functioned as a regional monopoly. All of this changed in 1995 with the beginning of deregulation in public utilities. Emphasis was now being placed on cost cutting and competitiveness.

The Information Systems Division of Avalon was always regarded as a "thorn in the side" of the company. The employees acted like prima donnas and refused to accept any of the principles of project management. Cost-cutting efforts at Avalon brought to the surface the problem that the majority of the work in the Information Systems Division could be outsourced at a significantly cheaper price than performing the work internally. Management believed project management could make the division more competitive, but would employees now be willing to accept the project management approach?

According to a spokesperson for Avalon Power and Light:

> Two prior attempts to implement a standard application-development methodology had failed. Although our new director of information systems aggressively supported this third effort by mandating the use of a standard methodology and standard tools, significant obstacles were still present.
>
> The learning curve for the project management methodology was high, resulting in a tendency of the leads to impose their own interpretations on methodology tasks rather than learning the documented explanations. This resulted in an inconsistent interpretation of the methodology, which in turn produced inconsistencies when we tried to use previous estimates in estimating new projects.
>
> The necessity to update project plans in a timely manner was still not universally accepted. Inconsistency in reporting actual hours and finish dates resulted in inaccurate availabilities. Resources were not actually available when indicated on the departmental plan.
>
> Many team leads had not embraced the philosophy behind project management and did not really subscribe to its benefits. They were going through the motions, producing the correct deliverables, but managing their projects intuitively in parallel to the project plan rather than using the project plan to run their projects.
>
> Information systems management did not ask questions that required use of project management in reporting project status. Standard project management metrics were ignored in project status reports in favor of subjective assessments.

The Information Systems Division realized that its existence could very well be based upon how well and how fast it would be able to develop a mature project management system. By 1997, the sense of urgency for maturity in project management had permeated the entire Information Systems Division. When asked what benefits were achieved, the spokesperson remarked:

> The perception of structure and the ability to document proposals using techniques recognized outside of our organization has allowed Information Systems to successfully compete against external organizations for application development projects.
>
> Better resource management through elimination of the practice of "hoarding" preferred resources until another project needs staffing has allowed Information Systems to actually do more work with less people.

We are currently defining requirements for a follow-on project to the original project management implementation project. This project will address the lessons learned from our first two years. Training in project management concepts (as opposed to tools training) will be added to the existing curriculum. Increased emphasis will be placed on why it is necessary to accurately record time and task status. An attempt will be made to extend the use of project management to non–application-development areas, such as network communications and technical support. The applicability of our existing methodology to client-server development and Internet application development will be tested. We will also explore additional efficiencies such as direct input of task status by individual team members.

We now offer project management services as an option in our service-level agreements with our corporate "customers." One success story involved a project to implement a new corporate identity in which various components across the corporation were brought together. The project was able to cross department boundaries and maintain an aggressive schedule. The process of defining tasks and estimating their durations resulted in a better understanding of the requirements of the project. This in turn provided accurate estimates that drove significant decisions regarding the scope of the project in light of severe budget pressures. Project decisions tended to be based on sound business alternatives rather than raw intuition.

## 3.20 ROADWAY EXPRESS

In the spring of 1992, Roadway Express realized that its support systems (specifically information systems) had to be upgraded in order for Roadway Express to be well positioned for the twenty-first century. Mike Wickham, then president of Roadway Express, was a strong believer in continuous change. This was a necessity for his firm, because the rapid changes in technology mandated that reengineering efforts be an ongoing process. Several of the projects to be undertaken required a significantly larger number of resources than past projects had needed. Stronger interfacing between functional departments would also be required.

At the working levels of Roadway Express, knowledge of the principles and tools of project management was minimal at best in 1992. However, at the executive levels, knowledge of project management was excellent. This would prove to be highly beneficial. Roadway Express recognized the need to use project management on a two-year project that had executive visibility and support and that was deemed strategically critical to the company. Although the project required a full-time project manager, the company chose to appoint a line manager who was instructed to manage his line and the project at the same time for two years. The company did not use project management continuously, and the understanding of project management was extremely weak.

After three months, the line manager resigned his appointment as a project manager, citing too much stress and being unable to manage his line effectively while performing project duties. A second line manager was appointed on a part-time basis and, like his predecessor, he found it necessary to resign as project manager.

The company then assigned a third line manager, but this time released her from all line responsibility while managing the project. The project team and selected company personnel were provided with project management training. The president of the company realized the dangers of quick implementation, especially on a project of this magnitude, but was willing to accept the risk.

After three months, the project manager complained that some of her team members were very unhappy with the pressures of project management and were threatening to resign from the company if necessary simply to get away from project management. But when asked about the project's status, the project manager stated that the project had met every deliverable and milestone thus far. It was quickly apparent to the president, Mike Wickham, and other officers of the company that project management was functioning as expected. The emphasis now was how to "stroke" the disgruntled employees and convince them of the importance of their work and how much the company appreciated their efforts.

To quell the fears of the employees, the president assumed the role of the project sponsor and made it quite apparent that project management was here to stay at Roadway Express. The president brought in training programs on project management and appeared at each training program.

The reinforcement by the president and his visible support permeated all levels of the company. By June of 1993, less than eight months after the first official use of project management, Roadway Express had climbed further along the ladder to maturity in project management than most other companies accomplish in two to three years due to the visible support of senior management.

Senior management quickly realized that project management and information systems management could be effectively integrated into a single methodology. Mike Wickham correctly recognized that the quicker he could convince his line managers to support the project management methodology, the quicker they would achieve maturity. According to Mike Wickham, President of Roadway Express at that time (and later chairman of the board):

> Project management, no matter how sophisticated or how well trained, cannot function effectively unless all management is committed to a successful project outcome. Before we put our current process in place, we actively involved all those line managers who thought it was their job to figure out all of the reasons a system would never work! Now, the steering committee says, "This is the project. Get behind it and see that it works." It is a much more efficient use of resources when everyone is focused on the same goal.

## 3.21 DEFCON CORPORATION

A defense contractor that wishes to remain nameless (we call it Defcon Corporation) had survived for almost 20 years on fixed-price, lump-sum government contracts. A characteristic of a fixed-price contract is that the customer does not audit your books, costs, or perhaps even your project management system. As a result, the company managed its projects rather loosely between 1967 and 1987. As long as deliveries were on time, the capabilities of the project management system were never questioned.

By 1987, the government-subcontracting environment had changed. There were several reasons for this:

- The DoD was undergoing restructuring.
- There were cutbacks in DoD spending, and the cutbacks were predicted to get worse.
- The DoD was giving out more and more cost-reimbursable contracts.
- The DoD was pressuring contractors to restructure from a traditional to a product-oriented organizational form.
- The DoD was pressuring contractors to reduce costs, especially the overhead rates.
- The DoD was demanding higher-quality products.
- The DoD was now requiring in its proposals that companies demonstrate higher-quality project management practices.

Simply to survive, Defcon had to bid on cost-reimbursable contracts. Internally, this mandated two critical changes. First, the organization had to go to more formal rather than informal project management. Second, the organization had to learn how to use and report earned-value measurement. In order to be looked upon favorably by the government for the award of a cost-reimbursable contract, a company must have its earned-value cost control/reporting system validated by the government.

A manager within one such company that was struggling made the following comments on how "survival" had forced the organization to climb the ladder to maturity over the past 10 years:

> Formal project management began with the award of the first major government program. There was a requirement to report costs by contract line item and to report variances at specific contract levels. A validated system was obtained to give us the flexibility to submit proposals on government programs where cost schedule reporting was a requirement of the RFP (request for proposal).
>
> We have previous experience in PERT (program evaluation and review technique) networking, work breakdown structures, and program office organizations. Management was also used to working in a structured format because of our customer's requirements for program reviews. After system validation in 1987, it took six months to a year to properly train and develop the skills needed by cost account managers and work package supervisors. As you move along in a program, there is the need to retrain and review project management requirements with the entire organization.
>
> We visited other companies and divisions of our own company that had prior experience in project management. We sent people to seminars and classes held by experts in the field. We conducted internal training classes and wrote policies and procedures to assist employees with the process of project management. Later we purchased canned reporting packages to reduce the cost of internal programming of systems.
>
> We established dedicated teams to a contract/program. We have program office organizations for large programs to follow through and coordinate information to internal management and our customer. We adjusted our systems and reports to meet both our internal and external customers' needs.
>
> Implementation of integrated systems will provide data on a timelier basis. These data will allow management to react quickly to solve the problem and minimize the cost impact.

Project management has allowed us to better understand costs and variances by contract/program. It provides us with timely data and makes tracking of schedule issues, budget issues, and earned values more manageable. Project management has given us visibility into the programs that is useful in implementing cost reductions and process improvements. Having a validated system allows us to remain competitive for bidding on those programs that require formal cost schedule control systems.

## 3.22 KOMBS ENGINEERING

The company described in the last section was very fortunate to have identified the crises and taken the time to react properly. Some companies are not so fortunate. Although the next two companies appear to be outdated, there are valuable lessons that can be learned about what not to do when embarking on the path to maturity. Consider the Michigan-based Kombs Engineering (name of the company is disguised at company's request).

In June 1993, Kombs Engineering had grown to a company with $25 million in sales. The business base consisted of two contracts with the Department of Energy (DoE), one for $15 million and one for $8 million. The remaining $2 million consisted of a variety of smaller jobs for $15,000 to $50,000 each.

The larger contract with the DoE was a five-year contract for $15 million per year. The contract was awarded in 1988 and was up for renewal in 1993. The DoE had made it clear that, although it was very pleased with the technical performance of Kombs, the follow-on contract must go through competitive bidding by law. Marketing intelligence indicated that the DoE intended to spend $10 million per year for five years on the follow-on contract with a tentative award date of October 1993. On June 21, 1993, the solicitation for proposal was received at Kombs. The technical requirements of the proposal request were not considered to be a problem for Kombs. There was no question in anyone's mind that on technical merit alone Kombs would win the contract. The more serious problem was that the DoE required a separate section in the proposal on how Kombs would manage the $10 million/year project as well as a complete description of how the project management system at Kombs functioned.

When Kombs won the original bid in 1988, there had been no project management requirement. All projects at Kombs were accomplished through the traditional organizational structure. Only line managers acted as project leaders.

In July 1993, Kombs hired a consultant to train the entire organization in project management. The consultant also worked closely with the proposal team in responding to the DoE project management requirements. The proposal was submitted to the DoE during the second week of August. In September 1993, the DoE provided Kombs with a list of questions concerning its proposal. More than 95 percent of the questions involved project management. Kombs responded to all questions.

In October 1993, Kombs received notification that it would not be granted the contract. During a postaward conference, the DoE stated that it had no "faith" in the Kombs project management system. Kombs Engineering is no longer in business.

Kombs Engineering is an excellent case study to give students in project management classes. It shows what happens when a subcontractor does not recognize how smart the customer has become in project management. Had Kombs been in close contact with its customers, the company would have had five years rather than one month to develop a mature project management system.

## 3.23 WILLIAMS MACHINE TOOL COMPANY

The strength of a culture can not only prevent a firm from recognizing that a change is necessary but also block the implementation of the change even after need for it is finally realized. Such was the situation at Williams Machine Tool Company (another disguised case).

For 75 years, the Williams Machine Tool Company had provided quality products to its clients, becoming the third largest U.S.-based machine tool company by 1980. The company was highly profitable and had an extremely low employee turnover rate. Pay and benefits were excellent.

Between 1970 and 1980, the company's profits soared to record levels. The company's success was due to one product line of standard manufacturing machine tools. Williams spent most of its time and effort looking for ways to improve its "bread and butter" product line rather than to develop new products. The product line was so successful that other companies were willing to modify their production lines around these machine tools, rather than asking Williams for major modifications to the machine tools.

By 1980, Williams Company was extremely complacent, expecting this phenomenal success with one product line to continue for 20 to 25 more years. The recession of 1979–1983 forced management to realign its thinking. Cutbacks in production had decreased the demand for the standard machine tools. More and more customers were asking either for major modifications to the standard machine tools or for a completely new product design.

The marketplace was changing and senior management recognized that a new strategic focus was necessary. However, attempts to convince lower-level management and the workforce, especially engineering, of this need were meeting strong resistance. The company's employees, many of them with over 20 years of employment at Williams Company, refused to recognize this change, believing that the glory days of yore would return at the end of the recession.

In 1986, the company was sold to Crock Engineering. Crock had an experienced machine tool division of its own and understood the machine tool business. Williams Company was allowed to operate as a separate entity from 1985 to 1986. By 1986, red ink had appeared on the Williams Company balance sheet. Crock replaced all of the Williams senior managers with its own personnel. Crock then announced to all employees that Williams would become a specialty machine tool manufacturer and the "good old days" would never return. Customer demand for specialty products had increased threefold in just the last 12 months alone. Crock made it clear that employees who would not support this new direction would be replaced.

The new senior management at Williams Company recognized that 85 years of traditional management had come to an end for a company now committed to specialty products. The company culture was about to change, spearheaded by project management, concurrent engineering, and total quality management.

Senior management's commitment to project management was apparent by the time and money spent in educating the employees. Unfortunately, the seasoned 20+ year veterans still would not support the new culture. Recognizing the problems, management provided continuous and visible support for project management in addition to hiring a project management consultant to work with the people. The consultant worked with Williams from 1986 to 1991.

From 1986 to 1991, the Williams Division of Crock Engineering experienced losses in 24 consecutive quarters. The quarter ending March 31, 1992, was the first profitable quarter in over six years. Much of the credit was given to the performance and maturity of the project management system. In May 1992, the Williams Division was sold. More than 80 percent of the employees lost their jobs when the company was relocated over 1500 miles away.

Williams Machine Tool Company did not realize until too late that the business base had changed from production-driven to project-driven. Living in the past is acceptable only if you want to be a historian. But for businesses to survive, especially in a highly competitive environment, they must look ahead and recognize that change is inevitable.

# Project Management Methodologies

## 4.0 INTRODUCTION

In Chapter 1 we described the life-cycle phases for achieving maturity in project management. The fourth phase was the growth phase, which included the following:

- Establish life-cycle phases.
- Develop a project management methodology.
- Base the methodology upon effective planning.
- Minimize scope changes and scope creep.
- Select the appropriate software to support the methodology.

The importance of a good methodology cannot be understated. Not only will it improve your performance during project execution, but it will also allow for better customer relations and customer confidence. Good methodologies can also lead to sole-source or single-source procurement contracts.

Creating a workable methodology for project management is no easy task. One of the biggest mistakes made is developing a different methodology for each type of project. Another is failing to integrate the project management methodology and project management tools into a single process, if possible. When companies develop project management methodologies and tools in tandem, two benefits emerge. First, the work is accomplished with fewer scope changes. Second, the processes are designed to create minimal disturbance to ongoing business operations.

This chapter discusses the components of a project management methodology and some of the most widely used project management tools. Detailed examples of methodologies at work are also included.

## 4.1 EXCELLENCE DEFINED

Excellence in project management is often regarded as a continuous stream of successfully managed projects. Without a project management methodology, repetitive successfully completed projects may be difficult to achieve.

Today, everyone seems to agree somewhat on the necessity for a project management methodology. However, there is still disagreement on the definition of excellence in project management, the same way that companies have different definitions for project success. In this section, we will discuss some of the different definitions of excellence in project management.

Some definitions of excellence can be quite simple and achieve the same purpose as complex definitions. According to a spokesperson from Motorola[1]:

Excellence in project management can be defined as:

- Strict adherence to scheduling practices
- Regular senior management oversight
- Formal requirements change control
- Formal issue and risk tracking
- Formal resource tracking
- Formal cost tracking

A spokesperson from AT&T defined excellence at AT&T as:

Excellence [in project management] is defined as a consistent Project Management Methodology applied to all projects across the organization, continued recognition by our customers, and high customer satisfaction. Also our project management excellence is a key selling factor for our sales teams. This results in repeat business from our customers. In addition there is internal acknowledgement that project management is value-added and a must have.

Doug Bolzman, Consultant Architect, PMP®, ITIL expert at Hewlett-Packard, discusses his view of excellence in project management:

Excellence is rated, not by managing the pieces, but by understanding how the pieces fit together, support each other's dependencies, and provide value to the business. If project management only does what it is asked to do, such as manage 300 individual projects in the next quarter, it is providing a low value-added function that basically is

---

1. H. Kerzner, *Project Management Best Practices: Achieving Global Excellence*, Wiley, Hoboken, NJ, 2006, p. 136.

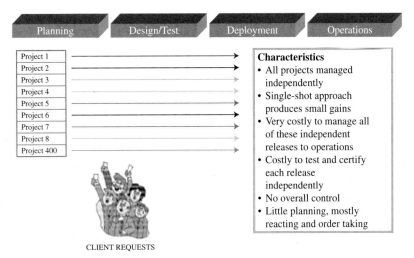

**Figure 4–1.** Release management stages.

**Figure 4–2.** Release management stages: bundling requests.

the "pack mule" that is needed, but only does what it is asked—and no more. Figures 4–1 and 4–2 demonstrate that if mapping project management to a company's overall release management framework, each project is managed independently with the characteristics shown

Using the same release framework and the same client requests, project management disciplines can understand the nature of the requirements and provide a valuable service to bundle the same types of requests (projects) to generate a forecast of the work, which will assist the company in balancing its financials, expectations, and resources. This function can be done within the PMO.

At DTE Energy, excellence in project management is defined using the project management methodology. According to Jason Schulist, manager—continuous improvement, Operating Systems Strategy Group:

> DTE Energy leverages a four-gate/nine-step project management model for continuous improvement. (See Figures 4–3 and 4–4.) In gate1 the project lead clearly defines the metric that will measure the success in the project. In gate 2 of the project, after the ideal state has been defined, the project lead gains approval from the champion confirming not only the appropriate metrics but also the target.
>
> This target for success is maintained throughout the project and the project does not close (gate 4) unless the target is achieved. If the project does not achieve the target through the committed actions, the project reverts to gate 2 where the ideal state is revisited and further actions are determined.

Allan Dutch, PMP®, senior project manager, software engineering, methods and staffing at DTE Energy, believes that:

> A DTE Energy information technology (IT) project exhibiting success and excellence in project management is one where the project manager directs her or his team much like a

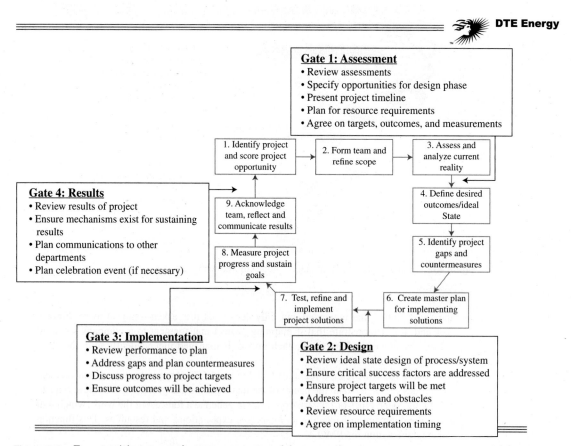

**Figure 4–3.** Four-gate/nine-step project management model.

**Figure 4–4.** Four-gate/nine-step model and tool selection process map.

conductor directs her or his orchestra. Business value is demonstrable and recognized by the business unit while interactions with support organizations and infrastructure requirements are coordinated smoothly and according to plan.

## 4.2 RECOGNIZING THE NEED FOR METHODOLOGY DEVELOPMENT

Simply having a project management methodology and following it do not lead to success and excellence in project management. The need for improvements in the system may be critical. External factors can have a strong influence on the success or failure of a company's project management methodology. Change is a given in the current business climate, and there is no sign that the future will be any different. The rapid changes in technology that have driven changes in project management over the past two decades are not likely to subside. Another trend, the increasing sophistication of consumers and clients, is likely to continue, not go away. Cost and quality control have become virtually the same issue in many industries. Other external factors include rapid mergers and acquisitions and real-time communications.

Project management methodologies are organic processes and need to change as the organization changes in response to the ever-evolving business climate. Such changes, however, require that managers on all levels be committed to the changes and develop a vision that calls for the development of project management systems along with the rest of the organization's other business systems.

Today, companies are managing their business by projects. This is true for both non–project-driven and project-driven organizations. Virtually all activities in an organization can be treated as some sort of project. Therefore, it is only fitting that well-managed companies regard a project management methodology as a way to manage the entire business rather than just projects. Business processes and project management processes will be merged together as the project manager is viewed as the manager of part of a business rather than just the manager of a project.

Developing a standard project management methodology is not for every company. For companies with small or short-term projects, such formal systems may not be cost-effective or appropriate. However, for companies with large or ongoing projects, developing a workable project management system is mandatory.

For example, a company that manufactures home fixtures had several project development protocols in place. When they decided to begin using project management systematically, the complexity of the company's current methods became apparent. The company had multiple system development methodologies based on the type of project. This became awkward for employees who had to struggle with a different methodology for each project. The company then opted to create a general, all-purpose methodology for all projects. The new methodology had flexibility built into it. According to one spokesman for the company:

> Our project management approach, by design, is not linked to a specific systems development methodology. Because we believe that it is better to use a (standard) systems development methodology than to decide which one to use, we have begun development of a

guideline systems development methodology specific for our organization. We have now developed prerequisites for project success. These include:

- A well-patterned methodology
- A clear set of objectives
- Well-understood expectations
- Thorough problem definition

During the late 1980s, merger mania hit the banking community. With the lowering of costs due to economies of scale and the resulting increased competitiveness, the banking community recognized the importance of using project management for mergers and acquisitions. The quicker the combined cultures became one, the less the impact on the corporation's bottom line.

The need for a good methodology became apparent, according to a spokesperson at one bank:

> The intent of this methodology is to make the process of managing projects more effective: from proposal to prioritization to approval through implementation. This methodology is not tailored to specific types or classifications of projects, such as system development efforts or hardware installations. Instead, it is a commonsense approach to assist in prioritizing and implementing successful efforts of any jurisdiction.

In 1996, the information services (IS) division of one bank formed an IS reengineering team to focus on developing and deploying processes and tools associated with project management and system development. The mission of the IS reengineering team was to improve performance of IS projects, resulting in increased productivity, and improved cycle time, quality, and satisfaction of the projects' customers.

According to a spokesperson at the bank, the process began as follows:

> Information from both current and previous methodologies used by the bank was reviewed, and the best practices of all these previous efforts were incorporated into this document. Regardless of the source, project methodology phases are somewhat standard fare. All projects follow the same steps, with the complexity, size, and type of project dictating to what extent the methodology must be followed. What this methodology emphasizes are project controls and the tie of deliverables and controls to accomplishing the goals.

To determine the weaknesses associated with past project management methodologies, the IS reengineering team conducted various focus groups. These focus groups concluded that there was:

- Lack of management commitment
- Lack of a feedback mechanism for project managers to determine the updates and revisions needed to the methodology
- Lack of adaptable methodologies for the organization
- Lack of training curriculum for project managers on the methodology
- Lack of focus on consistent and periodic communication on the methodology deployment progress
- Lack of focus on the project management tools and techniques

Based on this feedback, the IS reengineering team successfully developed and deployed a project management and system development methodology. Beginning June 1996 through December 1996, the target audience of 300 project managers became aware and applied a project management methodology and standard tool (MS Project).

The bank did an outstanding job of creating a methodology that reflects guidelines rather than policies and provides procedures that can easily be adapted on any project in the bank. Below the selected components of the project management methodology are discussed.

**Organizing**

With any project, you need to define what needs to be accomplished and decide how the project is going to achieve those objectives. Each project begins with an idea, vision, or business opportunity, a starting point that must be tied to the organization's business objectives. The project charter is the foundation of the project and forms the contract with the parties involved. It includes a statement of business needs, an agreement of what the project is committed to deliver, an identification of project dependencies, the roles and responsibilities of the team members involved, and the standards for how project budget and project management should be approached. The project charter defines the boundaries of the project.

**Planning**

Once the project boundaries are defined, sufficient information must be gathered to support the goals and objectives and to limit risk and minimize issues. This component of project management should generate sufficient information to clearly establish the deliverables that need to be completed, define the specific tasks that will ensure completion of these deliverables, and outline the proper level of resources. Each deliverable affects whether or not each phase of the project will meet its goals, budget, quality, and schedule. For simplicity's sake, some projects take a four-phase approach:

- *Proposal:* Project initiation and definition.
- *Planning:* Project planning and requirements definition.
- *Development:* Requirement development, testing, and training.
- *Implementation:* Rollout of developed requirements for daily operation.

Each phase contains review points to help ensure that project expectations and quality deliverables are achieved. It is important to identify the reviewers for the project as early as possible to ensure the proper balance of involvement from subject matter experts and management.

**Managing**

Throughout the project, management and control of the process must be maintained. This is the opportunity for the project manager and team to evaluate the project, assess project performance, and control the development of the deliverables. During the project, the following areas should be managed and controlled:

- Evaluate daily progress of project tasks and deliverables by measuring budget, quality, and cycle time.
- Adjust day-to-day project assignments and deliverables in reaction to immediate variances, issues, and problems.

- Proactively resolve project issues and changes to control scope creep.
- Aim for client satisfaction.
- Set up periodic and structured reviews of the deliverables.
- Establish a centralized project control file.

Two essential mechanisms for successfully managing projects are solid status-reporting procedures and issues and change management procedures. Status reporting is necessary for keeping the project on course and in good health. The status report should include the following:

- Major accomplishment to date
- Planned accomplishments for the next period
- Project progress summary:
  - Percent of effort hours consumed
  - Percent of budget costs consumed
  - Percent of project schedule consumed
- Project cost summary (budget versus actual)
- Project issues and concerns
- Impact to project quality
- Management action items

Issues-and-change management protects project momentum while providing flexibility. Project issues are matters that require decisions to be made by the project manager, project team, or management. Management of project issues needs to be defined and properly communicated to the project team to ensure the appropriate level of issue tracking and monitoring. This same principle relates to change management because inevitably the scope of a project will be subject to some type of change. Any change management on the project that impacts the cost, schedule, deliverables, and dependent projects is reported to management. Reporting of issue and change management should be summarized in the status report denoting the number of open and closed items of each. This assists management in evaluating the project health.

Simply having a project management methodology and using it does not lead to maturity and excellence in project management. There must exist a "need" for improving the system so it moves toward maturity. Project management systems can change as the organization changes. However, management must be committed to the change and have the vision to let project management systems evolve with the organization.

## 4.3 ENTERPRISE PROJECT MANAGEMENT METHODOLOGIES

Most companies today seem to recognize the need for one or more project management methodologies but either create the wrong methodologies or misuse the methodologies that have been created. Many times, companies rush into the development or purchasing of a methodology without any understanding of the need for one other than the fact that their competitors have a methodology. Jason Charvat states[2]:

> Using project management methodologies is a business strategy allowing companies to maximize the project's value to the organization. The methodologies must evolve and be "tweaked" to accommodate a company's changing focus or direction. It is almost a mind-set, a way that reshapes entire organizational processes: sales and marketing, product design, planning, deployment, recruitment, finance, and operations support. It presents a radical cultural shift for many organizations. As industries and companies change, so must their methodologies. If not, they're losing the point.

Methodologies are a set of forms, guidelines, templates, and checklists that can be applied to a specific project or situation. It may not be possible to create a single enterprise-wide methodology that can be applied to each and every project. Some companies have been successful doing this, but there are still many companies that successfully maintain more than one methodology. Unless the project manager is capable of tailoring the enterprise project management methodology to his or her needs, more than one methodology may be necessary.

There are several reasons why good intentions often go astray. At the executive levels, methodologies can fail if the executives have a poor understanding of what a methodology is and believe that a methodology is:

- A quick fix
- A silver bullet
- A temporary solution
- A cookbook approach for project success[3]

At the working levels, methodologies can also fail if they:

- Are abstract and high level
- Contain insufficient narratives to support these methodologies
- Are not functional or do not address crucial areas
- Ignore the industry standards and best practices
- Look impressive but lack real integration into the business
- Use nonstandard project conventions and terminology
- Compete for similar resources without addressing this problem
- Don't have any performance metrics
- Take too long to complete because of bureaucracy and administration[4]

Deciding on the type of methodology is not an easy task. There are many factors to consider, such as:

- The overall company strategy—how competitive are we as a company?
- The size of the project team and/or scope to be managed

2. J. Charvat, *Project Management Methodologies*, Hoboken, NJ: Wiley, 2003, p. 2.

3. Ibid., p. 4.

4. Ibid., p. 5.

- The priority of the project
- How critical the project is to the company
- How flexible the methodology and its components are[5]

Project management methodologies are created around the project management maturity level of the company and the corporate culture. If the company is reasonably mature in project management and has a culture that fosters cooperation, effective communications, teamwork, and trust, then a highly flexible methodology can be created based upon guidelines, forms, checklists, and templates. Project managers can pick and choose the parts of the methodology that are appropriate for a particular client. Organizations that do not possess either of these two characteristics rely heavily upon methodologies constructed with rigid policies and procedures, thus creating significant paperwork requirements with accompanying cost increases and removing the flexibility that the project manager needs for adapting the methodology to the needs of a specific client.

Jason Charvat describes these two types as light methodologies and heavy methodologies.[6]

**Light Methodologies**

Ever-increasing technological complexities, project delays, and changing client requirements brought about a small revolution in the world of development methodologies. A totally new breed of methodology—which is agile and adaptive and involves the client every part of the way—is starting to emerge. Many of the heavyweight methodologists were resistant to the introduction of these "lightweight" or "agile" methodologies (Fowler, 2001[7]). These methodologies use an informal communication style. Unlike heavyweight methodologies, lightweight projects have only a few rules, practices, and documents. Projects are designed and built on face-to-face discussions, meetings, and the flow of information to the clients. The immediate difference of using light methodologies is that they are much less documentation oriented, usually emphasizing a smaller amount of documentation for the project.

**Heavy Methodologies**

The traditional project management methodologies (i.e., SDLC approach) are considered bureaucratic or "predictive" in nature and have resulted in many unsuccessful projects. These heavy methodologies are becoming less popular. These methodologies are so laborious that the whole pace of design, development, and deployment slows down—and nothing gets done. Project managers tend to predict every milestone because they want to foresee every technical detail (i.e., software code or engineering detail). This leads managers to start demanding many types of specifications, plans, reports, checkpoints, and schedules. Heavy methodologies attempt to plan a large part of a project in great detail over a long span of time. This works well until things start changing, and the project managers inherently try to resist change.

Enterprise project management methodologies can enhance the project planning process as well as providing some degree of standardization and consistency. Companies

---

5. Ibid., p. 66.

6. Ibid., pp. 102–104.

7. M. Fowler, *The New Methodology, Thought Works*, 2001. Available: www.martinfowler.com/articles.

have come to the realization that enterprise project management methodologies work best if the methodology is based upon templates rather than rigid policies and procedures. The International Institute for Learning has created a Unified Project Management Methodology (UPMM™) with templates categorized according to the *PMBOK® Guide*, 4th edition, Areas of Knowledge[8]:

*Communication:*
Project Charter
Project Procedures Document
Project Change Requests Log
Project Status Report
PM Quality Assurance Report
Procurement Management Summary
Project Issues Log
Project Management Plan
Project Performance Report

*Cost:*
Project Schedule
Risk Response Plan and Register
Work Breakdown Structure (WBS)
Work Package
Cost Estimates Document
Project Budget
Project Budget Checklist

*Human Resources:*
Project Charter
Work Breakdown Structure (WBS)
Communications Management Plan
Project Organization Chart
Project Team Directory
Responsibility Assignment Matrix (RAM)
Project Management Plan
Project Procedures Document
Kick-Off Meeting Checklist
Project Team Performance Assessment
Project Manager Performance Assessment

*Integration:*
Project Procedures Overview
Project Proposal
Communications Management Plan

---

8. Unified Project Management Methodology (UPMM™) is registered, copyrighted and owned by International Institute for Learning, Inc., © 2009; reproduced by permission.

Procurement Plan
Project Budget
Project Procedures Document
Project Schedule
Responsibility Assignment Matrix (RAM)
Risk Response Plan and Register
Scope Statement
Work Breakdown Structure (WBS)
Project Management Plan
Project Change Requests Log
Project Issues Log
Project Management Plan Changes Log
Project Performance Report
Lessons Learned Document
Project Performance Feedback
Product Acceptance Document
Project Charter
Closing Process Assessment Checklist
Project Archives Report

*Procurement:*
Project Charter
Scope Statement
Work Breakdown Structure (WBS)
Procurement Plan
Procurement Planning Checklist
Procurement Statement of Work (SOW)
Request for Proposal Document Outline
Project Change Requests Log
Contract Formation Checklist
Procurement Management Summary

*Quality:*
Project Charter
Project Procedures Overview
Work Quality Plan
Project Management Plan
Work Breakdown Structure (WBS)
PM Quality Assurance Report
Lessons Learned Document
Project Performance Feedback
Project Team Performance Assessment
PM Process Improvement Document

*Risk:*
Procurement Plan
Project Charter

Project Procedures Document
Work Breakdown Structure (WBS)
Risk Response Plan and Register

*Scope:*
Project Scope Statement
Work Breakdown Structure (WBS)
Work Package
Project Charter

*Time:*
Activity Duration Estimating Worksheet
Cost Estimates Document
Risk Response Plan and Register Medium
Work Breakdown Structure (WBS)
Work Package
Project Schedule
Project Schedule Review Checklist

## 4.4 BENEFITS OF A STANDARD METHODOLOGY

For companies that understand the importance of a standard methodology, the benefits are numerous. These benefits can be classified as both short- and long-term benefits. Short-term benefits were described by one company as:

- Decreased cycle time and lower costs
- Realistic plans with greater possibilities of meeting time frames
- Better communications as to "what" is expected from groups and "when"
- Feedback: lessons learned

These short-term benefits focus on KPIs or, simply stated, the execution of project management. Long-term benefits seem to focus more upon critical success factors (CSFs) and customer satisfaction. Long-term benefits of development and execution of a world-class methodology include:

- Faster "time to market" through better scope control
- Lower overall program risk
- Better risk management, which leads to better decision-making
- Greater customer satisfaction and trust, which lead to increased business and expanded responsibilities for the tier 1 suppliers
- Emphasis on customer satisfaction and value-added rather than internal competition between functional groups
- Customer treating the contractor as a "partner" rather than as a commodity
- Contractor assisting the customer during strategic planning activities

Perhaps the largest benefit of a world-class methodology is the acceptance and recognition by your customers. If one of your critically important customers develops its own methodology, that customer could "force" you to accept it and use it in order to remain a supplier. But if you can show that your methodology is superior or equal to the customer's, your methodology will be accepted, and an atmosphere of trust will prevail.

One contractor recently found that its customer had so much faith in and respect for its methodology that the contractor was invited to participate in the customer's strategic planning activities. The contractor found itself treated as a partner rather than as a commodity or just another supplier. This resulted in sole-source procurement contracts for the contractor.

Developing a standard methodology that encompasses the majority of a company's projects and is accepted by the entire organization is a difficult undertaking. The hardest part might very well be making sure that the methodology supports both the corporate culture and the goals and objectives set forth by management. Methodologies that require changes to a corporate culture may not be well accepted by the organization. Nonsupportive cultures can destroy even seemingly good project management methodologies.

During the 1980s and 1990s, several consulting companies developed their own project management methodologies, most frequently for information systems projects, and then pressured their clients into purchasing the methodology rather than helping their clients develop a methodology more suited to the clients' needs. Although there may have been some successes, there appeared to be significantly more failures than successes. A hospital purchased a $130,000 project management methodology with the belief that this would be the solution to its information system needs. Unfortunately, senior management made the purchasing decision without consulting the workers who would be using the system. In the end, the package was never used.

Another company purchased a similar package, discovering too late that the package was inflexible and the organization, specifically the corporate culture, would need to change to use the project management methodology effectively. The vendor later admitted that the best results would occur if no changes were made to the methodology.

These types of methodologies are extremely rigid and based on policies and procedures. The ability to custom design the methodology to specific projects and cultures was nonexistent, and eventually these methodologies fell by the wayside—but after the vendors made significant profits. Good methodologies must be flexible.

## 4.5 CRITICAL COMPONENTS

It is almost impossible to become a world-class company with regard to project management without having a world-class methodology. Years ago, perhaps only a few companies really had world-class methodologies. Today, because of the need for survival and stiffening competition, there are numerous companies with good methodologies.

The characteristics of a world-class methodology include:

- Maximum of six life-cycle phases
- Life-cycle phases overlap

- End-of-phase gate reviews
- Integration with other processes
- Continuous improvement (i.e., hear the voice of the customer)
- Customer oriented (interface with customer's methodology)
- Companywide acceptance
- Use of templates (level 3 WBS)
- Critical path scheduling (level 3 WBS)
- Simplistic, standard bar chart reporting (standard software)
- Minimization of paperwork

Generally speaking, each life-cycle phase of a project management methodology requires paperwork, control points, and perhaps special administrative requirements. Having too few life-cycle phases is an invitation for disaster, and having too many life-cycle phases may drive up administrative and control costs. Most companies prefer a maximum of six life-cycle phases.

Historically, life-cycle phases were sequential in nature. However, because of the necessity for schedule compression, life-cycle phases today will overlap. The amount of overlap will be dependent upon the magnitude of the risks the project manager will take. The more the overlap, the greater the risk. Mistakes made during overlapping activities are usually more costly to correct than mistakes during sequential activities. Overlapping life-cycle phases requires excellent upfront planning.

End-of-phase gate reviews are critical for control purposes and verification of interim milestones. With overlapping life-cycle phases, there are still gate reviews at the end of each phase, but they are supported by intermediate reviews during the life-cycle phases.

World-class project management methodologies are integrated with other management processes such as change management, risk management, total quality management, and concurrent engineering. This produces a synergistic effect, which minimizes paperwork, minimizes the total number of resources committed to the project, and allows the organization to perform capacity planning to determine the maximum workload that the organization can endure.

World-class methodologies are continuously enhanced through KPI reviews, lessons-learned updates, benchmarking, and customer recommendations. The methodology itself could become the channel of communication between the customer and contractor. Effective methodologies foster customer trust and minimize customer interference in the project.

Project management methodologies must be easy for workers to use as well as covering most of the situations that can arise on a project. Perhaps the best way is to have the methodology placed in a manual that is user friendly.

Excellent methodologies try to make it easier to plan and schedule projects. This is accomplished by using templates for the top three levels of the WBS. Simply stated, using WBS level 3 templates, standardized reporting with standardized terminology exists. The differences between projects will appear at the lower levels (i.e., levels 4–6) of the WBS. This also leads to a minimization of paperwork.

Today, companies seem to be promoting the use of the project charter concept as a component of a methodology, but not all companies create the project charter at the same

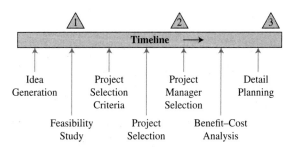

**Figure 4–5.** When to prepare the charter.

point in the project life cycle, as shown in Figure 4–5. The three triangles in Figure 4–5 show possible locations where the charter can be prepared:

- In the first triangle, the charter is prepared immediately after the feasibility study is completed. At this point, the charter contains the results of the feasibility study as well as documentation of any assumptions and constraints that were considered. The charter is then revisited and updated once this project is selected.
- In the second triangle, which seems to be the preferred method, the charter is prepared after the project is selected and the project manager has been assigned. The charter includes the authority granted to the project manager, but for this project only.
- In the third method, the charter is prepared after detail planning is completed. The charter contains the detailed plan. Management will not sign the charter until after detail planning is approved by senior management. Then, and only then, does the company officially sanction the project. Once management signs the charter, the charter becomes a legal agreement between the project manager and all involved line managers as to what deliverables will be met and when.

## 4.6 SAP[9]

**Project Quality Gates— Structured Approach to Ensure Project success**
Project quality is paramount in delivery of SAP projects; this fact is reflected in structured approach to quality management for SAP solution delivery—Quality Built In. The staple of the quality approach in SAP projects is the execution of formal project quality gates. The quality gates are defined in the ASAP project delivery methodology that SAP and its customers and partners use for project planning, management and delivery. Each project type has predetermined number of quality gates executed at key milestones of the project as shown in Figure 4–6.

SAP believes that the quality gates are essential for success of any project regardless of deployment strategy—like traditional or agile. The Q-Gates are integrated not only into our delivery methodology, but they are also coded into our delivery policies and internal

9. ©2013 by SAP. All rights reserved. Reproduced by permission. Material in this section has been provided by Jan Musil, Global Lead of Project Management Practice, SAP Field Services, SAP America, Inc.

**Figure 4–6.** Project quality gates are defined in the Project Management Plan and are set at critical stages in the project lifecycle.

systems. The results of each quality gate are recorded in the corporate project management information system and they are regularly reviewed and reported on. Dedicated quality team in SAP has accountability for management of the quality gates, review of project health and follow-up with project manager, stakeholders and leadership.

The project quality gates in the ASAP methodology provide clear guidance to project managers, stakeholders and project teams on how to structure and perform the quality gate review. During each quality gate the QA manager assesses completeness and quality of each deliverable produced in the project according to predefined quality gate checklist that includes not only deliverable name, but also detailed acceptance criteria. Each deliverable in the Q-Gate checklist is marked as either mandatory or optional for completion of the quality gate. Upon completion of the Q-Gate the QA manager assesses pass/fail score for the Q-Gate and proposes follow-up plan to take corrective actions addressing the deficiencies identified in this process.

The formal quality built in process has been shown to have positive impact on customer satisfaction, improved overall project portfolio health and positively impacted revenue.

**ASAP Methodology – Structured, Repeatable, Prescriptive Way to Deliver SAP Projects and Innovate Project Delivery**

SAP delivery projects follow structured, repeatable and prescriptive methodology for implementation ASAP. The ASAP Methodology for Implementation is SAP's content rich methodology for assisting with the implementation and/or upgrade of SAP solutions across industries and customer environments. Built on experience from thousands of SAP projects, ASAP provides content, tools, and best practices that help consultants to deliver consistent and successful results across industries and customer environments.

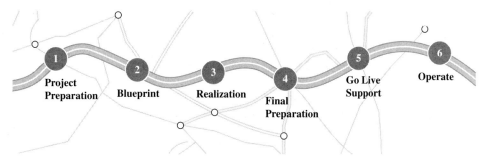

**Figure 4–7.** The phases of ASAP.

| Project Preparation | Blueprint | Realization | Final Preparation | Go-Live Support | Run |
|---|---|---|---|---|---|
| Conduct initial planning and preparation; define project goals, scope, and objectives; identify and train team members | Map value drivers to implementation scope; refine business requirements; document the to-be processes; define functional solution design; identify added functional, technical requirements; get business sign-off on requirements/design | Build and test a complete test and business environment; develop training material and user manuals; obtain business approval | Prepare system for production release; migrate/cut over data; execute training; and prepare the organization internal/external) for Go-Live | Execute new systems and processes; monitor business process results, and monitor the production environment; build Center of Excellence for support and enhancements | Run the new system; apply additional SAP operations standards to optimize system operation |

**Ensure the delivery of value throughout the solution life cycle ("Value Delivery")**

**Figure 4–8.** The ASAP methodology elements.

The six phases of ASAP provide support throughout the life cycle of SAP solutions. Underlying these phases is a series of value delivery checks to make sure that the solution, as implemented, delivers the expected value. Figure 4–7 illustrates the phases of ASAP.

The methodology covers key aspects of SAP implementation from Project Management guidance structured around PMI *PMBOK® Guide* through business process design, business value management, application lifecycle management, organizational change management, technical solution management, data management and other topics important for delivery of SAP solutions.

The ASAP methodology is not pure Project Management methodology, but instead it combines all key elements the project team needs to cover in order to deliver successful projects. This is shown below in Figure 4–8.

The ASAP methodology is designed in a way that enables flexibility and scalability from smaller projects like single consulting services delivery to more complex delivery of global deployments in multinational corporations. This flexible design allows us to use the methodology as a foundation for creation of all consulting services. Each engineered service leverages the common WBS of the ASAP methodology to define clearly the work that is performed, roles and skills that are required to deliver service and also detail about sourcing of the roles from within the organization.

This approach helps us achieve commonality between the services in areas that are not core expertise of service owners (like project management), it lowers the cost of service creation and simplifies the adoption process.

Thanks to the use of common taxonomy based on ASAP in creation of services the SAP projects can be assembled from individual engineered services and delivered in "assembled-to-order" approach rather than designed from scratch. SAP has been recognized by Technology Services Industry Association (TSIA) in 2012 for its innovative approach in services delivery with the SAP Advanced Delivery Management approach that is built on principles of common modular services that can be assembled and reused in different projects. See Figure 4–9.

With this approach SAP customers take advantage of prebuild services and content and lower the cost of deployment [and] complexity of projects and minimize the risk of deployment projects. The engineered services and rapid deployment solutions are used in early stages of the project to establish a baseline solution that is later enhanced in series of iterative incremental builds using the agile techniques. This innovative approach to project delivery has significantly changed the delivery of projects and requires our project managers to adopt their skills to this innovative way of project execution. One example is that project managers needed to learn how to structure and run projects with the iterative agile techniques to design, configure or develop customer-specific extensions, which is substantially different from traditional management of projects where solution is built from scratch.

Figure 4–9.

The common methodology and its taxonomy is not only great enabler for project delivery, but it also helped SAP innovate the way solutions are delivered and deployed.

## 4.7 CASSIDIAN: INTEGRATED MULTILEVEL SCHEDULES

Perhaps the single most important benefit of a good methodology is the ability to create integrated multilevel schedules for all stakeholders. The remainder of this section has been provided by Cassidian: ©2013 by Cassidian. Reproduced by permission. All rights reserved.

**Why Integrated Multilevel Schedules?**

The Integrated Multilevel Scheduling practice is aimed at providing each level of management of the project, from the Customer and/or the Company management, down to the Project Management, and down to the Work Package management, with consistent schedule baselines, consistent progress measurement and consistent estimates to completion.

This practice is widely used for large and complex projects.

### Definition of Integrated Multilevel Schedules

Project Integrated Multilevel Scheduling is aimed divided into three levels of management:

- Master Schedule (Level 1): Customer and/or the Company management,
- Project Summary Schedule (Level 2): Project management, and
- Detailed Schedule (Level 3): Work Package management

By default, the different levels of schedules are self-sufficient, reflecting the delegation of management responsibility in the project:

- Master Schedule is owned by the Project Manager
- Project Summary Schedule is owned by the Project Manager (for large project, the project manager may be supported by the Project Management Office)
- Detailed Schedules are owned by Work Package Managers

In addition, to deliver and maintain a multilevel project schedule, links shall be performed between the different levels to provide an Integrated Multilevel Project Schedule.

### Prerequisites to Prepare and to Deliver Integrated Multilevel Schedules

Before developing Integrated Multilevel Schedules, the following deliverables shall be completed and performed:

*Define the project scope through*
- Collect the requirements,
- Define the Product Breakdown Structure with all internal and external deliverables,
- Develop the Work Breakdown Structure—think **DELIVERABLES**.

- Specify Work Packages with a special focus on inputs needed and outputs expected including acceptance criteria of the deliverables, and
- Plan a review of all the Work Packages descriptions with all main stakeholders of the project, the objective is to share and control the consistency between inputs and outputs of the different Work Packages.
- Estimate target duration, work, cost and skills for each Work Package,
- Manage risks and opportunities to define and plan mitigation actions and to identify where buffers shall be added in the plans.

### Principles of Integrated Multilevel Schedules

The following principles shall be followed to ensure a success Integrated Multilevel Schedules:

- The *Master Schedule* (also called Level 1 Plan) shall deliver a synthetic view (one page) on the project schedule reflecting the *major milestones* (contractual milestones, major customer furnished Items or equipments (CFI / CFE) with critical impact on contractual deliverables), the *dependencies* between major milestones and summaries of the main phases, the *contractual dates* (contract commitments) and the current contract status (major milestones passed and current progress with forecasted dates).
- The *Project Summary Schedule* (also called Level 2 Plan) shall cover the complete WBS of the project and shall be organised according to the WBS structure. It shall provide the dependencies between all the WP of the project and the dependencies between the WP deliverables and the major milestones of the project.
- The *Detailed Schedule* (also called Level 3 Plan) shall deliver the detailed schedule of each work package, which shall be broken-down in activities, each activity shall conduct to a deliverable and shall be broken down in elementary tasks with resource assigned.

In addition, between the different levels, Multilevel Schedule Links shall be implemented (See Figure 4–10):

- Between the Level 1 Plan and the Level 2 Plan, links should be defined to follow the adherence to the contractual milestones dates (customer commitments) and other key milestones dates. Thereof, all major milestones (MM) reported in the Level 1 Plan shall be linked to corresponding MM in the Level 2 Plan,
- Between the Level 2 Plan and the Level 3 Plan, links should be defined to follow the adherence to the project dates (project target dates), where each WP, the input(s) and output(s) defined in the Level 2 Plan should be linked to corresponding input(s) and output(s) of the Level 3 Plan.
- Two types of links should be used:
- Hard links (also called driving links) for inputs in order to align automatically the date of availability of the corresponding input in the lower level
- Soft links for outputs to inform on the forecasted date in the lower level of the output without impacting automatically the date defined in the upper level

### Monitor and Control of Integrated Multilevel Schedules

The monitoring and controlling of the different levels of schedule, including the consistency, checking should be performed within one month according to the following scheme:

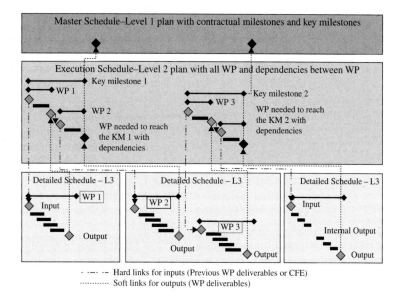

Figure 4–10. Example of Integrated Multilevel Schedule

1. Update of Level 3 schedules by each Work Package Manager based on current actual and new forecast according to ongoing Work Package progress.
2. Multi level consistency checking by the Project Management Office including the deviations and the identification of the impact over Key milestones.
3. Schedule Review among Project Management Office, Engineering Authority and Work Package Managers in order to identify actions plan to recover target dates

Validation by the Project Manager with the update of Level 3 and Level 2 schedules based on the decisions taken and update performance baseline if needed.

## 4.8 TECNICAS REUNIDAS[10]

**Open Book Estimate (OBE) as a Successful Contract Alternative to Execute Projects in the Oil & Gas Sector**

**Introduction**

As a result of the projected rate of energy demand growth, the oil and gas industry has a wide range of challenges and opportunities across different areas. Due to that the sector is proceeding since several years ago for the development of new facilities which in many cases are megaprojects.

10. ©2013 Tecnicas Reunidas. Reproduced by permission. All rights reserved. Material has been provided by Felipe Revenga López. Felipe Revenga López is the Chief Operations Officer of Tecnicas Reunidas since September 2008. He joined TR in 2002 as Project Director and then as Project Sponsor of a group of international strategic projects. He has large experience in EPC-LSTK and OBE-LSTK projects in the Oil & Gas Production Units, Refining, Petrochemical and Power Sector worldwide.
He is an Industrial Engineer (Specialty Chemicals) by the School of Industrial Engineers (ETSIIM) and currently finishing the Doctoral Program Engineering of Chemical and Biochemical Processes at the Polytechnic University of Madrid (ETSIIM).

The typical complete life cycle of a capital project in the Oil & Gas sector is focused in the overall stages that are showed in Figure 4–11. Understanding and managing these stages is crucial on the long-term success of the project.

LSTK and Cost Plus contracting are both very prevalent types of contracts within projects in the Oil & Gas industry. Depending upon the level of risk the Client of a project is willing to accept, budget constraints, and the Client's organization core competencies, will determine which method is best for their project.

A large amount of projects in this sector are performed under EPC-LSTK contracts; TECNICAS REUNIDAS major experience is based mainly in this type of projects that in general implies to manage the whole project and carry out the detailed engineering (in some cases is included the basic engineering or FEED in the scope of work), procure all the equipment and materials required, and then construct, precommission and start-up to deliver a functioning facility ready to operate. LSTK contracting tends to be riskiest, and all risks are assumed by EPC Contractor.

The Open Book Estimate (OBE) or Open Book Cost Estimate (OBCE) is an alternative to execute EPC Projects. With this type of contract the final purpose of the work is to define the total price of the Project in collaboration with the Client; the global costs of the Project are established in a transparent manner (Open Book).

### The Open Book Estimate (OBE)

The main purpose of this methodology is to build up an accurate EPC Price by applying some parameters previously agreed (between Client and the Contractor), the base cost through an OBE, the development of an extended front-end engineering and in some cases the placement of purchase orders for selected long lead and critical items to ensure the overall schedule of the project. The OBE will fix the project base cost and will become the basis for determining the lump sum EPC price for the project.

During an OBE Phase, Contractor develops a FEED and/or part of the detailed engineering under a reimbursable basis or lump sum price or alternatives, including a complete and open cost estimate of the plant. After an agreed period (usually between 6–12 months mainly dependent of accuracy grade, schedule and an others factors required by client) of engineering development and a Client and Contractor agreement on the base cost, the contract is changed or converted to an EPC LSTK Contract applying previously agreed multiplying factors.

**Cost Estimate Methodology**      **Principal Cost Elements and Pricing Categories:**

The OBE usually is based on sufficient engineering development in accordance with the deliverables identified in OBE contract. These deliverables have the greatest degree of development that is possible under the normal progress of the project. Required deliverables are prepared and submitted to client prior completion of conversion phase.

The principal cost elements, as seen in Figure 4–12, that will comprise the OBE are addressed below. The OBE cost estimate shall include the total Scope of Work:

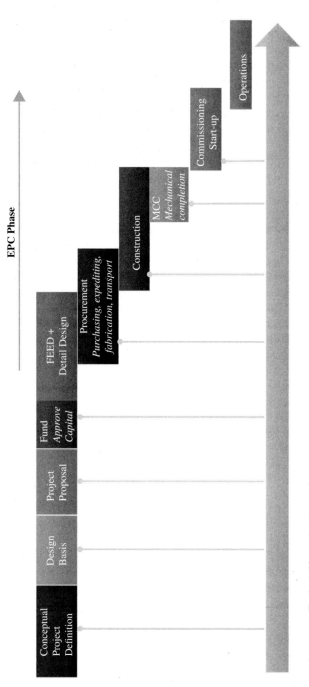

**Figure 4–11.** Typical life-cycle phases.

1. Detailed engineering, procurement and construction services
2. Supply of equipment, bulk materials and spare parts
3. Transport to construction site
4. Customs clearance
5. Construction and erection at site
6. Provision of subcontractors temporary construction facilities and services
7. Construction and precommissioning services
8. Commissioning and start-up services
9. Training services and vendor's assistance
10. Bonds, insurances, hedging charges
11. Other costs including third-party inspection and contractor insurance
12. Other . . .

### Cost Base

● An OBE procedure is developed during contract stage and implemented during project's OBE phase. All details on how to prepare an OBE have to be agreed and included as an annex in the Contract.
● Preagreement of allowances, growth, conditioning, technical design allowances, surplus and cut and waste.
● MTOs based on PDS, measured in P&ID and plot plans and estimates. All details on procedures to be agreed before OBE contract signature.

In executing projects under convertible basis, TR develops the OBE in parallel with the normal project execution ensuring that both activities can flow smoothly without interferences. During the OBE phase, in particular cases and if so agreed with client, TR can advance the procurement of main equipment and initiate negotiations for construction

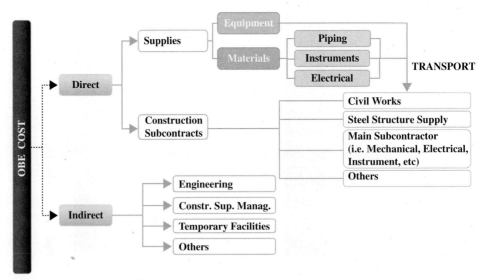

**Figure 4–12.** Typical cost elements.

subcontractors. The execution of these activities in advance facilitates the fulfillment of project schedule requirements.

This OBE phase of the project is jointly developed between client and TR. The OBE is fully transparent to client and the conversion to LSTK is easily agreed once the risk/reward element is fixed.

In Figure 4–13, we see the main steps and activities that are developed to achieve OBE goals and to convert to next phase of the project:

During OBE phase, cost saving ideas are developed in order to adjust the final cost estimation, for that, a special team of engineering specialist is appointed to work with both the Engineering Manager and the Estimating Manager, with the purpose to detect those areas where potential savings can be achieved by optimizing the design, without jeopardising Safety, Quality or Schedule. Any of these changes that could lead to cost saving are carefully evaluated from a technical point of view and, if the feasibility of the potential change is proven, the alternative solution, together with the cost saving impact evaluation will be forwarded to the CLIENTS for consideration of approval.

The EPC contract price is the result of the base cost multiplied by fixed percentages for fee and mark-ups related with equipment, bulk materials, construction and ancillary cost that are agreed between Client and Contractor. This price during the conversion phase is converted to lump sum price and thereafter remains fixed during the EPC-LSTK phase.

## Contracts

The typical models of contracts under this OBE alternative are:

- One Contract, two parts, OBE and EPC. Price for EPC part to be included at conversion
- Two Contracts, one OBE and other EPC: Both may be signed at the beginning or one at the beginning and the other at conversion.

- Obtain quotations for all equipment (for procurement)

- Preliminary Data Sheets When Necessary
- Develop Specific Drawings Valid only in the OBE Phase When Required
- Develop Cost Saving Ideas (Value Engineering)

- All Civil Works and Steel Structures Predesigned (Construction Driven)
- Complete Piping and Electrical and Instrument MTOs
- Main and Critical Equipment Designed
- Enquires Issued for 90% of Supplies
- Final Negotiations with Construction Subcontractors
- Conversion to EPC in a Established Month

**OBE COMPLETED CONVERSION PHASE**

OBE Cost Estimation

**Steps to Achieve OBE Target**

**Figure 4–13.** Steps to achieve OBE target.

The methodology of open book estimate is included in the contract.
In case of no conversion:

I. Contractual relationship disappears and both client and contractor break their commitment. Client may break if not interested. Consequences:
   - Six months new LSTK offer. 2-3 months plus evaluating the offers
   - Repeat FEED with different contractor
II. Contract provides mechanisms in the case of no agreement.
   - Continue contract on a service base (better LS Contract)
   - Agreement on partial conversion
   - Others as per contract agreement

### Advantages

An OBE + LSTK could optimize all project execution especially in cost and schedule.

- In terms of cost, Client and Contractor could together determine the project cost through an open book cost estimate because an estimation methodology, conversion conditions like multiplying factors are agreed, etc. also, transparency and mutual agreement and trust is share between clients and contractors. This model results in an accurate cost because unnecessary contingencies are avoided.
- On the other hand, this model results in schedule advantage because the bidding period is shortened or replaced by a conversion negotiation phase; an EPC stage is shortened because all the works developed during the extended feed and conversion stage. A representation of the schedule advantage is shown in Figure 4–14:

**Figure 4–14.** Typical schedule advantage.

In summary, the advantages in cost and schedule are:

| Cost | Schedule Reduction |
|---|---|
| • Develops and EPC estimate during 6 to 12 months. This provides a much better accuracy on the costs. <br> • Accurate prices based in real offers + an agreed conversion factor assures a fair juice to Client and Contractor. <br> • Time enough to develop the project and to avoid unnecessary contingencies. <br> • Application of cost saving to match project cost to Clients budget. <br> • Facilitates the possibility of funding, because a more accurate estimation. <br> • Risk are reduced and better controlled for common benefits both Client and Contractor | • **Short bidding period,** as cost estimate does not need to be as detailed <br> • **Shorten of the overall schedule** of the project as time for Extended FEED and EPC bidding is shortened dramatically. <br> • **Contract award procedure** much easier and shorter <br> • Some **long lead Items and critical equipment could be awarded or negotiated** |

## Close-out

Open Book Estimate (OBE) has been demonstrated to be a successful contract alternative for executing projects in the Oil & Gas Sector because it aligns Client and Contractor with the project's goals. Both are motivated to pursue the best cost estimation or project target cost and at the same time the schedule is optimized.

As was mentioned in the introduction, in the Oil & Gas Sector most of the current projects can be catalogued as megaprojects, where there are many risk associated also in a hot market with a high workload from suppliers, contractors, subcontractors, etc. Through an OBE alternative, the clients can better manage its risks through a more cooperative and agreed approach, where the risks are reduced during an accurate estimation, and then embraced rather than totally transferred to Contractors, and in this way project outcomes could be improved.

Técnicas Reunidas has developed and converted successfully 100% from OBE to LSTK-EPC projects.

## Acronyms

- **Client:** means the owner of the Oil & Gas Company
- **Contractor:** affiliated company responsible for performing the engineering, procurement and construction services
- **EPC:** Engineering, Procurement & Construction. Type of contract typical of industrial plant construction sector, comprising the provision of engineering services, procurement of materials and construction.
- **FEED:** Front-End Engineering Design means Basic Engineering which is conducted after completion of Conceptual Design or Feasibility Study. At this stage, before start of E.P.C (Engineering, Procurement and Construction), various studies take place to figure out technical issues and estimate rough investment cost.
- **LS Contract:** Lump Sum Contract implies that the Contractor agrees to do an specific project for a fixed price

- **LSTK Contract:** Lump Sum Contract + all systems are delivered to the client ready for operations.
- **MTOS:** Material take-offs of: Piping, Electrical, Instrumentation
- **OBE:** The Open Book Estimate (OBE) or Open Book Cost Estimate (OBCE)
- **PDS:** Software used for designing industrial plants through a multidisciplinary engineering activity
- **P&ID:** Process & Instrument Diagrams
- **TR:** Tecnicas Reunidas

## 4.9 TERADYNE: FROM MYTH TO REALITY

Sooner or later, all companies recognize the need for a project management methodology, perhaps even an enterprise-wide project management methodology (EPM). While recognition is easy, the actual development and ultimately implementation can be difficult if the company bases its decisions on myths rather than reality. In Table 4–1 are some of the characteristics of well-developed methodologies, at least in the author's opinion. The second column illustrates the myth, whereas the third column identifies what mature companies seem to do.

**TABLE 4–1.   DESIGN CHARACTERISTICS OF A GOOD METHODOLOGY**

| Characteristic | Myth | Reality |
| --- | --- | --- |
| Description of the Methodology | Excessive details are needed, perhaps hundreds of pages with illustrations, graphs, and tables | Description is a high-level framework, short and easy to read and understand; perhaps 25 pages or less |
| Readability | Excessive details and complex illustrations | Narrative descriptions supported mainly by flowcharts |
| Applicability | Need multiple methodologies based upon the types of programs, projects, deliverables, and functional areas | One high-level methodology can be used for both projects and programs |
| Alignment | Methodology should be aligned to the deliverables of the project or programs | Methodology should be aligned to the business and applicable to both projects and programs |
| Adaptability | Methodology must be based upon rigid policies and procedures | Methodology should have flexibility so that it can be adapted to all programs and projects and also be readily adaptable to necessary changes |
| *PMBOK® Guide* mapping | Methodologies have unique characteristics and are based upon rigid policies and procedures | The basic structure of the methodology should be mapped against the *PMBOK® Guide* |
| Role delineation | The methodology should clearly delineate the roles of all players; project managers, line managers, executives, and customers | Because of the need for flexibility and adaptability, only the role of the project manager needs to be delineated, specifically integration responsibility |
| Lessons learned | Methodology usage ends when the deliverables are met | Methodologies should have a final step to capture best practices and lessons learned |

A capital equipment producer for the electronics industry has developed a methodology that follows most of these reality characteristics. For example, the opening paragraph in the Teradyne's methodology states[11]:

> One purpose of this document is to present a high level description of the Program/ Project Manager's function within the product development cycle. A Program Manager is responsible for multiple related product aligned projects culminating in a major product deliverable where a Project Manager will focus on a single major product level deliverable. The purpose of the Program/Project Manager is to: "Partner with design leads in directing and integrating cross-functional project teams to ensure stakeholders objectives are met and deliver products in accordance with the approved project plan". . . . The value proposition for program management could be stated: A Program/Project Manager (PM) provides the organized, industry-aligned disciplined framework to integrate functional engineering and operational groups in driving product-based projects or programs to completion on-time and on-budget, while meeting market demands/needs/requirements. Another purpose of this document will be to present a high level description of the work instruction framework that Program Management uses in managing product development projects/programs. . . . Given the unique integrative role of the PM, this document will also attempt to link tie points with functional groups such as: marketing, engineering and operations.

The document also goes on to describe the role of the project and program managers, as well as the fact that the methodology applies to both programs and projects. In addition, the methodology also identifies the fact that the entire methodology may not be required on certain projects. This fact alone provides considerable flexibility in its use and illustrates the company's faith in the ability of their project/program managers.

> This document will pertain to all Programs and Projects managed by the Division Program Management Office (PMO). A Project Manager will focus on a single major product level deliverable where a Program Manager is responsible for multiple related product aligned projects culminating in a major product deliverable. Hereafter in this document the title Project Manager and Program Manager will be referenced using the acronym, PM, with the meaning taken from the context of the passage. *There will be variation in the level of effort that project or program managers will expend for any one project or program depending on multiple factors such as: project/program budget size, technical or schedule complexity.*
>
> The PMO has established the criteria to determine when a Project should be considered a Program. On the other end of the spectrum, it has also established criteria to define which project should use the "PM Lite" process, which does not require use of the PM tools & techniques identified in this document. Both definitions can be found in the "PM Lite" process document available on the PMO web page. This document will assume a situation where there is a need for a PM and the full range of standardized project tools/ techniques available would be in use.

---

11. The remaining quotes in this section were provided by Dr. Steve Lyons, Applications Project Manager at Teradyne, a capital equipment manufacturer in the electronics industry and adapted from their methodology for managing projects and programs. The material is reproduced with permission of Teradyne.

As mentioned previously, good methodologies focus on flowcharting that is easy to understand. Documents with complex charts and exhibits are not only difficult to follow but also make it difficult for the employees to follow and use. Complex or poorly constructed flowcharts can severely limit the project manager's flexibility. An excellent example of an easy-to-follow flowchart can be shown from the statement below taken from the Teradyne's methodology:

Figure 4–15 shows the alignment between product development and program management and the fact that heaviest PM involvement is in Path II, Phases 1 through 4. The company's functional groups and the program managers operate in a matrix, composite organization (*PMBOK® Guide* 2004, Figure 2.12). The program managers are responsible for the scope, schedule, and cost of a project but do not directly manage human resources. The program manager(s) generally work closely with a design lead(s) (the "two in a box" concept) in controlling the project's "triple-constraints" of scope, schedule, and cost. Note that path II, phase 1 can be divided into optional A and B segments. This would typically be recommended for large platform projects. Phase 2 is also divided into segments A & B on all projects.

The two circles in Figure 4–15 represent Dr. Deming's cycle of continuous improvement. This shows that quality is embedded in the process flow and the company's commitment to quality and the total quality management processes.

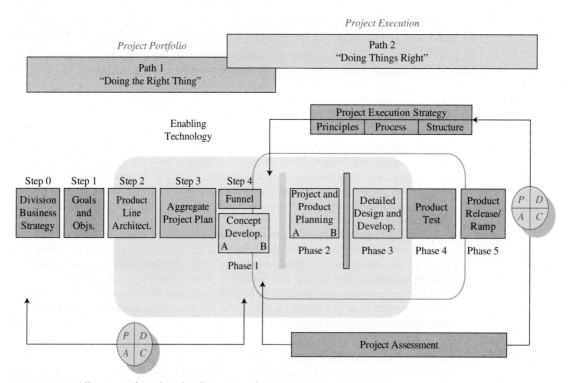

**Figure 4–15.** Alignment of product development and program management.

**TABLE 4–2.   MAPPING OF PHASE DELIVERABLES TO *PMBOK GUIDE* PROCESSES**

| Description | *PMBOK® Guide* Process Groups | *PMBOK® Guide* Process Knowledge Areas |
|---|---|---|
| Concept development | Initiation<br>Planning | Integration<br>Scope<br>Cost<br>Human resources |
| Project and product planning | Planning<br>Monitoring and control | Integration<br>Scope<br>Time<br>Cost<br>Quality<br>Human resources<br>Communications<br>Risk<br>Procurement |
| Detail design and development | Execute<br>Monitor and control | Integration<br>Scope<br>Time<br>Cost<br>Quality<br>Human resources<br>Communications<br>Risk |
| Product test and verification | Execute<br>Monitor and control | Integration<br>Scope<br>Time<br>Cost<br>Quality<br>Human resources<br>Communications<br>Risk |
| Product release and ramp | Closing | Integration<br>Procurement |

Another characteristic of many companies is the alignment of their methodology to the *PMBOK® Guide* processes. According to Teradyne's methodology:

> Table 4–2 shows the mapping of the phase deliverables to *PMBOK® Guide* processes. Phase 2 (planning) may require the use of all 9 process knowledge areas and is generally considered to be the phase requiring the largest time commitment from the PM because of the amount of actual "pick and shovel" work. For example, during Phase 2 the PM creates the detailed schedule, budget, resource plan, and risk management plans. Phases 3 and 4 require continued PM effort because in Phase 2 the work is planned but in Phases 3 and 4 you work the plan. In reality no project ever goes according to plan so the PM role of monitoring, adjusting, and re-planning (within the scope, schedule, and cost constraints) is challenging. PM involvement with Phase 5 is minimal at this time although end of project assessments do prove invaluable in a capturing the "lessons learned."

Finally, methodologies undergo changes and enhancements. Just like with configuration management and scope change control, the revisions are tracked for traceability. This is shown below in Table 4–3 which has been adapted from Teradyne's methodology.

**TABLE 4–3.   HISTORY OF REVISIONS**

| Date | Page(s) | Section | Comments | Rev. |
|------|---------|---------|----------|------|
| 3/15/2009 | 6–8 | 2.13 | Expansion of Roles and Responsibilities | 1.0 |
|  |  |  |  |  |
|  |  |  |  |  |
|  |  |  |  |  |
|  |  |  |  |  |
|  |  |  |  |  |
|  |  |  |  |  |
|  |  |  |  |  |
|  |  |  |  |  |

## 4.10  SLALOM CONSULTING: PROJECT MANAGEMENT FUNCTIONS

Over the years, companies have come to accept the *PMBOK® Guide* as the gospel of project management. But we must remember that the *PMBOK® Guide* is still just a guide and not necessarily the actual body of knowledge. Creating a perfect body of knowledge would have to be based upon the size, nature, and complexity of a firm's projects; the type of industry in which they compete; and the percentage of work within the firm that must utilize project management. Some project management educators argue that there are three ways to manage projects: the right way, the wrong way, and the *PMBOK® Guide* way. Although companies create project management methodologies based upon the *PMBOK® Guide*, the methodologies rarely follow it exactly. A large percentage of the areas of knowledge may be used, but the focus could be more heavily aligned to the domain areas rather than the areas of knowledge. Processes may be more important than knowledge areas. As an example, consider the following comments from Carl Manello, PMP, practice director—delivery effectiveness, Slalom Consulting:

> For all that project managers are responsible for, depending on their role (e.g., small or large projects, business focused programs, corporate mergers/acquisitions), there are several models for representing what each needs to know to become an expert in the practice of project management. One model, the *PMBOK® Guide*, breaks this tacit knowledge into nine basic knowledge areas. A model that I developed has further dissected the knowledge areas into twelve project management functions (PM*f*). There are scores of other models crowding the internet. Whether nine or twelve areas, the models help us to define what we as project managers do.
>
> In my experience, all project managers should know about and be experienced in the project management functions defined in my model. The basic PM*f* (see Figure 4–16) represents the functions as I have defined them for a project/program manager. The PM*f* are robust and change as the project management focus changes (i.e., project managers' functions are different from those of an Enterprise Program Office).[12] If we leverage the PM*f*

---

12. For more information on the PM*f*, readers are referred to C. Manello's forthcoming book, *Value Project Management*.

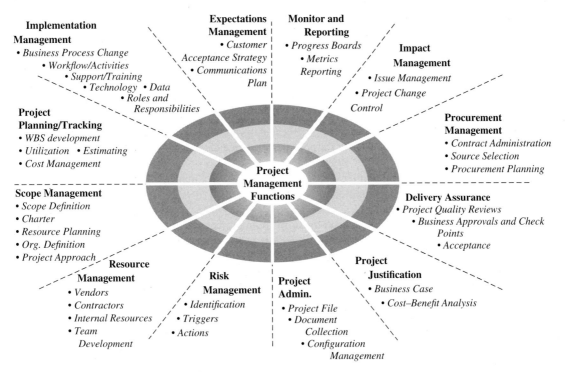

**Figure 4–16.** Project management functions.

model, these functions can be established as *standard practices* for project managers, but may have to be customized or adapted to a specific project or client. In my dealings with corporations—both large and small—I specifically avoid the term *best practice*, which has always been puzzling to me. If each company determines their own "best practices," then the term itself is diluted and loses its meaning. As an industry, project management cannot have hundreds of *best* practices. While it is certainly possible that the practices adopted within one enterprise may be *best*, it has been my experience that these are more often the "preferred practices" of the implementing organization or key leader at the time of implementation.

If we release ourselves from a focus on preferred, best or super-stupendous as a qualifier, the project management functional areas can be seen as key practice areas that project managers should be able to rally around. Each company must choose which functional areas are most important to them, define the right level or maturity to strive for, put plans in place to achieve mastery, reassess and continue on the never-ending path of evolving capabilities. Each organization will implement project management to the best of their ability and whether or not that is a true level of "best"—for which there is no globally defined standard—it will be the level most appropriate for that time and place.

Unfortunately, in my discussions with corporations, there appears to be an equality in the mind of many senior managers that the implementation of standard practices can only be realized with the implementation of project management tools (e.g., Clarity, PlanView, MS/Project). It has been my observation that companies that try to short-cut the establishment of process first and instead chase a tool to solve their project management challenges typically do not succeed with the implementation of any good practices, at least in a reasonable period of time. While I'm certain that this is not the case for all organizations—otherwise the tool vendors would be out of business—it should be clear that process cannot be ignored or substituted with a tool.

I have worked with a variety of industries. Within each industry I have seen companies that have spent millions of dollars on the acquisition, customization, implementation, education, and administration of project management tools. Each has had less than adequate results. A common reason for poor performance can be attributed to:

- Believing that the implementation of project management software is the same as the implementation of project management

There is certainly a need for software in project management. But the critical question is as follows: Should we select a project management software tool and then design our project management methodology around that tool, or should we design our methodology first and then select the appropriate tool to fit our methodology? The latter is certainly the better choice. Software is simply a tool. Projects are managed by people, not tools. People manage tools; tools do not manage people. A proper understanding of software and its capability is essential at the onset of project management implementation but should not be viewed as a replacement for good processes. Carl Manello states:

> As a consultant, I have had the opportunity [to] work with many companies and to observe multiple examples of tool implementations to support project management: both successes and failures. The life-cycle phases of the successful project management tool implementation projects are usually the same, but I do not think one can easily apply a repeatable time frame to a chronology to this effort. For example, the national implementation of a well-known software package as the tool of record for the financial replacement program for a major airline manufacturer—spanning several years and including over 800 full and part time resources—was not the same as the implementation within the engineering division of a major cell phone manufacturer. Each implementation is different. What is extensible, however, is the process by which an organization approaches the attainment of standard practices. One should begin by assessing where the organization is at in its capabilities (leveraging the PM*f*), determine where the definition, construction and implementation of process will support growth, select supporting toolsets, analyze, customize, and implement. It is crucial that *process* precede the *tools*.

## 4.11 SLALOM CONSULTING: REPLACING METHODOLOGIES WITH FRAMEWORKS

While project management methodologies seem to be in favor today, there is a growing trend toward replacing methodologies with framework models. Methodologies do have disadvantages and companies that see one and only one methodology, namely their own, may have a difficult time recognizing the pitfalls. But management consultants that have had the exposure to a variety of methodologies in various industries often have a much better understanding of the limitations. Carl Manello, PMP, practice director—delivery effectiveness, Slalom Consulting, states that:

> In the past twenty years I have seen a fair share of detailed methodologies. There are of course great success stories around methodologies, especially when hordes of highly trained consultants are brought in to run enormous enterprise-wide initiatives. However, for the small garden variety project or those instances when training an entire project team on how to use the methodology is unrealistic, methodologies can cause as many problems as they solve.
>
> While working at a large manufacturing company in the late 80's I had the privilege of working with a team of folks from the Technology Center and the Enterprise Program Office (of which I was a member). We leveraged a vendor framework, adopted it to our needs and put it in place for both IT and R&D projects. In my first stint as a Solution Lead for Program and Project Management with a different consultancy, I had the opportunity to take the approach of *frameworks* to other clients. Now, at Slalom, I am continuing to advocate frameworks over detailed methodologies. Slalom Consulting does not have a proprietary PM methodology. Instead we leverage frameworks.
>
> The difference is that methodologies tend to be defined as a series of inflexible steps that must be completed, usually in sequential order. "Frameworks" on the other hand, outline steps that may be followed. The difference is that the steps of a framework may be implemented differently per project, may be passed over, or may be implemented at varying degrees of detail. At one of my prior employers, I created a PM life-cycle framework and gave it an internal marketing tag of "Flexibly Applying a Rigor Process." The marketing phrase was developed not only to help sell the concept of a new process but also to help illustrate the key difference between a framework and a methodological tome. The framework's process was designed and built with a significant amount of detail (e.g., project steps, deliverables, metrics, approval processes). However the implementation rigor varied depending on the type of project, size, criticality, etc. Large projects had to adhere to the PM life-cycle with exacting detail, while small projects had the flexibility to omit specific portions of the approach. Not all projects are created equally.
>
> At one of my former clients, there is an exhaustive methodology defined and in place, which employs scores of templates, numerous processes and multiple governance teams that each own different pieces of the process. The challenge of this implementation of *methodology* is its inflexibility. Project managers spend too much time trying to navigate the process and are distracted or prevented from focusing on their PM activities. They spend more time doing project management process related work than adding value to their projects as project managers. This in turn forces the client to hire more people to do "project management" activities (having some project managers to project manage and some to do methodology compliance), thereby increasing the time spent on project management. This extra time spent cannot be shown to have a direct impact on the project or the financial benefits, and in fact erodes whatever initial benefits the programs claimed to be able to realize. Similarly, there is insufficient training, coaching, documentation or other support available to help project managers navigate their way through the rigorous process. Not only do project managers spend an inordinate amount of time complying with what they understand, they must also chase around the organization looking for the right governance office to provide them help on what to do. If the current enterprise-wide system replacement program should not do well, the project management methodology itself (which should have been enabling success) will certainly be one of the key elements in the downfall.

## 4.12 LIFE-CYCLE PHASES

Determining the best number of life-cycle phases can be difficult when developing a project management methodology. As an example, let's consider IT. During the 1980s, with the explosion in software, many IT consulting companies came on the scene with the development of IT methodologies using systems development life-cycle (SDLC) phases. The consultants promise their client phenomenal results if the client purchases the package along with the accompanying training and consulting efforts. Then, after spending hundreds of thousands of dollars, the client reads the fine print that states that the methodology must be used as is, and no customization of the methodology will take place. In other words, the client must change their company to fit the methodology rather than vice versa. Most of the IT consultancies that adopted this approach no longer exist.

For an individual company, agreeing on the number of life-cycle phases may be difficult at first. But when an agreement is finally reached, all employees should live by the same phases. However, for today's IT consulting companies, the concept of one-package-fits-all will not work. Whatever methodology they create must have flexibility in it so that client customization is possible. In doing so, it may be better to focus on processes rather than phases, or possibly a framework approach that combines the best features of each. Carl Manello, PMP, practice director, delivery effectiveness, Slalom Consulting, states that:

> While a project management life-cycle parallels an SDLC, they are separate and distinct (especially because many projects have nothing to do with IT or "systems"). With some special client-specific exceptions, the standard phases are:
>
> 1. Project Conceptualization—where the project begins to take form (usually an informal unstructured process)
> 2. Project Initiation—an approved initiative begins its journey and begins to complete the rigor of the life-cycle
> 3. Analysis & Design—begin to decompose project scope into the meat of what is to be accomplished
> 4. Development/Test /Implement—standard SDLC phases, but tailored to meet non-IT projects as needed
> 5. Close Down/Turn Over—ending the "project" and initiation of the "ongoing maintenance" (as needed)
> 6. Benefits Realization Tracking—the woefully neglected phase where we prove the realization of benefits claimed back in conceptualization
>
> During one of my corporate lives, I helped define a project management life cycle that built upon these six core phases (see Figure 4–17). As a framework, phases should be malleable to the needs of the situation. Our Project Management Office was part of the IT Governance function, which was primarily the financial arm for IT. The equivalent of the IT CFO wanted to ensure we allowed sufficiently detailed processes to enable the activation (or administrative start-up) of a project. Activation represented setting up the right time tracking buckets, financial buckets and establishing service level agreements. Similarly, we added the Quality Assurance and Testing phase because there was a newly formed organization that focused on these functions and IT wanted to ensure their prominence in the minds of project teams. Lastly, to help erase the years of short-comings at implementation, I created a separate phase for implementation. This late phase detailed the roles, responsibilities,

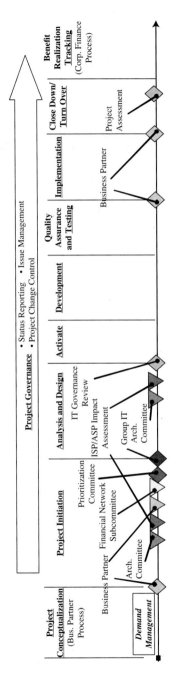

**Figure 4-17.** Extended PM framework.

deliverables and associated check-points that would enable us to manage a project past the "go live" date. As discussed above, each initiative flowed through this framework in a "flexible" way, leveraging the parts that made sense given the nature of the project.

## 4.13 EXPANDING LIFE-CYCLE PHASES[13]

Historically, we defined the first phase of a project as the Initiation Phase. This phase included bringing the project manager on board, handing him/her a budget and a schedule, and telling them to being project execution. Today, there is a preinitiation phase which Russ Archibald and his colleagues refer to as the Project Incubation/Feasibility Phase. In this phase, we look at the benefits of the project, the value expected at completion, whether sufficient and qualified resources are available, and the relative importance of the project compared to other projects that may be in the queue. It is possible that the project may never reach the Initiation Phase.

In the past, project management was expected to commence at the Initiation Phase because it was in this phase that the project manager was assigned. Today, project managers are expected to possess a much greater understanding of the business as a whole and companies have found it as beneficial to bring the project manager on board earlier than the Initiation Phase to assist in making business decisions rather than purely project decisions.

In the same context, we traditionally viewed the last life cycle phase as Project Closure. This includes the implementation of contractual closure, administrative closure and financial closure. After closure, the project manager would be reassigned to another project.

Today, we are including a Post-Project Evaluation Phase. Some companies refer to this as a Customer Satisfaction Management Phase. In this phase, selected members of the project team and sales/marketing personnel, as well as members from the governance committee, meet with the client to see what changes can be made to the methodology or processes used to execute the project, and what can be done differently on future projects for this client to improve even further the working relationship between client-contractor-stakeholders.

## 4.14 CHURCHILL DOWNS, INCORPORATED

Churchill Downs, Incorporated has created a project management methodology that clearly reflects its organization. According to Chuck Millhollan, director of program management:

> While we based our methodology on professional standards, we developed a graphic (and used terminology) understood by our industry to help with understanding and acceptance. For example, we have a structured investment request, approval and prioritization process. (See Figure 4–18). We used the analogy of bringing the thoroughbred into the paddock prior to race and then into the starting gate. The project, or race, is not run until the thoroughbred has entered the starting gate (approved business case and project prioritization).

---

13. Russell D. Archibald, Ivano Di Filippo, and Daniele Di Filippo have written an excellent paper on this topic, "The Six-Phase Comprehensive Project Life Cycle Model Including the Project Incubation/Feasibility Phase and the Post-Project Evaluation Phase." The paper is published in the *PM World Journal*, December, 2012.

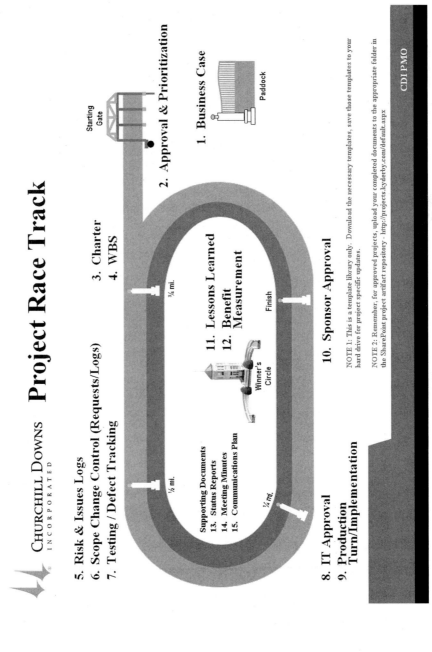

**Figure 4–18.** The Churchill Downs, Incorporated methodology.

## 4.15 INDRA: THE NEED FOR A METHODOLOGY

As mentioned previously in Chapter 3, the quest for excellence in project management is almost always accompanied by the development of a project management methodology. Such was the case at Indra. Indra defines excellence in project management as follows: "Excellence in project management is achieved by being able of repeatedly reaching the project targets, creating business opportunities, and improving the management process itself when managing the assigned projects." Enrique Sevilla Molina, PMP, formerly corporate PMO director, discusses the journey to excellence:

> A project management methodology was formally defined in the mid 90's based upon the experience gained in our major international contracts. The main problems we faced were related to the definition of the scope, the limits of the methodology, and the adoption of the correct strategy to spread this knowledge throughout the company. To solve these issues, our management chose to hire an external consulting company to act as a dynamic factor that boosted and drove the cultural change.
>
> Yes, the process was carefully sponsored from the beginning by senior executives and closely followed-up until its complete deployment in all areas of the company.
>
> The major milestones of the process have roughly been:

- Project management strategy decision................................................................mid 90's
- Methodology definition and documentation..........................................mid-late 90's
- Tools definition and preparation........................................................................late 90's
- Training process start...........................................................................................2000
- Risk management at department level................................................................2002
- PMP certification training start...........................................................................2004
- Risk management process defined at corporate level..........................................2007
- Program and portfolio management processes definition start.............................2008

> A PM methodology was developed in the early 90's and formalized during that decade. It has been eventually updated to cope with the company and the industry evolution. It is being used as a framework to develop and maintain the PMIS, and to train the PMs throughout the company.
>
> It is based on the project lifecycle and structured in the following two stages and six phases as shown in Figure 4–19:

> **Precontractual stage**
> Phase 1. Initiation
> Phase 2. Concept development, creation of offers and proposals
> Phase 3. Offer negotiation
>
> **Contractual stage**
> Phase 4. Project planning
> Phase 5. Execution, Monitoring and Control
> Phase 6. Closure

> The precontractual and contractual stages are both part of the project and its lifecycle. Most problems that appear during a project's lifespan originate during its definition and in the negotiation of its objectives, contents and scope with the customers. A proper management of the precontractual stage is the best way to prevent problems later on.

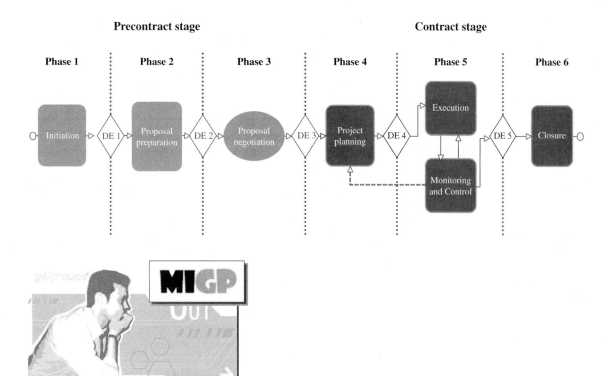

**Figure 4–19.** Project management life cycle.

At the end of each phase there is a specific result that will allow a key decision to be made, focusing and directing the actions of the next phase and thereby reducing the initial risks and uncertainties of the project.

The decision on the stages and phases was a decision mainly based upon the needs of our standard cycle of a project conception and development, and based on the most significant types of projects we were involved with.

Risk management processes are integrated into the methodology and into the corporate PM tools. An initial risk identification process is performed during the proposal phase, followed by a full risk management plan during the planning phase of the contract stage, and the subsequent monitoring processes during the execution phase of the project. QA and change control processes are considered main support processes in the methodology.

## 4.16 IMPLEMENTING THE METHODOLOGY

The physical existence of a methodology does not convert itself into a world-class methodology. Methodologies are nothing more than pieces of paper. What converts a standard methodology into a world-class methodology is the culture of the organization and the way the methodology is implemented.

The existence of a world-class methodology does not by itself constitute excellence in project management. The corporate-wide acceptance and use of it do lead to excellence. It is through excellence in execution that an average methodology becomes a world-class methodology.

One company developed an outstanding methodology for project management. About one-third of the company used the methodology and recognized its true long-term benefits. The other two-thirds of the company would not support the methodology. The president eventually restructured the organization and mandated the use of the methodology.

The importance of execution cannot be overestimated. One characteristic of companies with world-class project management methodologies is that they have world-class managers throughout their organization.

Rapid development of a world-class methodology mandates an executive champion, not merely an executive sponsor. Executive sponsors are predominantly on an as-needed basis. Executive champions, on the other hand, are hands-on executives who drive the development and implementation of the methodology from the top down. Most companies recognize the need for the executive champion. However, many companies fail to recognize that the executive champion position is a life-long experience. One Detroit company reassigned its executive champion after a few successes were realized using the methodology. As a result, no one was promoting continuous improvement to the methodology.

Good project management methodologies allow you to manage your customers and their expectations. If customers believe in your methodology, then they usually understand it when you tell them that no further scope changes are possible once you enter a specific life-cycle phase. One automotive subcontractor carried the concept of trust to its extreme. The contractor invited the customers to attend the contractor's end-of-phase review meetings. This fostered extreme trust between the customer and the contractor. However, the customer was asked to leave during the last 15 minutes of the end-of-phase review meetings when project finances were being discussed.

Project management methodologies are an "organic" process, which implies that they are subject to changes and improvements. Typical areas of methodology improvement might include:

- Improved interfacing with suppliers
- Improved interfacing with customers
- Better explanation of subprocesses
- Clearer definition of milestones
- Clearer role delineation of senior management
- Recognition of need for additional templates
- Recognition of need for additional metrics
- Template development for steering committee involvement
- Enhancement of the project management guidebook
- Ways to educate customers on how the methodology works
- Ways of shortening baseline review meetings

## 4.17 IMPLEMENTATION BLUNDERS

Even though companies recognize the driving forces which indicate a need for project management improvement, the actual decision to make an investment to do it may not happen until some crisis occurs or a significant amount of red ink appears on the company's balance sheet. Recognizing a need is a lot easier than doing something about it because doing it requires time and money. Too often, executives procrastinate giving the go-ahead in hopes that a miracle will occur and project management improvements will not be necessary. And while they procrastinate, the situation often deteriorates further. Consider the following comments from Carl Manello, PMP, practice director delivery effectiveness, Slalom Consulting:

> I find that the greatest motivation for my clients to invest in improvements is how they view the impact of project management on their initiatives. When their track record at driving large scale business initiatives is less than stellar (lacking sufficient process, methods, tools or skills), they begin to understand the need to invest. Unrealized project implementations, blown budgets and poor quality all speak loudly and capture the attention of senior executive management. The challenge is instead to arrest executive attention before millions of dollars are squandered.
>
> At first, many corporations are unlikely to want to invest in improving PM infrastructure like the PM*f*. "There are *real* projects with hard-core benefits to be realized instead." However, after these same organizations begin to struggle, understand their weaknesses and the need for improvement in basic project management, they begin to focus on those improvements.

Delayed investment in project management capabilities is just one of many blunders. Another common blunder, which can occur in even the best companies, is the failure to treat project management as a profession. In some companies, project management is a part-time activity to be accomplished in addition to one's primary role. The career path opportunities come from the primary role, not through project management. In other companies, project management may be regarded merely as a specialized skill in the use of scheduling tools. Carl Manello continues:

> While the PMI has done a super job, especially in the last 10 years, advocating project management as a specialized skill that should be left to the professionals, I find that many companies still believe project management is a skill, not a profession. Whether in marketing or engineering organizations, someone is often randomly assigned to be the project manager, regardless of their training, demonstrated skill level or capabilities as a project manager. This lack of attention to project management as a profession may be one of the contributing factors to projects around the world which continue to perform poorly. Too many projects do not have qualified experienced project managers at the helm.

## 4.18 OVERCOMING DEVELOPMENT AND IMPLEMENTATION BARRIERS

Making the decision that the company needs a project management methodology is a lot easier than actually implementing it. There are several barriers and problems that surface well after the design and implementation team begins their quest. Typical problem areas include:

- Should we develop our own methodology or benchmark best practices from other companies and try to use their methodology in our company?
- Can we get the entire organization to agree upon a singular methodology for all types of projects or must we have multiple methodologies?
- If we develop multiple methodologies, how easy or difficult will it be for continuous improvement efforts to take place?
- How should we handle a situation where only part of the company sees a benefit in using this methodology and the rest of the company wants to do its own thing?
- How do we convince the employees that project management is a strategic competency and the project management methodology is a process to support this strategic competency?
- For multinational companies, how do we get all worldwide organizations to use the same methodology? Must it be intranet based?

These are typical questions that plague companies during the methodology development process. These challenges can be overcome, and with great success, as illustrated by the companies identified in the following sections.

## 4.19 PROJECT MANAGEMENT TOOLS

Project management methodologies require software support systems. As little as five years ago, many of the companies described in this book had virtually no or limited project management capabilities. How did these companies implement project management so fast? The answer came with the explosion of personal computer-based software for project planning, estimating, scheduling, and control. These were critical for methodology development.

Until the late 1980s, the project management tools in use were software packages designed for project scheduling only. The most prominent were:

- Program evaluation and review technique (PERT)
- Arrow diagramming method (ADM)
- Precedence diagramming method (PDM)

These three networking and scheduling techniques provided project managers with computer capabilities that far surpassed the bar charts and milestone charts that had been in use. The three software programs proved invaluable at the time:

- They formed the basis for all planning and prediction and provided management with the ability to plan for the best possible use of resources to achieve a given goal within schedule and budget constraints.
- They provided visibility and enabled management to control one-of-a-kind programs.

- They helped management handle the uncertainties involved in programs by answering such questions as how time delays influence project completion, where slack exists among elements, and which elements are crucial to meeting the completion date. This feature gave managers a means for evaluating alternatives.
- They provided a basis for obtaining the necessary facts for decision making.
- They utilized a so-called time network analysis as the basic method of determining manpower, material, and capital requirements as well as providing a means for checking progress.
- They provided the basic structure for reporting information.

Unfortunately, scheduling techniques cannot replace planning. And scheduling techniques are only as good as the quality of the information that goes into the plan. Criticisms of the three scheduling techniques in the 1980s included the following:

- Time, labor, and intensive effort were required to use them.
- The ability of upper-level management to contribute to decision making may have been reduced.
- Functional ownership of the estimates was reduced.
- Historical data for estimating time and cost were lost.
- The assumption of uninvited resources was inappropriate.
- The amount of detail required made full use of the scheduling tools inappropriate.

Advancements in the memory capabilities of mainframe computer systems during the 1990s eventually made it possible to overcome many of the deficiencies in the three scheduling techniques being used in project management in the 1970s and 1980s. There emerged an abundance of mainframe software that combined scheduling techniques with both planning and estimating capabilities. Estimating then could include historical databases, which were stored in the mainframe memory files. Computer programs also proved useful in resource allocation. The lessons learned from previous projects could also be stored in historical files. This improved future planning as well as estimating processes.

The drawback was that mainframe project management software packages were very expensive and user unfriendly. The mainframe packages were deemed more appropriate for large projects in aerospace, defense, and large construction. For small and medium-sized companies, the benefits did not warrant the investment.

The effective use of project management software of any kind requires that project teams and managers first understand the principles of project management. All too often, an organization purchases a mainframe package without training its employees in how to use it in the context of project management.

For example, in 1986, a large, nationally recognized hospital purchased a $130,000 mainframe software package. The employees in the hospital's information systems department were told to use the package for planning and reporting the status of all projects. Less than 10 percent of the organization's employees were given any training in project management. Training people in the use of software without first training them in project management principles proved disastrous. The morale of the organization hit an all-time low point, and eventually no one even used the expensive software.

Generally speaking, mainframe software packages are more difficult to implement and use than smaller personal computer–based packages. The reason? Mainframe packages require that everyone use the same package, often in the same way. A postmortem study conducted at the hospital identified the following common difficulties during the implementation of its mainframe package:

- Upper-level managers sometimes did not like the reality of the results.
- Upper-level managers did not use the packages for planning, budgeting, and decision making.
- Day-to-day project planners sometimes did not use the packages for their own projects.
- Some upper-level managers sometimes did not demonstrate support and commitment to training.
- Clear, concise reports were lacking.
- Mainframe packages did not always provide for immediate turnaround of information.
- The hospital had no project management standards in place prior to the implementation of the new software.
- Implementation highlighted the inexperience of some middle managers in project planning and in applying organizational skills.
- Neither the business environment nor the organization's structure supported the hospital project management/planning needs.
- Sufficient/extensive resources (staff, equipment, etc.) were required.
- The business entity did not determine the extent of and appropriate use of the systems within the organization.
- Some employees viewed the system as a substitute for the extensive interpersonal skills required of the project manager.
- Software implementation did not succeed because the hospital's employees did not have sufficient training in project management principles.

Today, project managers have a large array of personal computer–based software available for planning, scheduling, and controlling projects. Packages such as Microsoft Project have almost the same capabilities as mainframe packages. Microsoft Project can import data from other programs for planning and estimating and then facilitate the difficult tasks of tracking and controlling multiple projects.

The simplicity of personal computer–based packages and their user friendliness have been especially valuable in small and medium-sized companies. The packages are so affordable that even the smallest of companies can master project management and adopt a goal of reaching project management excellence.

Clearly, even the most sophisticated software package can never be a substitute for competent project leadership. By themselves, such packages cannot identify or correct task-related problems. But they can be terrific tools for the project manager to use in tracking the many interrelated variables and tasks that come into play in contemporary project management. Specific examples of such capabilities include the following:

- Project data summary; expenditure, timing, and activity data
- Project management and business graphics capabilities

- Data management and reporting capabilities
- Critical-path analyses
- Customized as well as standardized reporting formats
- Multiproject tracking
- Subnetworking
- Impact analysis
- Early-warning systems
- Online analyses of recovering alternatives
- Graphical presentations of cost, time, and activity data
- Resource planning and analyses
- Cost and variance analyses
- Multiple calendars
- Resource leveling

Figure 4–20 shows that right now approximately 95 percent of the project management software focus on planning, scheduling, and controlling projects. In the future, we can expect more software to be created for the initiation of a project and the closure of a project.

Perhaps the biggest problems today with methodologies are that companies are not taking advantage of the full capabilities of the tools they are purchasing. A perfect example of this is with the earned-value measurement (EVM) system. Companies appear reluctant to use EVM probably for fear of the financial reality that the EVM numbers will show.

The intent behind EVM is to make sure that the person leading the project is functioning as a project manager rather than a project monitor. Project monitors simply record numbers and report them to a higher authority for decision making. The project manager, on the other hand, measures the variances from the baselines, develops contingency plans, gets the plans approved, implements the plans, and measures the new variances to see if the improvements worked. Without full use of an EVM system, it may be difficult to fully manage a project.

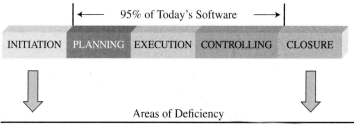

**Figure 4–20.** Project management software.

Enrique Sevilla Molina, PMP, formerly corporate PMO director at Indra, states:

> Earned valued techniques are used during project execution. Earned value measurement is performed regularly by the project manager making use of the corporate tool.
>
> Variances are analyzed monthly at a work package level and also at the project level.
>
> If corrective action is necessary, it is initiated by the project manager who analyses a set of project indicators (risk status, overall cost, schedule and margin deviations, project cash flow, . . .), and it is followed by the expert judgment by the area controller who also considers the analysis of the information provided by the project team. If it is the case, a specific Executive report can be requested from the PM for providing more detailed information on deviations by work package and the status of the program. Once the overall picture of the project is clear, a decision is taken on the need for corrective actions.

Perhaps EVM is the best approach for determining progress, status, and eventually forecasts of where we will end up. The critical component of EVM is earned value, which is the amount of work that has physically been completed and measured in either hours or dollars. All too often, companies accept EVM as a way of life without understanding the complexities in using EVM. Memorizing the 12–15 equations necessary to use EVM is easy. The hard part is capturing the data that go into the equations. Carl Manello, PMP, practice director, delivery effectiveness, Slalom Consulting, states that:

> Earned Value (EV) is not for the beginner. Conceptually a straightforward measure of work done/remaining to be done, the effort needed to collect the necessary data and to make the calculations is no simple undertaking. Many of my clients have done such a poor job getting a handle on their time capture, and don't have sufficient enough project plans, that capturing information at required levels for EV is almost impossible. After working with numerous Fortune 250 companies, I have found only one that tracked SPI/CPI.[14] The overhead required to collect data on that project, perform quality assurance and maintain the information accurately in a project management tool required significant overhead (the program employed more than a dozen people on the program whose sole accountability was data entry, data accuracy and reporting through the project tool).
>
> For the less mature projects, I have often used a variation of earned value. This method is based on the completion of deliverables. Using simple spreadsheets, I am able to track deliverables (assigned different values for draft or final), milestones and key activities. The spreadsheet generates graphs that represent how much work has been delivered, the cost associated with the scheduled time spent, and the earned value actuals compared to the plan (see Figures 4–21 through 4–23). Each deliverable is set against a planned schedule for completion and credit is only realized with any delivered value when it is complete (i.e., something 90% complete gets a big fat zero in this model. There are many

---

14. SPI is the schedule performance index and CPI is the cost performance index. SPI and CPI identify the trends or direction in which cost and schedule are heading. During executive briefings on project status, it is often the case that SPI and CPI are the first two numbers discussed.

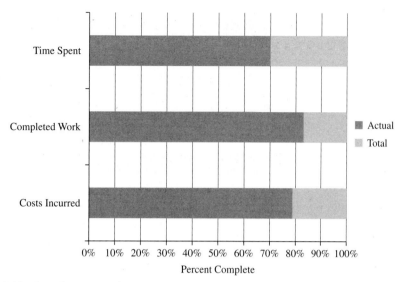

**Figure 4–21.** Actual versus total costs.

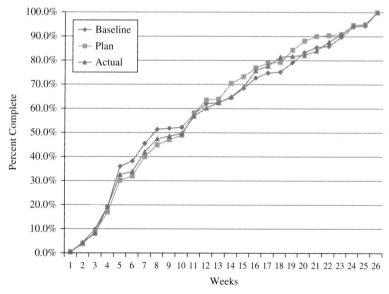

**Figure 4–22.** Earned-value actuals compared to the plan.

circumstances where this over-simplistic approach is not appropriate and a partially completed deliverable *is* of value. In this implementation, however, the choice is all or none). While this tool also requires some overhead to create the initial list based on the WBS, track completion and update status to reflect progress, the overhead is in the order of hours instead of multiple full-time equivalents.

| Status | | | |
|---|---|---|---|
| Deliverables | Schedule | Budget | Overall |
| 10% | 12% | 9% | |
| O | O | O | O |
| Key: | O On Target | ◎ At Risk | ● In Danger |

| Key Indicators | | | |
|---|---|---|---|
| Scope | Delivery | Cost | Staffing |
| ⇧ | ⇔ | ⇧ | ⇔ |
| Key: | ⇔ Holding Steady | ⇧ Increasing | ⇩ Decreasing |

**Figure 4–23.** Status and indicator directions.

## 4.20 WÄRTSILÄ: RECOGNIZING THE NEED FOR SUPPORTING TOOLS[15]

Although we have always had a strong passion for engines at Wärtsilä, we are now much more than an engine company. Today, professional project management has become essential for our continuing success due to the bigger and more complex marine and power plant projects we deliver.

**Excellent Project Management—A Prerequisite to Customer Satisfaction**

The Wärtsilä Project Management Office (WPMO) was established in 2007 to develop a project management culture, processes, competences and tools that would guarantee our customers receive the satisfaction they deserve.

One of the first things we did was to conduct a detailed project management analysis in order to identify improvement areas. At that time, we didn't have any software available for project and portfolio management. Therefore one of the first actions the WPMO took was to initiate a global improvement program called Gateway" to develop and implement a set of project and project portfolio management processes with a supporting application.

According to the Program Owner Antti Kämi, the starting point for Gateway and the reason why Wärtsilä needed to improve project management even further was as follows: "Professional project management was seen as truly essential for our profitability, competitiveness and for providing value to our Customers."

---

15. Material has been provided by Wärtsilä Project Management Office (WPMO). Copyright to Wärtsilä Corporation 2013. Reproduced by permission.

## Projects Divided into Three Categories

To achieve as many of the expected benefits as possible, it was decided that relevant parts of the unified processes of the new tool should be used in all divisions and in all three project categories in the company:

- Customer delivery projects
- Operational development projects
- Product and solution development projects

Using this new approach meant that thousands of projects could be managed with the new tool, involving approximately 2,000 people in project management.

## Good Project Management Practices

Today we have unified business processes (gate models) in use throughout Wärtsilä with harmonised guidelines and terminology. Additionally, we maintain this resource through a professional training and certification path for project management.

As with all projects of this magnitude, there have been challenges to face on the way, especially when developing both the way-of-working and the software in parallel. The varying project management maturity levels within the company have also proven to be challenging. On the upside, a continuous and active dialogue around project management is now in place, experiences are openly exchanged between divisions and project categories, and the work gives a true feeling of "One Wärtsilä."

In several project management areas we can already see improvements and benefits, especially in Portfolio Management and Resource Management.

Currently we use the new application as a project database for our Research & Development portfolio planning. This enables the projects to be arranged in portfolios, which means that there is a more structured follow-up process. This in turn leads to better transparency and visibility in projects, easier and quicker responses to stakeholder enquiries and more efficient project reporting on the whole.

First-class resource management is important today since Information Management resources are used in operational development projects throughout the company. Having a shared software tool ensures good resource availability, transparency for managing and monitoring the workload as well as reliable facts for good planning.

Further benefits with a common project and portfolio management tool include the possibility to record and utilize lessons learned, the ability to collaborate and to have information easily available for all project team members.

## Tools Really Make a Difference in Project Management

In a nut shell, this is what Gateway at Wärtsilä is all about: to work out and apply a more effective way to plan and run projects and a common tool to help us gather, handle and share project-related information. And by doing this, ensure that both internal and external customers are satisfied.

## 4.21 TECH MAHINDRA LIMITED: PROJECT
## PROCESS MONITORING

In the previous section we discussed some of the tools that are used with project management methodologies. EVM and scheduling tools are common to most methodologies. Some companies design tools that are unique to their organization and then capitalize on their expected benefits. Such is the situation with Tech Mahindra Limited. The remainder of the information in this section has been provided by Krishna Gali, Quality Group, Tech Mahindra Limited, and Anu Khendry, Tech Mahindra Limited Learning World, Tech Mahindra Limited.

Tech Mahindra Limited is a software services company that conducts its business through software projects. These projects need to comply with critical product and process parameters as defined by the customers. Project process monitoring (PPM) is an assurance activity that enables project managers, their teams, and senior management to understand these critical parameters and the related compliance levels of the project. PPM is a mandatory activity for all fixed-bid and high-impact/critical projects, which are defined based on certain criteria of size, effort, and customer focus. (See Figures 4–24 and 4–25, which contain two parts of PPM.)

*The objectives of PPM are to:*
- Provide early alerts on potential project problems with suggestions for corrective and preventive measures
- Identify and accelerate process improvements in projects
- Identify best practices and share them across the company

*The unique features of the PPM process include:*
- Assessment on 12 key project focus areas with predefined parameters
- Quantitative rating of the project on a scale of 100
- Automation and integration with other project management tools at Tech Mahindra Limited.
- Process assessment aligned to CMMI and International Organization for Standardization (ISO) 9001

*The PPM process includes:*
- Self-assessment by the project manager on all focus areas and parameters that are applicable to the project.
- Review and validation by quality assurance on project compliance. This is done by assessing project artifacts against applicable product/process requirements. At the end of this step the project has a compliance rating on a scale of 100 and a red/amber/green status based on this rating.
- Reports of project ratings to senior management.
- The deficiencies observed during PPM are followed up in the subsequent month.

The PPM process has enabled identification of right project issues at the right time for immediate resolution. PPM reports depict real-time data on project performance. Process

**Figure 4–24.** Project process monitoring.

**Context diagram on PPMs…**

**Figure 4–25.** Project process monitoring.

automation has also helped in conducting aging analysis on project issues and their tracking toward closure.

The PPM reviews highlight the impact of noncompliance on the three basic project parameters: quality, cost, and time. The success of this process is evident and it has become a key input supporting all business performance reviews at the corporate level with quantitative project issue reporting, month after month.

**Customer One**                    In the life cycle of a software project in Tech Mahindra Limited, software work products are created, reviewed, and tested in multiple phases, such as unit testing, integration testing, and system testing. The focus during these phases is on the detection of defects and fulfillment of customer requirements.

The final phase of testing prior to delivery is user acceptance testing (UAT). During this phase, project teams review, inspect, and test the final deliverable from the customer perspective. Any oversight at this time can lead to defect leakage to the customer, resulting in rework and customer dissatisfaction. This not only impacts efficient and effective project delivery but also impacts the overall relationship with the customer. The Customer One (C1) process in Tech Mahindra Limited ensures that the final deliverables are as per the expectations of the customer and will lead to customer delight. (See Figure 4–26.)

*The objectives of the C1 process are:*
- External validation of the final project deliverables from the perspective of the customer
- Go/no-go decision on the release of project deliverables to the customer

**Figure 4–26.**   C1: assessment components driving contractual commitments.

The focus of C1 assessment is to validate that the project deliverables not only meet the customer's explicit specifications and requirements but also conform to the implied requirements and intended business value. C1 thus helps to prevent any unpleasant surprises for the customer and subsequent escalations on the quality of delivery.

Projects are selected for C1 assessment based on project criticality to the customer, pricing model, and project type. The assessment is conducted by a team of experts with relevant competencies/skills. The C1 assessment team is supported with detailed guidelines on multiple focus areas, ranging from requirements, technology/tools and technical architecture, delivery methodology, testing, and predeployment activities.

The C1 assessments for any projects are broadly categorized into two components:

1. Assessment of solution effectiveness against SOW/requirements
2. Assessment of project deliverables against business requirements and specifications

*The assessment involves:*
- Performing pseudo-customer role
- Bringing customers' view in the assessment of solution/product/services
- Focusing on contractual commitments (SOW)
- Unearthing potential risks

The C1 assessment process is automated and integrated with other project management tools in Tech Mahindra Limited. The success of this process has been seen through a significant increase in customer appreciations on quality of deliverables.

## 4.22 TECH MAHINDRA LIMITED: CUSTOMER DELIGHT INDEX FOR PROJECTS

What's more important to the average company: maintaining alliances with existing clients or seeking out new clients? We could argue that both are equally important. However, because of the competitive forces in the marketplace, the edge goes to customer management and customer relations efforts. There is a tendency today to add in a life-cycle phase after contract closure entitled "customer satisfaction management" or "customer relations management." The purpose of this phase is to review how well satisfied the customer was throughout the project, with the end results, and with the flow of information provided by your project management methodology. In this phase, you are basically asking the customer, "What improvements would you like us to make prior to the next project we perform for you?" The customer may request more involvement in scope changes, a different format for status reporting or even better ways to track the status of their project through your project management methodology.

While this approach of having a life-cycle phase called customer satisfaction management has merit, it does have a downside risk that customer satisfaction is looked at only at the end of the project. Good companies track customer satisfaction throughout the project, not merely at the end. Tech Mahindra Limited has found a rather interesting way to do this. The remaining information has been provided by Hirdesh Singhal and Anu Khendry, School of Program and Project Management, Tech Mahindra Limited.

Tech Mahindra Limited is a software services company that manages projects that have been outsourced to it by its customers. Customer satisfaction or delight plays a critical role in getting repeat business from customers and this is important for the sustenance of Tech Mahindra Limited. Also, retaining a customer costs significantly less than acquiring a new customer. The customer delight index (CDI) framework has been developed in Tech Mahindra Limited to continuously measure and improve customer delight.

The CDI is measured for each project in Tech Mahindra Limited. (See Figure 4–27.) CDI is essentially the project manager's perspective on what he or she perceives about customer perception for the services provided depending on the communications and interactions with the stakeholders at the customer place. It helps senior management or investors of projects to proactively take actions and support the projects if required. For project managers, it helps in sending right signals proactively to the management that they need their support and intervention.

*The goals of CDI:*
- Assess the health of projects and take proactive steps to avoid customer escalations.
- Track project critical issues that directly affect customer delight to closure.
- Identify and record reasons for customer delight.

The CDI value is entered based on the perspective of the project manager on "how the customer is feeling about the project." It is arrived at based on customer feedback and interactions. Project managers need to enter the CDI, project issues, and/or reasons for delight on an agreed time frequency. Program managers review the status of the CDI on the same frequency.

There are four types of status for CDI (See Figure 4–28):

If the CDI status is *delighted,* the project is expected to have one or more reasons for customer delight. Project managers need to record the reason(s) and save the same.

If a project manager fails to enter the CDI within the specified period of one week, then the status is recorded as "Data Not Entered" (represented by "gray" color) for the week.

If the CDI status is *dissatisfied*, the project is expected to have a minimum of one open issue that reflects the reason for customer dissatisfaction. Project managers need to record the issue and save the issue.

After having updated the issues/reasons for delight for the project, the appropriate CDI rating is selected and the CDI status is saved (Figure 4–29).

**Project Issues**

Project issues need to be mapped against the predefined set of categories and subcategories. When the CDI status is other than delighted, project managers need to specify the project issues using the categories and subcategories.

It is mandatory for a project to have a minimum of one open issue that is the likely cause for customer dissatisfaction when the CDI status of the project is dissatisfied.

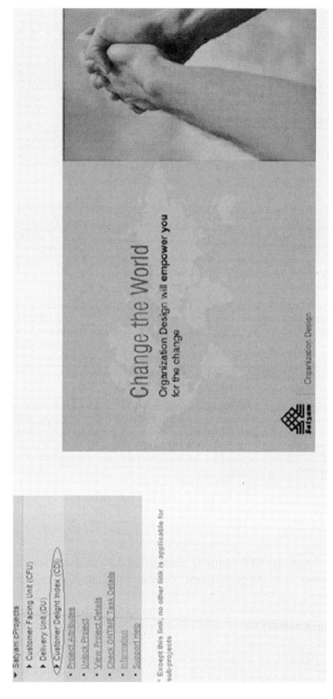

**Figure 4–27.** Customer delight index process.

| | |
|---|---|
| ◉ | Not Entered |
| ☒ | Dissatisfied |
| △ | Satisfied |
| ▢ | Delighted |

**Figure 4–28.** Summary categories of satisfaction.

| Parameter\ Interpretation | Delighted | Satisfied | Dissatisfied |
|---|---|---|---|
| Scope | Fulfill scope with higher business value solution/ service | Fulfill scope with in acceptable levels of solution/service performance | Scope not fulfilled |
| Quality | Zero-defect products/ services delivered | Products/services delivered with in customer acceptable level of defects | Products/services delivered with defects beyond customer acceptable level of defects |
| Schedule | Ahead of schedule/ turnaround time | On time/mutually acceptable delay | Delayed |
| CDI rating for the project can be selected based on a combination of the above factors. CDI rating shall be the lowest of the interpretations of all the above parameters put together. For example, if the interpretation for Scope for a given project is "Delighted" while that of "Quality" and "Schedule" is "Satisfied", the overall CDI rating for the project shall be "Satisfied". | | | |

**Figure 4–29.** Guidelines for CDI rating selection.

Projects where the CDI status is *satisfied* or delighted can also have issues that either have a potential to impact customer delight or are internal issues that need to be tracked to closure.

A project that has the CDI status as dissatisfied will not be allowed to move to satisfied or delighted without all the open issues being closed (Figure 4–30).

**Project Reasons**

In cases where the CDI status is delighted and/or satisfied, project managers need to enter the project reasons that have helped them delight the customers. In the same screen as Customer Delight, the project manager needs to enter the project reasons against the options that are listed. Based on the selection, the reason field will be enabled.

| Category | Sub category | Category | Sub category |
|---|---|---|---|
| **1. Customer Management** | Customer Communication | **4. Human Resources** | Competency |
| | Scope | | Attrition |
| | Customer Style | | Productivity |
| | Customer Escalations | | Motivation |
| | Customer Sign-off | | Attitude |
| | Customer Expectation | | Teaming |
| | | | Resource Availability |
| **2. Project Management** | Project Schedule | | Domain Knowledge |
| | Estimation | | |
| | Collaboration | **5. Infrastructure and Services** | Equipment |
| | Communication | | Office Space |
| | Process Compliance | | Communication Facilities |
| | Project Ownership | | Network Connectivity and Bandwidth |
| | Risks | | Travel and Boarding |
| | Project Tracking and Control | | Immigration |
| | Project Planning | | |
| | | **6. Financial** | Project Contract |
| **3. Software Engineering** | Requirements | | PO/SOW |
| | Design | | Invoicing |
| | Coding | | Project Cost |
| | Testing | | Project Profitability |
| | Delivery | | |
| | Deployment | | |
| | Technology | | |
| | Software Quality | | |
| | Configuration Management | | |
| | Environment (Development, Testing,...) | | |

**Figure 4–30.** Detailed categories of satisfaction.

## 4.23 GENERAL MOTORS POWERTRAIN GROUP

For companies with small or short-term projects, project management methodologies may not be cost-effective or appropriate. For companies with large projects, however, a workable methodology is mandatory. General Motors Powertrain Group is another example of a large company achieving excellence in project management. The company's business is based primarily on internal projects, although some contract projects are taken on for external customers. The size of the group's projects ranges from $100 million to $1.5 billion. Based in Pontiac, Michigan, the GM Powertrain Group developed and implemented a four-phase project management methodology that has become the core process for its business. The company decided to go to project management in order to get its products out to the market faster. According to Michael Mutchler, former vice president and group executive:

> The primary expectation I have from a product-focused organization is effective execution. This comprehends disciplined and effective product program development, implementation, and day-to-day operations. Product teams were formed to create an environment in which leaders could gain a better understanding of market and customer needs, to foster systems thinking and cross-functional, interdependent behavior, and to

enable all employees to understand their role in executing GM Powertrain strategies and delivering outstanding products. This organizational strategy is aimed at enabling a large organization to be responsive and to deliver quality products that customers want and can afford.

The program management process at GM Powertrain is based upon common templates, checklists, and systems. The following lists several elements that were common across all GM Powertrain programs during the 1990s:

- Charter and contract
- Program team organizational structure with defined roles and responsibilities
- Program plans, timing schedules, and logic networks
- Program-level and part-level tracking systems
- Four-phase product development process
- Change management process

Two critical elements of the GM Powertrain methodology are the program charter and program contract. The program charter defines the scope of the program with measurable objectives, including:

- Business purpose
- Strategic objective
- Results sought from the program
- Engineering and capital budget
- Program timing

The program contract specifies how the program will fulfill the charter. The contract becomes a shared understanding of what the program team will deliver and what the GM Powertrain staff will provide to the team in terms of resources, support, and so on.

Although the information here on GM Powertrain may appear somewhat dated, it does show that GM was several years ahead of most companies in the development of an enterprise project management methodology. GM has made significant changes to its methodology since then. What GM accomplished more than a decade ago many companies are just beginning to develop. Today, GM uses the above-mentioned methodology for new product development and has a second methodology for software projects.

## 4.24 ERICSSON TELECOM AB

General Motors Corporation and the bank were examples of project management methodologies that were internal to the organization (i.e., internal customers). For Ericsson Telecom AB, the problem is more complicated. The majority of Ericsson's projects are for external customers, and Ericsson has divisions all over the world. Can a methodology be developed to satisfy these worldwide constraints?

In 1989, Ericsson Telecom AB developed a project management methodology called PROPS.[16] Although it was initially intended for use at Business Area Public Telecommunications for technical development projects, it has been applied and appreciated throughout Ericsson worldwide, in all kinds of projects. In the author's opinion, PROPS is one of the most successful methodologies in the world.

New users and new fields of applications have increased the demands on PROPS. Users provide lessons-learned feedback so that their shared experiences can be used to update PROPS. In 1994, a second generation of PROPS was developed, including applications for small projects, concurrent engineering projects, and cross-functional projects and featuring improvements intended to increase quality on projects.

PROPS is generic in nature and can be used in all types of organizations, which strengthens Ericsson's ability to run projects successfully throughout the world. PROPS can be used on all types of projects, including product development, organizational development, construction, marketing, single projects, large and small projects, and cross-functional projects.

PROPS focuses on business, which means devoting all operative activities to customer satisfaction and securing profitability through effective use of company resources. PROPS uses a tollgate concept and project sponsorship to ensure that projects are initiated and procured in a business-oriented manner and that the benefits for the customer as well as for Ericsson are considered.

The PROPS model is extremely generic, which adds flexibility to its application to each project. The four cornerstones of the generic project model are:

- Tollgates
- The project model
- The work models
- Milestones

Tollgates are superordinate decision points in a project at which formal decisions are made concerning the aims and execution of the project, according to a concept held in common throughout the company. In PROPS, five tollgates constitute the backbone of the model. The function and position of the tollgates are standardized for all types of projects. Thus, the use of PROPS will ensure that the corporate tollgate model for Ericsson is implemented and applied.

The project sponsor makes the tollgate decision and takes the overall business responsibility for the entire project and its outcome. A tollgate decision must be well prepared. The tollgate decision procedure includes assessment and preparation of an executive summary, which provides the project sponsor with a basis for the decision. The project and its outcome must be evaluated from different aspects: the project's status, its use of resources,

---

16. The definition of the acronym PROPS is in Swedish. For simplicity sake, it is referred to as PROPS throughout this book.

and the expected benefit to the customer and to Ericsson. At the five tollgates, the following decisions are made:

- Decision on start of project feasibility study
- Decision on execution of the project
- Decision on continued execution, confirmation of the project or revision of limits, implementation of design
- Decision on making use of the final project results, handover to customer, limited introduction on the market
- Decision on project conclusion

The project model describes which project management activities to perform and which project documents to prepare from the initiation of a prestudy to the project's conclusion. The project sponsor orders the project and makes the tollgate decisions while most of the other activities described in the project model are the responsibility of the project manager. The project model is divided into four phases: prestudy, feasibility study, execution, and conclusion phases.

The purpose of the prestudy phase is to assess feasibility from technical and commercial viewpoints based on the expressed and unexpressed requirements and needs of external and internal customers. During the prestudy phase a set of alternative solutions is formulated. A rough estimate is made of the time schedule and amount of work needed for the project's various implementation alternatives.

The purpose of the feasibility study phase is to form a good basis for the future project and prepare for the successful execution of the project. During the feasibility study, different realization alternatives and their potential consequences are analyzed, as well as their potential capacity to fulfill requirements. The project goals and strategies are defined, project plans are prepared, and the risks involved are assessed. Contract negotiations are initiated, and the project organization is defined at the comprehensive level.

The purpose of the execution phase is to execute the project as planned with respect to time, costs, and characteristics in order to attain the project goals and meet the customer's requirements. Technical work is executed by the line organization according to the processes and working methods that have been decided on. Project work is actively controlled; that is, the project's progress is continuously checked and the necessary action taken to keep the project on track.

The purpose of the conclusion phase is to break up the project organization, to compile a record of the experiences gained, and to see to it that all outstanding matters are taken care of. During the conclusion phase, the resources placed at the project's disposal are phased out, and measures are suggested for improving the project model, the work models, and the processes.

Besides describing the activities that will be performed to arrive at a specific result, the work model also includes definitions of the milestones. However, to get a complete description of the work in a specific project, one or more work models should be defined and linked to the general project model. A work model combined with the general project model is a PROPS application. If there are no suitable work models described for a project, it is the project manager's responsibility to define activities and milestones so that the project plan can be followed and the project actively controlled.

A milestone is an intermediate objective that defines an important, measurable event in the project and represents a result that must be achieved at that point. Milestones link the work models to the project model. Clearly defined milestones are essential for monitoring progress, especially in large and/or long-term projects. Besides providing a way of structuring the time schedule, milestones will give early warning of potential delays. Milestones also help to make the project's progress visible to the project members and the project sponsor. Before each milestone is reached, a milestone review is performed within the project in order to check the results achieved against the milestone criteria. The project manager is responsible for the milestone review.

Ericsson's worldwide success can be partially attributed to the acceptance and use of the PROPS model. Ericsson has shown that success can be achieved with even the simplest of models and without the development of rigid policies and procedures.

## 4.25 INDRA: CLOSING THE PROJECT[17]

**Closing the Project**

In a technological company like Indra, with projects being managed to develop, manufacture and maintain complex hardware and software systems, an immature project closure can be, if not well treated, a cause of great losses on efficiency.

Projects usually require a curve of effort with its peak at the beginning and half of its project life cycle. See Figure 4–31. Or in other words, from the project manager point of view, planning and monitoring and control are the phases that require more of her attention.

During planning stage, project manager works toward clear goals. At the same time planning depends on established commitments either with the sponsor or the customer. While in monitoring and control project manager's attention is focused on coordinating team efforts to achieve project milestones, identifying variances to baselines and protecting the project from changes, which really take most of her time.

This is not the case at the end of the project: when commitments are fulfilled most of the pressure on PM is released. This occasionally makes that the last of the milestones (project closure) is not properly achieved, as PM attention and effort has dropped and even it is possible that a new assignment be waiting for the project manager, so she is released to start the new responsibility without properly closing the previous one.

In the context of an organization like Indra, whose main business is delivering project results to its customers, we intend to organize our resources in the most efficient manner, giving effective response to all commitments with our customers in a business improvement framework.

Performing a good project closure may be a little motivator and can be rated by PMs as a simple and administrative task. Therefore it could be forgotten that if we don't pay attention to it the opportunity to consolidate the efficiencies that were got in the project, that

17. ©2013 Indra. Reproduced by permission. All rights reserved. Material on Indra has been provided by Alfredo Vázquez Díaz, PMP, Director, Corporate Project Management Office

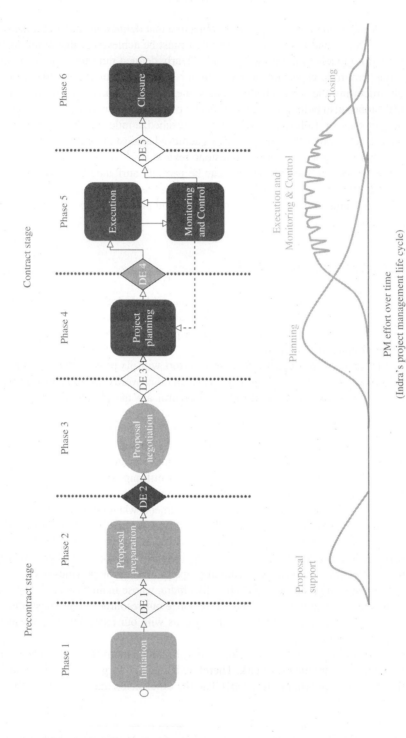

**Figure 4–31.** Indra's project management life cycle.

benefit can be lost for the organization, particularly in the management of the scope and resources (which, by the way, are the main values used in the calculation of productivity).

If we focus on management of the scope, if project closure is not well done there is a risk that acceptance agreements of the deliverables may tend to be diluted, reopened or reinterpreted by the customer. This happens if the project end is not well settled and confused and mixed with warranty period.

Let me show this: customer's needs after a new system is implanted may be changing, and that can make the interpretation of the requirements to evolve, losing traceability with the initial project scope and its former conditions of validation. The person at the client that performs the requirements validation on deliverables may change her perception as time passes without a formal closure. That way, the customer may try to relocate new needs back on the project instead of placing them in a project extension, as it should be.

Especially advisable is paying particular attention to the efforts dedicated to customer requirements acceptance when managing a project based on agile models, that are so fashionable today. The dynamic constant of micro delivery -validation can place the problem of poorly executed project closure in dimensions hardly treatable.

If we focus on management of resources, several organizational roadblocks can happen, being the most frequent the prevention of resources being released from our project to others. It also happens that not released resources lower the productivity gained during the project. Another negative effect is the lack of methodological focus during a not properly executed project closure, being carried away by the reaction of change events and incidents. This nonpreventive approach is a like a Trojan horse with respect to the assumption of changes, scope, improvements and responsibilities that were properly negotiated by the project manager during other stages of the project, and that at project closure could be put again at stake.

This concern has led Indra to improve project closure practices by implementing in the PMIS of a group of facilitators:

- Possibility of early beginning of project closure activities, by overlapping this phase with the previous (e.g., in cases where scope is cut or there is a planned closure date)
- Use the information that the PMIS has accumulated from the project along its life to help identify situations that could prevent formal closure
- Indicators and reporting associated with project closure, offering at the same time the project closure status
- Linking lessons learned which allow search in previous experiential knowledge; this could affect closing process (and to others)

The Spanish saying "close the door [or] the cat will escape" shows with ease the risks that the organization faces if project closure is not properly done. If we don't close the door (the project) the cat escapes, or in other words, risks that were controlled have an ultimate chance to happen. If the project manager doesn't carefully perform project closure, she may be adding great risks to the project that had had under control during earlier phases in which her effort and attention were high.

## 4.26  REPSOL: THE REPSOL E&P GIP© METHODOLOGY—THE PROJECT QUALITY MANAGEMENT PROCESS APPLIED TO DECISION MAKING[18]

The "Gestión Integrada de Proyectos" (GIP©) is the integrated project management methodology in Repsol Exploration and Production general direction (Repsol E&P). The GIP© is based on the stage gated concept.

The GIP© establishes a set of deliverables for each of the disciplines involved in an E&P project; in total there are 330 deliverables to cover from the Exploration Phase to Development Project handover. Each derivable is described in a "card," that includes its description, a guideline and a best in class example. The GIP© is available to all employees via the GIP Space© in the Repsol intranet through an interactive deliverables map.

The GIP© considers four phases. See Figure 4–32. The phases are in fact four projects in sequence where the PMI *PMBOK® Guide* project process groups; namely initiating, planning, executing, planning, monitoring and control and closing are applied. At the end of each phase there is a closing process and the initiation of the next phase.

The main deliverable of the Visualization Phase is a list of integrated development options, technical and economically feasible. These options include production profiles and the associated facilities wells. The Visualization Phase is a period of high creativity where all technically feasible options should be detected, to be analyzed and prioritize in the next phase. The phrase that defines the effort during the Visualization phase is "When a man has put a limit on what he will do, he has put a limit on what he can do." (Charles M. Schwab).

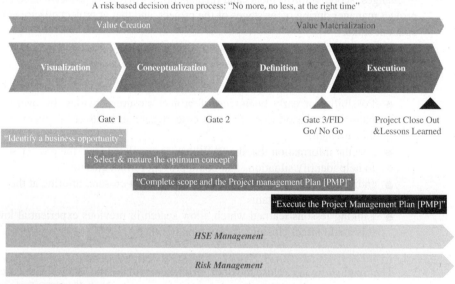

**Figure 4–32.**   The GIP© concept.

---

During the Conceptualization Phase the previously detected development options are analyzed and new options may appear. Then they are prioritized and only one option is selected. Once selected, the option is matured until the information is [great] enough to produce a Cost Estimate Class 4 according to AACEi. This is the phase where maximum value can be captured in a project. The phrase that defines best the philosophy behind the conceptualization effort is "Dress me slowly as I'm in a hurry." (Napoleon Bonaparte)

If the Conceptualization Phase is key because it is the period that captures the maximum project value, the Definition phase is also key because it is the phase that sets project scope and baseline and where the Project Management Plan (PMP) is completed. In this case the phrase that defines the philosophy behind the Definition activities is "Give me six hours to chop down a tree and I will spend the first four sharpening the axe." (Abraham Lincoln)

The Execution Phase includes all the activities planned and included in the PMP. The phase's final product is the project infrastructure: wells, facilities, buildings, etc. The phase is complete and closed when the infrastructure is handover to the "client" who is the Asset Manager. In this case the phrase that defines the activities is "Success is dependent on effort." (Sophocles)

The GIP© manages the risks associated with uncertainties along the project lifecycle, administering commitments and expenditures. These uncertainties may be related to reservoir modeling maturity, drilling and facilities scope to name a few. The level of uncertainty through a project life cycle can be visualized by mapping the Development Project Cost Estimate Class according to AACEi at each phase. See Figure 4–33. These uncertainty levels depend on the scope of the development options to develop a specific crude or gas reservoir. These cost estimate uncertainties do not cover the uncertainties associated to the reservoir knowledge, that will have to be assessed and considered in the surface facilities and wells' configuration and designs.

Like most of the stage gate methodology, The GIP©, establishes four possible decisions at each gate. See Figure 4–34. The possible decision outcomes are:

- continue to next phase
- recycle back because level of uncertainty is too high for the next commitment and more analysis, modeling, engineering, etc., is required.
- cancel
- shelve until conditions favor the development of the project

Decisions at each gate are made by gatekeepers' committees that will endorse or approve commitments and funds according their delegations levels in the company. Each gate is fed by Technical Review performed by a multidiscipline and integrated team of experts, where all E&P disciplines are represented. This integrated team reviews all deliverables developed during the evaluated stage according to the map in the GIP Space©. Other experts such as Fiscal and Legal, also assess the project. The assessment is made available to the gatekeepers in an integrated document and presentation called the Decision Support Document or DSD.

The Technical Reviews made at the end of each phase are, in fact, Quality Control events, where the level of maturity and completeness of main deliverables are checked and deviations identified. See Figure 4–35.

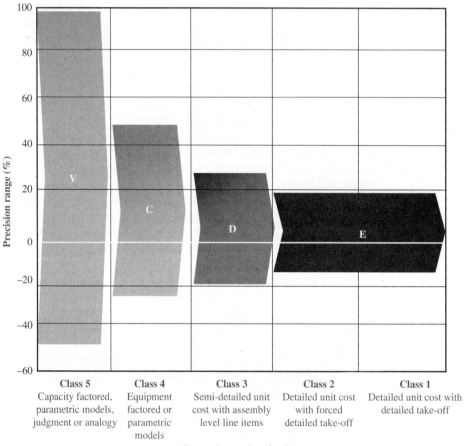

**Figure 4–33.**  Cost estimating precision according to GIP© phases.

**Figure 4–34.**  GIP© decision gates.

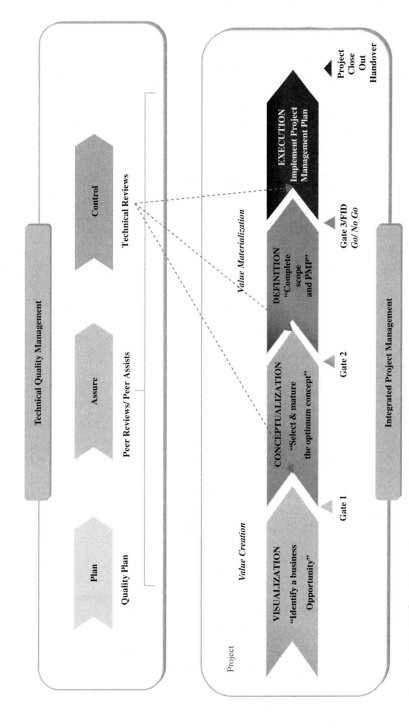

**Figure 4–35.** GIP© and the quality process for decision making.

At the beginning a project, officially after Gate 1, a Quality Events Map is established and agreed to between the Project Manager and the E&P Project Management Office. This map is incorporated in the Project Quality Plan and its milestones integrated in the Project Master Schedule. The Quality Events Map 1 identifies the Quality Control milestones associated to the decision making an also Quality Assurance events.

The Quality Events Map is project specific, meaning that it will depend on the investment level, complexity, company exposure or a combination. This Quality Events Map include the Quality Assurance events that provide confidence that the Quality Control events will be successful.

The Quality in Repsol E&P can be represented by a four-level pyramid. The first level corresponds to contractors and suppliers, the second to project teams, the third to the decision making and the fourth to corporate policies and norms. See Figure 4–36.

To better understand the philosophy behind this pyramid, it is important to consider that in an Exploration and Production Project most of the activities are contracted and goods are always supplied by third parties. The project teams literally plan and supervise the work executed by others.

The contractors and suppliers will execute the quality process to guaranty that services and goods are delivered according to the contracted requirements.

At the project team level, the execution of the quality process is less intense and requires less resources than the quality activities performed by contractors and suppliers. Quality Assurance activities are aimed to provide confidence that the contractors and suppliers are following or using the agreed specifications, procedures having the proper people profiles, etc., and will deliver services and goods according to offered requirements. Examples of this are shop inspections during fabrication and auditing to engineering

**Figure 4–36.**  The GIP© quality pyramid.

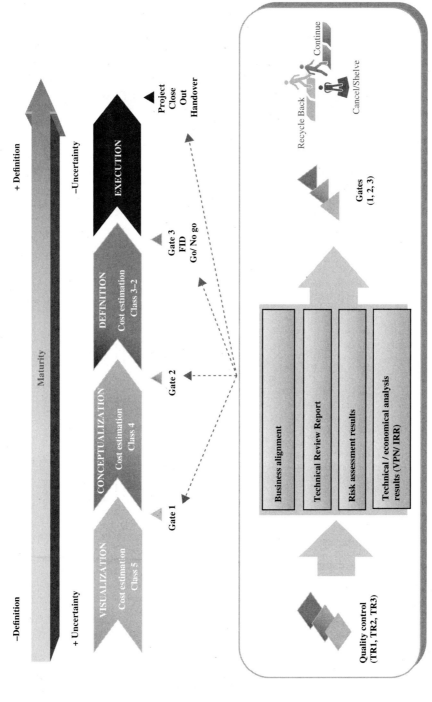

**Figure 4-37.** GIP© model and the decision-making process.

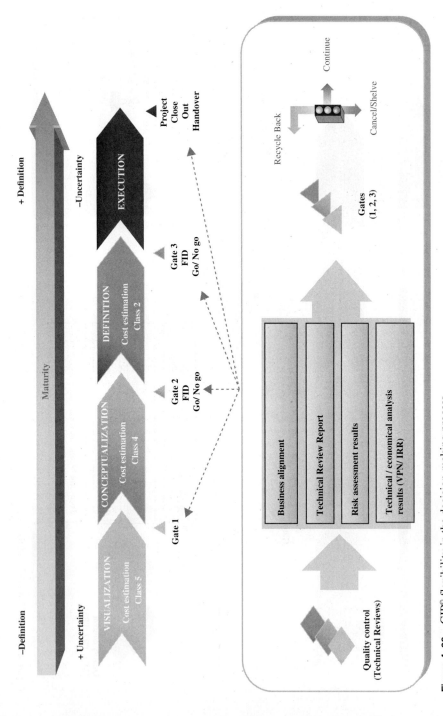

**Figure 4–38.** GIP© flexibility in the decision-making process.

products while being developed. Quality Control activities are associated to determine if delivered products are to be accepted or rejected.

During the preparation of the Project Quality plan the project team, establishes the deliverable requirements, assess the criticality of each project area and its deliverables, their associated contracts, materials and equipment, and establishes the number, timing and depth of the audits and inspections associated to both Quality Assurance and Control.

The third level of the pyramid corresponds to the quality associated to the inputs to the decision makers or gatekeepers. The "main clients" of the outcome of the third-level Quality Process are the gatekeepers that will use the results of the Technical Reviews, to assess the project value and its risks at the decision point; and then decide to proceed or not to the next phase. See Figure 4–37. The other client that is benefited with the third level Quality Process is the project team, because quality events will detect deviations that, once corrected, will assure or improve the project value. It is important to understand that the third-level Quality Process does not replace the Quality Process at the Project Teams level, it complements it.

The GIP© allows controlled flexibility in the decision-making process. For example, in certain cases, there are benefits in taking final investments decisions earlier than the gate 3. This means that certain parts of the project may not be mature enough to have the information to produce a Class 2 Cost Estimate according to AACEi. See Figure 4–38.

In this case an analysis is performed an analysis to assess the risks produced by the uncertainties associated to the low maturity, then plan and sometimes immediately perform response actions. The residual probability and impacts are reflected in the project main variables and considered in its value analysis.

When the Project Team and the project sponsor detect an opportunity to be exploited by taking a decision in an early stage with a low maturity project, as seen in Figure 4–39, the results of the risk analysis are discussed at the Technical Review and response actions confirmed. During the Technical Review additional risks could be detected and response actions be established and incorporated in the PMP. The Decision Support Document is submitted to the gatekeepers for approval after the endorsement of the multidiscipline integrated team of experts. See Figure 4–40.

**Positive:**

- Capturing opportunity earlier
- Complying with contractual commitments
- Capturing sales market
- Accelerating contracting and purchasing to assure project deadlines (schedule)
- Accelerating income

Positive effects

+

Response actions

**Negative:**

- Technical uncertainties when specifying equipment and materials
- High probability of rework
- Possible high level of uncertainty when developing cost estimate and schedule
- Uncertainties in project value
- Possible project failure

Negative effects

**Figure 4–39.** GIP© taking decisions with low maturity project inputs.

**Figure 4–40.** GIP© process for decision making with low-maturity project inputs.

## 4.27 ROCKWELL AUTOMATION: QUEST FOR A COMMON PROCESS[19]

Rockwell Automation was formed by bringing two major automation companies together in the late 1980s. These two companies, Allen-Bradley and Reliance Electric, were the foundation of what is now Rockwell Automation. Over the years, Rockwell Automation has continued to acquire leading automation suppliers as a growth strategy and also as a way to bring new advanced automation technologies into the company. In 2005, as Rockwell Automation was planning the rollout of a new SAP business system, we recognized the need for a new "common" product development process that would be defined based on company best practices combined with what was considered the industry's best practices for product development. This effort resulted in a "common product development" (CPD) process that was defined in a way to allow for enterprisewide adoption. This is shown in Figure 4–41. This means that 16 different product businesses ranging from high-volume component suppliers to complex continuous process control systems solution suppliers all use the same high-level process framework for their new product developments.

The resulting process is made up of six phases with a stage-gate review after each phase. The six phases are:

19. This section on Rockwell Automation was provided by James C. Brown, PgMP, PMP, OPM3 AC, MPM, CIPM, CSP, CSSMBB, Formerly Director, A&S Enterprise Program Management Office; Karen Wojala, Manager, Business Planning; and Matt Stibora, Lean Enterprise Manager.

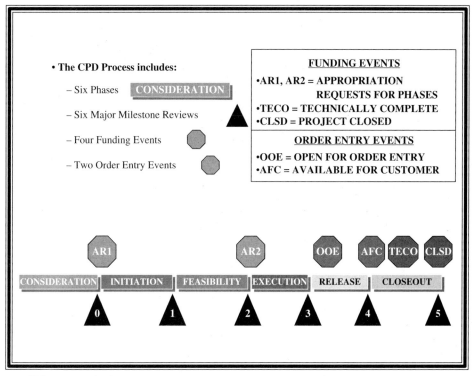

**Figure 4–41.** CPD process: basic concepts.

*Consideration*
- To develop a high-level business case and project proposal to justify AR1 funding for the execution of initiation and feasibility phase activities.

*Initiation*
- To refine the high-level business case document (BCD) created in the consideration phase into a set of customer requirements sufficient for the project team to create solution concepts, product requirement, and functional requirement documents in the feasibility phase. (See Figure 4–42.)

*Feasibility*
- To evaluate solution concepts to address the customer requirements from the initiation phase.
- To define the product requirements and functional requirements.
- To complete all project planning and scheduling to update the project plan for all activities and resources required to complete the execution, release, and closeout phases of the project.
- To develop a BCD that justifies the investment required to execute the project plan for the solution concept chosen. (See Figure 4–43.)

**Figure 4–42.** Initiation phase summary.

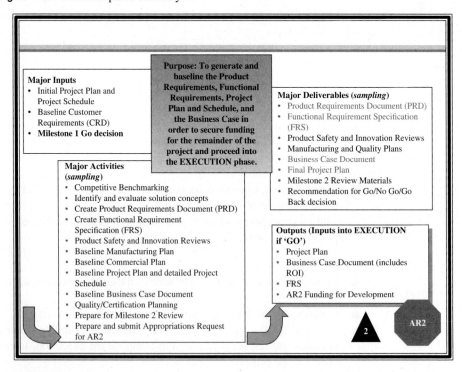

**Figure 4–43.** Feasibility phase summary.

*Execution*
- To develop the product or service according to the baselined functional requirement specification (FRS) from the feasibility phase; performing the necessary reviews; making approved requirement and/or design changes as the project progresses.

*Release*
- To finalize all test, certification, and other product verification documentation.
- To build and validate pilot production.
- To open for order entry and execute the commercial launch.

*Close*
- To position the product for transition to continuation/sustaining engineering; documentation cleanup, postmortems, lessons learned, record retention, and complete all financial transactions.

Our goal was to achieve a rapid, repeatable framework that consistently results in high-quality output. A major focus of the team that produced this new process was to drive the product businesses to be more disciplined in how innovation was embraced when deciding what projects proposals were funded and which ones were not. There were too many examples of projects receiving management support and funding without meeting a set of minimum criteria that would result in a higher probability of commercial success. Investment proposals were not always based on an ideation process that was driven by our customers.

We found examples of funded projects enjoying support and funding without any customer-driven commercial basis. The justification for these projects was based on new interesting technologies, investing in product family coverage for the sake of coverage without real market demand, providing niche solutions with limited potential driven by a single customer, etc.

To solve this problem the team's original focus was on two aspects of what was defined as best practice. There are numerous theories that attempt to describe the best way to capture customer needs and use them as the basis for creating effective new product concepts. Our goal was to understand the customer problems before we produced solution concepts and product solutions. We accomplished this goal by breaking apart an existing tool called the Marketing Requirements Document (MRD) and process used by Product Management into two tools.

We wanted the product owners to understand the market and target customer's problems before they considered solutions. By breaking the MRD into two deliverables, the first (Customer Requirements Document) focused on the market need and customer problems and the second (Product Requirements Document) focused on the solution concepts and product requirements, and locating these tools and activities in separate phases divided by a management stage gate review, we forced our product managers to break away from the "continuously evolve the product death march" that we were on.

Of course, accepted practice and company culture is hard to break so governance is critical in driving change. This simple step is the beginning of what will be a significant improvement in the new product development practices of this company.

The driving force behind management's commitment to implement this new process and to drive the cultural change was the vision of a common consistent methodology for new product development across the enterprise. This consistency was prioritized from the top (direct management involvement in the stage-gate reviews) down, in order to realize benefit as soon as possible.

All too often, businesses were forced to deny funding for strategic projects due to the never ending incremental product improvements that just kept coming. By forcing business management to approve each project's passage from one phase to the next, we pushed the visibility of every project, every resource and every dollar up to the decision makers who wrestled with trying to find dollars to fund the real game changers. This visibility made it easier for the business owners to kill projects with questionable returns, or to delay a project in order to free up critical resources. Once management began to see the returns from these decisions in the form of product introductions that really moved the needle, we began to focus on the fine tuning of the process and methodologies employed. (See Figure 4–44.)

The stage-gate review is the most important event of a project. Previously, under the old way of executing a project, these reviews were informal and haphazard. Teams were able to continue spending and even overspending without any real fear of cancellation. This new process ensures that every dependent organization is represented at the appropriate review and given the chance to agree or disagree with the project manager that all deliverables are available. The intent is to have the go/no-go decision made by both the primary organization responsible for the deliverables during the previous phase and the primary organization responsible for the deliverables in the subsequent phase. Both these organizations are required at each stage-gate review. If done correctly, we will be able to avoid surprises during the later phases by ensuring transparency during the earlier phases.

Once a project manager is assigned and a project team is formed, the importance of well-defined deliverables that are easy to locate and use became evident. The 12 months following the original launch of the new process was spent continuously improving the phase definitions, procedure documents, deliverable templates, and governance policies. There is a fine line between rigor and burden; the trick is to push this line hard to ensure rigorous implementation without slowing the progress of the project team down.

At the end of the feasibility phase, as the project enters the execution phase with requested funding secured, the project plan becomes the bible. The project plan drives all

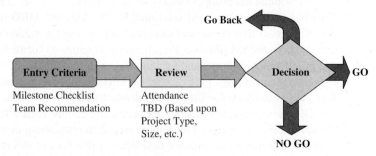

**Figure 4–44.**  The decision-making process.

activities through the execution phase, release phase, and finally project close. Any issue that the project team is faced with that requires a change in course must be recognized in an updated project plan. At the conclusion of the project, the plan must represent what actually occurred.

Prior to any new product being released for customer shipment, all impacted stakeholders must agree that the product is ready before giving the final approval.

There is a dimension of introducing an end-to-end process that has been assumed but must be mentioned. The company is built from many related but very different product businesses. Each business segment was at a different maturity level relative to all aspects of product development, even the existence of a formal project management organization.

Project managers are instrumental in the execution of a product development process. If consistency, transparency, and risk mitigation are important to a business, and they are to Rockwell Automation, then a formal well-recognized and managed project management entity is paramount.

Rockwell Automation is pursuing the discipline of project management at all levels in its organization.

## 4.28 SHERWIN-WILLIAMS

There are several ways that a company can develop a methodology for project management. Outsourcing the development process to another company can be beneficial. Some companies have template methodologies that can be used as a basis for developing their own methodology. This can be beneficial if the template methodology has enough flexibility to be adaptable to their organization. The downside is that this approach may have the disadvantage that the end result may not fit the needs of the organization or the company's culture. Hiring outside consultants may improve the situation a little, but the end result may still be the same unfavorable result as well as being more costly. This approach may require keeping contractors on the payroll for a long time such that they can fully understand the company's culture and the way it does business.

Benchmarking may be effective, but by the time the benchmarking is completed, the company could have begun the development of its own methodology. Another downside risk of benchmarking is that the company may not be able to get all of the needed information or the supporting information to make the methodology work.

Companies that develop their own methodology internally seem to have greater success, especially if they incorporate their own best practices and lessons learned from other activities. This occurred in most of the companies in this book.

The information below was provided by The Sherwin-Williams Company.

**Company Background**     The Sherwin-Williams Company engages in the development, manufacture, distribution, and sale of paints, coatings, and related products to professional, industrial, commercial, and retail customers in North and South America, the United Kingdom, Europe, China, and India. It operates in four segments: paint stores, consumer, Latin America, and global finishes. The paint stores segment sells paint, coatings, and related products to end-use customers. This segment markets and sells Sherwin-Williams branded architectural paints and coatings, industrial and marine products, and original equipment manufacturer product finishes and related items. As of December 31, 2008, it operated 3346 paint stores. The consumer segment engages in the development, manufacture, and distribution of paints, coatings, and related products to third-party customers and the paint stores segment. The Latin American and global finishes segments develop, license, manufacture, distribute, and sell architectural paint and coatings, industrial and marine products, automotive finishes and refinish products, and original equipment manufacturer coatings and related products. These segments also license certain technology and trade names as well as distribute Sherwin-Williams branded products through a network of 541 company-operated branches, direct sales staff, and outside sales representatives to retailers, dealers, jobbers, licensees, and various third-party distributors. The company was founded in 1866 and is headquartered in Cleveland, Ohio.

The Corporate Information Technology (IT) Department for The Sherwin-Williams Company provides shared services support for the three operating divisions, described above, that make up the organization.

**Case Study Background**  During the summer of 2002, the Corporate IT Department engaged in activities surrounding the conversion of international, interstate, intrastate, and local telecommunications services from the company's present voice telecommunications carrier to a new carrier. Project management disciplines and best practices, using a structured project management methodology, were utilized on this project, ultimately leading to a successful project outcome.

The project was implemented using a phased approach consisting of the major phases as described below. The phases were established to include many of the principles stated in the *PMBOK® Guide* and also included many of the best practices that had been developed previously at The Sherwin-Williams Company. The phases could overlap, if necessary, allowing for a gradual evolvement from one phase to the next. The overlapping also allowed the company to accelerate schedules, if need be, but possibly at an additional risk. Project reviews were held at the end of each phase to determine the feasibility of moving forward into the next phase, to make "go/no-go" decisions, to evaluate existing and future risks, and to determine if course corrections are needed.

- *Initiate*: The first phase is the initiate phase where the project team is formed, a project kickoff meeting is held, needs and requirements are identified, and roles and responsibilities are defined.
- *Planning*: The planning phase is the next phase and is regarded by most project managers as the most important phase. Most of the project's effort is expended in the planning phase, and it is believed that the appropriate time and effort invested in this phase ensure the development of a solid foundation for the project. Management wholeheartedly supports the efforts put forth in this phase because this is where many of the best practices have occurred. Also, a solid foundation in this phase allows for remaining phases of the project to be accomplished more

efficiently, giving senior management a higher degree of confidence in the ability of project managers to produce the desired deliverables and meet customer expectations.

A series of meetings are typically held throughout this phase to identify at the lowest level the project needs, requirements, expectations, processes, and activities/steps for the processes. The results of these meetings are several deliverables, including a needs and requirements document, a project plan, a risk management plan, an issue log, and an action item list. Additional documents maintained include quality management and change management plans. Together these documents provide management with an overview of the entire project and the effort involved to accomplish the goal of transitioning services by the target date established by management.

- *Execution:* The third phase in implementation is execution. This phase is evolved into gradually once the majority of planning has been completed. All activities outlined in the processes during the planning phase come to fruition at this time as actual communication line orders begin to take place as well as the installation of equipment where necessary. Services begin to be transitioned by the division/segment and implementation moves forward aggressively for this project due to a stringent timeframe. It is of vital importance that activities in this phase be monitored closely in order to facilitate the proactive identification of issues that may negatively impact the timeline, cost, quality, or resources of the project.

To facilitate monitoring and control of the project, weekly status meetings were held with the vendor and the project team, as well as short internal daily meetings to review activities planned for each day. Ad hoc meetings also occurred as necessary.

- *Closure:* The final phase of the project is closure. In this phase, there is typically a closure meeting to identify any remaining open issues and to determine the level of client satisfaction. This phase also included any "clean-up" from the project, administrative closeout, the communication of postimplementation support procedures, and a review of lessons learned.

Best practices that worked notably well for The Sherwin-Williams Company included the establishment of success criteria, consisting of project objectives and a needs/requirements analysis, regular communications both within the project team and with stakeholders, dedicated resources, defined roles and responsibilities, knowledge transfer between cross-functional teams, teamwork, the development of a fun, synergistic working environment, and a review of lessons learned.

One of the best practices in project management is that maturity and excellence in project management can occur quickly when senior management not only actively supports project management but also articulates to the organization their vision of where they expect project management to be in the future. This vision can motivate the organization to excel, and best practices improvements to a project management methodology seem to occur at a rapid rate. Such was the case at The Sherwin-Williams Company. Tom Lucas, chief information officer at The Sherwin-Williams Company, comments on his vision for The Sherwin-Williams Company:

> The future of project management at The Sherwin-Williams Company includes the integration of project management disciplines and best practices, combined with portfolio management techniques, to deliver high value project results on a consistent basis. The Sherwin-Williams Company anticipates that the use of a PMO will not only instill the best practices of project management as core competencies, but also aid in the growth of the organization's project management maturity.
>
> One goal has been to unify the goals and objectives of individual departments by applying a universal yet flexible project management framework in pursuit of better across-the-board results. We have made significant strides in this regard. The Sherwin-Williams Company desires to learn from past successes, as well as mistakes, make processes more efficient, and develop people's skills and talents to work more effectively through the establishment of standardized procedures within the company. Above all, we must demonstrate real business value in using professional project management.
>
> While project management professionals may reside in multiple operating units, so as to be as close as possible to our internal clients, our intent is to have a core group of project management professionals that would be the standards setting body and provide for best practices identification and sharing.
>
> We have all managed projects at one time or another, but few of us are capable of being Project Managers. Herein lies one of the biggest impediments to implementing professional project management. We can have the best-trained project managers, we can have all the right process in place, we can use all the right words, yet the PMO will either fail or be only a shell of what it can be. Staff and management have a hard time appreciating the power, and improved results, of a professionally managed project. Until staff and management become involved themselves, until they feel it, until they personally see the results, the distinction between managing a project and project management is just semantics.
>
> The difference between managing projects and professional project management is like the difference between getting across the lake in a rowboat versus a racing boat. Both will get you across the lake but the rowboat is a long and painful process. But how do people know until you give them a ride?
>
> The 2002 telecom case study was just such a ride. While the focus of the case study discussion was to articulate the mechanics of the PMO process, the real story is the direct per share profitability improvement resulting from this successful initiative. In addition, there was legitimate concern from the business on the potential impact this change may have on our internal clients and external customers should something go wrong during the transition. The professional project management that was used gave everyone the cautious optimism to proceed and the results made the staff and management "believers" in the process.
>
> Project by project, success by success, a cultural transition is in process. As we demonstrate improved business results because of professional project management we are able to offer services to a wider audience and are able to take on projects outside of IT where the PMO got its start.
>
> By staying focused on business results, by staying close to our clients so we understand their needs well, and by constantly challenging ourselves to improve our underlying processes our PMO services are maturing more and more every day. It becomes a fun ride for everybody.

## 4.29 MEDICAL MUTUAL[20]

Some companies have found that some of the readily available methodologies that can be purchased or leased have enough flexibility to satisfy their needs. This is particularly true in the IT area. The following information was provided by Dan Halicki, IT Administration Coordinator at Medical Mutual.

**Introduction**

Industry analysts believe that a quarter of all IT projects are delivered on time, and fewer are within budget. In response to this problem, various project management approaches have emerged with the primary objective of ensuring project success.

With the advent of healthcare reform Medical Mutual sought to reinforce its strategy to deliver the appropriate combination of benefit design, health management and other services to optimize the value of their dollars spent on healthcare. To help achieve this goal we have adjusted our IT project management philosophy to find the proper balance between rigorous methodology requirements and the realities of containing administrative project overhead.

Systems development at Medical Mutual has followed traditional project management principles for over twenty years, including a commitment to the Project Management Institute's (PMI) standards for project management. In 2005, a Project Management Practices Team was established to provide governance and ensure the project management methodology and standards have kept pace with subsequent iterations of the PMBOK®. The use of traditional project management principles, PMI concepts and this governance team has contributed to overall project success and helped the IT Division fulfill our mission to be the best provider of information technology to Medical Mutual through the delivery of solutions with exceptional value.

What, then, are the best practices that Medical Mutual follows to establish project priorities, infuse flexibility and bring the right amount of discipline to its information technology project management approach? The following discusses the techniques and deliverables that constitute the standard project management elements of initiation, planning, execution, controlling and monitoring, and closing. Additionally, it is important to look beyond the elements of the methodology itself to include the support system that makes it all work, such as, executive management buy-in, training and follow-up activity.

The Executive IT Steering Committee provides a Demand Management process and ensures that IT project initiatives are aligned with the strategic objectives of the organization.

Medical Mutual uses a demand management process that requires collaboration between business area management and IT planning to effectively align business vision with technical solutions. Corporate Strategic Planning establishes the Company goals that guide our IT strategy. To achieve the desired goals, related project proposals are drafted

20. The material in this section was taken from A. Wallen and D. Halicki, *Medical Mutual – Project Management Meets Healthcare Reform: Bringing the Right Amount of Discipline to a Project.* © 2013 Medical Mutual. All rights reserved. Reproduced by permission.

using a standardized project charter format that identifies how the proposed project aligns with specific value business drivers and strategic business drivers to warrant consideration. The long-standing business model of health insurance is undergoing significant change and it is imperative now more than ever that scarce IT resources are utilized on the most strategic projects.

### Provide On-Going Project Management Training throughout the Enterprise

The term "project management" can mean different things to different people. An instructor-led class explains how Information Technology applies their improved "Project Management Methodology" to ensure that all project participants, including the customer community, have a common understanding of the process, especially stakeholder participation and joint ownership. With duration of 24 hours over three days, the class provides instruction and a hands-on workshop on the major components of the project management process, e.g. project initiation, planning, execution, monitoring and control, and project closure. A project management handbook (PM Handbook) will be deployed in 2013 to help coach existing employees and instruct new ones on the project management philosophy in use at Medical Mutual.

### Structure Every Project in a Consistent Manner, Including Scope, Responsibilities, Risks, and High-Level Milestones

Central to the project management process is the creation of the project charter, which documents and formalizes the agreement of all concerned parties, including business sponsors, project manager and project team members. The charter spells out what will be accomplished and when, who is responsible for getting it done, known risks and the course of action to mitigate the risk, quality assurance measures, quantifiable measures of success and the major milestones to be met.

### Communicate and Update Project Plans and Status on an On-Going Basis, Using On-Line Tools When Appropriate

Once an approved project charter is in place, the project plan becomes the next focus of attention. The plan and the regular status reports that document progress versus the plan are the primary communication vehicles for keeping management and staff informed. Medical Mutual built and maintains a SharePoint-based an online real-time project status center to keep project planning, issues, change control and project information current and readily available throughout the organization. Key Performance information is also maintained and available on this system as an ongoing source of IT service level measures.

### Maintain an Official Project Issues List, Including Who Is Responsible, What Are the Potential Impacts, and How the Issue Was Resolved

As problems arise during the execution of a project, it is vital to maintain an accurate issues log to properly document the problems and what is being done about them. In project meetings, the issues log serves as a meaningful tool to prevent unproductively rehashing old information and assures concerned areas that their problems have not been forgotten. Along with the charter and the regular progress reports, the issues log can be accessed via the project status center and critical issues are highlighted in the project's status reporting dashboard.

### Use a Formal Change Control Process, Including an Executive Steering Committee to Resolve Major Changes

As issues arise that could require additional resources, the nemesis known as "scope creep" can begin to surface and impede a smooth-running project. A formal change control process is essential to help the project manager with the predicament of pleasing the customer or going over budget or missing milestones. The Medical Mutual change control process (request and assessment) can be accessed via the project status center and critical change control is highlighted in the project's status reporting dashboard.

### Institute Governance to Periodically Review the Project Management Process and Selected Projects to Determine How Well the Methodology Is Working and to Identify Opportunities for Improvement

With the establishment of project management standards and the sizeable investment in training, it makes sense to step back occasionally and assess if the overall objectives are being met. Where projects ran into difficulty, would a different approach have helped? Where projects went well, what aspects of the project execution ought to be shared by other projects? These are the kinds of questions that our IT Project Management Practices team is positioned to address. Benchmark efforts are also valuable to assess progress over time.

### Conclude Every Major Project with an Open Lessons Learned Presentation of the Results to Share Knowledge Gained, Demonstrate New Technology, and Gain Official Closure

Many organizations rush through the project conclusion phase in order to get on with the next assignment. However, taking the time to fully document accomplishments, outstanding issues, deferred deliverables, and lessons learned is a worthwhile exercise that gives proper credit for successful completions, reduces misunderstandings over omissions and facilitates transitions to follow-up projects. On a quarterly basis, utilizing webcast technology, IT management and lead staff meet to hear presentations on the Lessons Learned in recent projects and those presentations are retained for future reference as needed.

### Adapt to Meet the Business Needs of the Customer

Now more than ever, the rapidly evolving healthcare insurance environment dictates that Medical Mutual use more flexible software development methodologies to fulfill business requirements. Methodologies may include development models (sequential, incremental and iterative) used in conjunction with one or more techniques (agile, prototyping, object-oriented). Website development appears to be particularly well suited to an agile methodology where time to market is a key concern. Regardless of the model or technique used in the development effort, delivery of a successful project is dependent on the project management framework of common processes, e.g. initiation, planning, execution, monitor and control, and closure. This framework provides common vocabulary for project managers and allows all stakeholders to understand the project phases more clearly.

**Conclusion**

One of the major reasons projects lose control is because project managers operate reactively rather than anticipating possible stumbling blocks. When the stumbling block finally appears, they are hesitant to make effective decisions, which often results in situations that can kill a project. Medical Mutual' experience has been that the aforementioned best practices help foster a more proactive environment in partnership with the customer, resulting in successful project delivery more consistently. Moreover, the outlined approach strikes a balance that requires minimum project overhead while obtaining maximum benefit.

## 4.30 HOLCIM[21]

Holcim is one of the world's leading suppliers of cement and aggregates (crushed stone, gravel and sand) as well as further activities such as ready-mix concrete and asphalt including services. The Holcim Group holds majority and minority interests in more than 70 countries on all continents. The cement industry is a capital-intensive, long term investment business. Executing projects safely, within expected costs, time targets and quality targets is crucial for the long term financial success of such companies. Due to the structure of the company and limited available resources, projects are initiated and executed locally with a central support organization that provides the know-how in technology concepts, planning and quality control.

**Project Management Methodologies**

The Project Management Approach (PMA) is the Holcim standard methodology for all projects, and is based on PMI standards.

In the late '90s, Holcim decided to develop a standard and structured methodology in managing projects and it was introduced across the company for developing and executing all types of projects. This methodology consists of 5 phases and 25 steps covering the entire Project Management Life Cycle. It includes most of the 10 Knowledge areas of PMI's *PMBOK® Guide*, and it is supported by a series of project templates. Special emphasis is put on the definition, closing and evaluation phases, to enable knowledge sharing within the Holcim Group.

Holcim has Specific Project Management Standards for investment projects (CAPEX projects)

In order to maintain and improve its business results, over the last decade Holcim has been investing on average 50% of its cash flow from operations in so-called CAPEX projects; projects to maintain, improve and expand Holcim's production facilities. As a

---

21. ©2013 by Holcim. Reproduced by permission. All rights reserved. Material in this section has been provided by Roberto Nores, Head of CAPEX Project Management Office, Holcim Technology Ltd. - Holcim Group

further step in standardization, Holcim decided to expand PMA for application on CAPEX Projects, incorporating the most important and crucial tasks and deliverables. Four new methodologies were developed and implemented. They consist of the same 5 phases of PMA, but the number of steps has been extended to between 50 and 200; each one is supported by tools and guidelines. The new methodologies are:

- ProMap: Methodology for CAPEX projects in the Cement business. It is divided in 2 sub methodologies, ProMap for Small projects (investments below 2 million CHF) and ProMap for Medium and Large projects (investments above 2 million CHF)
- AggPro: Methodology for CAPEX projects in the Aggregates business.
- RMXPro: Methodology for CAPEX projects in the Ready-mixed Concrete business.
- AsphPro: Methodology for CAPEX projects in the Asphalt business.

Since most Project Managers know PMA, they can apply the CAPEX-specific methodologies and work steps with limited additional training.

**Project Portfolio Management** As a consequence of its efforts in Project Management, Holcim recognized the need to establish clear links between its yearly, high-level planning (business planning and budgeting processes) and the daily, operational execution of projects. An average the Holcim Group Company runs between 100 and 300 projects each year. There was a need to optimize project prioritization, cash flows, resource allocation and portfolio monitoring. Consequently, a Project Portfolio Management methodology was established that supports top management in each country to manage its annual Project Portfolio.

The Project Portfolio Management activities are supported by a common software that supports most of the Project Portfolio Management processes. So far this software is in place 60% of the Holcim operations, and it has good acceptance from employees and management.

**Project Management Training in Holcim** Throughout the years, Holcim has established a series of Project Management training programs to increase the level of competency of Project Managers. Currently Holcim works with a 3-level project management training concept as follows:

Local training programs
> PMA: this training aims at ensuring that project managers and teams understand the standard PMA and the basic Project Management principles behind it.
> ProMap: this training aims at ensuring CAPEX project managers understand the additional steps and tools of this approach.

Regional training programs:
> PM seminars and certification programs: these programs are designed for Project Managers dealing with the most important and complex projects in each country. The programs cover both hard Project Management skills (scheduling, Earned

Value Management, etc.) as well as soft competencies (leadership, stakeholder engagement, etc.). The courses include several pre- and post-seminar activities to maximize the impact of the new skills on their projects. The courses also facilitate the creation of networks of Project Managers in geographical regions (Asia, Europe, Latin America, etc.).

Global training:

Large CAPEX Project Management Seminar: this training serves as an exchange platform where Project Managers and key stakeholders of largest CAPEX projects get together to exchange experiences and discuss best practices. The training is held at a construction site of the largest and most complex projects of Holcim.

**Project Management Offices at Holcim**

Throughout the years, Holcim has been establishing Project Management Offices (PMOs) to support its efforts in Project Management. Back in 1999, the first global Project Management Office was put in place to develop project management methodologies and competencies. It focused on developing and implementing the standard Project Management Approach, providing a set of standard tools, developing competencies of personnel, provide mentoring/coaching for project managers and promoting Project Management within the organization.

In the late 2000s Holcim began setting up global, functional PMOs in the areas with the largest quantity and investments in projects: IT and CAPEX. These PMOs focus on:

- Monitoring & controlling project performance: reporting project status to upper management, monitoring project Key Performance Indicators, etc.
- Multiproject Management: coordinating between projects, identifying, selecting and prioritizing new projects, allocating resources between projects, etc.
- Organizational Learning: Managing archives of project documentation, conducting post-project reviews and project audits, documenting and implementing lessons learned, documenting and implementing risk management processes

The last PMO set up at group level focuses on strategic aspects of the business, in particular supporting the implementation of the key performance improvement program of Holcim, the so-called Holcim Leadership Journey, that aims at adding CHF 1.5 billion in profits by 2014 in order to improve the Return On Invested Capital after tax to at least 8%. The Holcim Leadership Journey PMO focuses on providing advice to upper management, participating in strategic and financial planning, monitoring achievement of project benefits and financial targets, etc.

**Results Achieved**

Back in 1999, the Holcim Executive Committee made PMA a worldwide standard for its operations. Today, PMA and the CAPEX specific Project Management methodologies are the common language for Project Management within the Group. The majority of the 5000+ Project Managers around the world make use of these approaches, adapting it to the nature of each individual project.

*The application of these methodologies helped to:*
- reduce time to execute projects.
- reduce cost overruns as well as costs for investments
- improve capability to execute projects
- improve collaboration and team work

These efforts in Project Management are helping the Holcim Group to maintain its position as one of the leaders in manufacturing cement and in the construction materials sector.

## 4.31 WESTFIELD GROUP

Developing a project management methodology may not be as complicated as one believes. There are activities that a company can do to make the process of methodology development easier. The information in this section has been provided by Janet Kungl, PMP, program manager, Westfield Group.

There are four essential building blocks that, if present, allow for the development of an effective project management methodology in a reasonably short period of time:

- Recognition and support at the executive level of the need for project management expertise
- Establishment of a project-focused organization with a vision of how the project management discipline will be integrated in the organization
- Leveraging the *PMBOK® Guide* for methodology development
- Commitment to developing and/or hiring skilled project managers

Westfield Insurance (westfieldinsurance.com), an insurance, banking and related financial services group of businesses headquartered in Westfield Center, Ohio, recognized in the late 1990s that the needs of the business had changed with regard to information technology and information systems projects. In order to meet those needs, senior management recognized that project management capabilities were required as well as an organization that was designed to "partner" with the business and specialize in providing high-quality business and technology solutions. Westfield reorganized its information systems department to include a delivery unit focused on project delivery via the use of "virtual teams" comprised of technology and business team members.

In order to build the project management capability, one of Westfield's project managers was assigned the project of developing the project management methodology. The approach was to use the *PMBOK® Guide* as a starting point and customize the processes to reflect Westfield's culture and organizational structure. A key component of the ongoing support for the methodology was having the other project managers assigned to work on developing the processes.

The methodology starts with an explanation of the project phases (see Figure 4–45). Within each of these phases, the activities to be performed are defined and cross-referenced according to the nine knowledge areas of the *PMBOK® Guide*. An example of this

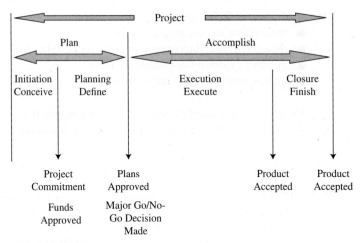

**Figure 4–45.**  Life-cycle phases.

| Initiation Phase | | | | | | | |
|---|---|---|---|---|---|---|---|
| Human Resource Management | | Project Startup Process | Project Manager Sourcing and Onboarding | Steering Committee Selection Process | | High-Level Resourcing | |
| Procurement Management | | | | Project Set-up | | | |
| Communications Management | | | | High-Level Comm Plan | | | Project Presentation Process |
| Risk Management | | | | | High-Level Risk Assessment | | |
| Scope Management | | | | | High-Level Scoping | | |
| Quality Management | | | | Initiate Customer Satisfaction Form for Team | | | Collect Lessons Learned |
| Cost Management | | | | | High-Level Cost/Benefit Analysis | | |
| Time Management | | | | | High-Level Scheduling | | |
| Integration Management | | Corporate Prioritization Process | Time → | | | Project Charter | |

**Figure 4–46.**  Initiation phase activities.

is shown in Figure 4–46. Each of the activities in Figure 4–47 can then be exploded into a flowchart showing the detailed activities needed for accomplishment of the deliverables for this activity. For example, the "Project Startup Process" activity in Figure 4–46 is cross-listed under Human Resource Management. The detailed activities are shown by the flowchart in Figure 4–47 Artifact templates and examples were also developed and are accessible in an electronic standard project folder. A continuous improvement process for the methodology was also implemented. Process improvements include the development of project success metrics, formalized project reviews, and defining the integration points with other processes [e.g., software development life cycle (SDLC), architecture governance, and portfolio planning]. Over time, experienced project managers were recruited from other organizations to supplement internally developed project managers. To support their professional development, a project manager forum was established in 2006. This biweekly meeting allows the project managers to share best practices and lessons learned with their peers. A project management experience path was documented in 2007 to identify key experiences and competencies that influence a project manager's career trajectory and define the path into, through, and out of the project management career track (see Figure 4–48).

The project management function gained acceptance by successfully delivering projects. Increased support from all levels of the organization is evident as more business customers request partnerships with teams led by project managers. Putting the right building

**Figure 4–47.** Project startup process.

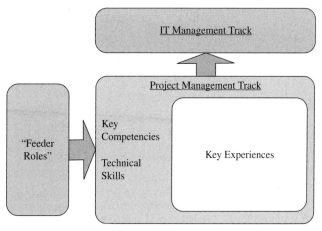

**Figure 4–48.** The project manager experience path.

blocks in place created the foundation necessary for moving ahead. Currently, Westfield is experiencing increased success in delivery and improved overall results with key business initiatives. These results validate the value of project management. Building toward excellence in project management continues at Westfield Insurance.

## 4.32 HEWLETT-PACKARD

Doug Bolzman, Consultant Architect, PMP®, ITIL expert at Hewlett-Packard, discusses the foundation for a project management methodology:

Many clients have a project management methodology and in addition, most companies have several methodologies that include project management disciplines. HP has been successful to integrate the definitive PM method into the other methodologies to allow for leveraging of PM standards and tools without duplication.

For the management of an enterprise IT environment, we have developed a client facing framework titled the Information Technology Enterprise Management (ITEM). This framework assists the client to map their strategic direction into feasible releases. ITEM is a preintegrated framework of three models and a methodology as illustrated in Figure 4–49.

Prior to the Release of ITIL the 2007 edition, we identified a gap for managing the lifecycle of a service and developed what we called back then, the Release Management Methodology which consisted of 4 Stages (Planning, Integration, Deployment and Operations). Each Stage consists of Phases - Activities - Tasks. This is shown in Table 4–4.

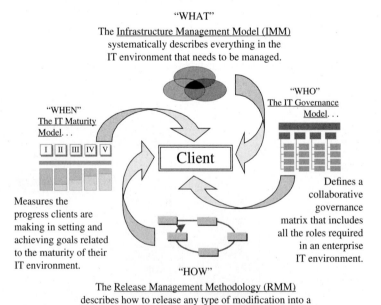

"WHAT"
The Infrastructure Management Model (IMM)
systematically describes everything in the
IT environment that needs to be managed.

"WHO"
The IT Governance
Model. . .

"WHEN"
The IT Maturity
Model. . .

I  II  III  IV  V

Client

Measures the
progress clients are
making in setting and
achieving goals related
to the maturity of their
IT environment.

Defines a
collaborative
governance
matrix that includes
all the roles required
in an enterprise
IT environment.

"HOW"
The Release Management Methodology (RMM)
describes how to release any type of modification into a
production IT environment in a structured, predictable manner.

**Figure 4–49.** An integrated approach to an integrated solution Activity phase mapping.

**TABLE 4–4. RELEASE METHODOLOGY STAGES**

| Stage | Description |
|---|---|
| Planning | The environment that is used to establish and manage the vision and strategic direction of the enterprise IT environment and proactively define the content and schedule of all IT releases. To provide a common means for the client and service providers to clearly and accurately plan the enterprise IT environment and manage all aspects of planning, estimating a release, and setting appropriate client expectations as to what each release will deliver. |
| Integration | The environment that is used to finalize the design of a planned infrastructure release and perform all of the required testing and client validation, preparing the release to be deployed to the user community. To provide a common means for the client, EDS, and service providers to clearly and accurately validate the accuracy, security, and content of each release and to finalize all development. To provide the client a clear and accurate portrait of the outcome of the release and to set proper expectations of the deployment and operations activities, costs, and schedules. |
| Deployment | The process that is used by the organization to implement new releases of the enterprise IT design (business, support, and technical components) to a target environment. To provide a common means for the client, EDS, and service providers to clearly and accurately schedule, deploy, and turn over to production the updated environment. |
| Operations | The production environment that is used to sustain and maintain the IT components and configurable items that are part of the enterprise IT environment. To provide a stable IT environment that is required by the IT users to support their business roles and responsibilities. |

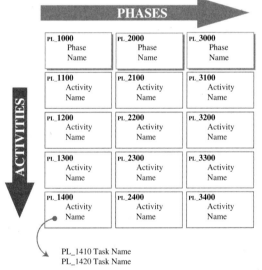

**Description:**
The sequential hierarchy of events should there be no need to deviate. This demonstrates the highest percentage of occurrence when the process is executed. Main intent for this chart is to expose the tasks subordinate to each activity.

**Value:**
• Each work element can be identified and mapped internal or external to the component
• All activities divided into tasks (justified)
• All work is represented once (even if executed many times)

**Structure:**
• All work elements have a unique identifier
  – Phase numbered by the thousands
  – Activity numbered by the thousands
  – Task numbered by the tens
• All work elements preceded by Component Name Identifier (PL_1000; PL_1100)
• Activity and Task Names kept relatively short; descriptive of work complete, not sentences
• All tasks make up the scope of the activity
• This is NOT a relationship model

**Figure 4–50.** Activity phase mapping.

Since then, ITIL has introduced the lifecycle approach in the 2007 and 2011 editions. Their stages, Service Strategy, Service Design, Service Transition, Service Operation and Continual Service Operation are almost an exact match to our RMM stages. Since the rollout of 2007, we have maintained much of our collateral, since the new ITIL editions did not provide a comprehensive work breakdown structure. We simply matched our WBS to the overall goals and contents of the ITIL lifecycle framework. That is an illustration of how we continually evolve our practices and continually apply industry advances.

Basically we used the 9x9 rule as shown in Figure 4–50. If there are more than 9 phases across a stage and more than 9 activities to each phase (81 total units of work), then the scope will become too large, and our numbering scheme will become unstable.

## 4.33 DTE ENERGY

Henry Campbell, principal analyst, Customer Service Project Management Office and Time Menke, PMP, senior continuous improvement expert, DO Performance Management at DTE Energy, discusses the growth of project management at DTE Energy:

> DTE Energy has created a common approach to ensure projects are implemented in a consistent manner. In the past, multiple project management methodologies were used throughout the enterprise. Several business units (specifically, the Customer Service (CS), Distribution Operations (DO), Gas Operations (GO), Operating Systems Strategy Group (OSSG), and Information Technology Services (ITS) groups) agreed that a "common" approach was needed.
>
> The "CI Project Portfolio Management Process" (Figure 4–51) was developed utilizing internal and external best practices. This approach was developed to provide a common framework for Continuous Improvement projects, however other types of projects (capital, ad-hoc, etc.) have also benefited from its structure. The process contains seven project management phases:

> I.  Identification
>
>     Potential projects are identified from a variety of sources including (but not limited to):
>
> ● Corrective Action Requests
> ● After Action Reviews
> ● Benchmarking
> ● Swarms
>
>     Projects deemed "Just Do It (JDI)" move past the Selection phase, directly into Initiation.

> II. Selection
>
>     Potential projects are then ranked according to established Project Selection Criteria, business cases are created if applicable, the project is registered with the appropriate organization and work is started on the project charter.

> III.   Initiation
>
>     The Initiation phase includes an assessment of the current state of business and why a change is needed. The Project Charter is refined and a decision to proceed is made. (Figure 4–52).

**Figure 4–51.** CI project portfolio management process.

# PROJECT CHARTER: Version                     Date:

## Project Information and Gate Sign-Off

| | Gate One through Four Sign Off- Initial and Date at Review | | | | | | |
|---|---|---|---|---|---|---|---|
| | Assess | | Design | | Implement | | Results |
| | G1 | Date | G2 | Date | G3 | Date | G4 | Date |
| Project Name | | | | | | | |
| Team Lead | | | | | | | |
| Champion | | | | | | | |
| Process Owner | | | | | | | |

| Business Impact Analysis Reviewed- G1 | Yes | No |
|---|---|---|

**Problem Statement-G1**

**Current State- G1**

| Metrics | Baseline- G1 | Target- G2 | Current- G3 | Final - G4 | Future State- G2 |
|---|---|---|---|---|---|
| | | | | | |
| | | | | | |
| | | | | | |
| | | | | | |

| Team Members- G1 | Role | % Time |
|---|---|---|
| | SME | |
| | Core Team | |
| | | |
| | | |

**Gaps- G2**

### Four Milestone Dates- G1

| Planned Date- | | Target Date | Rev 1 | Rev 2 | Rev 3 |
|---|---|---|---|---|---|
| Gate One | Assessment | | | | |
| Gate Two | Design | | | | |
| Gate Three | Implementation | | | | |
| Gate Four | Results | | | | |

**Results Summary- G4**

| Cost and Resource Assessment Reviewed- G1 | Yes | No |
|---|---|---|

**In Scope Out of Scope- G1**

**Figure 4-52.** Project charter template.

288

The Project Charter contains:

- Project Description and Purpose
- Team Members and Roles
- Timeline
- Case for Change
- Assumptions, Risks and Challenges
- Current State Identification
- Ideal State Identification
- Gaps Assessment (Current and Ideal States)
- Major Milestones
- Metrics (Baseline and Target)

## IV. Planning

The Planning phase establishes the desired state of the change with an emphasis on process, technical and cultural (re)design. Project managers are given several tools to assist with the overall project planning. One of the tools is the "Project Plan Template" (Figure 4–53). The template contains a standard planning methodology for Customer Service Projects. It also provides project managers with standard lag and lead times for certain resource intensive tasks. Each project manager customizes the template respective to their initiative.

The Project Plan Template contains:

- Key Project Activities
- Project Plan Elements (Common Activities, Start and Finish Dates, and Assigned Resources)
- High Level Milestones

## V. V/VI. Execution/Controlling

The actual process and procedural changes, technical coding, training, and necessary "course corrections" take place during the "Executing and Controlling" phases. If a project change is required that is related to scope, time, or financial resources, the project manager completes a "Change Request" (Figure 4–54) form. In addition, the project manager provides a bi-weekly status, used to populate the "Project Performance Board" (Figure 4–55).

The Project Performance Board tracks:

- Project Information
- Project Name and ID
- Project Manager Name
- Percent Complete
- Phase Completion (Gate Approvals)
- Issues

## VI. Closeout

During the "Closeout" phase, the project manager qualifies the actual project deliverables against the plan. The project is formally closed when the desired targets are attained. The project manager completes the "Project Closeout Report" as part of the closeout process.

| Activity ID | Activity Name | Performance % Complete | Start | Finish | Activity Owner |
|---|---|---|---|---|---|
| CSPMv3 Project Plan Template | | 0% | 27-Jan-... | 27-Jan-... | |
| CSPMv3.0 Benefits | | 0% | 27-Jan-... | 27-Jan-... | |
| CSPMv3.Gate 1 Initiation / Assessment | | 0% | 27-Jan-... | 27-Jan-... | |
| A3050 | Review Project Management Process and Methodology | 0% | 27-Jan-... | 27-Jan-... | |
| A1030 | Identify Project Scope | 0% | 27-Jan-... | 27-Jan-... | |
| A1040 | Secure Project Resources / Form Team | 0% | 27-Jan-... | 27-Jan-... | |
| A1050 | Assess Current Processes | 0% | 27-Jan-... | 27-Jan-... | |
| A1070 | Milestone: Gate 1 Approval | 0% | | 27-Jan-... | |
| A1005 | Project Charter | 0% | 27-Jan-... | 27-Jan-... | |
| CSPMv3.Gate 2 Planning - Design | | 0% | 27-Jan-... | 27-Jan-... | |
| A1080 | Complete Change Management Form | 0% | 27-Jan-... | 27-Jan-... | |
| A1090 | Develop Project Workplan | 0% | 27-Jan-... | 27-Jan-... | |
| A1170 | Milestone: Gate 2 Approval | 0% | | 27-Jan-... | |
| CSPMv3.Gate 2.2 Process | | 0% | 27-Jan-... | 27-Jan-... | |
| A1110 | Define Desired State of Processes | 0% | 27-Jan-... | 27-Jan-... | |
| A1120 | Conduct Gap Assessment | 0% | 27-Jan-... | 27-Jan-... | |
| A1130 | Develop Countermeaures / Prespecifications / Embedded Tests | 0% | 27-Jan-... | 27-Jan-... | |
| CSPMv3.Gate 2.1 Technology | | 0% | 27-Jan-... | 27-Jan-... | |
| A1140 | Develop Technical Requirements | 0% | 27-Jan-... | 27-Jan-... | |
| A1150 | Review Technical Requirements with IT | 0% | 27-Jan-... | 27-Jan-... | |
| A1160 | Design (IT) | 0% | 27-Jan-... | 27-Jan-... | |
| CSPMv3.Gate 3 Execution / Implementation | | 0% | 27-Jan-... | 27-Jan-... | |
| A1480 | Conduct After Action Review | 0% | 27-Jan-... | 27-Jan-... | |
| A1490 | Milestone: Gate 3 Approval | 0% | | 27-Jan-... | |
| CSPMv3.Gate 3.5 Training / Communications | | 0% | 27-Jan-... | 27-Jan-... | |
| A3060 | Implement Communication Plan | 0% | 27-Jan-... | 27-Jan-... | |
| CSPMv3.Gate 3.5.1 Develop Training and Communications Plan | | 0% | 27-Jan-... | 27-Jan-... | |
| A3180 | Content | 0% | 27-Jan-... | 27-Jan-... | |
| A3190 | Media(s) | 0% | 27-Jan-... | 27-Jan-... | |
| A3200 | Schedule | 0% | 27-Jan-... | 27-Jan-... | |
| A3210 | Logistics | 0% | 27-Jan-... | 27-Jan-... | |
| CSPMv3.Gate 3.5.2 Organization Rollout of Training | | 0% | 27-Jan-... | 27-Jan-... | |
| A3070 | Revenue Management & Protection | 0% | 27-Jan-... | 27-Jan-... | |
| A3080 | Field Operations | 0% | 27-Jan-... | 27-Jan-... | |
| A3090 | Billing | 0% | 27-Jan-... | 27-Jan-... | |
| A3100 | Data Acquisition | 0% | 27-Jan-... | 27-Jan-... | |
| A3110 | Consumer Affairs | 0% | 27-Jan-... | 27-Jan-... | |
| A3120 | Gas Operations | 0% | 27-Jan-... | 27-Jan-... | |
| A3130 | Other Organizations | 0% | 27-Jan-... | 27-Jan-... | |
| CSPMv3.Gate 3.5.2.1 Call Center | | 0% | 27-Jan-... | 27-Jan-... | |
| A3140 | Notify the CCC Board of Project Scope / Hi Level Training Requirements | 0% | 27-Jan-... | 27-Jan-... | |
| A3150 | Arrange for Project SME's to provide floor support (if necessary) | 0% | 27-Jan-... | 27-Jan-... | |
| A3160 | Conduct Training (or communicate) for Non Customer Reps | 0% | 27-Jan-... | 27-Jan-... | |
| A3170 | Conduct Training for Customer Reps | 0% | 27-Jan-... | 27-Jan-... | |
| CSPMv3.Gate 3.1 Process | | 0% | 27-Jan-... | 27-Jan-... | |
| A1180 | Develop / Modify Processes | 0% | 27-Jan-... | 27-Jan-... | |
| A1190 | Develop/Modify Procedures (SWI's) | 0% | 27-Jan-... | 27-Jan-... | |
| A1200 | Develop Implementation Plan | 0% | 27-Jan-... | 27-Jan-... | |
| CSPMv3.Gate 3.2 Technology | | 0% | 27-Jan-... | 27-Jan-... | |
| A1220 | Development (IT) | 0% | 27-Jan-... | 27-Jan-... | |
| A1230 | System Testing (IT) | 0% | 27-Jan-... | 27-Jan-... | |
| A1240 | Conduct User Accceptance Testing (UAT | 0% | 27-Jan-... | 27-Jan-... | |
| A1235 | Develop UAT Test Plan (Scenarios) | 0% | 27-Jan-... | 27-Jan-... | |
| A1241 | Report any UAT defects | 0% | 27-Jan-... | 27-Jan-... | |
| A1242 | Re-test fixed defects | 0% | 27-Jan-... | 27-Jan-... | |
| CSPMv3.Gate 3.4 Conduct Pilot (Placeholder, if necessary) | | 0% | 27-Jan-... | 27-Jan-... | |
| A1430 | Design | 0% | 27-Jan-... | 27-Jan-... | |
| A1440 | Develop | 0% | 27-Jan-... | 27-Jan-... | |
| A1450 | Implement | 0% | 27-Jan-... | 27-Jan-... | |
| A1470 | Modify Processes (Implement pilot learnings) | 0% | 27-Jan-... | 27-Jan-... | |
| CSPMv3.Gate 4 Closeout / Results | | 0% | 27-Jan-... | 27-Jan-... | |
| A1500 | Develop Sustainability Plan (if necessary) | 0% | 27-Jan-... | 27-Jan-... | |
| A1510 | Conduct Project After Action Review | 0% | 27-Jan-... | 27-Jan-... | |
| A1520 | Monitor Implementation Results | 0% | 27-Jan-... | 27-Jan-... | |
| A1540 | Make adjustments to implementation (if necessary) | 0% | 27-Jan-... | 27-Jan-... | |
| A1550 | Acknowledge and Release Project Resources | 0% | 27-Jan-... | 27-Jan-... | |
| A1660 | Prepare Project Closeout Document | 0% | 27-Jan-... | 27-Jan-... | |
| A1580 | Milestone: Gate 4 Approval / Project Complete | 0% | | 27-Jan-... | |

**Figure 4–53.** Project plan template.

| Customer Service Program Management | Project Change Template |
|---|---|

| Request Date | &lt;date&gt; | Organization | |
|---|---|---|---|

| *Project Manager* | &lt;pm name&gt; | *Project Number* | &lt;date&gt; |
|---|---|---|---|
| *Project Name* | &lt;project name&gt; | *Director/Sponsor* | &lt;date&gt; |

| Priority | High ☐ Medium ☐ Low ☐ | Business Case Impacted? | Yes ☐ No ☐ |
|---|---|---|---|
| Re-Baseline Required? | Yes ☐ No ☐ | | |

**Describe How this Change Impacts the Project**

o Include affected budget IT dollars

o State as change amount and change percent, if possible.

For example: Project cost increased by x dollars which represents y percent of approved project cost.

Scope: ☐ _____

Time: ☐ _____

Cost: ☐ _____

Other: ☐ _____

**Benefit of Change Request**

*Tracking Information—To be completed by Project Management Office*

| Additional Impact Analysis Required: Yes ☐ No ☐ | Review Date: |
|---|---|
| | Reviewer: |

Issues/Concerns with Change Request:

*Sponsor Approval – To be completed by Project Management Office*

| Approved ☐ Rejected ☐ | Approval Date: |
|---|---|

Reason:

**Attachments:** *Please attach: (a) revised cost estimate; (b) proposed schedule, as needed.*

**APPROVALS**

_____  Project Sponsor

_____  Director of CSPMO

_____  Project Manager

**Figure 4–54.** Project change template.

Customer Service Project
Management
Project Scorecard

| Project Name / Contact | Start Date / End Date | % Complete | Gate / Phase* | | Description | Comments | Project Type | 2009 Savings Target | Actual | Full Benefit Due Date | Assigned Resources |
|---|---|---|---|---|---|---|---|---|---|---|---|
| Customer Service Program Management | | | | | | | | | | | |
| Customer Billing | | | | | | | | | | | |
| Data Acquisition | | | | | | | | | | | |
| Customer Care Center | | | | | | | | | | | |
| Consumer Affairs | | | | | | | | | | | |

Cancelled: Project was Cancelled
Deferred: The Project has been Postponed until
Planned: Charter in progress
Active: Project is currently in progress
Implemented: Project has completed Gate 3

*OSSG 4 Gate = PMI Phase
Gate 1 Assessment = Initiation
Gate 2 Design = Planning
Gate 3 Implementation = Execution
Gate 4 Sustain = Closeout

**Figure 4–55.** Project performance board.

The Closeout Report contains:

- Project Summary
- Objectives Achieved
- Objectives Not Met
- Measures/Metrics
- Lessons Learned
- Further Actions

Every phase of the Common Project Management Process ends with a "Gate Approval" from the project sponsor. If approval is granted, the project formally moves to the next phase. Where appropriate, each business unit agrees to implement projects utilizing this common approach. While the process is the same across DTE Energy, the project management procedures are tailored to the needs of each specific department or organization.

## 4.34 ALSTOM[22]

**Introduction**

The context and environment of integrated system projects have become increasingly challenging due to the global and complex structure of these projects. Project integration is a key process assisting to align stakeholders and to efficiently manage a project through various dimensions. In modern project management, these dimensions include, in addition to quality, costs, and deadline (QCD); financial, environmental, political, health and safety aspects.

**V-cycle Methodology in Project Management of Industrial Integrated Systems:**

The V-cycle is an operational methodology of project integration. This chapter describes how this methodology can be applied to large and complex industrial projects.

### Customer Focus

V-cycle methodologies are increasingly used in project management to ensure an effective realization of projects. Military and aerospace industries were the initiators of the implementation of this management model, building complex integrated systems based on innovative products and technologies.

The project cycle (see Figure 4–56) assists with the process of transforming customer requirements (Input) into expected values/deliverables (Output). This corresponds also to the project cycle of industrial projects building integrated systems. This project cycle can be supported by a V-cycle methodology.

The implementation of the V-Cycle reinforces the role of the customer in a project cycle by continuously recalling both, the initial requirements and the expected final value. Monitoring check points, called gates, increases customer visibility and involvement

22. ©2013 by ALSTOM. Reproduced by permission. All rights reserved. Material provided by Lahcen Zeggoud, ALSTOM Group Programme Manager

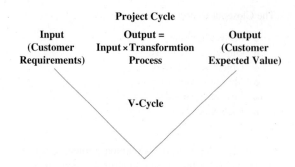

**Figure 4–56.**   V-cycle is a customer-to-customer process (C2C).

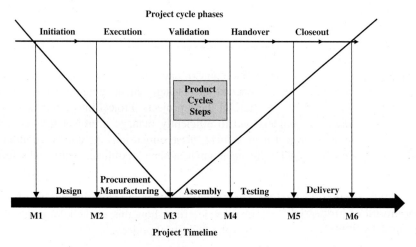

M1: Design start; M2: Procurement/Manufacturing start; M3:
Assembly start; M4: Testing start;  M5: Delivery  start; M6: Closeout

**Figure 4–57.**   Alignment of Project Management Phase and product steps.

during the project execution and validation. Consequently, the application of the V-cycle methodology helps project teams and organizations to be more customer focused.

## Alignment of Project and Product Cycles

The generic steps of a product development cycle and the generic phases of a project management cycle are shown on the V-cycle (See Figure 4–57). The alignment of the cycles, steps and phases ensures a controlled execution of a project. Therefore, the V-cycle assists project stakeholders and project managers to have a common understanding of the various stages of their projects and of the key check points to assess the physical progress, the risks and the deliverables.

## Monitoring of Milestones and Key Dates

Milestones and key dates of industrial projects are generally related to major events of the project, such as notice to proceed (NTP), delivery on site, start of revenue service or commercial operations, but can also be related to physical progress, like design validation, main purchase orders approval, factory tests report acceptance. Some of these milestones and key dates are used to declare revenues or, as payment milestones with customers or suppliers.

The milestones M1 to M6 shown in Figure 4–57 correspond to critical events where key decisions need to be taken. These milestones are generally included in the critical path of the project. The monitoring of V-cycle milestones improves the solidity and the consistency of the physical progress.

## Gate Reviews

V-cycle gates represent check points to assess the progress of the project. This is shown in Figure 4–58. These gates are implemented through formal project reviews with the participation of key stakeholders. They are monitored by an independent decision board (assessors) deciding on the Go/No Go. For enhanced transparency and alignment of key stakeholders, the gate review process is generally open to external stakeholders.

The gate reviews use the risk management process through assessment of all project activities against checklists. Critical risks or deliverables of a gate are assessed by KO questions; if declared no OK, the gate will be declared No Go.

## Technology and Product Freeze Risks

The V-cycle model can be applied to technology, products and systems cycles. A technology V-cycle ends with the technology fully validated and mature to be used in a product. A product V-cycle ends with the product fully validated and mature to be used in an integrated system cycle. Both Technology and Product cycles are managed, generally, within R&D projects or programs. See Figure 4–59.

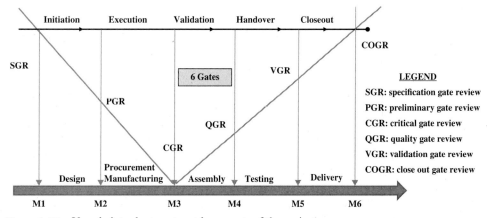

**Figure 4–58.** V-cycle introduces gates at key events of the project.

**Figure 4–59.** Technology, product and system cycle.

The freeze (the upper portion of Figure 4–59) of a technology or a product is considered the close out of the cycle. This close out specifies and validates the performance level, the costs, and the implementation limits as well as associated risks related to the technology or product. This freeze is, therefore, a crucial decision to perform a controlled system project. Generally, the lack of a technology or product cycle close out introduces high level of uncertainties in the scope definition, the time line and the risk management of system projects.

Most of industrial integrated system projects or programs using immature and not frozen technologies or products (the lower portion of Figure 4–59) are out of control from a planning, cost and risk point of view. Their baselines are difficult to establish and to validate by stakeholders during the project initiation phase. In addition, margin and/or schedule slippages often occur as consequence, leading to contractual claims and disputes between stakeholders. In some cases, these disputes can only be resolved through arbitration or termination of the contract.

### Gates Synchronization

The management of interfaces is one of the key challenges in project integration management. In large and complex projects usually several interfaces need to be managed due to the large number of work packages allocated to project stakeholders. A top-down synchronization applies to specification/requirement gates (left leg of the V-cycle), a bottom-up approach applies to assembly and validation gates (right leg of the V-cycle).

Project specifications have to be frozen with the customer in a specification gate review (SGRP) before freezing the work package specifications with stakeholders

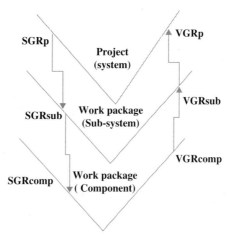

**Figure 4–60.** Synchronization process of V-cycles through gates alignment.

(SGRsub). The sub-system work package is validated (VGRsub) before its integration to system project and validation (VGRp). See Figure 4–60.

The synchronization of V-cycles gates allows the alignment of stakeholders in terms of physical progress and interface management at all stages of the project.

## 4.35 CASSIDIAN: GOLDEN RULES IN PROJECT MANAGEMENT[23]

All of the companies discussed in this chapter have excellent methodologies for project management. When a company captures best practices in project management and relates it to the methodology, then the company can develop "Golden Rules in Project Management." The remainder of this section has been provided by Cassidian. Cassidian provides excellent examples of how project management should work for the benefits and value to be achieved. This material is also representative of what the outcome may be from the journey to excellence in project management as discussed in Chapter 3.

**Why Golden Rules for Project Management?** The Golden Rules for Project Management are the top-level element of the Programme Management over project life cycle. They are developed under the requirement to be easily understandable and applicable for every project/programme. The system of having Golden Rules for Programme Management provides a set of rules which are common for all projects and programmes.

They form a basic body of regulations which have to be followed without exception. They are written in a way that they fit both the highly complex, high-volume, high-risk projects [and] the low-budget [ones] as such resource limited projects.

---

They describe the major areas where all projects have to cover the same level of performance in order to increase the quality of the projects/programmes. Following these rules shall lead to a common basic quality standard which in the end shall help to avoid any critical deficiencies.

Finally, these Golden Rules shall help to increase the overall project/programme excellence with respect to "time – cost – quality" requirements and are the backbone of the process "Program & Project Management."

**Golden Rules for Project Management**

*Golden Rule:* The Project Manager is **fully responsible** for the project in terms of cost, cash, time and quality (as indicated in the Project Charter) and is actively supported by his/her sponsor in the line management.

*Target:* Empowerment of Project Manager and clear definition of responsibilities within Proposal and Execution Team against other functional areas, but also obliging the Project Manager to fulfill his responsibility.

*Golden Rule:* An adequately qualified **Project Manager** shall be assigned by the Organization, actively involved during proposal preparation and contract negotiation.

*Target:* Foster active co-operation of proposal and project execution responsible in order to increase transparency and lossless transfer from initiating to planning phase.

*Golden Rule:* Project Management responsibility, based on the **contract related baselines and project charter** (e.g., cost, scope and schedule, prefunding agreements, final project categorization), shall be handed over officially from the bid team to the project team within maximum 10 working days after contract signature.

*Target:* Fast transfer, full set of documents available, long-term target is fixed time period. By being within these 10 days, proof is gained that proposal team provided adequate quality

*Golden Rule:* The Project Manager shall establish a formal, realistic and measureable **integrated project plan** during the project planning phase not later than three months after contract signature. A performance measurement baseline (cost, scope and schedule) shall be established, against which the project progress is measured and reported on a monthly basis.

*Target:* Ensure integrated planning including project schedule, major & minor milestones, milestone/dependencies, cost planning, resource planning and baselines. Establish Earned Value Management as the basis to monitor and to track project planning based on an initial baseline.

*Golden Rule:* The Proposal Manager (before contract signature) and afterwards the Project Manager (in the Project Execution) own the **risks and opportunities** of the project and ensure they are managed proactively. The Chief Engineer supports in that task by taking responsibility for the technical risks and opportunities.

*Target:* Ensure proper risk and opportunity management within the projects following defined rules and regulations from project initiation.

*Golden Rule:* The project scope and targets shall be managed with **focus on customer requirements**, and a dedicated Requirements Management shall be established in order to avoid scope creep. In addition change & configuration management shall be fully in place prior to Project Execution

*Target:* Increase of Requirements Management both against gold-plated and non-compliances. Strong focus on customer requirements necessary as well as definition of project scope and targets through complete project life cycle. Enhance of early customer acceptance of project requirements in order to avoid future misunderstanding.

*Golden Rule:* The Project Manager shall establish **communications** (formalized within Project Communication Plan) to facilitate working as an **I**ntegrated **P**roject **T**eam and ensure an optimum information flow within the team.

*Target:* Everyone within project team knows their ways of communication and their internal or external interfaces. Reporting is clearly establish and for any project stakeholders.

## 4.36 WHEN TRADITIONAL METHODOLOGIES MAY NOT WORK

While methodologies serve a viable purpose, traditional methodologies may not work well when projects become distressed and rapid action is required to save a failing project. In such a case, other factors may become more important than following the traditional life-cycle phases.

**Insights about Recovering Troubled Projects and Programs**

Today's projects and programs have become so complex that, on a daily basis, effective project recovery techniques may be required regardless of the country you are in or the business base of your company. We must be willing to face different or unforeseen situations that are not related to citizenship, the language we speak, or the experiences we all have. We are affected by a multitude of internal and external risks that can become real issues during project execution.

Below some insights from Dr. Alexandre Sörensen Ghisolfi, who for many years has collaborated with the International Institute for Learning and faced these challenges in the global project management community. Following are some of his best practices and things to think about.

Project recovery can be source of many ideas and lessons learned. When projects require recovery, they are normally accompanied by conflicts, disagreements and even fights.

When projects are in trouble, you will probably get a better understanding of who the people really are, and whether they are truly committed to the organization. In other words, when things go well we can easily see smiles and we often know the best side of people. When things go bad, a different individual usually emerges.

In this kind of environment we learned that recovery usually requires a team of experts and effective project management leadership for recovery to be successful. Not all project managers have skills in recovery project management techniques. Trust in both the project manager and the solution is probably the most essential criteria for recovery to work. Furthermore, dedicated project management teams are usually essential.

When conditions indicate that failure may be imminent, we must to be able to clearly change the cultural aspects where bad feelings can and will surface. In this way, we may be able to create a new culture conducive to recovery.

Recovery project management teams are composed of senior people, professional experts, young people with new ideas, new talent that recognizes that successful recovery may benefit their career goals, and recovery project managers with leadership skills. They must all work together such that we can convert a bad feeling environment to an environment where people believe we can still bring back the project and deliver what is required. If the team succeeds successfully, this will, even more, make the team proud of their accomplishments. This can develop a great buy-in feeling that remains over a lifetime.

During recovery, we must consider two different environments:

● Human behaviors
● Application of technical expertise

**Human Behaviors**        Each project that is in trouble has very different scenarios and alternatives. The recovery process depends about your experience and ability to find solutions. It also depends on how well you can influence the different stakeholders to bring them to an agreed upon vision where they recognize that the game can be won. For successful solutions that are based upon the different contributions of the team members, the project manager needs to know how to extract the best from them or influence them to achieve what is expected by the leader.

But before you start to identify and evaluate different alternatives, it would be good to consider some different aspects related to human behaviors which strongly influences the output. For simplicity sake, we will not talk about politics, hierarchy, knowledge and other aspects that influence human behaviors, but we suggest you at least clearly understand what kind of company and team you have. We can try to understand it through the study of organizational maturity.

Do you have a mature team? Is the company likewise mature in project management?

*Some best practices you should always try to put in place:*
● The organizational maturity of a company and/or team will directly affect the outcomes, so the more mature and professional team you have, the better the capability to recover failing programs. The best practice is to first deeply analyze company and team member maturity; having results at your disposal, you will be able to identify gaps, issues and conditions that may require change. After the maturity analysis, and having maturity reports in your hands, it will be necessary to prepare a recovery plan and show the project sponsors justification why some important actions must urgently take place. When working in matrix organizations, we can face important difficulties related to resources that are not directly part of our project organizational structure and which may have different interests in the outcome of the project.

*Additional best practices would certainly be:*
● Remind people about the necessity for change. If required, let's remind people every day about our mission and daily tasks.
● Give training to people as appropriate and act as a role model for the attitudes that are necessary. It is probably the best way to improve team and organizational maturity.

- Empower team members; make our challenge, their challenge.
- Make sure the entire team is deeply committed to the challenge of bringing back good results.

Ensure that the communications processes are effective. You can communicate less or even more, but at the end of the day communication must be effective. It depends again about your company and team maturity and about how much we are committed to the project. Truly committed team members will focus on communication. Communication must naturally flow.

When talking about processes and work, flexibility is important. But on other side, discipline to delivery key critical activities must be in place. Again, here the team organizational structure, departments, and suppliers interests can have a huge impact for good and bad things to happen.

Human behavior is probably more challenging than the application of the required technical expertise.

The study of organizational and team maturity can point use to a more secure way of proceeding. Since faster and better performance needs to be in place quickly, you cannot fail again; actions must be effective.

## Application of Technical Expertise

On the technical side, when recovering projects, it is may be even more important to clearly know or define priorities.

You will probably put emphasis on the quality criteria (i.e., quality acceptance criteria) of the products you need to delivery; undoubtedly you cannot sacrifice quality. The equilibrium of constraints will depend upon your negotiation abilities, as well the contractual conditions you may have with your customers.

You may not be able to delivery everything that is required by the project because sacrifices must be made. Perhaps you will delivery results differently when compared to the project baselines. A recovery plan needs to be in place immediately.

What to sacrifice? Costs? Project costs? Product costs? Timing? Downgrade specifications? Change product/project delivery strategies? Communication downgrading? Documentation writing, presentations?

Many factors can significantly impact success and increase the issues when having trouble projects. Here are some additional best practices you may try to put in place:

- You will probably be secure that emphasis in risk management is a must have condition. In troubled projects, risk management becomes even more important. When you encourage the team to perform a plan driven by risk management, the team is already defining priorities, which are the ones resulting from risk analysis. Risk management, as a holistic vision, can drive everything else around, such as: scope, time, team organization, skills, communications, etc.
- Put the best people you have on the most difficult activities first.
- Put emphasis on critical activities to shorten their respective durations.

- Avoid bringing new people on board that lack sufficient experience; but on the other hand, you could bring on board new people where you need to change the cultural aspects and/or interests in place.
- Avoid conflict of interests; we cannot lose time or waste resources solving unnecessary issues. Work even harder with your sponsor to gain their support.
- Adaptation of best practices available in the project scenario is a key point also. You must find out a way, often "out of the box," to make your team perform things that possibly never have been in practice before. Challenge your team members; ask them what they think and how we can start to work in different ways. In this way you are developing the buy-in feeling.
- You will probably look for quick wins. You will soon observe that some best practices the team tried to apply have been useful and some other ones not so well received. Replace or adopt best practices with no previous adoption and not necessarily applicable to your project, by other ones where you can quickly have satisfactory results. Fast identification of what is working and what is not working is crucial to recover the time lost.

Successfully recovering bad projects can be an amazing experience, and when you have a team with the proper mindset in place, it can significantly contribute to increasing enterprise and team project management maturity. Great results can be achieved on future projects by preventing them from entering into critical situations that can lead to failure.

# 5 Integrated Processes

## 5.0 INTRODUCTION

Companies that have become extremely successful in project management have done so by performing strategic planning for project management. These companies are not happy with just matching the competition. Instead, they opt to exceed the performance of their competitors. To do this on a continuous basis requires processes and methodologies that promote continuous rather than sporadic success.

Figure 5–1 identifies the hexagon of excellence. The six components identified in the hexagon of excellence are the areas where the companies excellent in project management exceed their competitors. Each of these six areas is discussed in Chapters 5–10. We begin with integrated processes.

**Figure 5–1.** Six components of excellence. Source: Reprinted from H. Kerzner: In Search of Excellence in Project Management, Hoboken, NJ: Wiley, 1998, p. 14.

## 5.1 UNDERSTANDING INTEGRATED MANAGEMENT PROCESSES _____

As we discussed in Chapter 1, since 1985 several new management processes (e.g., concurrent engineering) have supported the acceptance of project management. The most important complementary management processes and the years they were introduced are listed below:

- 1985: Total quality management (TQM)
- 1990: Concurrent engineering
- 1992: Employee empowerment and self-directed teams
- 1993: Reengineering
- 1994: Life-cycle costing
- 1995: Change management
- 1996: Risk management
- 1997: 1998: Project offices and centers of excellence
- 1999: Co-located teams
- 2000: Multinational teams
- 2001: Maturity models
- 2002: Strategic planning for project management
- 2003: Intranet status reporting
- 2004: Capacity-planning models
- 2005: Six Sigma integration with project management
- 2006: Virtual project management teams
- 2007: Lean/agile project management
- 2008: Best practices libraries
- 2009: Project management business process certification
- 2010: Managing complex projects
- 2011: Governance by committees
- 2012: Competing constraints using a value component
- 2013: Advances in metrics management

The integration of project management with these other management processes is key to achieving sustainable excellence. Not every company uses every process all the time. Companies choose the processes that work the best for them. However, whichever processes are selected, they are combined and integrated into the project management methodology. Previously, we stated that companies with world-class methodologies try to employ a single, standard methodology based upon integrated processes. This includes business processes as well as project management–related processes.

The ability to integrate processes is based on which processes the company decides to implement. For example, if a company implemented a stage gate model for project management, the company might find it an easy task to integrate new processes such as concurrent engineering. The only precondition would be that the new processes were not treated as independent functions but were designed from the onset to be part of a project management system already in place. The four-phase model used by the General Motors Powertrain Group and the PROPS model used at Ericsson Telecom AB readily allow for the assimilation of additional business and management processes.

Previously, we stated that project managers today are viewed as managing part of a business rather than just a project. Therefore, project managers must understand the business and the processes to support the business as well as the processes to support the project. Companies such as Visteon and Johnson Controls understand this quite well and have integrated business processes either into or with their project management methodology.

This chapter discusses each of the management processes listed and how the processes enhance project management. We then look at how some of the integrated management processes have succeeded using actual case studies.

## 5.2 EVOLUTION OF COMPLEMENTARY PROJECT MANAGEMENT PROCESSES

Since 1985, several new management processes have evolved in parallel with project management. Of these processes, TQM and concurrent engineering are the most relevant. Companies that reach excellence are the quickest to recognize the synergy among the many management options available today. Companies that reach maturity and excellence the quickest are those that recognize that certain processes feed on one another. As an example, consider the seven points listed below. Are these seven concepts part of a project management methodology?

- Teamwork
- Strategic integration
- Continuous improvement
- Respect for people
- Customer focus
- Management by fact
- Structured problem solving

These seven concepts are actually the basis of Sprint's TQM process. They could just as easily have been facets of a project management methodology.

During the 1990s, Kodak taught a course entitled Quality Leadership. The five principles of Kodak's quality leadership program included:

**Customer Focus**

"We will focus on our customers, both internal and external, whose inputs drive the design of products and services. The quality of our products and services is determined solely by these customers."

**Management Leadership**

"We will demonstrate, at all levels, visible leadership in managing by these principles."

**Teamwork**

"We will work together, combining our ideas and skills to improve the quality of our work. We will reinforce and reward quality improvement contributions."

**Figure 5–2.** Totally uncoupled processes.

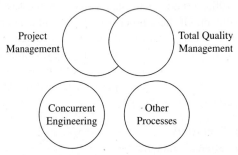

**Figure 5–3.** Partially integrated processes.

### Analytical Approach

"We will use statistical methods to control and improve our processes. Data-based analyses will direct our decisions."

### Continuous Improvement

"We will actively pursue quality improvement through a continuous cycle that focuses on planning, implementing, and verifying of improvements in key processes."

Had we looked at just the left column, we could argue that these are the principles of project management as well.

Figure 5–2 shows what happens when an organization does not integrate its processes. The result is totally uncoupled processes. Companies with separate methodologies for each process may end up with duplication of effort, possibly duplication of resources, and even duplication of facilities. Although there are several processes in Figure 5–2, we will focus on project management, TQM, and concurrent engineering only.

As companies begin recognizing the synergistic effects of putting several of these processes under a single methodology, the first two processes to become partially coupled are project management and TQM, as shown in Figure 5–3. As the benefits of synergy and integration become apparent, organizations choose to integrate all of these processes, as shown in Figure 5–4.

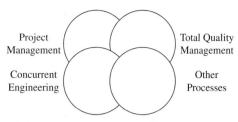

**Figure 5–4.** Totally integrated processes.

Excellent companies are able to recognize the need for new processes and integrate them quickly into existing management structures. During the early 1990s, integrating project management with TQM and concurrent engineering was emphasized. Since the middle 1990s, two other processes have become important in addition: risk management and change management. Neither of these processes is new; it's the emphasis that's new.

During the late 1990s, Steve Gregerson, formerly vice president for product development at Metzeler Automotive Profile System, described the integrated processes in its methodology[1]:

> Our organization has developed a standard methodology based on global best practices within our organization and on customer requirements and expectations. This methodology also meets the requirements of ISO 9000. Our process incorporates seven gateways that require specific deliverables listed on a single sheet of paper. Some of these deliverables have a procedure and in many cases a defined format. These guidelines, checklists, forms, and procedures are the backbone of our project management structure and also serve to capture lessons learned for the next program. This methodology is incorporated into all aspects of our business systems, including risk management, concurrent engineering, advanced quality planning, feasibility analysis, design review process, and so on.

Clearly, Metzeler sees the integration and compatibility of project management systems and business systems. Another example of integrated processes is the methodology employed by Nortel. During the late 1990s, Bob Mansbridge, then vice president, supply chain management at Nortel Networks, believed[2]:

> Nortel Networks project management is integrated with the supply chain. Project management's role in managing projects is now well understood as a series of integrated processes within the total supply chain pipeline. Total quality management (TQM) in Nortel Networks is defined by pipeline metrics. These metrics have resulted from customer and external views of "best-in-class" achievements. These metrics are layered and provide connected indicators to both the executive and the working levels. The project manager's role is to work with all areas of the supply chain and to optimize the results to the benefit of the project at hand. With a standard process implemented globally, including the monthly

---

1. H. Kerzner, *Advanced Project Management: Best Practices on Implementation*, Hoboken, NJ: Wiley, 2000, p. 188.
2. Ibid.

review of pipeline metrics by project management and business units, the implementation of "best practices" becomes more controlled, measurable, and meaningful.

The importance of integrating risk management is finally being recognized. According to Frank T. Anbari, professor of project management, Drexel University:

> By definition, projects are risky endeavors. They aim to create new and unique products, services, and processes that did not exist in the past. Therefore, careful management of project risk is imperative to repeatable success. Quantitative methods play an important role in risk management. There is no substitute for profound knowledge of these tools.

Risk management has been a primary focus among health-care organizations for decades, for obvious reasons, as well as financial institutions and the legal profession. Today, in organizations of all kinds, risk management keeps us from pushing our problems downstream in the hope of finding an easy solution later on or of the problem simply going away by itself. Change management as a complement to project management is used to control the adverse effects of scope creep: increased costs (sometimes double or triple the original budget) and delayed schedules. With change management processes in place as part of the overall project management system, changes in the scope of the original project can be treated as separate projects or subprojects so that the objectives of the original project are not lost.

Today, excellent companies integrate five main management processes (see Figure 5–5):

- Project management
- Total quality management
- Risk management
- Concurrent engineering
- Change management

Self-managed work teams, employee empowerment, reengineering, and life-cycle costing are also combined with project management in some companies. We briefly discuss these less widely used processes after we have discussed the more commonly used ones.

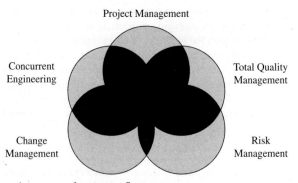

**Figure 5–5.** Integrated processes for twenty-first century.

## 5.3 ZURICH AMERICA INSURANCE COMPANY[3]

One of the benefits of having integrated processes is that it allows for more comprehensive and realistic contingency planning. Kathleen Cavanaugh states:

As we know, simply put, the goal of all PMOs is to deliver projects on time and on budget. There is increased scrutiny placed on projects these days and many companies have enacted a protective governance gauntlet to help ensure the right projects come to fruition. With the time to market push/shove, it is easy for projects budgets and end dates to be estimated, even "promised" too soon much to the unease of the Project Manager. To help alleviate this potential heartbreak, the IT PMO at Zurich American Insurance Company has implemented a project contingency process for both duration and dollars.

The contingency process helps to mitigate the risk of known unknowns within the scope of a project. Since we are part of the global Zurich Financial Services Group we have a rather strict governance process that takes time and money to navigate. Once you make it through the gauntlet you don't really want to go back for re-approvals, date extension authorizations, etc., so, proper planning for risks and project changes is imperative.

The contingency process uses a determination matrix that considers particular risk factors such as resources and technology complexity just to name a couple. It is designed to help the PM assign the appropriate amount of contingency needed for both dollars and duration. The concept is nothing new but is still not widely accepted as a necessity.

The goal is not only to protect us from time consuming project governance submissions, but also to move away from padding individual estimates. Before this contingency approach was introduced, some Project Managers buried contingency within their estimates. The major disadvantage of this is that because the contingency is "hidden," there is no systematic process to release funds back into the project funding pool as risks lessen throughout the project. Now, contingency is kept separate in the project plan so that we can have better insight into where and why original estimates were off. Only then can we find the root cause and improve upon the estimating approach.

One important thing to note is that the customer takes part in determining the need for contingency so they have a good understanding of why it is so critical to the success of the project. The process makes contingency transparent and once the customer understands that contingency dollars cannot be used without their acknowledgement, they are more open to understanding the inherent changes that occur during the life of a project.

Contingency should be actively managed as the project progresses. Each month when project risks are re-assessed and risk probability decreases, the amount of contingency should be adjusted accordingly in both budget and schedule. It is expected that contingency would be released from the project if it is determined to no longer be needed based on the updated risk assessment. This allows money to be available for other efforts in the company.

To summarize, the bullet points below are an outline of the steps taken to effectively use the contingency process.

*Plan:* The IT Project Manager works with the Business Project Manager and project team using the determination matrix to calculate the appropriate contingency percent when the project is ready to go for full funding.

*Document/Communicate:* PM updates the project plan and Usage Log to document and communicate the contingency figures.

*Approve:* The Sponsor is responsible for providing sign-off before contingency is used.

---

3. Material provided by Kathleen Cavanaugh, PMP, Zurich—ZNA PMO Lead

*Manage:* The Project Manager maintains the Usage Log as contingency is consumed and other contingency data is updated each month along with the risk assessment.

*Release:* Contingency funds are released back into project funding pool as risks decrease.

Overall, this process addresses the age old problem of project work starting before everything that needs to be known is known and gives PMs a fighting chance to deliver on-time and on-budget in this ever changing environment. Because after all, change happens.

## 5.4 TOTAL QUALITY MANAGEMENT

During the past decade, the concept of TQM has revolutionized the operations and manufacturing functions of many companies. Companies have learned quickly that project management principles and systems can be used to support and administer TQM programs and vice versa. Ultimately excellent companies have completely integrated the two complementary systems.

The emphasis in TQM is on addressing quality issues in total systems. Quality, however, is never an end goal. Total quality management systems run continuously and concurrently in every area in which a company does business. Their goal is to bring to market products of better and better quality and not just of the same quality as last year or the year before.

Total quality management was founded on the principles advocated by W. Edwards Deming, Joseph M. Juran, and Phillip B. Crosby. Deming is famous for his role in turning postwar Japan into a dominant force in the world economy. Total quality management processes are based on Deming's simple plan–do–check–act cycle.

The cycle fits completely with project management principles. To fulfill the goals of any project, first you plan what you're going to do, then you do it. Next, you check on what you did. You fix what didn't work, and then you execute what you set out to do. But the cycle doesn't end with the output. Deming's cycle works as a continuous-improvement system, too. When the project is complete, you examine the lessons learned in its planning and execution. Then you incorporate those lessons into the process and begin the plan–do–check–act cycle all over again on a new project.

Total quality management also is based on three other important elements: customer focus, process thinking, and variation reduction. Does that remind you of project management principles? It should. The plan–do–check–act cycle can be used to identify, validate, and implement best practices in project management.

One of the characteristics of companies that have won the prestigious Malcolm Baldrige Award is that each has an excellent project management system. Companies such as Motorola, Armstrong World Industries, General Motors, Kodak, Xerox, and IBM use integrated TQM and project management systems.

In the mid-1990s, during a live videoconference on the subject, "How to Achieve Maturity in Project Management," Dave Kandt, Group vice president for quality and program management at Johnson Controls during the time of the videoconference, commented on the reasons behind Johnson Controls' astounding success:

> We came into project management a little differently than some companies. We have combined project management and TQC (total quality control) or total quality management. Our first design and development projects in the mid-1980s led us to believe that

our functional departments were working pretty well separately, but we needed to have some systems to bring them together. And, of course, a lot of what project management is about is getting work to flow horizontally through the company. What we did first was to contact Dr. Norman Feigenbaum, who is the granddaddy of TQC in North America, who helped us establish some systems that linked together the whole company. Dr. Feigenbaum looked at quality in the broadest sense: quality of products, quality of systems, quality of deliverables, and, of course, the quality of projects and new product launches. A key part of these systems included project management systems that addressed product introduction and the product introduction process. Integral to this was project management training, which was required to deliver these systems.

We began with our executive office, and once we had explained the principles and philosophies of project management to these people, we moved to the management of plants, engineering managers, analysts, purchasing people, and of course project managers. Only once the foundation was laid did we proceed with actual project management and with defining the role and responsibility so that the entire company would understand their role in project management once these people began to work. Just the understanding allowed us to move to a matrix organization and eventually to a stand-alone project management department. So how well did that work? Subsequently, since the mid-1980s, we have grown from 2 or 3 projects to roughly 50 in North America and Europe. We have grown from 2 or 3 project managers to 35. I don't believe it would have been possible to manage this growth or bring home this many projects without project management systems and procedures and people with understanding at the highest levels of the company.

In the early 1990s we found that we were having some success in Europe, and we won our first design and development project there. And with that project, we carried to Europe not only project managers and engineering managers who understood these principles but also the systems and training we incorporated in North America. So we had a company-wide integrated approach to project management. What we've learned in these last 10 years that is the most important to us, I believe, is that you begin with the systems and the understanding of what you want the various people to do in the company across all functional barriers, then bring in project management training, and last implement project management.

Of course, the people we selected for project management were absolutely critical, and we selected the right people. You mentioned the importance of project managers understanding business, and the people that we put in these positions are very carefully chosen. Typically, they have a technical background, a marketing background, and a business and financial background. It is very hard to find these people, but we find that they have the necessary cross-functional understanding to be able to be successful in this business.

At Johnson Controls, project management and TQM were developed concurrently. Dave Kandt was asked during the same videoconference whether companies must have a solid TQM culture in place before they attempt the development of a project management program. He said:

I don't think that is necessary. The reason why I say that is that companies like Johnson Controls are more the exception than the rule of implementing TQM and project management together. I know companies that were reasonably mature in project management and then ISO 9000 came along, and because they had project management in place in a reasonably mature fashion, it was an easier process for them to implement ISO 9000 and TQM. There is no question that having TQM in place at the same time or even first would make it a little easier, but what we've learned during the recession is that if you want to compete in Europe and you want to follow ISO 9000 guidelines, TQM must be implemented. And using project management as the vehicle for that implementation quite often works quite well.

There is also the question of whether or not successful project management can exist within the ISO 9000 environment. According to Dave Kandt:

Not only is project management consistent with ISO 9000, a lot of the systems that ISO 9000 require are crucial to project management's success. If you don't have a good quality system, engineering change system, and other things that ISO requires, the project manager is going to struggle in trying to accomplish and execute that project. Further, I think it's interesting that companies that are working to install and deploy ISO 9000, if they are being successful, are probably utilizing project management techniques. Each of the different elements of ISO requires training, and sometimes the creation of systems inside the company that can all be scheduled, teams that can be assigned, deliverables that can be established, tracked, and monitored, and reports that go to senior management. That's exactly how we installed TQC at Johnson Controls, and I see ISO 9000 as having a very similar thrust and intent.

While the principles of TQM still exist, the importance of Six Sigma concepts has grown. According to Eric Alan Johnson and Jeffrey Alan Neal[4]:

**Total Quality Management**

In addition to the TQM PDCA cycle, the continuous improvement DMAIC (Define, Measure, Analyze, Improve, and Control) model can be used to improve the effectiveness of project management. This model has been successfully employed for Six Sigma and Lean Enterprise process improvement, but the basic tenets of its structured, data enabled problem solving methodology can also be employed to improve the success of project management.

By assessing data collected on both project successes and root cause of project failures, the DMAIC model can be used to improve and refine both the management of projects and the ultimate quality of products produced.

In the define phase, specific project definition and requirements are based on data gathered from the customer and on historical project performance. Gathering as much information as possible in these areas allows the project manager to concentrate on what is truly important to the customer while reviewing past performance in order to avoid the problems of and continue to propagate the successes of past projects. In the define stage, available data on the people, processes, and suppliers is reviewed to determine their ability to meet the cost, quality and schedule requirements of the project. The define phase, in short, should assess not only the requirements of the customer, but should also assess the capability of your system to meet those requirements. Both of these assessments must

---

4. Eric Alan Johnson, Satellite Control Network Contract Deputy Program Director, AFSCN, and the winner of the 2006 Kerzner Project Manager of the Year Award; and Jeffrey Alan Neal, Blackbelt/Lean Expert and Lecturer, Quantitative Methods, University of Colorado, Colorado Springs.

be based on data gathered by a dedicated measurement system. Additionally, the define stage should establish the metrics to be used during projects execution to monitor and control project progress. These metrics will be continually evaluated during the measure and analysis phase (these DMAIC phase are concurrent with the PMI phases of project management).

The next phase of the DMAIC model, measure, data (the metrics identified in the define stage) from the measurement system is continually reviewed during project execution to ensure that the project is being effectively managed. The same data metrics used in the define stage should be updated with specific project data to determine how well the project is progressing. The continual assessment of project performance, based on data gathered during the execution phase, is the key to data enabled project management.

During the continual measurement of the progress of the project, it is likely that some of these key metrics will indicate problems either occurring (present issues) or likely to occur (leading indicators). These issues must be addressed if the project is to execute on time, and on budget to meeting requirements. This is where the analysis aspect of the DMAIC model becomes a critical aspect of project management. The analysis of data is an entire field onto itself. Numerous books and articles have addressed the problem of how to assess data, but the main objective remains. The objective of data analysis is to turn data into usable information from which to base project decisions.

The methods of data analysis are specific to the data type and to the specific questions to be answered. The first step (after the data have been gathered) is to use descriptive techniques to get an overall picture of the data. This overall picture should include a measure of central tendency (i.e., mean), and a measure of variation such as standard deviation. Additionally, graphical tools such as histograms and Pareto charts are useful in summarizing and displaying information. Tests of significance and confidence interval development are useful in determining if the results of the analysis are statistically significant and for estimating the likelihood of obtaining a similar result.

In the continual monitoring of processes, control charts are commonly used tools to assess the state of stability of processes and to determine if the variation is significant enough to warrant additional investigation. In addition, control charts provide a basis for determining if the type of variation is special cause or common cause. This distinction is critical in the determination of the appropriate corrective actions that may need to be taken.

To provide a basis for the identification of potential root causes for project performance issues, tools such Failure Modes and Effects Analysis and the Fishbone (also known as the Ishkawa) diagram can be used to initiate and document the organized thought process needed to separate main causes of non-conformities from contributing causes.

If the data meets the statistical condition required, such tests as Analysis of Variance (ANOVA) and regression analysis can be extremely useful in quantifying and forecasting process and project performance. Because ANOVA (the General Linear Model) can be used to test for mean differences of two or more factors or levels, ANOVA can be used to identify important independent variables for various project dependent variables. Various regression models (simple linear, multiple linear and binary) can be used to quantify the different effects of independent variable on critical dependent variable that are key to project success.

In short, this phase uses the data to conduct an in-depth and exhaustive root cause investigation to find the critical issue that was responsible for the project execution problem and effects upon the project if left uncorrected.

The next phase involves the process correction and improvement that addressed the root cause identified in the previous phase. This is corrective action (fix the problem you are facing) and preventive action (make sure it or one like it doesn't come back). So, once the root cause has been identified, both corrective and preventive process improvement actions can be taken to address current project execution and to prevent the reoccurrence of that particular issue in future projects. To ensure that current projects do not fall victim to that problem recently identified and that future projects avoid the mistakes of the past, a control plan is implemented to monitor and control projects. The cycle is repeated for all project management issues.

The continuing monitoring of project status and metrics along with their continual analysis and correction is an ongoing process and constitutes the control phase of the project. During this phase, the key measurements instituted during the initiation phase are used to track project performance against requirements. When the root cause of each project problem is analyzed, this root cause and the subsequent corrective and preventive action are entered into a "lessons learned" database. This allows for consistent problem resolution actions to be taken. The database is also then used to identify potential project risks and institute a priori mitigation actions.

## Risk/Opportunity Management Using Six Sigma Tools and Probabilistic Models

Risk/Opportunity Management is one of the critical, if not the most, tools in a project or program managers tool box—regardless of contract type. Typically projects/programs focus on the potential impact and/or probability of a risk occurrence. While these are very critical factors to developing a good risk mitigation plan, the adroit ability of the project team to *detect* the risk will have the greatest impact on successful project execution. If you can't detect the risk, then your ability to manage it, will always be reactive. The undetectable risk is a greater threat to execution, than the high-probability or high-impact factors. This is where using one of the six sigma tools Failure Modes and Effects Analysis (FMEA) can be very effective. The FMEA tool can help a project team evaluate—risk detection. Focusing on risk detection will help the team think "out of the box" in proposing, planning or executing a successful project.

Example: If your project/program has a risk that has a significant probability of occurrence, then it is probably not really a risk —it is an issue/problem. If the impact is great and the probability is low, then you will keep an eye on this, but not usually spend Management Reserve (MR) to mitigate. However, if the risk has a high impact or probability, but has a low level of detectability, the results could be devastating.

The other side of managing a project/program is the lack of focus on opportunity identification and management. If a project team is only risk management focused, they may miss looking at the projects potential opportunities. Opportunities need to be evaluated with the same rigor as risks. The same level of focus in the areas of impact, probability AND the ability to *recognize* the opportunity must occur for a project team. The FMEA is also. . . . very useful for opportunity recognition and management. Sometimes undetectable risks will occur, but the ability to recognize and realize opportunities can counter this risk impacts. The use of opportunity recognition can have the greatest impacts on Fixed-Price projects where saving costs can increase the projects profit margin.

If a project has risk schedule, how can we quantify that risk? One method is through the use probabilistic modeling. Probabilistic modeling of your schedule can help you forecast the likelihood of achieving all your milestones within your period of performance. If the risk of achieving your schedule is too high, you can use these models to perform "what if" analysis until the risk factors, can be brought to acceptable levels. This analysis should be done BEFORE the project is baselined or (ideally) during the proposal phase.

The key to successful implementation of this strategy is a relational database of information which will allow you to build the most realistic probabilistic model possible. This information must be gathered on a wide variety of projects so that information on projects of similar size and scope/complexity can be evaluated. It must be integrated with "lessons learned" from these other projects in order to build the best probabilistic model to mitigate your schedule risks. Always remember that a model is only as good as the information used to build it.

## 5.5 CONCURRENT ENGINEERING

The need to shorten product development time has always plagued U.S. companies. During favorable economic conditions, corporations have deployed massive amounts of resources to address the problem of long development times. During economic downturns, however, not only are resources scarce, but time becomes a critical constraint. Today, the principles of concurrent engineering have been almost universally adopted as the ideal solution to the problem.

Concurrent engineering requires performing the various steps and processes in managing a project in tandem rather than in sequence. This means that engineering, research and development, production, and marketing all are involved at the beginning of a project, before any work has been done. That is not always easy, and it can create risks as the project is carried through. Superior project planning is needed to avoid increasing the level of risk later in the project. The most serious risks are delays in bringing product to market and costs when rework is needed as a result of poor planning. Improved planning is essential to project management, so it is no surprise that excellent companies integrate concurrent engineering and project management systems.

Chrysler (now DaimlerChrysler) Motors used concurrent engineering with project management to go from concept to market with the Viper sports car in less than three years. Concurrent engineering may well be the strongest driving force behind the increased acceptance of modern project management.

## 5.6 RISK MANAGEMENT

Risk management is an organized means of identifying and measuring risk and developing, selecting, and managing options for handling those risks. Throughout this book, I have emphasized that tomorrow's project managers will need superior business skills in assessing and managing risk. This includes both project risks and business risks. Project managers in the past were not equipped to quantify risks, respond to risks, develop contingency plans, or keep lessons-learned records. They were forced to go to senior managers for advice on what to do when risky situations developed. Now senior managers are empowering project managers to make risk-related decisions, and that requires a project manager with solid business skills as well as technical knowledge.

Preparing a project plan is based on history. Simply stated: What have we learned from the past? Risk management encourages us to look at the future and anticipate what can go wrong and then to develop contingency strategies to mitigate these risks.

We have performed risk management in the past, but only financial and scheduling risk management. To mitigate a financial risk, we increased the project's budget.

To mitigate a scheduling risk, we added more time to the schedule. But in the 1990s, technical risks became critical. Simply adding into the plan more time and money is not the solution to mitigate technical risks. Technical risk management addresses two primary questions:

- Can we develop the technology within the imposed constraints?
- If we do develop the technology, what is the risk of obsolescence, and when might we expect it to occur?

To address these technical risks, effective risk management strategies are needed based upon technical forecasting. On the surface, it might seem that making risk management an integral part of project planning should be relatively easy. Just identify and address risk factors before they get out of hand. Unfortunately, the reverse is likely to be the norm, at least for the foreseeable future.

For years, companies provided lip service to risk management and adopted the attitude that we should simply live with it. Very little was published on how to develop a structure risk management process. The disaster with the Space Shuttle Challenger in January 1986 created a great awakening on the importance of effective risk management.[5]

Risk management today has become so important that companies are establishing separate risk management organizations within the company. However, many companies have been using risk management functional units for years, and yet this concept has gone unnoticed. The following is an overview of the program management methodology of the risk management department of an international manufacturer headquartered in Ohio. This department has been in operation for approximately 25 years.

> The risk management department is part of the financial discipline of the company and ultimately reports to the treasurer, who reports to the chief financial officer. The overall objective of the department is to coordinate the protection of the company's assets. The primary means of meeting that objective is eliminating or reducing potential losses through loss prevention programs. The department works very closely with the internal environmental health and safety department. Additionally, it utilizes outside loss control experts to assist the company's divisions in loss prevention.
>
> One method employed by the company to insure the entire corporation's involvement in the risk management process is to hold its divisions responsible for any specific losses up to a designated self-insured retention level. If there is a significant loss, the division must absorb it and its impact on their bottom-line profit margin. This directly involves the divisions in both loss prevention and claims management. When a claim does occur, risk management maintains regular contact with division personnel to establish protocol on the claim and reserves and ultimate resolution.
>
> The company does purchase insurance above designated retention levels. As with the direct claims, the insurance premiums are allocated to its divisions. These premiums are calculated based upon sales volume and claim loss history, with the most significant percentage being allocated to claim loss history.
>
> Each of the company's locations must maintain a business continuity plan for its site. This plan is reviewed by risk management and is audited by the internal audit and environmental health and safety department.

---

5. The case study "The Space Shuttle Challenger Disaster" appears in H. Kerzner, *Project Management Case Studies,*, 4th ed. Hoboken, NJ: Wiley, 2013, p. 447.

Risk management is an integral part of the corporation's operations as evidenced by its involvement in the due diligence process for acquisitions or divestitures. It is involved at the onset of the process, not at the end, and provides a detailed written report of findings as well as an oral presentation to group management.

Customer service is part of the company's corporate charter. Customers served by risk management are the company's divisions. The department's management style with its customers is one of consensus building and not one of mandating. This is exemplified by the company's use of several worker's compensation third-party administrators (TPAs) in states where it is self-insured. Administratively, it would be much easier to utilize one nationwide TPA. However, using strong regional TPAs with offices in states where divisions operate provides knowledgeable assistance with specific state laws to the divisions. This approach has worked very well for this company that recognizes the need for the individual state expertise.

The importance of risk management is now apparent worldwide. The principles of risk management can be applied to all aspects of a business, not just projects. Once a company begins using risk management practices, the company can always identify other applications for the risk management processes.

For multinational companies that are project-driven, risk management takes on paramount importance. Not all companies, especially in undeveloped countries, have an understanding of risk management or its importance. These countries sometimes view risk management as an overmanagement expense on a project.

Consider the following scenario. As your organization gets better and better at project management, your customers begin giving you more and more work. You're now getting contracts for turnkey projects, or complete-solution projects. Before, all you had to do was deliver the product on time and you were through. Now you are responsible for project installation and startup as well, sometimes even for ongoing customer service. Because the customers no longer use their own resources on the project, they worry less about how you're handling your project management system.

Alternatively, you could be working for third-world clients who haven't yet developed their own systems. One hundred percent of the risk for such projects is yours, especially as projects grow more complex (see Figure 5–6). Welcome to the twenty-first century!

One subcontractor received a contract to install components in a customer's new plant. The construction of the plant would be completed by a specific date. After construction was completed, the contractor would install the equipment, perform testing, and then start

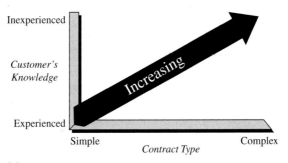

**Figure 5–6.** Future risks.

up. The subcontractor would not be allowed to bill for products or services until after a successful startup. There was also a penalty clause for late delivery.

The contractor delivered the components to the customer on time, but the components were placed in a warehouse because plant construction had been delayed. The contractor now had a cash flow problem and potential penalty payments because of external dependencies that sat on the critical path. In other words, the contractor's schedule was being controlled by actions of others. Had the project manager performed business risk management rather than just technical risk management, these risks could have been reduced.

For the global project manager, risk management takes on a new dimension. What happens if the culture in the country with which you are working neither understands risk management nor has any risk management process? What happens if employees are afraid to surface bad news or identify potential problems? What happens if the project's constraints of time, cost, and quality/performance are meaningless to the local workers?

## 5.7 WÄRTSILÄ: THE NEED FOR PROACTIVE RISK MANAGEMENT[6]

**Proactive Project Risk Management in Wärtsilä Power Plant Projects**

At Wärtsilä, project risk management has traditionally been much about identifying and planning. We have seen that this now needs to be expanded to cover reflection and proactive action-taking in today's complex projects. Risks need to be tackled upfront before they occur and potentially jeopardize the objectives of projects. We will now briefly present what we have done in this respect.

How uncertainty and risk are handled in projects very much depends on experience. It can be said that many project managers only deal with risk and uncertainty as a result of what they are actually experiencing in their projects. Experienced project managers, however, can recognize risks far ahead before they become issues. Likewise, opportunities and positive uncertainties can be recognized easier by experienced project managers. However, the recognition of opportunities is not only restricted to experience, since a willingness to take risks is also required. In many cases, a change of mindset is required from project managers to be able to accomplish this.

As large projects are becoming increasingly complex to manage today, it is essential that the project manager has enough experience to have an accurate perception of what is involved. Besides the project itself, it is of huge importance to know about the location, the customer and the environment. Failure to have the knowledge or experience about these issues in advance will cause major problems, making the project more complex and challenging than necessary. In order to avoid such pitfalls, it is important to use the combined knowledge, experience and creativity of the whole project team. Although risk management is the project manager's responsibility, it is not solely his/her task. The whole project team needs to share this responsibility.

This brings us to the importance of having a lessons learned database with information that has been shared among the project teams. Such a database is an important resource for

---

6. Material has been provided by Wärtsilä Project Management Office (WPMO). Copyright to Wärtsilä Corporation 2013. Reproduced by permission.

a new project manager or other team member joining the force. Likewise, when a project team accepts a new project type, or a project in a totally new location, it is beneficial to be able to access the knowledge about similar cases. In light of this, a lessons learned database is being implemented where all this knowledge and experience can be shared.

We have seen that knowledge and experience play an important role in managing risk, uncertainty and other factors in projects. However, proactive risk management is something that is not always easy to implement, since it depends on so many people's different perceptions of it. A lot of communication is needed in order to gain a common understanding of what the organization needs regarding risk and uncertainty, as well as a clear understanding of the potential benefits they bring into the organization. Proactive risk management is not only about identifying, qualifying and quantifying the risks; it is much more. The utilization of the risk management process is all about having the maturity to use the previously learned experience and knowledge to prevent risks from occurring in the first place, as well as the confidence to take the necessary actions to be able to encourage positive opportunities to develop.

A project team needs a tool for project risk management where upcoming events, both foreseen and unforeseen, can be continuously followed. A risk management process tool does not need to be complex. The most important aspect is the way it is utilized in the organization. We see that in this case the statement: *"the simpler the better"* describes quite well what is needed.

The proactive risk management process taken into use at Wärtsilä consists of three different phases (Figure 5–7). Firstly a project classification should be done to define the complexity of the project. Thereafter, the risk process itself will be managed as a continuous process throughout the whole project lifecycle. In addition, lessons learned should be recorded on risks where the actions taken significantly differed from the planned response. At Wärtsilä we have implemented this entire process in one common PM tool used by all PM teams and management.

The classification process will provide important information for the risk identification steps. The intention of the process is to encourage project managers to think about the project and define both where the complexity of the project is situated as well as provide an input for the risk management process identification. It must describe the project from an objective point of view. One of the core added values that project classification brings to project management is by defining the needed resources for resource allocation.

The risk management process basically also relies on it being one continuous process. All the same elements that were used in the classification process are implemented in this process. The traditional risk management process described in *PMBOK® Guide* (2008) has been used as the basis for the new risk management process.

In order for a proactive risk management process to become successful, it is vital that the project team makes full use of it. Ignorance of risks and uncertainties will cause large problems within project management when they materialize unexpectedly and become harmful issues.

A good communication system needs to be created in order to implement a uniform risk management process among the project teams. In addition, training should be given on how to use the risk management process in order to improve the understanding of how the proactive risk process can be utilized and to gain an appreciation as to why it is so important.

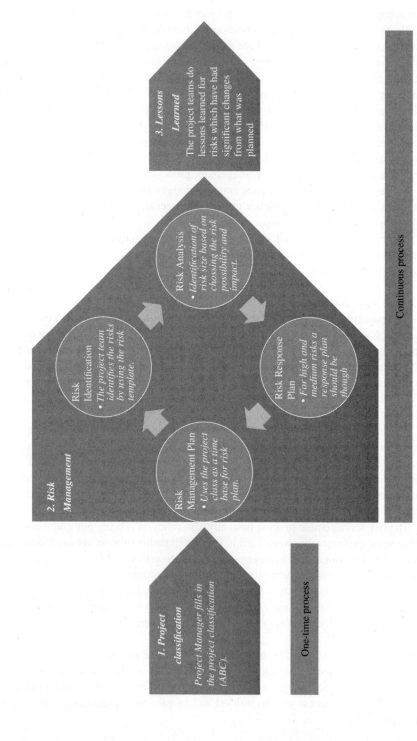

**1. Project classification**

*Project Manager fills in the project classification (ABC).*

**2. Risk Management**

Risk Identification
- *The project team identifies the risks by using the risk template.*

Risk Analysis
- *Identification of risk size based on choosing the risk possibility and impact.*

Risk Response Plan
- *For high and medium risks a response plan should be though*

Risk Management Plan
- *Uses the project class as a time base for risk plan.*

**3. Lessons Learned**

The project teams do lessons learned for risks which have had significant changes from what was planned

One-time process

Continuous process

**Figure 5–7.** Proactive project risk management process in Wärtsilä power plants.

## 5.8 ILLUMINAT: EFFECTIVE RISK MANAGEMENT[7]

ILLUMINAT (Trinidad & Tobago) Limited (ILLUMINAT) is one of the leading providers of Information, Communications and Technology (ICT) products and services in Trinidad and Tobago (T&T) and the wider Caribbean Region. ILLUMINAT is part of the Neal and Massy Group of companies, a conglomerate operating in the majority of English-speaking countries in the Caribbean. Much of ILLUMINAT's business with its external clients is managed through projects using a standardised project management methodology, where a Project Management Office (PMO) has been established and the use of the PMO's methodology, templates, tools and processes are strongly encouraged across the organisation. According to KPMG (1997) and Nieto-Rodriguez and Evrard (2004), organisational project management capability is increasingly being seen as a key to competitive advantage and therefore worthy of investment in its development. Risk Management (RM) is a key component of effective project management and as such, it is one of the approaches used to support the strategic goals of the organisation.

**Business Context**

Figure 5–8 represents the business context in which ILLUMINAT operates and which impacts the strategy used to integrate the portfolio, programme and project risk management approaches with the overall organizational strategy:

The concept of RM is embedded within ILLUMINAT where its application is evidenced at the portfolio, programme and project management levels throughout the organisation. The governance structures in place for effective RM include the following key dimensions as shown in Figure 5–9:

**Figure 5–8.** ILLUMINAT's business context.

---

7. Information has been provided by Ms. Kerdell Brereton, Manager, Project Management Services Unit, ILLUMINAT(Trinidad & Tobago) Limited

**PORTFOLIO**

- Programmes
- Projects

**PROCESSES**

- Systems
- Infrastructure
- Business Continuity Plan (human and physical resources)

**PEOPLE**

- ILLUMINAT resources
- Clients
- Suppliers
- Strategic Alliances (Business Partnerships, Industry affiliations etc.)

**Figure 5–9.** ILLUMINAT's key risk management dimensions.

**Figure 5–10.** Stages of risk management in ILLUMINAT.

These key dimensions are used to assess the risks in business opportunities for projects from inception. The evaluation includes risks to commercial viability, technical feasibility and organisational capacity to name a few. The external and internal environments are also taken into consideration and all risks are aggregated and reviewed holistically in order to inform the final decisions for ILLUMINAT's portfolio.

**Risk Management Approach**     Figure 5–10 is a high-level view of the risk management governance structure in place to manage risk within ILLUMINAT where it is managed in three (3) stages:

### Stage: Preprogramme/ Project Award (Request for Proposal)

- Assemble team including Sales resources, Technical Subject Matter Experts, Project Management resource, Executive/Senior Manager.
- Define Technical solution and degree of integration (single or multiple suppliers) required to put solution together.

- Review all aspects of programme/project inclusive of customer and supplier profiles, timeframe for delivery (milestone dates), currency to be used, logistics required (e.g. single or multiple countries involved), storage, potential causes for delay and the impact on the project etc.
- Identify and assess each risk and define feasible mitigation actions that can be taken to eliminate or reduce the impact of the risks (if they materialise). These actions plans are integrated into the contractual agreements (client and supplier), commercial agreements, the project implementation plan and other working arrangements (e.g., insurance) which are encapsulated within the final proposal.
- Review the proposal and once approved by Executive Management, release to the client for signoff.

## Stage: Implementation of Programme/Project

Once the business is won and the programme/project initiated, RM is actively managed at these levels as follows:

- Create plan to define governance approaches for RM taking into consideration the specific nuances of the programme/project.
- Conduct deeper RM analysis to identify risks that may impact the project at all stages (internal and external factors).
- Evaluate each risk using relevant data and expert knowledge to define and implement effective mitigation (and contingency) plans as well as prioritise the risks.
- Continuously monitor risks and implement corrective action and/ or contingency plans where applicable.
- Include risk analysis of risks with the priority of "Very High" and "High" in regular reports to key stakeholders.

Risk analysis is performed from three main perspectives: strategic, commercial and technical. In order to conduct this activity efficiently and effectively, Programme/Project Owners are assigned who are charged with the responsibility of ensuring that all arrangements are in place to make certain that the goals and objectives of the programme/project are achieved while significantly minimizing any negative impacts. The Programme/Project Owner is typically a senior resource with the authority to make requisite decisions and is a key escalation point to address issues as they arise.

## Stage: Post Programme/Project

Upon completion of the programme/project, clients are polled to identify their overall satisfaction with ILLUMINAT's services as well as any areas for improvement. This is done through Customer Satisfaction Surveys and Project Impact Assessment Surveys. The feedback from these surveys is then analysed and where applicable and feasible, used by the organisation to improve the relevant areas identified.

**Risk Management Reporting**

At the **portfolio management level** a Risk Profile for each opportunity is created and this is done through two (2) stages at the time of proposal:

- Using the organisation's *"Delegation of Authority"* policy, all proposals are reviewed and signed off depending on value of project within the organisation [Business Unit (BU) Head, Chief Executive Officer (CEO), Chairman, Group CEO). This is to ensure that the level of scrutiny (which increases with value of project) required to comprehensively assess the inherent risks is in place.
- A document is produced (Project Outline) which details the risks involved, the internal and external factors that can have an impact, the mitigating strategies to be used and an evaluation of the likelihood of winning the business. If the risk profile of the opportunity is deemed acceptable, the decision will be made to proceed. The inverse may occur if the risk profile is deemed unacceptable and the organisation may choose not to bid for the opportunity.

Once the business is won, the risk profile is taken into consideration in the overall planning of the projects where reporting on projects that meet specific criteria is done on a monthly basis. These reports are reviewed by Executive Management in monthly meetings of the Board of Directors where the status of the projects and the risks in particular, are reviewed.

At the **program/ project levels**, risk reporting is done on a more frequent basis where risks are reviewed at various stage gates, particularly when moving from one phase to another:

- Initiation – further identification and in-depth analysis of risks
- Planning – development and ratification of Risk Management Plan
- Execution – implementation of risk response plans
- Monitoring and Control – implementation of corrective action if required
- Close-out – retiring of risks

Reporting is done during project team meetings and if required, meetings are convened specifically to focus on risks at critical stages in the programme/project. Reporting is also included in formal status reports which are distributed to key stakeholders at the frequency documented within the programme/project's Communication Plan (at a minimum monthly). Additionally, for those programmes/projects that meet specific criteria, Summary Reports are submitted to the Board of Directors for monthly review (as noted above).

## Lessons Learned

ILLUMINAT is a knowledge-based organisation where one of the goals is to assist customers in making their businesses successful. As such, effective risk management and the application of risk-related lessons learned in the implementation of programmes and projects in the portfolio are key components to achieving this.

Consequently, as part of the overall risk management governance, programmes/projects are reviewed and common denominators are noted in the approaches used and the effectiveness of the outcomes. These are then used for process improvements within the organisation (e.g., definition of robust standardized processes and criteria for evaluating strategic business partners).

## Conclusion

Risk Management is entrenched into the organisational culture at ILLUMINAT and has been recognized as one of the key strategies in ensuring the completion of successful programmes/projects. This is demonstrated by the thoughts expressed by ILLUMINAT's CEO, Mr. David Belgrave:

> I think it (Risk Management) is a very positive thing. It ensures that we give sufficient thought to a project before we enter it, including the fact that we may decide not to enter the project itself in the worst case. But at the very least, we assess the risk of the project, we propose mitigating actions, we ensure that we place the degree of attention and the degree of rigour that's required to ensure success on the project. This action impacts on time, on money, on the number and type of resources required and ultimately customer service. The whole process of Risk Assessment in approaching projects allows for a more professional job and a more profitable job in the end. Through techniques that we have utilised in project management, we were able to anticipate some of the issues well before they occurred and take corrective action to keep programme/project on track and that's a major benefit to the organization.

As such, the risk management approaches used at ILLUMINAT have been seen to add value to the efficient and effective management of the portfolio of projects and through continuous learning, enhancement of these approaches can be done in order to achieve the organisation's strategic goals and ensure growth performance.

## 5.9 INDRA: WHEN A RISK BECOMES REALITY (ISSUE MANAGEMENT)[8]

In a company like Indra, with thousands of active projects geographically distributed, a solid and continuous risk management practice is vital. This is shown in Figure 5–11. Actions that might contain the impact that risks can have on the results of the project are planned and monitored throughout the project life cycle.

The percentage of projects with risk plans and risk registers is very high in Indra. However we at Corporate PMO noted that these high figures didn't prevent some risks from happening and even worse, that other unknowns-unknowns risks showed up as issues that project managers haven't been able to identify in their plans.

However an issue is not in the future, it is now already affecting project milestones or schedule, or certain WBS elements, or the committed budget or the quality level of the project. For that reason an issue usually demands immediate response and its resolution must be addressed as quickly and effectively as possible to avoid affecting other areas of the project.

We think that analyzing the relationship between risks and issues is essential for an integrated approach to risk management. By a pre-mortem review of risks and their related issues, their sorting and classification, and by analyzing why those risks were not well addressed in earlier planning stages, we seek to understand the causes that originated them and to determine earlier risk screening criteria.

In a second stage we need to know their actual effects on the project and whether the solutions proposed to contain the effects have been effective. That will allow us to create a database and to identify some lessons learned to help us to early identify issue-prone risks and prevent them from appearing in future projects.

---

8. Material provided by Alfredo Vázquez Díaz, PMP, Director, Corporate Project Management Office.

At proposal preparation phase:

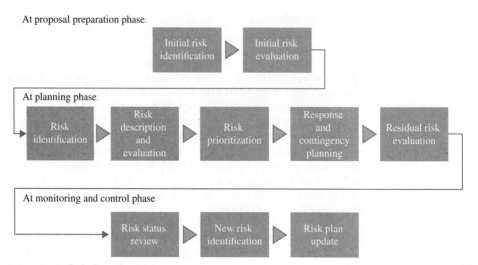

Figure 5–11.   Indra's project risk management process.

Not all issues are equal or affect the organization the same way. Their impact depend on the project size or volume, its internal or external visibility, complexity, variances on initial economic forecast, time needed to put it on track or their impact on the image of the company.

Having all those considerations up front, we have decided to focus our efforts on the projects with more critic issues. Those are considered projects that require close surveillance, and for those projects identifying which is the source of the issue, its immediate effects and the follow-up of the action plan are key. To allow this we created a new functionality in our PMIS called Issue Registry. This is embedded within our PMIS Issue Management module.

The Issue Registry works like this: When a PMO or a user responsible for project control analyzes a project and detects it is having serious problems, the project manager is required to complete detailed information in the Issue Registry module. This can be triggered automatically through a red alert in the PMIS (See Figure 5–12.)

The project manager must describe the existing issues in her project, the action plans to cope with them and the recovery target that must be obtained to get the project back on track. Once fulfilled, and to avoid this being a static snapshot of the project, the project manager is expected to update the information every reporting period from that moment on, indicating action plan updates and project status with regards to the initial targets.

What we intend to achieve with the Issue Registry? We want to focus our attention and our efforts on:

● Detecting which problems and issues have emerged from previously identified risks and which not
● Get an homogeneous classification and typification of projects that had become seriously affected by issues
● Track the effectiveness of the issue action plans
● Automated and systematic reporting to business management on those issues and their projects from different perspectives (business unit, solution, type of project, geographic location. . .)

**Figure 5–12.** Indra's project risk monitoring in the PMIS.

Historically the efforts of our organization had been focused on the management of risks, leaving the management of issues and problems as a secondary process, not connected with risk management. Issue management was more dependent on the involvement and proactiveness of the project manager, for that reason it was heterogeneously approached, usually in an internal project context and without a close monitoring and follow-up by the organization, as it was the case with risk management.

With this new feature, the ICU Registry, the registration and follow-up of ICU projects and the traceability between risks and issues can be made from the PMIS. To reverse actual dynamics we will take a first step by learning from already occurred issues and problems, and based on that, and in a second stage we will focus on prevention of recurrent issues, those that have been registered, diagnosed and solved by other projects managers, obtaining a invaluable information for other project managers to learn from other's experience in their own future similar projects.

Inside the problem is the solution and only knowing the problem, we can solve and avoid it.

## 5.10 THE FAILURE OF RISK MANAGEMENT

There are numerous reasons why risk management can fail. Typical reasons might include:

● The inability to perform risk management effectively
● Not being able to identify the risks
● Not being able to measure the uncertainty of occurrence

- Not being able to predict the impact, whether favorable or unfavorable
- Having an insufficient budget for risk management work
- Having team members that do not understand the importance of risk management
- Fear that identification of the true risks could result in the cancellation of the project
- Fear that whoever identifies critical risks will get unfavorable recognition
- Peer pressure from colleagues and superiors that want to see this project completed regardless of the risks

All of these failures occur during the execution of the project. These failures we seem to understand and can correct with proper education and budget allocations for risk management activities. But perhaps the worst failures occur when people refuse to even consider risk management because of some preconceived notion about its usefulness of importance to the project. David Dunham discusses some of the reasons why people avoid risk management on new product development (NPD) projects[9]

Discussing risk in new product development certainly seems to be a difficult thing to do. Despite the fact that the high-risk nature of new product development is built into the corporate psyche, many corporations still take a fatalistic approach toward managing the risk. Reasons for not being anxious to dwell on risk differ depending on the chair in which you are sitting.

**Program Manager**

- Spending time on risk assessment and management is counter to the action culture of many corporations. "Risk management does not create an asset," to quote one executive.
- Management feels that the learning can/should be done in the market.

**Program Manager**

- There is a natural aversion among developers to focus on the downside.
- Highlighting risk is counterintuitive for development teams who want to promote the opportunity when competing for NPD funding.

## 5.11 DEFINING MATURITY USING RISK MANAGEMENT

For years, project management maturity was measured by how frequently we were able to meet the project's triple constraints of time, cost, and performance or scope. Today, we are beginning to measure maturity in components, such as the areas of knowledge in the *PMBOK® Guide*. Maturity is now measured in stages and components, such as how well we perform scope management, time management, risk management, and other areas of knowledge. Gregory Githens believes that the way we handle risk management can be an indicator of organizational maturity[10]:

---

9. D. J. Dunham, "Risk Management: The Program Manager's Perspective," in P. Belliveau, A. Griffin, and S. Somermeyer, *The PDMA Toolbook for New Product Development*, Hoboken, NJ: Wiley, 2002, p. 382.

10. G. D. Githens, "How to Assess and Manage Risk in NPD: A Team-Based Approach," in P. Belliveau, A. Griffin, and S. Somermeyer, *The PDMA Toolbook for New Product Development*, Hoboken, NJ: Wiley, 2002, p. 208.

Some firms have more capability to manage risk well, and these firms are the most consistent in their growth and profitability. Perhaps the simplest test for examining risk management maturity is to examine the level of authority given to the [New Product Development] NPD program [project] manager: If authority is high, then the organization is probably positioning itself well to manage risks, but if authority is low, then the blinders may be on. Another test is the use of checklists: if ticking off a checklist is the sole company response to risk, then organizational maturity is low. Risk management provides. . . [an] excellent lens by which to evaluate a firm's ability to integrate and balance strategic intent with operations.

Many firms ignore risk management because they have not seen the need for it. They perceive their industry as stable and mostly focus on their competitive rivals and operational challenges. . . . By addressing risk at the project level, you encourage the organization to surface additional strategic concerns.

Top NPD firms have a sophisticated capability for risk management, and they will "book" a project plan, pay attention to the details of product scope and project scope, use risk management tools such as computer simulations and principle-based negotiation, and document their plans and assumptions. These more mature firms are the ones that will consider risk in establishing project baselines and contracts. For example, Nortel uses a concept called "out of bounds" that provides the NPD program managers with the freedom to make trade-offs in time, performance, cost and other factors. Risk analysis and management is an important tool.

Less mature firms typically establish a due date and pay attention to little else (and in my experience, this is the majority of firms). Firms that use the decision rule "Hit the launch date" default to passive acceptance—hiding the risk instead of managing it. Firefighting and crisis management characterize their organizational culture, and their strategic performance is inconsistent. These firms are like the mythological character Icarus: They fly high but come crashing down because they ignored easily recognizable risk events.

## 5.12 BOEING AIRCRAFT COMPANY

As companies become successful in project management, risk management becomes a structured process that is performed continuously throughout the life cycle of the project. The two most common factors supporting the need for continuous risk management is how long the project lasts and how much money is at stake. For example, consider Boeing's aircraft projects. Designing and delivering a new plane might require 10 years and a financial investment of more than $15 billion.

From an academic perspective, Table 5–1 shows the characteristics of risks at the Boeing Aircraft Company. (The table does not mean to imply that risks are mutually exclusive of each other, nor does it imply that these are the only risks.) New technologies can appease customers, but production risks increase because the learning curve is lengthened with new technology compared to accepted technology. The learning curve can be lengthened further when features are custom designed for individual customers. In addition, the loss of suppliers over the life of a plane can affect the level of technical and production risk. The relationships among these risks require the use of a risk management matrix and continued risk assessment.

**TABLE 5–1.  RISK CATEGORIES AT BOEING[11]**

| Type of Risk | Risk Description | Risk Mitigation Strategy |
|---|---|---|
| Financial | Upfront funding and payback period based upon number of planes sold | Funding by life-cycle phases<br>Continuous financial risk management<br>Sharing risks with subcontractors<br>Risk reevaluation based upon sales commitments |
| Market | Forecasting customers' expectations on cost, configuration, and amenities based upon a 30- to 40-year life of a plane | Close customer contact and input<br>Willingness to custom design per customer<br>Development of a baseline design that allows for customization |
| Technical | Because of the long lifetime for a plane, must forecast technology and its impact on cost, safety, reliability, and maintainability | A structured change management process<br>Use of proven technology rather than high-risk technology<br>Parallel product improvement and new product development processes |
| Production | Coordination of manufacturing and assembly of a large number of subcontractors without impacting cost, schedule, quality, or safety | Close working relationships with subcontractors<br>A structured change management process<br>Lessons learned from other new airplane programs<br>Use of learning curves |

## 5.13 CHANGE MANAGEMENT

Companies use change management to control both internally generated changes and customer-driven changes in the scope of projects. Most companies establish a configuration control board or change control board to regulate changes. For customer-driven changes, the customer participates as a member of the configuration control board. The configuration control board addresses the following four questions at a minimum:

- What is the cost of the change?
- What is the impact of the change on project schedules?
- What added value does the change represent for the customer or end user?
- What are the risks?

The benefit of developing a change management process is that it allows you to manage your customer. When your customer initiates a change request, you must be able to predict immediately the impact of the change on schedule, safety, cost, and technical performance. This information must be transmitted to the customer immediately, especially if your methodology is such that no further changes are possible because of the life-cycle phase you have entered. Educating your customer as to how your methodology works is critical in getting customer buy-in for your recommendations during the scope change process.

Risk management and change management function together. Risks generate changes that, in turn, create new risks. For example, consider a company in which the project manager is given the responsibility for developing a new product. Management usually establishes a launch date even before the project is started. Management wants the income stream from the project to begin on a certain date to offset the development costs. Project managers view

11. The information in this section on how Boeing might characterize risks on a new airplane project is the author's opinion and not necessarily Boeing's official opinion.

executives as their customers during new project development, but the executives view their customers as the stockholders who expect a revenue stream from the new product. When the launch date is not met, surprises result in heads rolling, usually executive heads first.

In the previous edition of the book, we stated that Asea, Brown and Boveri had developed excellent processes for risk management, so it is understandable that it also has structured change management processes. In companies excellent in project management, risk management and change management occur continuously throughout the life cycle of the project. The impact on product quality, cost, and timing is continuously updated and reported to management as quickly as possible. The goal is always to minimize the number and extent of surprises.

## 5.14 OTHER MANAGEMENT PROCESSES

Employee empowerment and self-directed work teams took the business world by storm during the early 1990s. With growing emphasis on customer satisfaction, it made sense to empower those closest to the customer—the order service people, nurses, clerks, and so on—to take action in resolving customers' complaints. A logical extension of employee empowerment is the self-managed work team. A self-directed work team is a group of employees with given day-to-day responsibility for managing themselves and the work they perform. This includes the responsibility for handling resources and solving problems.

Some call empowerment a basis for the next industrial revolution, and it is true that many internationally known corporations have established self-directed work teams. Such corporations include Esso, Lockheed-Martin, Honeywell, and Weyerhauser. Time will tell whether these concepts turn out to be a trend or only a fad.

Reengineering a corporation is another term for downsizing the organization with the (often unfortunate) belief that the same amount of work can be performed with fewer people, at lower cost, and in a shorter period of time. Because project management proposes getting more done in less time with fewer people, it seems only practical to implement project management as part of reengineering. It still is not certain that downsizing executed at the same time as the implementation of project management works, but project-driven organizations seem to consider it successful.

Life-cycle costing was first used in military organizations. Simply stated, life-cycle costing requires that decisions made during the R&D process be evaluated against the total life-cycle cost of the system. Life-cycle costs are the total cost of the organization for the ownership and acquisition of the product over its full life.

## 5.15 HEWLETT-PACKARD

Information technology enterprise management (ITEM) integrates all of the project management disciplines along with other IT disciplines such as engineering, testing, and model office. Many projects start with a charter and scope that make a project manager view the work with a definitive start and definitive end. (See Figure 5–13.) If this is applied to the relative stages of a release, they simply stretch the project management process groups across

**Figure 5–13.** Typical application of the *PMBOK® Guide*.

**Figure 5–14.** Correct application of the *PMBOK® Guide*.

all of the release. That makes the suppliers of the project try to understand if designing and building are part of execution or if deployment of the project is the execution of the project.

As stated in the *PMBOK® Guide*, project management process groups should be repeated for each release stage. This promotes other project management strategies, such as roll wave planning and resource balancing. With this framework view, other capabilities can be applied to each stage, along with the project management capabilities (Figure 5–14).

## 5.16 EARNED-VALUE MEASUREMENT

An integral part of most project management methodologies is the ability to perform earned-value measurement. Earned-value measurement was created so that project managers would manage projects rather than merely monitor results. Even though some companies do not use earned-value measurement on a formal basis, core concepts such as variance analysis and reporting are being used. As an example, Keith Kingston, PMP®, Manager of Program Management at Motorola, states:

> Variances to schedule performance are analyzed weekly or biweekly but not as part of a formal earned value approach. Greater than three days of schedule variation on any internal deliverable requires analysis and a mitigation plan. Any variances that impact meeting a customer-required date requires analysis and a mitigation plan.

## 5.17 DTE ENERGY

One of the characteristics of integrated processes is that it must include an integration with a project management information system capable of reporting earned-value measurement. Kizzmett Collins, PMP, senior project manager, software engineering, methods, and staffing at DTE Energy, describes the integration with earned-value measurement:

Getting Started Beginning in 2001, the Information Technology Services' (ITS) project management office (PMO) sponsored several projects that helped to advance and develop earned-value analysis (EVA) understanding and knowledge among management and project managers.

The implementation of the Primavera Teamplay project management suite easily enabled the tracking of EVA metrics for individual projects and project portfolios. Among other sources of education, TeamPlay product training introduced ITS project managers and management to EVA metrics. The PMO developed processes and reports to aid project managers and management in the analysis and reporting of EVA metrics. ITS contracted Quentin W. Flemming to conduct EVA courses, which increased project managers' and ITS management's understanding of EVA.

**DTE Energy's ITS Earned-Value Analysis Journey**

Reporting EVA Metrics EVA metrics (such as SPI, CPI, CV, SV, and EAC) are reported weekly in project status reports and in the ITS project portfolio. The project manager provides additional status commentary for variances from target that exceed $\pm 10\%$.

During the ITS planning and management table (PMT) meetings, project mangers report project status, usually on a monthly basis. The PMT reviews EVA metrics and discusses variances and indicators as warranted. Other triggers that may necessitate a review of the projects' EVA metrics include:

- A project is at 20 percent of original estimated duration.
- A significant phase has ended.
- The project manager presents issues, risks, or changes.

Linking Rewards to EVA CPI Metric In 2003, the ITS Organization began linking rewards to the CPI metric. The CPI metric results are now included in both project manager performance reviews and the ITS Organizational Scorecard.

The ITS Organizational Scorecard is tied to the corporate Rewarding Employee Plan (REP). REP pays employees bonuses that are based on achieving the corporate and organizational goals. The ITS Organization's CPI metric is the aggregate of the CPI of each project in the project portfolio. As a result, all ITS employees have a monetary stake in the success of each project.

Project manager performance goals include the CPI metric for all projects within their area of responsibility. CPI results greater than 0.95 and less than 1.05 exceed performance expectations. Each project manager's performance review is linked to their merit increase.

Opportunities for Improvement The ITS Organization introduced EVA to the corporation through the ITS scorecard. Both within ITS and across the corporation, further training is needed to expand our shared understanding of EVA. Internally, ITS can take the next step in further understanding what the EVA metrics indicate. For example, we have an opportunity to be more deliberate in allocating dollars elsewhere when a project's CPI indicates cost issues or problems.

# 6 Culture

## 6.0 INTRODUCTION

Perhaps the most significant characteristic of companies that are excellent in project management is their culture. Successful implementation of project management creates an organization and culture that can change rapidly because of the demands of each project and yet adapt quickly to a constantly changing dynamic environment, perhaps at the same time. Successful companies have to cope with change in real time and live with the potential disorder that comes with it.

Change is inevitable in project-driven organizations. As such, excellent companies have come to the realization that competitive success can be achieved only if the organization has achieved a culture that promotes the necessary behavior. Corporate cultures cannot be changed overnight. The time frame is normally years. Also, if as few as one executive refuses to support a potentially good project management culture, disaster can result.

In the early days of project management, a small aerospace company had to develop a project management culture in order to survive. The change was rapid. Unfortunately, the vice president for engineering refused to buy into the new culture. Prior to the acceptance of project management, the power base in the organization had been engineering. All decisions were either instigated or approved by engineering. How could the organization get the vice president to buy into the new culture?

The president realized the problem but was stymied for a practical solution. Getting rid of the vice president was one alternative, but not practical because of his previous successes and technical know-how. The corporation was awarded a two-year project that was strategically important to the company. The vice president was then temporarily assigned as the project manager and removed from his position as vice president for engineering. At the completion of the project, the vice president was assigned to fill the newly created position of vice president of project management.

## 6.1 CREATION OF A CORPORATE CULTURE

Corporate cultures may take a long time to create and put into place but can be torn down overnight. Corporate cultures for project management are based upon organizational behavior, not processes. Corporate cultures reflect the goals, beliefs, and aspirations of senior management. It may take years for the building blocks to be in place for a good culture to exist, but it can be torn down quickly through the personal whims of one executive who refuses to support project management.

Project management cultures can exist within any organizational structure. The speed at which the culture matures, however, may be based upon the size of the company, the size and nature of the projects, and the type of customer, whether it be internal or external. Project management is a culture, not policies and procedures. As a result, it may not be possible to benchmark a project management culture. What works well in one company may not work equally well in another.

Good corporate cultures can also foster better relations with the customer, especially external clients. As an example, one company developed a culture of always being honest in reporting the results of testing accomplished for external customers. The customers, in turn, began treating the contractor as a partner and routinely shared proprietary information so that the customers and the contractor could help each other.

Within the excellent companies, the process of project management evolves into a behavioral culture based upon multiple-boss reporting. The significance of multiple-boss reporting cannot be overstated. There is a mistaken belief that project management can be benchmarked from one company to another. Benchmarking is the process of continuously comparing and measuring against an organization anywhere in the world in order to gain information that will help your organization improve its performance and competitive position. Competitive benchmarking is where one benchmarks organizational performance against the performance of competing organizations. Process benchmarking is the benchmarking of discrete processes against organizations with performance leadership in these processes.

Since a project management culture is a behavioral culture, benchmarking works best if we benchmark best practices, which are leadership, management, or operational methods that lead to superior performance. Because of the strong behavioral influence, it is almost impossible to transpose a project management culture from one company to another. As mentioned earlier, what works well in one company may not be appropriate or cost-effective in another company.

Strong cultures can form when project management is viewed as a profession and supported by senior management. A strong culture can also be viewed as a primary business differentiator. Strong cultures can focus on either a formal or informal project management approach. However, with the formation of any culture, there are always some barriers that must be overcome.

According to a spokesperson from AT&T:

Project Management is supported from the perspective that the PM is seen as a professional with specific job skills and responsibilities to perform as part of the project team. Does the PM get to pick and choose the team and have complete control over budget allocation? No. This is not practical in a large company with many projects competing for funding and subject matter experts in various functional organizations.

A formal Project Charter naming an individual as a PM is not always done, however, being designated with the role of Project Manager confers the power that comes with that role. In our movement from informal to more formal, it usually started with Project Planning and Time Management, and Scope Management came in a little bit later.

In recent memory PM has been supported, but there were barriers. The biggest barrier has been in convincing management that they do not have to continue managing all the projects. They can manage the project managers and let the PMs manage the projects. One thing that helps this is to move the PMs so that they are in the same work group, rather than scattered throughout the teams across the company, and have them be supervised by a strong proponent of PM. Another thing that has helped has been the PMCOE's execution of their mission to improve PM capabilities throughout the company, including impacting the corporate culture supporting PM.

Our success is attributable to a leadership view that led to creating a dedicated project management organization and culture that acknowledges the value of Project Management to the business. Our vision: Establish a global best in class Project Management discipline designed to maximize the customer experience and increase profitability for AT&T.

In good cultures, the role and responsibilities of the project manager is clearly identified. It is also supported by executive management and understood by everyone in the company. According to Enrique Sevilla Molina, formerly corporate PMO director at Indra:

Based on the historical background of our company and the practices we set in place to manage our projects, we found out that the project manager role constitutes a key factor for project success. Our project management theory and practice has been built to provide full support to the project manager when making decisions and, consequently, to give him (or her) full responsibility for project definition and execution.

We believe that he or she is not just the one that runs the project or the one that handles the budget or the schedule, but the one that "understand and look at their projects as if they were running their own business," as our CEO used to say, with an integrated approach to his/her job.

Our culture sets the priority on supporting the project managers in their job, helping them in the decision-making processes, and providing them with the needed tools and training to do their job. This approach allow for a certain degree of a not so strict formal processes. This allows the Project manager's responsibility and initiative to be displayed, but always under compliance with the framework and set of rules that allows for a solid accounting and results reporting.

We can say that project management has always been supported throughout the different stages of evolution of the company, and throughout the different business units, although some areas have been more reluctant in implementing changes in their established way of performing the job. One of the main barriers or drawbacks is the ability to use the same project management concepts for the different types of projects and products.

It is still a major concern in our training programs to try to explain how the framework and the methodology is applied to projects with a high degree of definition in scope and to projects with a lesser degree of definition (fuzzy projects).

## 6.2 CORPORATE VALUES

An important part of the culture in excellent companies is an established set of values that all employees abide by. The values go beyond the normal "standard practice" manuals and morality and ethics in dealing with customers. Ensuring that company values and project management are congruent is vital to the success of any project. In order to ensure this congruence of values, it is important that company goals, objectives, and values be well understood by all members of the project team.

Successful project management can flourish within any structure, no matter how terrible the structure looks on paper, but the culture within the organization must support the four basic values of project management:

- Cooperation
- Teamwork
- Trust
- Effective communication

## 6.3 TYPES OF CULTURES

There are different types of project management cultures, which vary according to the nature of the business, the amount of trust and cooperation, and the competitive environment. Typical types of cultures include:

- *Cooperative cultures:* These are based upon trust and effective communication, not only internally but externally as well.
- *Noncooperative cultures:* In these cultures, mistrust prevails. Employees worry more about themselves and their personal interests than what is best for the team, company, or customer.
- *Competitive cultures:* These cultures force project teams to compete with one another for valuable corporate resources. In these cultures, project managers often demand that the employees demonstrate more loyalty to the project than to their line manager. This can be disastrous when employees are working on multiple projects at the same time.
- *Isolated cultures:* These occur when a large organization allows functional units to develop their own project management cultures. This could also result in a culture-within-a-culture environment. This occurs within strategic business units.
- *Fragmented cultures:* Projects where part of the team is geographically separated from the rest of the team may result in a fragmented culture. Fragmented cultures

also occur on multinational projects, where the home office or corporate team may have a strong culture for project management, but the foreign team has no sustainable project management culture.

Cooperative cultures thrive on effective communications, trust, and cooperation. Decisions are made based upon the best interest of all of the stakeholders. Executive sponsorship is more passive than active, and very few problems ever go up to the executive levels for resolution. Projects are managed more informally than formally, with minimum documentation, and often with meetings held only as needed. This type of project management culture takes years to achieve and functions well during both favorable and unfavorable economic conditions.

Noncooperative cultures are reflections of senior management's inability to cooperate among themselves and possibly their inability to cooperate with the workforce. Respect is nonexistent. Noncooperative cultures can produce a good deliverable for the customer if one believes that the end justifies the means. However, this culture does not generate the number of project successes achievable with the cooperative culture.

Competitive cultures can be healthy in the short term, especially if there exists an abundance of work. Long-term effects are usually not favorable. An electronics firm would continuously bid on projects that required the cooperation of three departments. Management then implemented the unhealthy decision of allowing each of the three departments to bid on every job. Whichever department would be awarded the contract, the other two departments would be treated as subcontractors.

Management believed that this competitiveness was healthy. Unfortunately, the long-term results were disastrous. The three departments refused to talk to one another and the sharing of information stopped. In order to get the job done for the price quoted, the departments began outsourcing small amounts of work rather than using the other departments, which were more expensive. As more and more work was being outsourced, layoffs occurred. Management now realized the disadvantages of a competitive culture.

The type of culture can be impacted by the industry and the size and nature of the business. According to Eric Alan Johnson and Jeffrey Alan Neal[1]:

*Data-oriented culture:* The data-oriented culture (also known as the data driven culture and knowledge-based management) is characterized by leadership and project managers basing critical business actions on the results of quantitative methods. These methods include various tools and techniques such as descriptive and inferential statistics, hypothesis testing and modeling. This type of management culture is critically dependent on a consistent and accurate data collection system specifically designed to provide key performance measurements (metrics). A robust measurement system analysis program is needed to insure the accuracy and ultimate usability of the data.

This type of culture also employs visual management techniques to display key business and program objects to the entire work population. The intent of a visual management program is not only to display the progress and performance of the project, but to instill a

1. Eric Alan Johnson, Satellite Control Network Contract Deputy Program Director, AFSCN, and the winner of the 2006 Kerzner Project Manager of the Year Award; and Jeffrey Alan Neal, Blackbelt/Lean Expert and Lecturer, Quantitative Methods. University of Colorado, Colorado Springs.

sense of pride and ownership in the results with those who are ultimately responsible for project and program success . . . the employees themselves.

Also critical to the success of this type of management culture is the training required to implement the more technical aspects of such a system. In order to accurately collect, assess and enable accurate decision making the diverse types of data (both nominal and interval data), the organization needs specialists skilled in various data analysis and interpretation techniques.

## 6.4 CORPORATE CULTURES AT WORK

Cooperative cultures are based upon trust, communication, cooperation, and teamwork. As a result, the structure of the organization becomes unimportant. Restructuring a company simply to bring in project management will lead to disaster. Companies should be restructured for other reasons, such as getting closer to the customer.

Successful project management can occur within any structure, no matter how bad the structure appears on paper, if the culture within the organization promotes teamwork, cooperation, trust, and effective communications.

**Boeing**

In the early years of project management, the aerospace and defense contractors set up customer-focused project offices for specific customers such as the Air Force, Army, and Navy. One of the benefits of these project offices was the ability to create a specific working relationship and culture for that customer.

Developing a specific relationship or culture was justified because the projects often lasted for decades. It was like having a culture within a culture. When the projects disappeared and the project office was no longer needed, the culture within that project office might very well disappear as well.

Sometimes, one large project can require a permanent cultural change within a company. Such was the case at Boeing with the decision to design and build the Boeing 777 airplane. The Boeing 777 project would require new technology and a radical change in the way that people would be required to work together. The cultural change would permeate all levels of management, from the highest levels down to the workers on the shop floor. Table 6–1 shows some of the changes that took place.[2] The intent of Table 6–1 is to show that on large, long term projects, cultural change may be necessary.

As project management matures and the project manager is given more and more responsibility, project managers may be given the responsibility for wage and salary administration. However, even excellent companies are still struggling with this new approach. The first problem is that the project manager may not be on the management pay scale in the company but is being given the right to sign performance evaluations.

---

2. The Boeing 777 case study, "Phil Condit and the Boeing 777: From Design and Development to Production and Sales", appears in H. Kerzner, *Project Management Case Studies*, 4th Edition, Hoboken, NJ: Wiley, 2013, p. 97. The information presented in Table 6–1 is Harold Kerzner's interpretation of some of the changes that occurred, not necessarily Boeing's official opinion.

**TABLE 6–1.   CHANGES DUE TO BOEING 777 NEW AIRPLANE PROJECT**

| Situation | Previous New Airplane Projects | Boeing 777 |
|---|---|---|
| Executive communications | Secretive | Open |
| Communication flow | Vertical | Horizontal |
| Thinking process | Two dimensional | Three dimensional |
| Decision-making | Centralized | Decentralized |
| Empowerment | Managers | Down to factory workers |
| Project managers | Managers | Down to nonmanagers |
| Problem solving | Individual | Team |
| Performance reviews (of managers) | One way | Three ways |
| Human resources problem focus | Weak | Strong |
| Meetings style | Secretive | Open |
| Customer involvement | Very low | Very high |
| Core values | End result/quality | Leadership/participation/customer satisfaction |
| Speed of decisions | Slow | Fast |
| Life-cycle costing | Minimal | Extensive |
| Design flexibility | Minimal | Extensive |

The second problem is determining what method of evaluation should be used for union employees. This is probably the most serious problem, and the jury hasn't come in yet on what will and will not work. One reason why executives are a little reluctant to implement wage and salary administration that affects project management is because of the union involvement. This dramatically changes the picture, especially if a person on a project team decides that a union worker is considered to be promotable when in fact his or her line manager says, "No, that has to be based upon a union criterion." There is no black-and-white answer for the issue, and most companies have not even addressed the problem yet.

**Midwest Corporation (Disguised Company)**

The larger the company, the more difficult it is to establish a uniform project management culture across the entire company. Large companies have "pockets" of project management, each of which can mature at a different rate. A large Midwest corporation had one division that was outstanding in project management. The culture was strong, and everyone supported project management. This division won awards and recognition on its ability to manage projects successfully. Yet at the same time, a sister division was approximately five years behind the excellent division in maturity. During an audit of the sister division, the following problem areas were identified:

- Continuous process changes due to new technology
- Not enough time allocated for effort
- Too much outside interference (meetings, delays, etc.)
- Schedules laid out based upon assumptions that eventually change during execution of the project
- Imbalance of workforce
- Differing objectives among groups
- Use of a process that allows for no flexibility to "freelance"

- Inability to openly discuss issues without some people taking technical criticism as personal criticism
- Lack of quality planning, scheduling, and progress tracking
- No resource tracking
- Inheriting someone else's project and finding little or no supporting documentation
- Dealing with contract or agency management
- Changing or expanding project expectations
- Constantly changing deadlines
- Last-minute requirements changes
- People on projects having hidden agendas
- Scope of the project is unclear right from the beginning
- Dependence on resources without having control over them
- Finger pointing: "It's not my problem"
- No formal cost-estimating process
- Lack of understanding of a work breakdown structure
- Little or no customer focus
- Duplication of efforts
- Poor or lack of "voice of the customer" input on needs/wants
- Limited abilities of support people
- Lack of management direction
- No product/project champion
- Poorly run meetings
- People do not cooperate easily
- People taking offense at being asked to do the job they are expected to do, while their managers seek only to develop a high-quality product
- Some tasks without a known duration
- People who want to be involved but do not have the skills needed to solve the problem
- Dependencies: making sure that when specs change, other things that depend on it also change
- Dealing with daily fires without jeopardizing the scheduled work
- Overlapping assignments (three releases at once)
- Not having the right personnel assigned to the teams
- Disappearance of management support
- Work being started in "days from due date" mode, rather than in "as soon as possible" mode
- Turf protection among nonmanagement employees
- Risk management nonexistent
- Project scope creep (incremental changes that are viewed as "small" at the time but that add up to large increments)
- Ineffective communications with overseas activities
- Vague/changing responsibilities (who is driving the bus?)

Large companies tend to favor pockets of project management rather than a company-wide culture. However, there are situations in which a company must develop a companywide culture to remain competitive. Sometimes it is simply to remain a major competitor; other times it is to become a global company.

## 6.5 INDRA: BUILDING A COHESIVE CULTURE[3] _____

At Indra, the project manager role constitutes a key factor for project success. This is because running projects is a core part of our business. As such, company policies and practices are oriented to provide full support to project managers and to give them full responsibility on the project definition and execution. In the words of our former CEO: "Project managers must look at their projects as if they were running their own business."

This sentence distils the basis of the PM culture at Indra. It implies that a project manager must have an integrated approach to their job, not only focusing on main objectives tied to the triple constraint, taking care of schedule and cost baselines, but also having a business perspective and pushing to deliver results that will fulfill their business unit objectives (profitability, cost efficiency, development of resources, productivity, etc.) The PM foundations are shown in Figure 6–1.

The Corporate PMO provides support to around 3,300 project managers needing clear directions, missions, strategies, methodologies and a set of common tools and procedures to develop their job. We are responsible to develop and update IPMM, the Corporate PM Methodology. Based on that development, requirements for upgrading the company PMIS are defined and deployed. Ongoing support is provided both to business units and PM individuals, in terms of training and education, informal networking and participation in

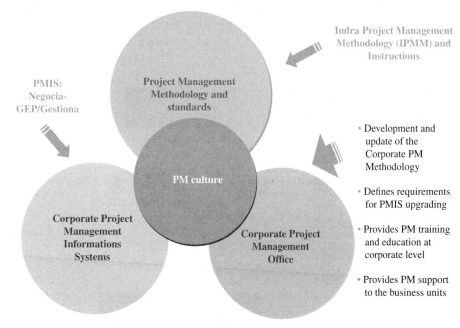

**Figure 6–1.** PM foundations.

_____

3. Material provided by Alfredo Vázquez Díaz, PMP, Director, Corporate Project Management Office, Indra

different initiatives related to PM that are required by the business units. Our final objective is building and consolidating a strong and recognizable PM culture within Indra, whatever the performing unit, geography or business sector.

In 2005 we started an internal PMP certification program for a small group of senior program and project managers and business unit managers. This certification program has been carried out yearly since then, and has become one of the most required training initiatives by project managers. Business managers carefully select the candidates that participate in the program.

In total more than 950 professionals have been through the PMP training process. We achieved the objective of counting on 500 certified PMPs by the end of 2012. As of today (May 2013), we count on 558 PMPs and have around 40 people more pending to run the test.

These figures wouldn't mean nothing without a context. For us achieving these figures mean that an important proportion of the most experienced and talented professionals at Indra are well trained in PM best practices. Taking into account that our PM methodology, IPMM, is aligned with the PMBOK Guide, then we could intuit that a certified PMP could easily spread out the knowledge and experience in PM best practices in her area of influence, be this her program, project or business unit. This is a way that works when it comes to settle a strong PM culture in all branches within the company. (See Figure 6–2.)

We started in 2008 having PMPs collaborating as internal trainers by delivering content of the course "Project Management at Indra," created by the Corporate PMO. This course explains IPMM Methodology and PM information systems. Thanks to this initiative, we are training our people in the PMBOK standard, and at the same time, the experience of the PMP trainer is used to provide a fitted project management context, using

**Figure 6–2.** People: Internal trainers.

projects and services that Indra provides to its customers as training examples. In fact, this collaboration has been a success, having win-win result for all participants:

- PMPs contribute to create a better PM culture, spreading best practices within the company and also getting PDUs to maintain their certification.
- Trainees connect directly with the content, without any interpretation that an external trainer could provide, as the teacher is a PMP who knows well which issues must be handled when it comes to manage a project in our company.
- HR training departments also win, because they can invest money in other areas that could need external trainers.
- Corporate PMO, which is supervising and supporting the consistency of the message being delivered in the training process

At the end is Indra as whole who benefits, because this PM course content has been put into e-learning format, was translated to English and Portuguese languages, and it has been included as a mandatory content in the PM training paths of every Indra company, wherever in the world this might be. (See Figure 6–3.)

In addition to this, at 2010, HR department made available to all employees one platform accessed from the intranet aimed to let people connect, share and learn from each other. This platform (named 'Sharing knowledge') has the look and feel of a social network and aims to support the informal exchange of knowledge and experiences between professionals. Its scope is corporate and local, and it helps to fast and easily deliver content on best practices and methodologies, management and technical issues, business information and has the possibility of creating groups and communities, and even broadcasting digital content and courses.

For us, Sharing Knowledge has been a powerful tool to get our PMs on the loop and in touch with the Corporate PMO and also to keep building PM culture. We created PMPnet,

**Figure 6–3.** "Project Management at Indra" course in e-learning platform.

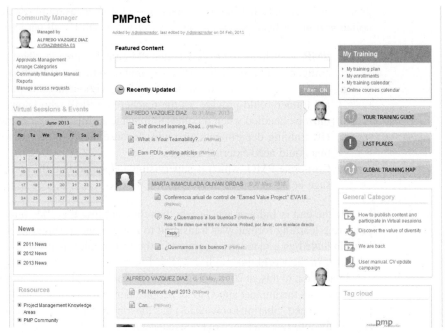

**Figure 6–4.** "PMPnet" in Sharing Knowledge tool.

(See Figure 6–4) for certified professionals at Indra who want to be in touch, be updated with any interesting initiative, activity or simply contribute with experiences and ideas.

Also a community of methodologies where all employees of the company have access include information relative to IPMM and its associated MIDAS libraries as seen in Figure 6–5. MIDAS is the Indra framework for different project life cycles (consulting, engineering, development, integration and services)

## 6.6  maxIT-VCS

When you work within your own company, you often have the luxury of time for changing the culture to be more supportive of project management. But when you work as a consultant and must perform miracles at a client that has a culture that refuses to support project management, life can be quite difficult. Marc Hirshfield, PMP, director, Project Management Office at maxIT-VCS, states:

> As a project management consultant working at a client site, the culture of the company many times does not typically support project management. Usually, the initial reaction to a PM is skepticism until they have a PM lead a project team and implement a successful project. Once they realize the importance of project management, including a well defined workplan, status reports, issues list, etc., their opinions change. In most cases, the culture allows project management to work on an informal basis. However, a PM can be successful by slowly introducing more formal methodologies to a project that create standards for future projects.

**START NOW**
MIDAS stands for methodology
for development, tailoring and services used by Indra

LISTINGS | ROLES-PROFILES | PLANNINGS | TOOLS | STANDARDS | RESOURCES

**Figure 6–5.** MIDAS framework.

One specific cultural barrier to an organization's acceptance of the project management function is the perception that project management is too time-consuming, requires too much detail, and that use of the project management tool methodologies overwhelm the project. With too much emphasis on the tool, this perception can become a reality. Projects work best, and are better accepted by management, if the PM focuses on communication as the goal, with the tool used as simply a way to keep large amounts of tasks organized and tracked into manageable pieces.

Another large cultural challenge in our industry is scope creep. Key stakeholders, including the management staff, may request new scope to the project, which puts the project at risk. One way to overcome this challenge is to have a scope document prepared and approved before the project begins. If anyone attempts to add scope, the PM can remind the group of the original plan (scope statement) and how adding new work would impact the budget and the timeline.

Lack of commitment by key stakeholders is also a cultural challenge in our industry. Sometimes people want to be included in the decision making process, but do not wish to attend the meetings. This causes unnecessary delays, which can impact the project's completion. One way to overcome this challenge is to meet with the key stakeholder(s) and personally ask them to attend the meeting, check their schedules in advance to ensure they are available or arrange to meet with them weekly to review all decisions and obtain their approval before proceeding with the project. Although, this may be additional work on the PM, it will ensure that unnecessary delays are avoided and support is provided.

We find some organizations that employ us have an expectation today for project manager certification, implying greater awareness of and desire for the professional project manager. This documentation is often requested as part of an RFP.

As project professionals, we realize the importance of managing a client's project with consideration of their culture: taking into account their decision-making processes, formal versus informal governance and meeting management, culture of communication, etc. A good project manager is able to adapt project management tools and techniques to fit the organizational culture of the client environment.

Internally, within the context of integrating two successful consulting firms with yet a third, we have embraced a strong commitment to a culture that supports and expects project management excellence. We are integrating project management and other methodologies among the companies, selecting the "best of the best" as our standards for tools and templates. We will then train consultants on those defined methodologies. We are developing our quality program based on specific expectations, so project managers have incentives to deliver to consistent standards. We recognize that projects often succeed or fail based on compliance to a solid project management foundation, and have resolved to provide the structure of methodologies, training and QA to support our consultants and our clients.

## 6.7 DFCU FINANCIAL[4]

At \$3.4 billion in assets, DFCU Financial is the largest credit union in Michigan and among the top 40 largest in the nation. With a 318.7% increase in net income since 2000, DFCU Financial has never done better, and effective project implementation has played a key role. At the root of this success story is a lesson in how to leverage what is best about your corporate culture. The story is also proof that staying true to core values is a sure way to sustain success over the long haul.

**1997–2005: Overcoming the Past**

Rolling back the clock to late 1997, I had just volunteered to be the Y2K project manager—the potential scope, scale and risk associated with this project scared most folks away. And with some justification—this was not a company known for its project successes. We made it through very well, however, and it taught me a lot about the DFCU Financial culture. We did not have a fancy methodology. We did not have business unit managers who were used to being formally and actively involved in projects. We did not even have many IT resources who were used to being personally responsible for specific deliverables. What we did have, however, was a shared core value to outstanding service—to doing whatever was necessary to get the job done well. It was amazing to me how effective that value was when combined with a well-chosen sampling of formal project management techniques.

Having tasted project management success, we attempted to establish a formal project management methodology—the theory being that if a little formal project management

---

worked well, lots more would be better. In spite of its bureaucratic beauty, this methodology did not ensure a successful core system conversion in mid-2000. We were back to the drawing board concerning project management and were facing a daunting list of required projects.

With the appointment of a new president in late 2000, DFCU Financial's executive team began to change. It did not take long for the new team to assess the cultural balance sheet. On the debit side, we faced several cultural challenges directly affecting project success:

- Lack of accountability for project execution
- Poor strategic planning and tactical prioritization
- Projects controlled almost exclusively by IT
- Project management overly bureaucratic
- Limited empowerment

On the plus side, our greatest strength was still our strong service culture. Tasked with analyzing the company's value proposition in the market place, former senior vice president of marketing, Lee Ann Mares, made the following observations:

> Through the stories that surfaced in focus groups with members and employees, it became very clear that this organization's legacy was extraordinary service. Confirming that the DFCU brand was all about service was the easy part. Making that generality accessible and actionable was tough. How do you break a high-minded concept like *outstanding service* into things that people can relate to in their day-to-day jobs? We came up with three crisp, clear Guiding Principles: Make Their Day; Make It Easy; and Be An Expert. Interestingly enough, these simple rules have not only given us a common language, but have helped us to keep moving the bar higher in so many ways. We then worked with line employees from across the organization to elaborate further on the Principles. The result was a list of 13 Brand Actions—things each of us can do to provide outstanding service (Table 6–2).

While we were busy defining our brand, we were also, of course, executing projects. Since 2000, we had improved our operational efficiency through countless process improvement projects. We replaced several key sub-systems. We launched new products and services and opened new branches. We also got better and better at project execution, due in large part to several specific changes we made in how we handled projects. When we looked closer at what these changes were, it was striking how remarkably congruent they were with our Guiding Principles and Brand Actions. As simple as it may sound, we got better at project management by truly living our brand.

## Brand Action—Responsibility

Project control was one of the first things changed. Historically, the IT department exclusively controlled most projects. The company's project managers even reported to the CIO. As former chief financial officer, Eric Schornhorst commented, "Most projects had weak or missing sponsorship on the business side. To better establish project responsibility, we moved the project managers out of IT, and we now assign them to work with a business unit manager for large-scale projects only. The project managers play more of an administrative and facilitating role, with the business unit manager actually providing project

**TABLE 6–2.   DFCU FINANCIAL BRAND ACTIONS**

**Make Their Day — • Make It Easy — • Be An Expert**

| | |
|---|---|
| Voice | We recognize team members as the key to the company's success, and each team member's role, contributions and voice are valued. |
| Promise | Our brand promise and its guiding principles are the foundation of DFCU Financial's uncompromising level of service. The promise and principles are the common goals we share and must be known and owned by all of us. |
| Goals | We communicate company objectives and key initiatives to all team members, and it is everyone's responsibility to know them. |
| Clarity | To create a participative working environment, we each have the right to clearly defined job expectations, training and resources to support job function and a voice in the planning and implementation of our work. |
| Teamwork | We have the responsibility to create a teamwork environment, supporting each other to meet the needs of our members. |
| Protect | We have the responsibility to protect the assets and information of the company and our members. |
| Respect | We are team members serving members, and as professionals, we treat our members and each other with respect. |
| Responsibility | We take responsibility to own issues and complaints until they are resolved or we find an appropriate resource to own them. |
| Empowerment | We are empowered with defined expectations for addressing and resolving member issues. |
| Attitude | We will bring a positive, "can do" attitude to work each day—it is my job! |
| Quality | We will use service quality standards in every interaction with our members or other departments to ensure satisfaction, loyalty and retention. |
| Image | We take pride in and support our professional image by following dress code guidelines. |
| Pride | We will be ambassadors for DFCU Financial by speaking positively about the company and communicate comments and concerns to the appropriate source. |

leadership." We updated our leadership curriculum, which all managers must complete, to include a very basic project management course, laying the foundation for further professional development in this area.

### Guiding Principle—Make It Easy

With project ownership more clearly established, we also simplified our project planning and tracking process. We began tracking all large corporate and divisional projects on a single spreadsheet that was reviewed by the executive team monthly (see Table 6–3 for the report headers). Project priority was tied directly to our strategic initiatives. Our limited resources were then applied to the most impactful and most critical projects. Eric Schornhorst commented, "Simplifying project management forms and processes has enabled us to focus more on identifying potential roadblocks and issues. We are much better at managing project risk."

### Brand Action—Goals

Chief information officer Vince Pittiglio recalled the legacy issue of IT over-commitment. "Without effective strategic and tactical planning, we used to manage more of a project

**TABLE 6–3.   DFCU FINANCIAL CORPORATE PROJECTS LIST REPORT HEADERS**

| Column Label | Column Contents |
|---|---|
| Priority | 1 = Board reported and/or top priority<br>2 = High priority<br>3 = Corporate priority, but can be delayed<br>4 = Business unit focused or completed as time permits |
| Project | Project Name |
| Description | Brief entry, especially for new initiatives |
| RequirementsDocument Status | R = Required<br>Y = Received<br>N/A = Not needed |
| Status | Phase (Discovery, Development, Implementation) and percentage completed for current phase |
| Business Owner | Business unit manager who owns the project |
| Project Manager | Person assigned to this role |
| Projected Delivery Time | The year/quarter targeted for delivery |
| Resources | Functional areas or specific staff involved |
| Project Notes | Brief narrative on major upcoming milestones or issues |

wish list than a true portfolio of key projects. We in IT would put our list of key infrastructure projects together each year. As the year progressed, individual managers would add new projects to our list. Often, many of these projects had little to do with what we were really trying to achieve strategically. We had more projects than we could do effectively, and to be honest, we often prioritized projects based on IT's convenience, rather than on what was best for the organization and our members." Focusing on key initiatives made it possible to say "no" to low priority projects that were non-value-added or simply not in our members' best interest. And the new measuring stick for project success was not merely whether the IT portion of the project was completed, but rather that the project met its larger objectives and contributed to the company's success as a whole.

## Brand Action—Teamwork

Historically, DFCU Financial was a strong functional organization. Cross-departmental collaboration was rare and occurred only under very specific conditions. This cultural dynamic did not provide an optimal environment for projects. The monthly project review meeting brought together the entire executive team to discuss all current and upcoming projects. The team decided which projects were in the best interest of the organization as a whole. This critical collaboration contributed to building much more effective, cross-functional project teams. We began to develop a good sense of when a specific team or department needed to get involved in a project. We also gained a much better understanding of the concept that we will succeed or fail together and began working together better than ever.

## Brand Action—Empowerment

As now retired chief operating officer Jerry Brandman pointed out, "our employees have always been positive and pleasant. But our employees were never encouraged to speak their minds, especially to management. This often had a direct negative impact on projects—people foresaw issues, but felt it was not their place to sound the alarm. A lot of the fear related to not wanting to get others 'in trouble.' We have been trying to make it comfortable for people to raise issues. If the emperor is naked, we want to hear about it! To make people visualize the obligation they have to speak up, I ask them to imagine they are riding on a train and that they believe they know something that could put the trip in jeopardy. They have an obligation to pull the cord and stop the train. This has not been easy for people, but we are making headway every day."

### Brand Action—Quality

At DFCU Financial project implementation in the past followed more of the big bang approach—implement everything all at once to everyone. When the planets aligned, success was possible. More often than not, however, things were not so smooth. Commented Jerry Brandman, "You have to have a process for rolling things out to your public. You also need to test the waters with a small-scale pilot whenever possible. This allows you to tweak and adjust your project in light of real feedback." Most employees have accounts at DFCU Financial, so we found we had a convenient pilot audience for major projects such as ATM to debit card conversions and the introduction of eStatements to ensure everything functioned correctly prior to launching to the entire membership.

Bottom line, the most significant best practice at DFCU Financial has been to be true to our core cultural value of providing extraordinary service. As we were working on defining this value and finding ways to make it actionable, we were also making changes to the way we approach project management that were very well aligned with our values. Our commitment to living our brand helped us:

- Move project responsibility from IT to the business units
- Simplify project management forms and processes
- Use project review meetings to set priorities and allocate resources more effectively
- Break down organizational barriers and encourage input on projects from individuals across the organization
- Improve project success through pilots and feedback

As president and CEO, Mark Shobe summarized back in 2005, "good things happen when you have integrity, when you do what you say you are going to do. The improvements we have made in handling projects have rather naturally come out of our collective commitment to really live up to our brand promise. Have we made a lot of progress in how we manage projects? Yes. Is everything where we want it to be? Not yet. Are we moving in the right direction? You bet. And we have a real good road map to get there."

**2005–2009: Poised for Growth**

So, how good was that road map? The preceding material was written in early 2005. By objective measures, fiscal years 2005 through 2008 were good ones for DFCU Financial (see Table 6–4). With over $2 billion in assets in late 2008, the credit union was ranking in the top 10 among its peers in the most important key measures.

So, from a purely financial perspective, DFCU was doing very well, especially given the global economic climate as 2009 began and the fact that DFCU was largely serving

**TABLE 6–4.  DFCU FINANCIAL RESULTS FOR QUARTER ENDING SEPTEMBER 30, 2008**

| | | Result | | Ranking | |
| --- | --- | --- | --- | --- | --- |
| Metric | DFCU | National Peer[1] Average | Regional Peer[2] Average | National Peers | Regional Peers |
| Return on Assets | 1.94% | 0.42% | 0.77% | 1 | 1 |
| Return on Equity | 13.98% | 4.23% | 7.19% | 2 | 2 |
| Efficiency Ratio | 49.57% | 65.41% | 69.31% | 5 | 3 |
| Capital/Assets | 13.91% | 9.82% | 10.69% | 2 | 9 |
| Total Assets | $2.0B | $3.4B | $1.0B | 39 | 5 |

[1] 50 largest credit unions as measured by total assets.
[2] Credit unions with at least $500 million in total assets in the states of Michigan, Pennsylvania, Ohio, Indiana, Illinois, Wisconsin and Minnesota.

members associated with the auto industry sector that is so much a part of the notoriously troubled Michigan economy.

The solid financials were the result of the current administration's efforts over an eight-year period to streamline and improve operations, clarify DFCU's brand and value proposition and initiate effective project selection and execution processes.

While these efforts were underway, the executive team and board of directors were evaluating a troubling metric – one that could undermine the company's ability to sustain its recent successes: DFCU's membership has been nearly flat for many years – a trend affecting nearly all U.S. credit unions. They therefore were focused on the critical, strategic issue of growth. And again DFCU's brand and guiding principles were helping shape the results.

## Brand Actions—Voice and Quality; Guiding Principle—Make Their Day

While DFCU's executive team and board of directors explored several growth options, work on selection of a new core processing system began in mid-2005. The core system conversion project was viewed as a strategic imperative for growth, regardless of DFCU's operating structure. The prior system conversion in 2000 suffered from poor execution that left behind lingering data and process issues that needed to be addressed. At the start of the conversion project, which began formally in January 2006, the conversion was targeted for October of that same year. From the outset, however, we ran into difficulties with the system vendor. The vendor was going through one of its largest client expansions ever and was having a hard time satisfying all of the demands of the conversion projects in its pipeline. The impact to us was noticeable – high turnover in key vendor project team members, poor quality deliverables and lack of responsiveness to conversion issues. Due to the poor quality of the data cuts, the project conversion date was at serious risk by June.

Due to its scope, the system conversion project was the only corporate project commissioned in 2006, and all attention was on it. It was not an easy task, therefore, to deliver the message that the project was in trouble. "Mark was well aware that we were having difficulties when we sat down to discuss whether we would have a smooth October conversion," commented chief information officer, Vince Pittiglio. "I have had to deliver bad news before to other bosses, but the talk I had with Mark was a lot different than those I had before." Our stated objective of this conversion was basically to do no harm. We all agreed that we could not put our members or our employees through the same type of conversion we experienced in July 2000. It had to be as close to a non-event as possible. According to Pittiglio, "I laid out the key issues we were facing and the fact that none of us on the project believed they could be resolved by the October date. If we kept to the original date, we believed we would negatively impact member experience." But our CEO Mark Shobe was very clear – he insisted that this project be a quality experience for both members and employees, as we all had agreed at the outset, and he was willing to go to the board to bump the date and to put other key initiatives on hold to ensure the conversion's success. The project team agreed upon a revised conversion date in early June 2007. According to conversion project manager and senior vice president, Martha Peters, "Though the team continued to face difficulties with the vendor, we worked hard and got it done without any major issues. It made a real difference to know that the chiefs and the board took our feedback seriously. In the end, it really was a non-event for most of our members and employees, exactly as we all wanted it to be. It was a tough decision to postpone the conversion, one that many companies are not willing to make. But it was the right decision to make – we really try to live our brand."

### Brand Action—Clarity and Teamwork

By the time the system conversion was completed in June 2007, we had really only engaged one project in the previous two years, albeit a large-scale project of strategic importance – which was consciously delayed by eight months. This project drained our resources, so little else was accomplished in those two years. "Coming out of the system conversion was this huge, pent-up project demand. And everyone thought that the issues facing their division were, of course, the most pressing," commented former chief financial officer, Eric Schornhorst. "We found out very quickly that our handy project tracking list and monthly corporate project meeting were insufficient tools for prioritizing how to deploy our scarce project resources." A small team was quickly assembled to put together a process for initiating and approving projects more effectively and consistently. A key objective for this team was to minimize bureaucracy, while trying to establish some useful structure, including a preliminary review of all new requests by the IT division. The output was a simple flow diagram that made the steps in the request and approval process clear to everyone (see Figure 6–6) and a form that integrated the instructions for each section. As the senior vice president of human resources related at the time, "My group was one of the first to use the new process. It was surprisingly well put together and easy to use. We made the pitch to replace our learning management system with a more robust, outsourced solution. It was one of the projects that made it on the 2008 list. To be honest, we were really at the point in our company's history where we needed a bit more discipline in this process. In years past, we independently advocated for our projects at budget time with our division heads. If we received budgetary approval, we viewed our project as 'on the list.' When it came time to actually execute, however, we often had trouble lining up the resources from all the different areas that needed to be involved, especially those in IT."

The new process contributed not only to clarity regarding corporate projects, but to teamwork as well. As Schornhorst had reflected in late 2008, "We reviewed the projects requested for 2009 using the new process. While we haven't completely addressed the backlog of projects created by the system conversion, we also have another project likely

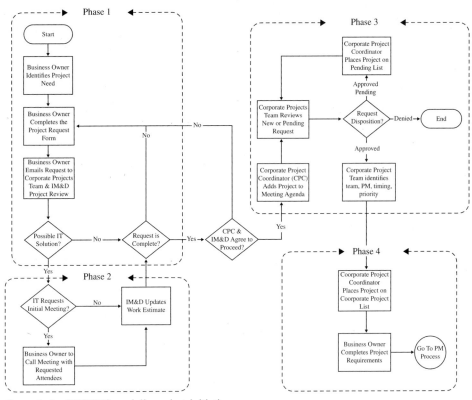

**Figure 6–6.** DFCU Financial's project initiation process.

to take over the majority of resources in 2009. This is not good news for areas that have not yet seen their projects addressed. What's interesting though is how little contention there was as we reviewed the docket for 2009 – and we had to put many important things on the back burner. I think that when you have everyone review the facts together in lock step, it's easier to get to a set of priorities that make sense for the organization and are mutually supported regardless of personal interest. It helps to bring the best out in all of us."

## So Where Was the Road Headed?

In early 2009, DFCU Financial was poised for membership growth. After a review of charter options, the board and executive team decided to remain a credit union, but to pursue other avenues for growth. To that end, as 2009 began, the board put forth for vote a proposal to DFCU's membership for a merger CapCom Credit Union that had 9 branch locations in the lower central and western areas of Michigan. According to former chief operating officer, Jerry Brandman, "We considered many different options to address our strategic vision for growth and expansion in membership – from mergers to various internal growth strategies. While we in the credit union industry currently face the same challenges as other financial institutions, we also are an industry known for service. And here at DFCU Financial we not only talk about service – we deliver. Our employees are happy and enthusiastic. They treat our members very well. We have been rated in the top 101 companies to work for in southeastern Michigan for 5 years in a row, based on feedback from our employees. And we have a member shops program that shows us how our service stacks up to industry benchmarks. We have consistently performed at the highest level of service relative to our peers. Merging with CapCom and shifting to a community charter will provide us the growth we need to ensure a bright future for our members and our employees. It will allow us to spread the DFCU brand to other geographic areas. We believe this proposal will be appealing to our members and will be successful. We believe that people want to be a part of our organization."

And for good reason. Also in early 2009, DFCU Financial paid out a $17 million patronage dividend to its members for the third consecutive year, for a total of more than $50 million being paid out since inception, during one of the worst economic times to hit the Motor City in decades. According to Keri Boyd, senior vice president of marketing at the time, "Since we are committed to remaining a credit union, we have looked at ways we could improve our value proposition to existing members and attract new members. The patronage dividend is the cornerstone of our approach to growing the business as a credit union." As Mark Shobe summed it up, "The board and I did not want to begin the payout until we were confident that we could sustain it over the years. It took hard work, some tough decisions, excellent project execution and diligence in our day-to-day operations to be in the position to share our success with our membership. The simple truth is that the driving force behind our success is our collective commitment to our brand." So, with a 2009 schedule of agreed-upon projects, a potential merger in the offing and some very, very satisfied members, how is the DFCU road map working? "Quite well, thank you!" replied Mark.

**2009–2013: Paying Dividends in More Ways Than One**

During this most recent period in DFCU Financial's history, we have made good progress on our strategic goal of growth. We completed the CapCom merger in late 2009, the same year we began due diligence for another merger with MidWest Financial Credit Union, based in Ann Arbor, Michigan. The MidWest merger was completed in early 2011. In January 2012 we opened a new branch we built in Novi, Michigan, and at the time of this writing we are building a new branch for the Ann Arbor market that is expected to open in May 2013. Table 6–5 summarizes the key growth metrics for the last four years.

Just as we set out to do, we have indeed grown the business through merger and new branch projects. But growth is not the only measure of success. If managed poorly, growth can have a deleterious effect on core financials, service and employee morale. So, how have we done?

During this same period of time, we have maintained a strong position financially when compared with our peers. Our financial strength has allowed us to continue to pay out an annual Patronage Dividend to our membership. In January 2013, we paid out our 7th dividend totaling $21.8 million, accounting for an accumulated total of $133.4 million since 2006. Member satisfaction has never been higher as measured through our member shops program, and in 2012 we were again in the top ranks on this element when benchmarked against our peers.

Not least importantly, we have continued to receive awards that recognize DFCU Financial as a premier employer – awards that are based solely on employee feedback

**TABLE 6–5.   DFCU FINANCIAL GROWTH METRICS**

|                       | 2000      | 2008      | 2009      | 2010      | 2011      | 2012      |
| --------------------- | --------- | --------- | --------- | --------- | --------- | --------- |
| **Number of Members** | 170,812   | 167,910   | 201,329   | 218,374   | 213,869   | 214,454   |
| **Number of Branches**| 6         | 12        | 12        | 21        | 22        | 23        |
| **Number of Employees**| 409      | 336       | 434       | 426       | 408       | 413       |
| **Assets in Billions**| $1.2      | $1.9      | $2.4      | $2.7      | $3.0      | $3.2      |

**TABLE 6–6.   DFCU FINANCIAL SUCCESS METRICS**

|                                       | 2000    | 2008    | 2009    | 2010    | 2011    | 2012     |
| ------------------------------------- | ------- | ------- | ------- | ------- | ------- | -------- |
|                                       |         | **Key Financial Metrics[1]** | | | | |
| **Return on Assets**                  | 1.01%   | 1.69%   | 1.25%   | 1.23%   | 1.38%   | 1.55%    |
| **Return on Equity**                  | 13.08%  | 12.06%  | 9.52%   | 9.53%   | 11.11%  | 12.36%   |
| **Efficiency Ratio**                  | 77.50%  | 50.84%  | 52.90%  | 54.73%  | 57.56%  | 52.26%   |
| **Capital/Assets**                    | 7.66%   | 14.00%  | 13.12%  | 12.94%  | 12.70%  | 12.49%   |
|                                       |         |         | **Special Patronage Dividend** | | | |
| **Total Payout in Millions**          | -       | $17.5   | $19.3   | $18.9   | $21.1   | $21.8    |
|                                       |         |         | **Member Satisfaction** | | | |
| **Member Shops 0 – 5 scale**          | -       | 4.80    | 4.86    | 4.89    | 4.96    | 4.97[2]  |
|                                       |         |         | **Employee Satisfaction** | | | |
| **Best & Brightest – Metro Detroit[3]** | -     | ✓       | ✓       | ✓       | ✓       | ✓        |
| **Best & Brightest – West MI**        | -       | -       | -       | -       | -       | ✓        |
| **Best & Brightest – National**       | -       | -       | -       | -       | -       | ✓        |

[1]Compared to our top 50 national peers, DFCU ranked 8th in ROA, 15th in ROE, 5th in efficiency and 5th in C/A as of 12/31/12.
[2]Versus a peer compare score of 4.82 for 2012. Note: Shops began in 2002 with a baseline of 4.05.
[3]DFCU has received the 101 Best & Brightest Companies to Work For-Metro Detroit for 8 consecutive years.

– such as the 101 Best and Brightest Companies to Work For and the *Detroit Free Press* "Top Workplace" awards.

We have also been recognized for our growth and geographic expansion over the last few years by being named a Michigan Economic "Bright Spot" by *Corp! Magazine*. These successes are summarized in Table 6–6.

## Brand Is Still How We Do It

When we grow by building new branches, the DFCU Financial brand extends rather naturally. In new branch projects, all elements of brand are well controlled – from the look and feel of the facility, to the training and on boarding of new employees, to even the project methodology we use. Growth by merger is not as organic. Ensuring that brand is protected through these projects is no small feat. According to president and CEO Mark Shobe, "In seeking potential merger partners, we look for organizations that are not only a good fit strategically, but which also ostensibly share our core values. But no matter how well suited the match appears to be, the biggest challenge in any merger project is culture."

## Brand Action—Goals and Clarity

CapCom was the first merger DFCU Financial completed during this period. It provided us access to two new geographical markets, but most importantly, it presented us with an opportunity for charter change. Originally a federally chartered credit union, DFCU Financial is now a state chartered, community credit union that has growth access to all counties in Michigan's lower peninsula. To change the charter required a vote of the membership. The membership charter vote was one of the most important tasks within the scope of the CapCom merger project. According senior vice president for strategic marketing and operations, Martha Peters, who was the CapCom merger project manager:

> For us to achieve what we set out to with this merger, we not only needed to successfully integrate systems and functional areas, we also needed to make a compelling argument to our existing membership base to change our business structure. In making both things happen, we leveraged our brand. We needed to assure our existing members that we would be the same DFCU they had relied upon – only better, while at the same time we had to demonstrate to our new CapCom colleagues how they and the members they historically served would benefit by adopting the DFCU Financial culture. To facilitate our success on this project, we were very careful about how we structured the project and communicated our goals. We established a steering committee, comprised of DFCU and CapCom executives, which was responsible for making all major project decisions and assessing and communicating the member and staff impact of these decisions. This structure helped us to keep a pulse on employee and member feedback and to address project risk very effectively.

All key objectives of this project were met – the charter change, the functional integration, the legal merger and the system integration – and, more importantly, the DFCU Financial brand came out a winner.

## Brand Actions—Respect and Responsibility; Guiding Principle—Make It Easy

As the CapCom merger was successfully wrapping up, I was assigned as project manager for the MidWest Financial merger. While charter change figured heavily in the CapCom merger, it was also a project where we learned a lot about how to do mergers. Since we conducted a formal lessons learned exercise coming out of the CapCom merger, I was able to re-use the elements that worked and focus on the areas that needed strengthening.

Being a small company, we do not have dedicated project management resources. Front-line and back office business unit managers and their teams are expected to participate as project team members, and often serve as project managers. In our merger projects, all managers are responsible for all functional integration tasks that relate to their business unit. They are essentially the project manager for their functional area. Managers whose operations are tightly coupled with the core computer system have even greater responsibility to ensure that data from the relinquishing system is safely and correctly converted into the surviving system. It is a daunting set of responsibilities.

One of the lessons learned from the CapCom merger is that these responsibilities were not clear to everyone, so we had varying degrees of compliance with project methodology. By the same token, we all learned together on that first merger what the big pieces were and how they fit together. To build on what we learned from CapCom, I worked with the executive team to develop a project charter that explained project scope, goals, structure and participant responsibilities and featured the review of this document at the kick off meeting. Then, to make this large-scale project more manageable for the business unit managers, I pulled together the small set of tools that everyone was required to use - our usual project database in Lotus Notes, a simple task list template in Excel and a simple status reporting template - and clearly articulated the rules of engagement: 1) the tools had to be proactively used, 2) tasks lists and status reports posted by deadline to the database and 3) the project manager for each area had to review their written, posted status report at each project meeting with the entire project team. I made myself readily available to anyone who needed help using the tools and was happy to see the extent to which we increased our project management competency through this project.

The methodology requirements and tool set we insisted upon project leads and participants using for the MidWest project paid big dividends for us, and not just for that merger project.

## Brand Actions—Empowerment, Responsibility, and Quality

With two mergers happening back to back and the first occurring not too long after our major core system conversion in 2007, our project backlog was growing by the day. A real moment of reckoning occurred in October 2010 when the MidWest merger was still in full swing. One of the issues that had arisen during the CapCom merger related to cash dispenser machines (CDMs). All original DFCU branches had CDMs serving each teller window. CapCom's branches did not. Rather than purchasing CDMs for CapCom, we rewrote dozens of cash operations procedures to accommodate branches with and without CDMs. While this solved the immediate issue, branch senior management did not like the complexity this inserted in operations. Also having other, more technical concerns about CDMs, the management team decided to evaluate de-implementing CDMs. They quietly worked on a pilot in one of the busier branches. They liked the result, and decided to move immediately to de-implementation system-wide. In short order, we had another project happening somewhat under the radar in the midst of a merger. It was time to pull the brake on the train. According to Steven Schulman, senior vice president of branch development, "There I was, recently promoted to SVP and my first big task was to get rid of our CDMs. No big deal, right? All of a sudden, I'm facing lots of questions from other managers – Have you thought about the impact to procedures? Are the units fully amortized? Is Facilties going to patch the teller cabinetry – do they have budget for that? Do we need to remove the CDM cash boxes from the system? And so on. I felt I was doing the right thing, but it became clear to me that to do it right way, I needed to follow our normal project management protocols. Lessons learned!"

The CDM de-implementation project was a watershed moment for us on a number of levels. It was absolutely the right thing to do and was probably not wise to put off for much longer. At that same time, all of us business unit managers had numerous projects of both large and small scale on our to do lists, and some of them, just like this one, really needed to proceed – in spite of the merger. So, for the benefit of all parties concerned, we really needed to push out our project methodology and tool set to a much broader audience. So, we got Steven the tools he needed to get his project done, and then spent 2011 not only finishing the MidWest merger, but also putting into place some improvements in how we manage projects and communicate about them. Though we still use dedicated databases for large scale projects

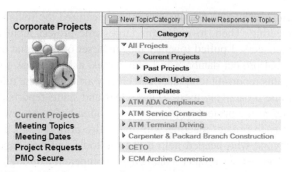

**Figure 6–7.** DFCU Finanical's corporate projects database.

like mergers, we rolled out the Corporate Projects database in Lotus Notes that contains project templates and provides a place to post documents for all active projects (see Figure 6–7).

In late 2011 we also introduced a formal Business Change Process designed to ensure that we avoid another surprise project like CDM de-implementation. The process is light on bureaucracy and requires business unit managers to start the conversation by putting in a quick ticket to a support database that describes the problem they need to solve or the project they believe needs to be engaged. We have a small business change committee, comprised of representatives from each division, that reviews the tickets each week to better understand the scope of each initiative and to determine which areas need to be involved. In some cases, business change tickets move on to corporate project stature, subject to our formal project management protocols. In most cases, however, the business change initiatives proceed as small-scale projects that are worked on as time permits, in deference to the approved corporate projects and the required recurring projects like core system updates and tax reporting.

Along with this process change, we created a new project list that summarizes all projects in play at any given point in time and allows management to understand not only project priorities, but where the resources across the company are currently committed (see Figure 6–8). The list is released monthly to all management who are encouraged to take the time to review what is on the list and to share what they feel is appropriate with their teams. The intent of new project list is to ensure better transparency on how resources are being deployed on projects and to build awareness on initiatives that will eventually impact staff and/or members.

## And the Brand Lives On

We have had significant financial success over the last several years, and the executive team is very focused on sustaining that success over the long haul. The DFCU Financial brand guides us in what we need to do, and in many cases, how and when we need to do it. Our brand has had a lot to do with shaping how we approach projects because it reflects how we think about ourselves and what we want to achieve. We will continue to make progress as we can on our strategy of managed growth. Our recently completed branch expansion study will inform our decisions for building new branches, and though the environment is changing, we continue to look for merger opportunities. At the same time, we are looking carefully for ways to improve, to take our products, services and processes to the next level. So, as we start 2013, we have plenty of projects and initiatives in the works that found their starting point in brand.

2013 Resource Planning

| # | Corporate Projects | Strategic Value | Impact | Size | Lead | J | F | M | A | M | J | J | A | S | O | N | D | Comments |
|---|---|---|---|---|---|---|---|---|---|---|---|---|---|---|---|---|---|---|
| 1 | IVR Replacement | Req'd | Member | Med | Peters | | | | | | | | | | | | | Launched 2/13; monitoring member feedback. |
| 2 | Mobile Banking & Security Alerts | High | Member | Med | Peters | | | | | | | | | | | | | Complete! |
| 3 | ATM Terminal Driving | High | Member | Large | Peters | | | | | | | | | | | | | Conversions finish in March. |
| 4 | MFA | Req'd | Member | Large | Dinn | | | | | | | | | | | | | Still testing error message. |
| 5 | ECM Phase 1 – BDS-OnBase Conversion | High | Staff | Large | Hershey | | | | | | | | | | | | | Training Mar 11-29; launch on April 1st. |
| 6 | Plymouth Rd Branch Construction | High | Member | Large | Winnik | | | | | | | | | | | | | All tasks on track. |
| 7 | Plymouth Rd Branch Opening | High | Member | Large | Kidwell | | | | | | | | | | | | | In full swing; early May opening. |
| 8 | IB RFP | High | Staff | Small | Kidwell | | | | | | | | | | | | | Vendor responses due 4/12. |
| 9 | Hardware Security Module | High | Staff | Small | Peters | | | | | | | | | | | | | Awaiting equipment delivery. |
| 10 | Public Website Infrastructure Upgrade | Req'd | Member | Large | Sweet | | | | | | | | | | | | | Add to April meeting. |
| | **Required Recurring Projects** | | | | | | | | | | | | | | | | | |
| 1 | Patronage Dividend | High | Member | Med | Kalinski | | | | | | | | | | | | | Jan 2013 payout complete! |
| 2 | 2012 Year End & Tax Reporting | Req'd | Member | Large | Kidwell | | | | | | | | | | | | | On track. |
| 3 | 2013 SDB Billing/Drilling | Req'd | Member | Med | Chiles | | | | | | | | | | | | | Collections mode. |
| 4 | 2013 Escheats | Req'd | Member | Med | Mahony | | | | | | | | | | | | | Project starting. |
| 5 | 2013 Fee Analysis | Med | Member | Med | Soullier | | | | | | | | | | | | | CETO replaces this for 2013. |
| 6 | 2013 Privacy Mailing | Req'd | Member | Small | Mallery | | | | | | | | | | | | | Awaiting US Senate vote. |
| 7 | Velocity | Req'd | Staff | Med | Cochran | X | X | | | | | | | | | | | Release buggy; awaiting 3.0; timing tbd. |
| 8 | C Cure | Req'd | Staff | Small | Winnik | | | | | | | | | | | | | GF & SF done; OB up next. |
| 9 | i-Vu | Req'd | Staff | Small | Winnik | | | | | | | | | | | | | Upgrade HVAC controls for FL |
| 10 | CETO | High | Member | Med | Kalinski | | | | | | | | | | | | | Testing, fee schedule & disclosure updates in process. |

| **Business Process Change Requests** | | | | |
|---|---|---|---|---|
| # | Project | Lead | Team | To – From | Status |
|---|---|---|---|---|---|
| 17 | Business Account Process Changes | Golles | Kidwell, Mallery, Belanger, Conway, Cameron 5 | 2013 | Suspend |
| 25 | Address/Multiple Statement Clean Up | Chiles | Soullier, Northup, Peters, Earhart, DCS | Mar 12 – Mar 13 | Open |
| 42 | Branch BCP Procedures & Forms | Golles | Branch, DCS, Fitzpatrick | Mar 12 – Mar 13 | Open |
| 55 | Branch Appointments Database (follow up from Account Switch project) | Golles | Hershey, Belanger, Mtg, MSC, Branch, ConsLend | 2013 | Suspend |
| 62 | eReceipt & Voucher Study (for later phase ECM project) | Kidwell | DCS, Marketing, IM&D | Aug 12 – Mar 13 | Open |
| 70 | FHLB1 Mortgage Loan Pledge Extract | Solarz | Core Resource Group, IM&D | Dec 12 – Mar 13 | Open |
| 71 | Replace Email Vendor with Exact Target | J Nelson | O'Gorman, Soullier, Buchanan, Moss, Moorhead | 2nd Q 2013 | Open |
| 72 | Member Retention Model | Soullier | Will work alone this first phase | Tbd | Open |
| 75 | Modify POD Fix to Remove Deceased Accounts | Kidwell | Northup, Blazo, Davis | March 1st Install | Open |
| 76 | Custom Notice for POS Declines | Buchanan | O'Gorman, DCS, Ops, Programming, DataMail | Jan – tbd 2013 | Open |
| 77 | Modifications to the Suggestions Database | Morgan | Belanger | Feb – tbd 2013 | Open |
| 78 | Gift Card PIN Rollout | Jackson | Moorhead, Kulik, Kidwell, T&D | Tbd | Open |

| **IM&D Infrastructure Projects** | | | | | |
|---|---|---|---|---|---|
| Project | Lead | Team | To – From | Status | % Done |
|---|---|---|---|---|---|
| MS SQL servers need DR plan* | Tim | ENG | Apr 12 – Jul 13 | Active | 25% |
| Upgrade lotus notes to current ver 8.53* | Doug | CSS | Jan 12 – Apr 13 | Active | 99% |
| Upgrade vmware to current ver 5.0* | Doug | CSS | Jan 12 – June 13 | Active | 50% |
| Network Upgrade (Fairlane datacenter) | Cyndi | Voice & Data | Jan – June 13 | Active | 40% |

*Become priority in the event of another merger

**Figure 6–8.** DFCU Finanical's resource planning document.

## Brand Action—Protect

Like other financial services organizations, DFCU Financial has been facing lower net interest revenue due to the repricing of our loan and investment portfolios caused by the protracted low rate environment and increasing pressure on fee income from what is described as 'consumer protection.' This has translated into a loss about about $6 million annually for DFCU over the last several years. Chief financial officer, Marv Elenbaas has been keeping the management team focused on a couple of key metrics that are extremely important to watch during times of tight margin compression: core net operating revenue growth and core non-interest expense growth. He also demonstrates the impact the environment is having on us by including in the monthly financials the average earning asset yield for the past five quarters, which went from 3.34% at the end of the 4th quarter 2011 to 2.66% at the end of the 4th quarter 2012. By building awareness about the impact the environment was having on our core financials, Marv was able to make the case and gain support for a project to perform a thorough review and analysis of our fee structures. But feeing members is a topic about which different areas in the credit union have very passionate, and often diametrically opposed, views. And these views are solidly rationalized against our brand. To ensure a productive, collaborative analysis and decision making process, Marv looked to an objective third party to facilitate the review. The key project deliverable was a set of fee change recommendations. According to Marv, "The fee changes agreed upon through this project are well supported and well understood by all parties and, most importantly, reflect a balance of our brand principles. The plan addresses the critical need for increasing our revenue stream over the next few years. At the same time, it ensures that DFCU continues to offer a solid value proposition to members. Not only have members benefited from the downward repricing of loan rates, we are still positioned as less aggressive on fees when benchmarked with our peers, allowing us to maintain our 'make their day' claim in the marketplace. It is not often that stakeholders from finance, customer service and marketing agree on a topic like fees, but we came pretty close."

### Guiding Principles—Make Their Day and Make It Easy

One of the major infrastructure projects engaged as we wrapped up the recent mergers was the implementation of a new, robust enterprise content management system. While the initial phases of this multiyear initiative are unremarkable, though foundational – replacing the existing document archive and converting content – the later phases are expected to delight employees and members by capturing signatures electronically on account paperwork, eliminating the handling and movement of paper, providing expanded self-service eDocuments to members and moving from paper vouchers and receipts on the teller line to eReceipts – things our employees and members have been requesting for a while as more and more businesses go green.

### Brand Action—Voice

To ensure that we sustain our position as a preferred employer and to strengthen employee retention, which is sure to be challenged as the Michigan economy improves, human resources launched a new and improved employee suggestion process at our annual employee meeting in January 2013. This solid commitment to listening to employee feedback will no doubt be the starting point for projects and initiatives designed to attract and retain the talent DFCU will need to take our business to the next level.

### Brand Action—Promise

2013 is also a year in which we will focus on assessing and improving internal customer service. According to Keri Boyd, executive vice president for corporate development, "Our front line staff who provide direct member service have been shopped for years – and their scores are impressive. But we have been asking ourselves: can we be even better? This questioning has led us to look more broadly at the service delivery process. When front line team members need help, they seek it from our back office areas – whether IT, operations, underwriting, collections. We want to ensure that those interactions are as supportive, positive and effective as possible. We are all part of the service value chain. Looking closely at internal service, we will be able to identify areas for improvement."

### Guiding Principle—Be an Expert

A critical component of outstanding member service is competence, hence the call to 'be an expert.' When we were going through our core system conversion, everything was new for every area, so documenting detailed procedures was an imperative component of that project. Policy and procedure loomed large in the merger projects as well since we needed to ensure new team members were clear on the DFCU way. According to delivery channel support supervisor, Kelly Kidwell, "We now have a very seasoned team, yet we are seeing some evidence of a literal reliance on our written knowledge base which sometimes gets in the way of effective problem solving and decision making when providing member service. This is a big concern for me as my area writes and maintains all operational policies and procedures and provides phone support to front line team members on operational matters." To begin to address this trend, Kelly initiated a "solve the right problem" campaign with her team and is coaching them to improve their listening and facilitation skills when providing support to the front line and is encouraging internal discussion about what is working and what is not. This initiative has already helped identify the need to overhaul our service delivery processes for deceased and fiduciary accounts, a project that will get underway in mid-2013.

### Guiding Principle—Make It Easy

While we were focused on system conversion and mergers, we did not have the band width to address our delivery channels. As Martha Peters, senior vice president for strategic marketing and operations, notes, "We had to keep some things constant. Mergers and conversions present huge changes by themselves. We could not risk *also* making channel changes, which are always disruptive to our members. So, needless to say, the last couple of years have been all about the channels. In 2012, I led projects to switch out our ATM terminal driving platform, replace our IVR and launch our mobile banking application. We are currently going through the RFP process to replace our internet banking channel. We are targeting that change for 2014. A key strategic requirement for all of these initiatives is to simplify our solutions and make it easier for us to change and improve. To do so, we need to move from our currently highly customized platforms to solutions that will allow us to adapt more rapidly to changes in technology and in the market place."

As DFCU Financial embarks on the next leg of its journey, the road map is more about honing expertise, improving the quality of service delivery and being successful and nimble in an increasingly wired world. And, as usual, the 'how' of our journey will be guided by our interest in sustaining and building upon our success and leveraging our brand to do so.

## 6.8 ILLUMINAT (TRINIDAD & TOBAGO) LIMITED

On paper, the merger of two or more companies may look easy. But if the companies have different cultures, and each culture has a different view of the benefits of project management, the actual merger may not produce the benefits expected. The remainder of this section has been provided by Cynthia James-Cramer, EMBA, PMP, Consultant, Project Management Services, ILLUMINAT (Trinidad & Tobago) Limited.

**Managing the PM Culture in a Merger**

A merger of companies always brings some inherent challenges to the newly formed organization. The methods employed in rising to these challenges would determine not only the success of the merger but also how soon the operations can reach a level of efficient operations.

When three entities of the ITC group (Information Technology & Communications) of the Neal & Massy Holdings company merged in the year 2001, September 11, the then CEO of ILLUMINAT, Mr. Keith Thomas, embarked on a thrust to introduce the organization to a Professional Services Methodology (Profserv) in which a project management methodology was embedded along with organizational learning and organizational adoption.

Of the three entities, only one company previously had a culture that supported project management because of its IT focus and had introduced the concept of a project office just prior to the merger. The tasks before the CEO were now to not only implement the Profserv methodology but also to institutionalize a project management culture. This was essential as the new company's mission was now focused on providing solutions which spanned the offerings of all three entities distinct from their previous undertakings of providing products and services.

### The Mandate

The CEO issued a mandate for adoption of the Profserv methodology to all employees inclusive of senior management who was held responsible for the success of its execution. A similar mandate was extended to a companywide adoption of a standard approach to project management.

### The Approach

First all managers, team leaders, and key department resources were sensitized to a project management culture and also shown what fallouts can emerge when such a culture does not exist. The evidence was apparent from a research done on a sample of key projects, which were executed without a project management methodology that clearly demonstrated where things went wrong and the impact on the success of these projects. The response and feedback from the employees were tremendous, which resulted in some employees' eagerness on the training up ahead. The organization itself was restructured to a team-based approach to services delivery and, where relevant, the supervisor and manager roles were renamed team leader.

The project office was now refocused to a project management office (PMO) with a mandate to establish and roll out a common project management methodology and to act as a repository for disseminating project management information as well as for maturing the culture in the short and long term. The organization's leadership showed its commitment by investing in staff training and complementing the existing staff with suitable, experienced project management recruitment.

### The Success

The merger challenges of mixed cultures, merged departments, merged functions, and varying levels of project management maturity or the absence of it were now bridged and tackled as a single test. The emerging success has been rewarding as the team approach has taken off. There is now a standardized approach to the whole services delivery cycle from winning the opportunity to implementing the solution. There has also been evidence of an increase in repeat business from existing customers and the opportunity for sole tendering.

Recently another research was done with a similar sample of projects to the first research, this time with projects executed using the project management methodology. The results spoke myriads to improved project success.

### Sustaining the PM Culture

Top management is continuing its investment in the project management culture by supporting the ongoing development of the PMO staff and its endeavors. The PMO, in keeping its visibility in the organization, has launched a team-based annual recognition award program and has started a quarterly newsletter, which is circulated through the entire organization. The newsletter reports on project inventory, project successes, and project management–focused tidbits. Along with these efforts, staff with a specific interest in the discipline are encouraged to pursue the basic project management certification, some have already done so, and others are taking up the challenge. In addition, new employees are introduced to the PMO through the new-employee orientation program. Yet another thrust is planned internal training for all key delivery personnel to reinforce the project management methodology.

**Conclusion**  Adopting a PM culture in a merger may look insurmountable at first, but by following best business practices of top management buy-in, involvement, and support, half of the job gets done. The rest lies with the PMO to stay visible by mentoring, teaching, and providing the necessary information and support to all arms of the organization until the transfer of the culture has been embedded.

In the words of the then CEO:

Getting started! The approach to adoption is perhaps the most critical aspect of any change process. Arguably more critical than the change itself, it must be very carefully thought through and implemented. What follows is the result of due deliberation on the part of management and feedback from staff focus groups.

Some believe that Adults accept new approaches best when they understand why. Therefore, we would begin with organization-wide sessions on the WHY.

This would be followed by the HOW.

## 6.9 DTE ENERGY

Tim Menke, PMP®, Senior Continuous Improvement Expert, DO Performance Management at DTE Energy, describes the culture at DTE Energy:

While the level of support for project management varies across departments, generally it is increasing overall. Several years ago it was common to find experienced practitioners managing large, high visibility projects using both internal and external methodologies. Today practitioners can be found across the company applying the tools and techniques to a variety of projects. In some cases this is a result of stakeholder expectations. In other cases, this is driven by the person managing the project. The Continuous Improvement Project Management Methodology is used by all Lean Six Sigma Black Belt Candidates at DTE Energy as it is. . . . required of all projects submitted in support of certification.

Both formal and informal project management approaches are used. Projects reporting progress to governance boards or stakeholder committees tend to be managed more formally. Most departments have implemented progress reporting templates to drive consistency. Projects managed within a Director's organization or below tend to be managed less formally. Recognizing all projects do not require the same level of project management rigor, our Continuous Improvement Project Management Methodology has been revised to "scale" to the particular project and stakeholder needs.

Challenges to implementing project and program management we have experienced include:

● Stakeholders driving scope creep
● Project Managers with varied experience levels
● Team members struggling to balancing operational and project needs

Ways we have overcome these barriers include:

- Peer-to-peer mentoring by experienced practitioners
- Employee development plan activities to improve proficiency
- Increased focus on managing the matrix reporting relationship

## 6.10 HEWLETT-PACKARD

Doug Bolzman, Consultant Architect, PMP®, ITIL Expert at HP, believes that:

In many cases, instituting project management an enabling is part of the cultural change required by an organization. When major improvements are needed, the culture is improved by instituting PM and PM cannot depend on the culture being there.

Implementing Enterprise wide releases/projects requires the culture to move the organization from a functional to matrix management, move the delivery from project centric to component centric and move the planning from tactical (emotional) to strategic (analytical) planning. This level of cultural change needs to be identified, designed and implemented.

We recently worked with a PMO for an organization that was primarily managing individual projects, had the typical dashboard in place to communicate status, and was basically an administrative function for the organization. We offered them the opportunity to be a "premier PMO" and that we could help them implement a number of best practices to allow them to:

1. Direct the IT Service Roadmap in terms of releases to the client

   a. In relationship to the timing of business needs
   b. In relationship to the compatibility / integration of the services/architectures

2. Manage/Integrate IT Service Releases

   a. For each Service/Delivery Team
   b. Utilizing the Service Management Lifecycle
   c. Manage integration and dependencies for each release

3. Own the Design & Implementation of Non-Standard Service Requests

   a. Projects, Onboarding, Applications, Employee Moves
   b. Establish a quarterly forecast and open Purchase Order to bill requests

4. Communicate to the IT Service Owner and/or Client on Service Status
5. Establish and Measure IT Service Lifecycles / Policies

   a. Establish the overall Service Management Lifecycle and continually improve
   b. Institute consistency for how services are delivered (i.e. Change Policy)
   c. Onboard Leveraged Service Providers to proper client & pharmaceutical standards

Since the PMO did not formally "recognize" ITIL as a best practice, they declined the opportunity to advance or mature their project management services and kept to their administrative practices. They had the opportunity to be a significant driving force for their company and provide value in the management of the overall IT direction but did not want to change, they desired status quo.

The next client we worked with had a PMO established with the same basic parameters of providing project support. As we presented this opportunity, they jumped for joy over the fact that someone was willing to help them be more valuable to the business and to get the out of "administration mode". So the moral to the story is, the opportunities are available (through the leveraging of best practices) and the people who will take advantage of the best practices are the ones who desire to provide service excellence.

## 6.11 BARRIERS TO IMPLEMENTING PROJECT MANAGEMENT IN EMERGING MARKETS

Growth in computer technology and virtual teams has made the world smaller. First world nations are flocking to emerging market nations to get access to the abundance of highly qualified human capital that is relatively inexpensive and want to participate in virtual project management teams. There is no question that there exists an ample supply of talent in these emerging market nations. These talented folks have a reasonable understanding of project management and some consider it an honor to work on virtual project teams.

But working on virtual project management teams may come with headaches. While the relative acceptance of project management appears at the working levels where the team members operate, further up in the hierarchy there might be resistance to the implementation and acceptance of project management. Because of the growth of project management worldwide, many executives openly provide "lip service" to its acceptance yet, behind the scenes, create significant barriers to prevent it from working properly. This creates significant hardships for those portions of the virtual teams in first world nations that must rely upon their other team members for support. The ultimate result might be frustrations stemming from poor information flow, extremely long decision-making processes, poor cost control, and an abundance of external dependencies that elongate schedules beyond the buyer's contractual dates. Simply stated, there are strong cultural issues that need to be considered. In this section, we will typically use the United States as an example of the first world nations.

Barriers to effective project management implementation exist worldwide, not merely in emerging market nations. But in emerging market nations, the barriers are more apparent. For simplicity sake, the barriers can be classified into four categories:

- Cultural barriers
- Status and political barriers
- Project management barriers
- Other barriers

**Culture**

A culture is a set of beliefs that people follow. Every company could have its own culture. Some companies may even have multiple cultures. Some cultures are strong while others are weak. In some emerging market nations, there exist national cultures that can be so strong that they dictate the corporate cultures. There are numerous factors that can influence the culture of an organization. Only those factors that can have an impact on the implementation and acceptance of project management are discussed here and include:

- Bureaucratic centralization of authority in the hands of a few
- Lack of meaningful or real executive sponsorship
- Importance of the organizational hierarchy
- Improper legal laws
- The potential for corruption

*Centralization of Authority:* Many countries maintain a culture in which very few people have the authority to make decisions. Decision-making rests in the hands of a few and it serves as a source of vast power. This factor exists in both privately held companies and governmental organizations. Project management advocates decentralization of authority and decision-making. In many countries, the senior-most level of management will never surrender their authority, power, or right to make decisions to project managers. In these countries an appointment to the senior levels of management is not necessarily based upon performance. Instead, it is based upon age, belonging to the right political party, and personal contacts within the government. The result can be executives that possess little knowledge of their own business or who lack the leadership capacity.

*Executive Sponsorship:* Project sponsorship might exist somewhere in the company but most certainly not at the executive levels. There are two reasons for this. First, senior managers know their limitations and may have absolutely no knowledge about the project. Therefore, they could be prone to making serious blunders that could become visible to the people that put them into these power positions. Second, and possibly most important, acting as an executive sponsor on a project that could fail could signal the end of the executive's political career. Therefore, sponsorship, if it exists at all, may be at a low level in the organizational hierarchy, and at a level where people are expendable if the project fails. The result is that project managers end up with sponsors who either cannot or will not help them in time of trouble.

*Organizational Hierarchy:* In the United States, project managers generally have the right to talk to anyone in the company to get information relative to the project. The intent is to get work to flow horizontally as well as vertically. In some emerging market nations, the project manager must follow the chain of command. The organizational hierarchy is sacred. Following the chain of command certainly elongates the decision-making process to the point where the project manager has no idea how long it will take to get access to needed information or for a decision to be made even though a sponsor exists. There is no mature infrastructure in place to support project management. The infrastructure exists to filter bad news from the executive levels and to justify the existence of each functional manager.

In the United States, the "buck" stops at the sponsor. Sponsors have ultimate decision-making authority and are expected to assist the project managers during a crisis. The role of the sponsor is clearly defined and may be described in detail in the enterprise project management methodology. But in some emerging market countries, even the sponsor might not be authorized to make a decision. Some decisions may need to go as high as a government minister. Simply stated, one does not know where and when the decision needs to be made and where it will be made. Also, in the United States reporting bad news ends up in the hands of the project sponsor. In some nations, the news may go as high as government ministers. Simply stated, you cannot be sure where project information will end up.

*Improper Legal Laws:* Not all laws in emerging market nations are viewed by other nations as being legal laws. Yet American project managers, partnering with these nations, must abide by these laws. As an example, procurement contracts may be awarded not to the most qualified supplier or to the lowest bidder but to any bidder that resides in a city that has a high unemployment level. As another example, some nations have laws that imply that bribes are an acceptable practice when awarding contracts. Some contracts might also be awarded to relatives and friends rather than the best qualified supplier.

*Potential for Corruption:* Corruption can and does exist in some countries and plays havoc on project managers that focus on the triple constraint. Project managers traditionally lay out a plan to meet the objectives and the triple constraint. Project managers also assume that everything will be done systematically and in an orderly manner, which assumes no corruption. But in some nations, there are potentially corrupt individuals or organizations that will do everything possible to stop or slow down the project, until they could benefit personally.

## Status and Politics

Status and politics are prevalent everywhere and can have a negative impact on project management. In some emerging market nations, status and politics actually sabotage project management and prevent it from working correctly. Factors that can affect project management include:

- Legal formalities and government constraints
- Insecurity at the executive levels
- Status consciousness
- Social obligations
- Internal politics
- Unemployment and poverty
- Attitude toward workers
- Inefficiencies
- Lack of dedication al all levels
- Lack of honesty

*Legal Formalities and Government Constraints:* Here in the United States, we believe that employees that perform poorly can be removed from the project or even fired. But in some emerging market nations, employees have the right to hold a job even if their performance is substandard. Having a job and a regular paycheck is a luxury. There are laws that clearly state under what conditions a worker can be fired, if at all.

There are also laws on the use of overtime. Overtime may not be allowed because paying someone to work overtime could eventually end up creating a new social class. Therefore, overtime may not be used as a means to maintain or accelerate a schedule that is in trouble.

*Insecurity:* Executives often feel insecurity more so than the managers beneath them because their positions may be the result of political appointments. As such project managers may be seen as the stars of the future and may be viewed as a threat to executives.

Allowing project managers who are working on highly successful projects to make presentations to the senior-most levels of management in the government could be mired. If the project is in trouble, then the project manager may be forced to make the presentation. Executives are afraid of project managers.

*Status Consciousness:* Corporate officers in emerging market nations are highly status conscious. They have a very real fear that the implementation of project management may force them to lose their status, yet they refuse to function as active project sponsors. Status often is accompanied by fringe benefits such as a company car and other special privileges.

*Social Obligations:* In emerging market nations, social obligations due to religious beliefs (and possibly superstitious beliefs) and politics may be more important than in first world nations. Social obligations are ways of maintaining alliances with those people that have put an executive or a project manager in power. As such, project managers may not be allowed to interface socially with certain groups. This could also be viewed as a threat to project management implementation.

*Internal Politics:* Internal politics exist in every company in the world. Before executives consider throwing their support behind a new approach such as project management, they worry about whether they will become stronger or weaker, have more or less authority, and have a greater or lesser chance for advancement. This is one of the reasons why only a small percentage of emerging market companies have PMOs. Whichever executive gets control of the PMO could become more powerful than other executives. In the United States, we have solved this problem by allowing several executives to have their own PMO. But in the emerging markets, this is viewed as excessive headcount.

*Unemployment and Government Constraints:* Virtually all executives understand project management and the accompanying benefits, yet they remain silent rather than visibly showing their support. One of the benefits of project management implementation is that it can make organizations more efficient to the point where fewer resources are needed to perform the required work. This can be a threat to an executive because, unless additional business can be found, efficiency can result in downsizing the company, reducing the executive's power and authority, increasing the unemployment level, and possibly increasing poverty in the community. Therefore, the increased efficiencies of project management could be looked upon unfavorably.

*Attitude toward Employees:* In some nations, employees might be viewed as stepping-stones to building an empire. Hiring three below-average workers to do the same work as two average workers is better for empire building, yet possibly at the expense of the project's budget and schedule. It is true, however, that finding adequate human resources may be difficult, but sometimes companies simply do not put forth a good effort in their search. Friends and family members may be hired first regardless of their qualifications. The problem is further complicated when one must find people with project management expertise.

*Inefficiencies:* Previously, we stated that companies might find it difficult to hire highly efficient people in project management. Not all people are efficient. Some people simply are not committed to their work even though they understand project management. Other people may get frustrated when they realize that they do not have the power, authority, or responsibility of their colleagues in first world countries. Sometimes, new hirers that want to be efficient workers are pressured by the culture to remain inefficient or else the individual's colleagues will be identified as poor workers. Peer pressure exists and can prevent people from demonstrating their true potential.

*Lack of Dedication:* It is hard to get people motivated when they believe they cannot lose their job. People are simply not dedicated to the triple constraint. Some people prefer to see schedules slip because it provides some degree of security for a longer period of time. There is also a lack of dedication for project closure. As a project begins to wind down, employees will begin looking for a home on some other project. They might even leave their current project prematurely, before the work is finished, to guarantee employment elsewhere.

*Honesty:* People working in emerging market countries have a tendency to hide things from fellow workers and project managers, especially bad news, either to keep their prestige or to retain their power and authority. This creates a huge barrier for project managers that rely upon timely information, whether good or bad, in order to manage the project successfully. Delays in reporting could waste valuable time when corrective action could have been taken.

## Implementation of Project Management

While culture, status, and politics can create barriers for any new management philosophy, there are other barriers that are directly related to project management, including:

- Cost of project management implementation
- Risks of implementation failure
- Cost of training and training limitations
- Need for sophistication
- Lack of closure on projects
- Work ethic
- Poor planning

*Cost of Implementation:* There is a cost associated with the implementation of project management. The company must purchase hardware and software, create a project management methodology, and develop project performance reporting techniques. This requires a significant financial expenditure, which the company might not be able to afford, and also requires significant resources to be tied up in implementation for an extended period of time. With limited resources, and the fact that the better resources would be required for implementation and removed from ongoing work, companies shy away from project management even though they know the benefits.

*Risk of Failure:* Even if a company is willing to invest the time and money for project management implementation, there is a significant risk that implementation will fail. And even if implementation is successful but projects begin to fail for any number of reasons, blame will be placed upon faulty implementation. Executives may find that their position in the hierarchy is now insecure once they have to explain the time and money expended for no real results. This is why some executives either refuse to accept or visibly support project management.

*Training Limitations:* Implementation of project management is difficult without training programs for the workers. This creates three additional problems. First, how much money must be allocated for training? Second, who will provide the training and what are the credentials of the trainers? Third, should people be released from project work to attend training classes? It is time consuming and expensive to train people in project management, whether it is project managers or team members that need to be trained. Adding together the cost of implementation and the cost of training might frighten executives from accepting project management.

*Need for Sophistication:* Project management requires sophistication, not only with the limited technology or tools that may be available but also in the ability of people to work together. This teamwork sophistication is generally lacking in emerging market countries. People may see no benefit in teamwork because others may be able to recognize their lack of competencies and mistakes. They have not been trained to work properly in teams and are not rewarded for their contribution to the team.

*Lack of Closure on Projects:* Employees are often afraid to be attached to the project at closure when lessons learned and best practices are captured. Lessons learned and best practices can be based upon what we did well and what we did poorly. Employees may not want to see anything in writing that indicates that best practices were discovered from their mistakes.

*Work Ethic:* In some nations, the inability to fire people creates a relatively poor work ethic which is contrary to effective project management practices. There is a lack of punctuality in coming to work and attending meetings. When people do show up at meetings, only good news is discussed in a group whereas bad news is discussed one-on-one. Communication skills are weak as is report writing. There is a lack of accountability because accountability means explaining your actions if things go bad.

*Poor Planning:* Poor planning is paramount in emerging market nations. There exists a lack of commitment to the planning process. Because of a lack of standards, perhaps attributed to the poor work ethic, estimating duration, effort, and cost is very difficult. The ultimate result of poor planning is an elongation of the schedule. For workers that are unsure about their next assignment, this can be viewed as job security at least for the short term.

## Other Barriers

There are other barriers that are too numerous to mention. However, some of the more important ones are shown below. These barriers are not necessarily universal in emerging market nations, and many of these barriers can be overcome.

- Currency conversion inefficiencies
- Inability to receive timely payments
- Superstitious beliefs
- Laws against importing and exporting intellectual property
- Lack of tolerance for the religious beliefs of virtual team partners
- Risk of sanctions by partners' governments
- Use of poor or outdated technologies

## Recommendations

Although we have painted a rather bleak picture, there are great future opportunities in these nations. Emerging market nations have an abundance of talent that is yet to be fully harvested. The true capabilities of these workers are still unknown. Virtual project management teams might be the starting point for the full implementation of project management.

As project management begins to grow, senior officers will recognize and accept the benefits of project management and see their business base increase. Partnerships and joint ventures using virtual teams will become more prevalent. The barriers that impede successful project management implementation will still exist, but we will begin to excel in how to live and work within the barriers and constraints imposed on the continually emerging virtual teams.

Greater opportunities are seen for the big emerging market economies. They are beginning to see more of the value of project management and have taken strides to expand its use. Some of the rapidly developing economies are even much more aggressive in providing the support needed for breaking many of the barriers addressed above. As more success stories emerge, the various economies will strengthen, become more connected, and start to fully utilize project management for what it really is.

# 7 Management Support

## 7.0 INTRODUCTION

As we saw in Chapter 6, senior managers are the architects of corporate culture. They are charged with making sure that their companies' cultures, once accepted, do not come apart. Visible management support is essential to maintaining a project management culture. And above all, the support must be continuous rather than sporadic.

This chapter examines the importance of management support in the creation and maintenance of project management cultures. Case studies illustrate the vital importance of employee empowerment and the project sponsor's role in the project management system.

## 7.1 VISIBLE SUPPORT FROM SENIOR MANAGERS

As project sponsors, senior managers provide support and encouragement to the project managers and the rest of the project team. Companies excellent in project management have the following characteristics:

- Senior managers maintain a hands-off approach, but they are available when problems come up.
- Senior managers expect to be supplied with concise project status reports.
- Senior managers practice empowerment.

- Senior managers decentralize project authority and decision-making.
- Senior managers expect project managers and their teams to suggest both alternatives and recommendations for solving problems, not just to identify the problems.

However, there is a fine line between effective sponsorship and overbearing sponsorship. Robert Hershock, former vice president at 3M, said it best during a videoconference on excellence in project management:

> Probably the most important thing is that they have to buy in from the top. There has to be leadership from the top, and the top has to be 100 percent supportive of this whole process. If you're a control freak, if you're someone who has high organizational skills and likes to dot all the i's and cross all the t's, this is going to be an uncomfortable process, because basically it's a messy process; you have to have a lot of fault tolerance here. But what management has to do is project the confidence that it has in the teams. It has to set the strategy and the guidelines and then give the teams the empowerment that they need in order to finish their job. The best thing that management can do after training the team is get out of the way.

To ensure their visibility, senior managers need to believe in walk-the-halls management. In this way, every employee will come to recognize the sponsor and realize that it is appropriate to approach the sponsor with questions. Walk-the-halls management also means that executive sponsors keep their doors open. It is important that everyone, including line managers and their employees, feels supported by the sponsor. Keeping an open door can occasionally lead to problems if employees attempt to go around lower-level managers by seeking a higher level of authority. But such instances are infrequent, and the sponsor can easily deflect the problems back to the appropriate manager.

## 7.2 PROJECT SPONSORSHIP

Executive project sponsors provide guidance for project managers and project teams. They are also responsible for making sure that the line managers who lead functional departments fulfill their commitments of resources to the projects underway. In addition, executive project sponsors maintain communication with customers.

The project sponsor usually is an upper-level manager who, in addition to his or her regular responsibilities, provides ongoing guidance to assigned projects. An executive might take on sponsorship for several concurrent projects. Sometimes, on lower-priority or maintenance projects, a middle-level manager may take on the project sponsor role. One organization I know of even prefers to assign middle managers instead of executives. The company believes this avoids the common problem of lack of line manager buy-in to projects (see Figure 7–1).

In some large, diversified corporations, senior managers do not have adequate time to invest in project sponsorship. In such cases, project sponsorship falls to the level below corporate senior management or to a committee.

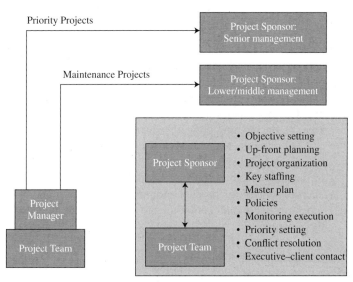

**Figure 7–1.** Roles of project sponsor. *Source:* Reprinted from H. Kerzner, *In Search of Excellence in Project Management*, Hoboken, NJ: Wiley, 1998, p. 159.

Some projects do not need project sponsors. Generally, sponsorship is required on large, complex projects involving a heavy commitment of resources. Large, complex projects also require a sponsor to integrate the activities of the functional lines, to dispel disruptive conflicts, and to maintain strong customer relations.

Consider one example of a project sponsor's support for a project. A project manager who was handling a project in an organization within the federal government decided that another position would be needed on his team if the project were to meet its completion deadline. He had already identified a young woman in the company who fit the qualifications he had outlined. But adding another full-time-equivalent position seemed impossible. The size of the government project office was constrained by a unit-manning document that dictated the number of positions available.

The project manager went to the project's executive sponsor for help. The executive sponsor worked with the organization's human resources and personnel management department to add the position requested. Within 30 days, the addition of the new position was approved. Without the sponsor's intervention, it would have taken the organization's bureaucracy months to approve the position, too late to affect the deadline.

In another example, the president of a medium-sized manufacturing company wanted to fill the role of sponsor on a special project. The project manager decided to use the president to the project's best advantage. He asked the president/sponsor to handle a critical situation. The president/sponsor flew to the company's headquarters and returned two days later with an authorization for a new tooling the project manager needed. The company ended up saving time on the project, and the project was completed four months earlier than originally scheduled.

**Sponsorship by Committee**     As companies grow, it sometimes becomes impossible to assign a senior manager to every project, and so committees act in the place of individual project sponsors. In fact, the recent trend has been toward committee sponsorship in many kinds of organizations. A project sponsorship committee usually is made up of a representative from every function of the company: engineering, marketing, and production. Committees may be temporary, when a committee is brought together to sponsor one time-limited project, or permanent, when a standing committee takes on the ongoing project sponsorship of new projects.

For example, General Motors Powertrain had achieved excellence in using committee sponsorship. Two key executives, the vice president of engineering and the vice president of operations, led the Office of Products and Operations, a group formed to oversee the management of all product programs. This group demonstrated visible executive-level program support and commitment to the entire organization. Their roles and responsibilities were to:

- Appoint the project manager and team as part of the charter process
- Address strategic issues
- Approve the program contract and test for sufficiency
- Assure program execution through regularly scheduled reviews with program managers

**Phases of Project Sponsorship**     The role of the project sponsor changes over the life cycle of a project. During the planning and initiation phases, the sponsor plays an active role in the following activities:

- Helping the project manager establish the objectives of the project
- Providing guidance to the project manager during the organization and staffing phases
- Explaining to the project manager what environmental or political factors might influence the project's execution
- Establishing the project's priority (working alone or with other company executives) and then informing the project manager about the project's priority in the company and the reason that priority was assigned
- Providing guidance to the project manager in establishing the policies and procedures for the project
- Functioning as the contact point for customers and clients

During the execution phase of a project, the sponsor must be very careful in deciding which problems require his or her guidance. Trying to get involved with every problem that comes up on a project will result in micromanagement. It will also undermine the project manager's authority and make it difficult for the executive to perform his or her regular responsibilities.

For short-term projects of two years or less, it is usually best that the project sponsor assignment is not changed over the duration of the project. For long-term projects of five years, more or less, different sponsors could be assigned for every phase of the project, if necessary. Choosing sponsors from among executives at the same corporate level works best, since sponsorship at the same level creates a "level" playing field, whereas at different levels, favoritism can occur.

Project sponsors need not come from the functional area where the majority of the project work will be completed. Some companies even go so far as assigning sponsors from line functions that have no vested interest in the project. Theoretically, this system promotes impartial decision making.

## Customer Relations

The role of executive project sponsors in customer relations depends on the type of organization (entirely project-driven or partially project-driven) and the type of customer (external or internal). Contractors working on large projects for external customers usually depend on executive project sponsors to keep the clients fully informed of progress on their projects. Customers with multimillion-dollar projects often keep an active eye on how their money is being spent. They are relieved to have an executive sponsor they can turn to for answers.

It is common practice for contractors heavily involved in competitive bidding for contracts to include both the project manager's and the executive project sponsor's resumes in proposals. All things being equal, the resumes may give one contractor a competitive advantage over another.

Customers prefer to have a direct path of communication open to their contractors' executive managers. One contractor identified the functions of the executive project sponsor as:

- Actively participating in the preliminary sales effort and contract negotiations
- Establishing and maintaining high-level client relationships
- Assisting project managers in getting the project underway (planning, staffing, and so forth)
- Maintaining current knowledge of major project activities
- Handling major contractual matters
- Interpreting company policies for project managers
- Helping project managers identify and solve significant problems
- Keeping general managers and client managers advised of significant problems with projects

## Decision Making

Imagine that project management is like car racing. A yellow flag is a warning to watch out for a problem. Yellow flags require action by the project manager or the line manager. There is nothing wrong with informing an executive about a yellow-flag problem as long as the project manager is not looking for the sponsor to solve the problem. Red flags, however, usually do require the sponsor's direct involvement. Red flags indicate problems that may affect the time, cost, and performance parameters of the project. So red flags need to be taken seriously and decisions need to be made collaboratively by the project manager and the project sponsor.

Serious problems sometimes result in serious conflicts. Disagreements between project managers and line managers are not unusual, and they require the thoughtful intervention of the executive project sponsor. First, the sponsor should make sure that the disagreement could not be solved without his or her help. Second, the sponsor needs to gather information from all sides and consider the alternatives being considered. Then, the sponsor must decide whether he or she is qualified to settle the dispute. Often, disputes are of a technical nature and require someone with the appropriate knowledge base to solve them. If the sponsor is unable to solve the problem, he or she will need to identify another source of authority that has the needed technical knowledge. Ultimately, a fair and appropriate solution can be shared by everyone involved. If there were no executive sponsor on the project, the disputing parties would be forced to go up the line of authority until they found a common superior to help them. Having executive project sponsors minimizes the number of people and the amount of time required to settle work disputes.

**Strategic Planning**    Executives are responsible for performing the company's strategic planning, and project managers are responsible for the operational planning on their assigned projects. Although the thought processes and time frames are different for the two types of planning, the strategic planning skills of executive sponsors can be useful to project managers. For projects that involve process or product development, sponsors can offer a special kind of market surveillance to identify new opportunities that might influence the long-term profitability of the organization. Furthermore, sponsors can gain a lot of strategically important knowledge from lower-level managers and employees. Who else knows better when the organization lacks the skill and knowledge base it needs to take on a new type of product? When the company needs to hire more technically skilled labor? What technical changes are likely to affect their industry?

## 7.3 EXCELLENCE IN PROJECT SPONSORSHIP

In excellent companies, the role of the sponsor is not to supervise the project manager but to make sure that the best interests of both the customer and the company are recognized. However, as the next two examples reveal, it is seldom possible to make executive decisions that appease everyone.

Franklin Engineering (a pseudonym) had a reputation for developing high-quality, innovative products. Unfortunately, the company paid a high price for its reputation: a large R&D budget. Fewer than 15 percent of the projects initiated by R&D led to the full commercialization of a product and the recovery of the research costs.

The company's senior managers decided to implement a policy that mandated that all R&D project sponsors periodically perform cost–benefit analyses on their projects. When a project's cost–benefit ratio failed to reach the levels prescribed in the policy, the project was canceled for the benefit of the whole company.

Initially, R&D personnel were unhappy to see their projects canceled, but they soon realized that early cancellation was better than investing large amounts in projects that were likely to fail. Eventually, the project managers and team members came to agree that it made no sense to waste resources that could be better used on more successful projects.

Within two years, the organization found itself working on more projects with a higher success rate but no addition to the R&D budget.

Another disguised case involves a California-based firm that designs and manufactures computer equipment. Let's call the company Design Solutions. The R&D group and the design group were loaded with talented individuals who believed that they could do the impossible and often did. These two powerful groups had little respect for the project managers and resented schedules because they thought schedules limited their creativity.

In June 1997, the company introduced two new products that made it onto the market barely ahead of the competition. The company had initially planned to introduce them by the end of 1996. The reason for the late releases: Projects had been delayed because of the project teams' desire to exceed the specifications required and not just meet them.

To help the company avoid similar delays in the future, the company decided to assign executive sponsors to every R&D project to make sure that the project teams adhered to standard management practices in the future. Some members of the teams tried to hide their successes with the rationale that they could do better. But the sponsor threatened to dismiss the employees, and they eventually relented.

The lessons in both cases are clear. Executive sponsorship actually can improve existing project management systems to better serve the interests of the company and its customers.

## 7.4 HEWLETT-PACKARD SPONSORSHIP IN ACTION

According to Doug Bolzman, consultant architect, PMP®, ITIL Expert at HP:

> Management supports the results of project management and if the PMO is in a position to understand the current environment and how to improve the results, management will listen. If management is given a problem statement regarding how projects are not being managed effectively, then they will be of little value.
>
> We share a story of when a fellow business associate came to us and asked why our Release Management Methodology was not formally implemented across the organization as a corporate method. This sponsor of ours had connections and scheduled a meeting with the CIO of our company (this was a few years after the turn of the century). We presented the overall methodology, the value to the organization, the success stories of what we achieved. His response was, "This is all fine and good, but what do you want me to do?" The sponsor of this meeting fell back into his chair and replied; "we are looking for 'management commitment' for this methodology". As the meeting concluded, I realized that if it were not for the good meal we had the night before after we flew into headquarters, the meeting would have been a total loss. The point to be made here is that we did not define what specifically we were looking for from the management team. Now when we are in the same situation, we explain the investments, the funding, the resource needs, the cultural change and the policies that they need to sanction and enforce. We describe their role and responsibilities for supporting the project, the practice, the PMO, or whatever we are looking for them to support.

## 7.5 ZURICH AMERICA INSURANCE COMPANY: IMPROVING STAKEHOLDER ENGAGEMENT

Kathleen Cavanaugh, PMP, Zurich—ZNA PMO lead, provides us with an insight of sponsors may help improve stakeholder engagement. Project lessons learned offer valuable insights into what's working and not working well within projects/programs. When lessons learned are rolled up to the portfolio level, the common problems faced by these projects/programs are revealed and trends appear that may otherwise have been overlooked.

One of the improvements opportunities identified from project lessons learned reviews in recent years at Zurich in North America (ZNA) was the need for a new project leadership role called the Business Lead. The aggregated project feedback showed that on large, transformational projects/programs stronger business engagement was a "must have" from the beginning of the effort all the way through post-implementation. It was also noted that middle layers of management can often pose the biggest alignment challenges for these types of large process and/or people related change projects/programs. The need was clear. Stakeholders at all levels must be kept informed, aligned with project goals and involved in the project/program throughout the life cycle. Without stakeholder commitment and ownership, the project/program will suffer from low acceptance/adoption rates or, worse, may be canceled altogether.

So, to help solve this problem, leaders from across the business worked to define the responsibilities in order to bring the Business Lead role to life. The role was first piloted to help prove the concept and to gain agreement prior to rollout. Guidelines were provided to help clarify the Business Lead responsibilities which originally included: aligning stakeholders, ensuring project goals continually support the business strategy, utilizing change management to prepare stakeholders for project impacts, ensuring post implementation support and benefits realization. However, the actual duties of the Business Lead are ultimately determined by the Project Sponsor and may vary from project to project.

The role itself was designed to complement and collaborate with the Project Sponsor and Project Lead roles creating a Project Leadership Team triangle. Together they form a strong partnership and join forces to make key decisions for the project/program. The Project Lead and Business Lead often work closely together throughout the project/program connecting the Business and IT worlds. To help ensure stakeholders stay engaged a variety of methods may be used, including focus group meetings, user councils, and multiple feedback mechanisms. Middle management should be a focal point for project/program communications since these are the people who can ensure end users hear and understand key messages. The Sponsor provides continual oversight and guidance for the project/program. If the project/program hits a roadblock, the escalation path usually goes straight to the Sponsor and/or Steering Committee for resolution.

To be effective, the person filling the Business Lead role needs to be someone at a fairly high level in the organization, and they must possess a unique blend of skills . . . skills that include cross-functional knowledge, the ability to challenge the status quo, and break down barriers and unite stakeholders. The role itself inherently calls for the chosen one to be strong in the art of change management and communication. But for this person to be truly successful, he/she must be allowed to devote himself/herself to the project/program especially on large, cross-functional efforts. We often find the role requires heavy involvement in the day-to-day activities of the project/program. So, in many cases the selected Business Lead relinquishes their current duties for at least 12 to 18 months in order to become officially 100% dedicated to this key project/program role. Convincing leadership to take someone out of a role and to dedicate them to a large transformational effort for this amount of time is not always an easy task. But, if the project/program is truly important to the organization then it's a small price to pay to help ensure a successful implementation. It is a big decision to make and one that is usually driven by the Sponsor who has the organizational awareness and knowledge to identify the right person for the role.

The value of the Business Lead role has grown significantly over the past couple of years and a community of practice is being formed so that Business Leads can share their experiences and lessons learned in order to continue improving their capabilities. The role is becoming a solid cornerstone of the Project Leadership Team for large, transformational projects/programs at ZNA. The collaboration and synergy between the Project Sponsor, Project Lead and Business Lead roles is the key to their success. This project leadership structure is proving to be a viable model to help ensure stakeholder engagement and alignment throughout the project/program life cycle, which translates to more successful results and improved end user satisfaction.

## 7.6 PROJECT GOVERNANCE

All projects have the potential of getting into trouble but, in general, project management can work well as long as the project's requirements do not impose severe pressure upon the project manager, and a project sponsor exists as an ally to assist the project manager when trouble does appear. Unfortunately, in today's chaotic environment, this pressure appears to be increasing because:

- Companies are accepting high risk and highly complex projects as a necessity for survival
- Customers are demanding low volume, high quality products with some degree of customization
- Project life cycles and new product development times are being compressed
- Enterprise environmental factors are having a greater impact on project execution
- Customers and stakeholders want to be more actively involved in the execution of projects
- Companies are developing strategic partnerships with suppliers, and each supplier can be at a different level of project management maturity
- Global competition has forced companies to accept projects from customers that are all at a different level of project management maturity and with different reporting requirements

These pressures tend to slow down the decision-making processes at a time when stakeholders want the projects and processes to be accelerated. One person, while acting as the project sponsor, may have neither the time nor capability to address all of these additional issues. The result will be a project slowdown and can occur because of:

- The project manager being expected to make decisions in areas where he/she has limited knowledge
- The project manager hesitating to accept full accountability and ownership for the projects
- Excessive layers of management being superimposed on top of the project management organization
- Risk management being pushed up to higher levels in the organization hierarchy resulting in delayed decisions
- The project manager demonstrating questionable leadership ability on some of the nontraditional projects

The problems resulting from these pressures may not be able to be resolved, at least easily and in a timely manner, by a single project sponsor. These problems can be resolved using effective project governance. Project governance is actually a framework by which decisions are made. Governance relates to decisions that define expectations, accountability, responsibility, the granting of power or verifying performance. Governance relates to consistent management, cohesive policies, processes and decision-making rights for a given area of responsibility. Governance enables efficient and effective decision making to take place.

Every project can have different governance even if each project uses the same enterprise project management methodology. The governance function can operate as a separate process or as part of project management leadership. Governance is not designed to replace project decision making, but to prevent undesirable decisions from being made.

Historically, governance was provided by a single project sponsor. Today, governance is a committee and can include representatives from each stakeholder's organization. Table 7–1 shows various governance approaches based upon the type of project team. The membership of the committee can change from project to project and industry to industry.

**TABLE 7–1. TYPES OF PROJECT GOVERNANCE**

| Structure | Description | Governance |
|---|---|---|
| Dispersed locally | Team members can be full time or part time. They are still attached administratively to their functional area. | Usually a single person is acting as the sponsor but may be an internal committee based upon the project's complexity. |
| Dispersed geographically | This is a virtual team. The project manager may never see some of the team members. Team members can be full time or part time. | Usually governance by committee and can include stakeholder membership. |
| Co-located | All of the team members are physically located in close proximity to the project manager. The project manager does not have any responsibility for wage and salary administration. | Usually a single person acting as the sponsor. |
| Projectized | This is similar to a co-located team but the project manager generally functions as a line manager and may have wage and salary responsibilities. | May be governance by committee based upon the size of the project and the number of strategic partners. |

The membership may also vary based upon the number of stakeholders and whether the project is for an internal or external client. On long term projects, membership can change throughout the project.

Governance on projects and programs sometimes fails because people confuse project governance with corporate governance. The result is that members of the committee are not sure what their role should be. Some of the major differences include:

- **Alignment:** Corporate governance focuses on how well the portfolio of projects is aligned to and satisfies overall business objectives. Project governance focuses on ways to keep a project on track.
- **Direction:** Corporate governance provides strategic direction with a focus on how project success will satisfy corporate objectives. Project governance is more operation direction with decisions based upon the predefined parameters on project scope, time, cost and functionality.
- **Dashboards:** Corporate governance dashboards are based upon financial, marketing and sales metrics. Project governance dashboards have operations metrics on time, cost, scope, quality, action items, risks, and deliverables.
- **Membership:** Corporate governance committees are composed of the senior most levels of management. Project government membership may include some membership from middle management.

Another reason why failure may occur is [that] members of the project or program governance group do not understand project or program management. This can lead to micromanagement by the governance committee. There is always the question of what decisions must be made by the governance committee and what decisions the project manager can make. In general, the project manager should have the authority for decisions related to actions necessary to maintain the baselines. Governance committees must have the authority to approve scope changes above a certain dollar value, and to make decisions necessary to align the project to corporate objectives and strategy.

## 7.7 TOKIO MARINE: EXCELLENCE IN PROJECT GOVERNANCE

**Executive Management Must Establish IT Governance: Tokio Marine Group**

By: Yuichi (Rich) Inaba, CISA, and Hiroyuki Shibuya[1]

Tokio Marine Group is a global corporate group engaged in a wide variety of insurance businesses. It consists of about 70 companies on five continents, including Tokio Marine and Nichido Fire Insurance (Japan), Philadelphia Insurance (US), Kiln (UK), and Tokio Marine Asia (Singapore).

In addition to Tokio Marine and Nichido Fire Insurance, which is the largest property and casualty insurance company in Japan, Tokio Marine Group has several other domestic companies in Japan, such as Tokio Marine and Nichido Life Insurance Co. Ltd, as well as service providers, such as Tokio Marine and Nichido Medical Service Co Ltd. and Tokio Marine and Nachido Facilities Inc.

1. Originally published in Cobit® Focus, Volume 1, January 2013. © 2013 ISACA. All rights reserved.

### Implementing IT Governance at Tokio Marine Group

Tokio Marine Holdings, which is responsible for establishing the group's IT governance approach, observed that the executive management of Tokio Marine Group companies believes that IT is an essential infrastructure for business management, and it hoped to strengthen company management by utilizing IT. However, some directors and executives had a negative impression of IT—that IT is difficult to understand, costs too much, and results in frequent system troubles and system development failures.

It is common for an organization's executive management to recognize the importance of system development but to put its development solely on the shoulders of the IT department. Other executives go even further, saying that the management or governance of IT is not anyone's business but the IT department's or chief information officers (CIO's). This line of thinking around IT is similar to the thought process that accounting is the job of the accounting department and handling personnel affairs is the role of the human resources department.

These are typical behaviors of organizations that fail to implement IT governance systems. Tokio Marine Holdings' executive management recognized that IT is not for IT's sake alone, but is a tool to strengthen business.

Tokio Marine Holdings' management recognized that there were various types of system development failures (e.g., development delays for the service-in date, projects being over budget). Even more frequently, the organization was finding requirement gaps—for instance, where after building a system, the business people say, "This is not the system that we asked you to build" or "The system that you built is not easy to use. It is useless for the business."

### Why the Requirement Gaps Occur

The process of system development is similar to that of a buildings construction. However, there is a distinct difference between the two: system development is not visible, whereas building construction is. Therefore, in system development, it is inevitable that there are recognition and communication gaps between business and IT.

### Tokio Marine Group's Solution for System Development Success

To fill these gaps, business and IT must communicate enough to minimize the gaps [between] A and C in Figure 7–2 and maximize a common understanding of B. The road to success for system development is to improve the quality of communication between business and IT.

Such communication cannot be reached or maintained in a one-sided relationship. Ideal communication is enabled only with as equal partnership between business and IT with appropriate roles and responsibilities mutually allocated.

This is the core concept of Tokio Marine Group's Application Owner System.

### Implementing the Application Owner System

Tokio Marine Holdings decided to implement the Application Owner System as a core concept of the Group It Governance System. Tokio Marine Holdings believes it is essential for the group companies to succeed in system development and to achieve the group's growth in the current business environment.

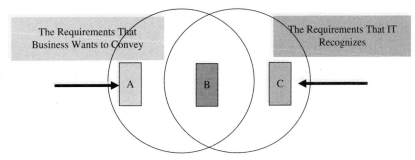

**Figure 7–2.** The Requirements Gap.

**Figure 7–3.** The Application Owner System in Tokio Marine Group.

The basic idea of the Application Owner System (Figure 7–3) is:

- Mutual cooperation between business and IT with proper check-and-balance functions, appropriately allocated responsibilities and shared objectives.
- Close communication between business and IT, each taking their own respective roles and responsibilities into account.

## Early Success in Tokio Marine and Nichido Fire Insurance

Tokio Marine and Nichido Fire Insurance Co. Ltd., the largest group company, implemented the Application Owner System in 2000. Implementation of the Application Owner System immediately reduced system troubles and problems by 80 percent. (See Figure 7–4.)

## Mind-Set of IT

Tokio Marine's mind-set is that only executive management can establish the enterprise's IT governance system. Thus, IT governance is the responsibility of executive management.

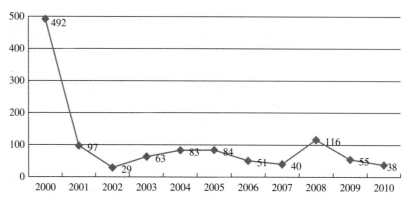

**Figure 7–4.** Number of system troubles.

Furthermore, the organization is of the mind-set that all employees, not only executive management, should understand the principle that strong IT systems cannot be realized by the IT department alone but require cooperation between business and IT. It is important that all employees recognize IT matters as their own, not as the matter of the IT function.

Establishing such a mind-set within the enterprise is a role of the executive management.

## Tokio Marine Group's IT Governance System

Characterized by the Application Owner System, Tokio Marine Holdings has introduced an IT governance framework, focused on the COBIT 4.1 framework, specifically the Plan and Organize (PO) domain.

*The main goals of the IT Governance Framework are:*
- Establishing basic policies for IT governance—Tokio Marine Holdings established the Basic Policies for IT Governance as the policies for the group's IT governance framework.
- Establishing guiding principles for IT governance—Tokio Marine Holdings defines seven principles as the guiding principles. (See Table 7–2.) These cover the five focus areas defined in the Board Briefing on IT Governance, particularly focusing on strategic alignment and value delivery. The seven principles are included in the Basic Policies for IT Governance. Tokio Marine Holdings thinks that the most important principle is the Application Owner System, which is stated as follows:

     In implementing the plan, it is important for the IT unit and the application owner units to cooperate with each other with proper checks and balance functions. Management shall clearly determine the appropriate sharing of roles between the IT unit and application owner units, secure human resources of adequate quality in both units, and establish a management system to assure that each unit will execute the plan according to its responsibilities.
- **Establishing a governance and management system for Tokio Marine Group**—Tokio Marine Holdings defines the governance and management system to be

**TABLE 7–2. SEVEN GUIDING PRINCIPLES**

| No. | Guiding Principles | Focus Area |
|---|---|---|
| 1 | Establish an IT strategic plan that enables management to achieve its business strategic plan, build the business processes for it, and develop an execution plan. | Strategic Alignment |
| 2 | In executing the plan, ensure that the IT unit and the application owner units cooperate with each other with proper check-and-balance functions. | Strategic Alignment |
| 3 | In the development or implementation of information systems, ensure that management scrutinizes the validity of the project plan from the standpoint of quality assurance, usability, commitment to service-in date, appropriate cost estimation and matching to the human resources availability. | Value Delivery |
| 4 | Ensure that the information systems are fully utilized by all staff in the company in order to achieve the objectives for the development or implementation of the information systems. | Value Delivery |
| 5 | Conduct appropriate IT resource management, including computer capacity management and human resources management. | Resource Management |
| 6 | Conduct appropriate risk management and information security management, and establish contingency plans for system faults in consideration of the accumulation of various risk factors in IT, such as high dependency of business processes on IT, centralization of important information and threats from wider use of the Internet. | Risk Management |
| 7 | Encourage the transparency of IT operations to be improved, and monitor their progress, which includes, for example, the progress of projects, the usage of IT resources and utilization of information systems. | Performance Measurement |

implemented in the group companies. It covers five domains and consists of three major components: establishment of the organizational structure, establishment of policies and standards, and execution of the plan, do, check, and act (PCDA) cycle for improvement. The governance and management system required for Tokio Marine Group companies is detailed in the Group IT Governance Standard.

- **Establishing an IT governance standard** (the definition of Tokio Marine's priority processes)—Tokio Marine Holdings has decided to utilize COBIT 4.1 to define the management system. However, the organization recognizes that it is difficult for relatively small group companies to implement matured processes for all COBIT 4.1 processes. To handle this concern, the organization focused on the minimal set and processes or more detailed control on objectives, which are essential for its group business in terms of IT governance and the most important controls for Tokio Marine Group.

In the IT Governance Standard, Tokio Marine Holdings defined the IT controls outlined in Figure 7–5 as priority for the Tokio Marine Group. The priority IT controls are defined as five domains, 14 processes, and 39 control objectives, which are selected processes from the 210 control objectives of COBIT 4.1. See Table 7–3.

The group companies are required to improve the priority controls to reach a maturity level 3, according to the COBIT Maturity Model, and report the progress of improvements of Tokio Marine Holdings.

### Toward the Future

Since the establishment of the IT governance system for Tokio Marine Group, Tokio Marine Holdings has extensively communicated not only with CIOs but also with CEOs

**Figure 7–5.** Tokio Marine group's priority controls.

TABLE 7–3.  TOKIO MARINE GROUP'S PRIORITY CONTROLS.

|    | Domain name | Id | Process Name |
|----|-------------|----|--------------|
| a. | Planning and organization | a1 | Annual IT planning |
|    | | a2 | Definitions of roles and responsibilities of the IT unit and application owner units |
|    | | a3 | Establishment of an IT steering committee |
| b. | Projects management | b1 | Management of development and implementation projects |
| c. | Change management | c1 | Change control |
| d. | Operations Management | d1 | Incident/problem management |
|    | | d2 | Vendor management |
|    | | d3 | Security management |
|    | | d4 | IT asset management |
|    | | d5 | Computer capacity management |
|    | | d6 | Disaster recovery and backup/restore |
| e. | Monitoring performance and return on investment | e1 | Annual IT review |
|    | | e2 | Monitoring the IT steering committee |
|    | | e3 | Monitoring project management, change management, and systems operation management |

and executive management of the group companies to ensure that they understand, agree on and take leadership for IT governance implementation.

Through these activities, the organization is confident that the core concept of IT governance has become better understood by management and good progress is being made as a result of the implementation of the application owner system in group companies. Tokio Marine Holdings will continue its evangelist mission to the group companies, realizing the benefit for the group business and giving value to stakeholders.

## Yuichi (Rich) Inaba, CISA

Is a senior consultant specialist in the area of IT governance, IT risk management and IT information security in the Tokio Marine and Nichido Systems Co. Ltd. (TMNS), a Tokio Marine Group company. Before transferring to TMNS, he had worked in the IT Planning Dept. of Tokio Marine Holdings Inc. and had engaged in establishing Tokio Marine Group's IT governance framework based on COBIT 4.1. His current responsibility is to implement and practice Tokio Marine Group's IT governance at TMNS. Inaba is a member of the ISACA Tokyo Chapter's Standards Committee and is currently engaged in translating COBIT 5 publications in Japanese.

### Hiroyuki Shibuya

Is an executive officer in charge of IT at Tokio Marine Holdings Inc. From 2000–2005, he led the innovation project from the IT side, which has totally reconstructed the insurance product lines, their business processes and the information systems of Tokio Marine and Nichido Fire Insurance Co. Ltd. To leverage his experience from this project as well as remediate other troubled development projects of group companies, he was named the general manager of the newly established IT planning department at Tokio Marine Holdings in July 2010. Since then, he has been leading the efforts to establish IT governance basic policies and standards to strengthen IT governance throughout the Tokio Marine Group.

## 7.8 EMPOWERMENT OF PROJECT MANAGERS

One of the biggest problems with assigning executive sponsors to work beside line managers and project managers is the possibility that the lower-ranking managers will feel threatened with a loss of authority. This problem is real and must be dealt with at the executive level. Frank Jackson, formerly a senior manager at MCI, believes in the idea that information is power:

> We did an audit of the teams to see if we were really making the progress that we thought or were kidding ourselves, and we got a surprising result. When we looked at the audit, we found out that 50 percent of middle management's time was spent in filtering information up and down the organization. When we had a sponsor, the information went from the team to the sponsor to the operating committee, and this created a real crisis in our middle management area.
>
> MCI has found its solution to this problem. If there is anyone who believes that just going and dropping into a team approach environment is an easy way to move, it's definitely not. Even within the companies that I'm involved with, it's very difficult for managers to give up the authoritative responsibilities that they have had. You just have to move into it, and we've got a system where we communicate within MCI, which is MCI mail. It's an electronic mail system. What it has enabled us to do as a company is bypass levels of management. Sometimes you get bogged down in communications, but it allows you to communicate throughout the ranks without anyone holding back information.

Not only do executives have the ability to drive project management to success, they also have the ability to create an environment that leads to project failure. According to Robert Hershock, former vice president at 3M:

> Most of the experiences that I had where projects failed, they failed because of management meddling. Either management wasn't 100 percent committed to the process, or management just bogged the whole process down with reports and a lot of other innuendos. The biggest failures I've seen anytime have been really because of management. Basically, there are two experiences where projects have failed to be successful. One is the management meddling where management cannot give up its decision-making capabilities, constantly going back to the team and saying you're doing this wrong or you're doing that wrong. The other side of it is when the team can't communicate its own objective. When it can't be focused, the scope continuously expands, and you get into project creep. The team just falls apart because it has lost its focus.

Project failure can often be a matter of false perceptions. Most executives believe that they have risen to the top of their organizations as solo performers. It is very difficult for them to change without feeling that they are giving up a tremendous amount of power, which traditionally is vested in the highest level of the company. To change this situation, it may be best to start small. As one executive observed:

> There are so many occasions where senior executives won't go to training and won't listen, but I think the proof is in the pudding. If you want to instill project management teams in your organizations, start small. If the company won't allow you to do it using the Nike theory of just jumping in and doing it, start small and prove to them one step at a time that they can gain success. Hold the team accountable for results—it proves itself.

It is also important for us to remember that executives can have valid reasons for micromanaging. One executive commented on why project management might not be working as planned in his company:

> We, the executives, wanted to empower the project managers and they, in turn, would empower their team members to make decisions as they relate to their project or function. Unfortunately, I do not feel that we (the executives) totally support decentralization of decision-making due to political concerns that stem from the lack of confidence we have in our project managers, who are not proactive and who have not demonstrated leadership capabilities.

In most organizations, senior managers start at a point where they trust only their fellow managers. As the project management system improves and a project management culture develops, senior managers come to trust project managers, even though they do not occupy positions high on the organizational chart. Empowerment does not happen overnight. It takes time and, unfortunately, a lot of companies never make it to full project manager empowerment.

## 7.9 MANAGEMENT SUPPORT AT WORK

Visible executive support is necessary for successful project management and the stability of a project management culture. But there is such a thing as too much visibility for senior managers. Take the following case example, for instance.

**Midline Bank**

Midline Bank (a pseudonym) is a medium-sized bank doing business in a large city in the Northwest. Executives at Midline realized that growth in the banking industry in the near future would be based on mergers and acquisitions and that Midline would need to take an aggressive stance to remain competitive. Financially, Midline was well prepared to acquire other small- and medium-sized banks to grow its organization.

The bank's information technology group was given the responsibility of developing an extensive and sophisticated software package to be used in evaluating the financial health of the banks targeted for acquisition. The software package required input from virtually every functional division of Midline. Coordination of the project was expected to be difficult.

Midline's culture was dominated by large, functional empires surrounded by impenetrable walls. The software project was the first in the bank's history to require cooperation and integration among the functional groups. A full-time project manager was assigned to direct the project.

Unfortunately, Midline's executives, managers, and employees knew little about the principles of project management. The executives did, however, recognize the need for executive sponsorship. A steering committee of five executives was assigned to provide support and guidance for the project manager, but none of the five understood project management. As a result, the steering committee interpreted its role as one of continuous daily direction of the project.

Each of the five executive sponsors asked for weekly personal briefings from the project manager, and each sponsor gave conflicting directions. Each executive had his or her own agenda for the project.

By the end of the project's second month, chaos took over. The project manager spent most of his time preparing status reports instead of managing the project. The executives changed the project's requirements frequently, and the organization had no change control process other than the steering committee's approval.

At the end of the fourth month, the project manager resigned and sought employment outside the company. One of the executives from the steering committee then took over the project manager's role, but only on a part-time basis. Ultimately, the project was taken over by two more project managers before it was complete, one year later than planned. The company learned a vital lesson: More sponsorship is not necessarily better than less.

## Contractco

Another disguised case involves a Kentucky-based company I'll call Contractco. Contractco is in the business of nuclear fusion testing. The company was in the process of bidding on a contract with the U.S. Department of Energy. The department required that the project manager be identified as part of the company's proposal and that a list of the project manager's duties and responsibilities be included. To impress the Department of Energy, the company assigned both the executive vice president and the vice president of engineering as cosponsors.

The DoE questioned the idea of dual sponsorship. It was apparent to the department that the company did not understand the concept of project sponsorship, because the roles and responsibilities of the two sponsors appeared to overlap. The department also questioned the necessity of having the executive vice president serve as a sponsor.

The contract was eventually awarded to another company. Contractco learned that a company should never underestimate the customer's knowledge of project management or project sponsorship.

## Health Care Associates

Health Care Associates (another pseudonym) provides health-care management services to both large and small companies in New England. The company partners with a chain of 23 hospitals in New England. More than 600 physicians are part of the professional team, and many of the physicians also serve as line managers at the company's branch offices. The physician-managers maintain their own private clinical practices as well.

It was the company's practice to use boilerplate proposals prepared by the marketing department to solicit new business. If a client were seriously interested in Health Care Associates' services, a customized proposal based on the client's needs would be prepared. Typically, the custom-designed process took as long as six months or even a full year.

Health Care Associates wanted to speed up the custom-designed proposal process and decided to adopt project management processes to accomplish that goal. In January 1994, the company decided that it could get a step ahead of its competition if it assigned a physician-manager as the project sponsor for every new proposal. The rationale was that the clients would be favorably impressed.

The pilot project for this approach was Sinco Energy (another pseudonym), a Boston-based company with 8600 employees working in 12 cities in New England. Health Care Associates promised Sinco that the health care package would be ready for implementation no later than June 1994.

The project was completed almost 60 days late and substantially over budget. Health Care Associates' senior managers privately interviewed each of the employees on the Sinco project to identify the cause of the project's failure. The employees had the following observations:

- Although the physicians had been given management training, they had a great deal of difficulty applying the principles of project management. As a result, the physicians ended up playing the role of invisible sponsor instead of actively participating in the project.
- Because they were practicing physicians, the physician sponsors were not fully committed to their role as project sponsors.
- Without strong sponsorship, there was no effective process in place to control scope creep.
- The physicians had not had authority over the line managers, who supplied the resources needed to complete a project successfully.

Health Care Associates' senior managers learned two lessons. First, not every manager is qualified to act as a project sponsor. Second, the project sponsors should be assigned on the basis of their ability to drive the project to success. Impressing the customer is not everything.

**maxIT-VCS**

There are various levels of support from management regarding project management methodologies [at maxIT-VCS]. Initially, management may not appreciate the value until they have a PM identify the deliverables, track the progress, provide real-time status reports, identify issues and risks and have someone actively oversee the project and maintain momentum.

Management support of project management in our industry is growing as the formal role of a project manager is becoming more important to meeting the needs of complex projects, but managers still need to embrace and empower the PM's. There are times the PM is not asked when changes occur (resource allocation as an example) or in some cases not supported as the person in charge of the initiative. Obviously culture and politics plays a factor in every industry, but in order for healthcare to meet the demands of the industry, management must continue to further invest and support the growth of their PMs.

At VCS, the entire leadership team (including our C-suite) is very supportive and actively involved in helping grow the PMO. The PMO resides as the central "hub" for providing our customer base a full service consulting service offering. This includes the ability to run a complex healthcare IT initiative with a project manager who has subject matter expertise on both the operations and technical side of the house along with a published methodology and process to follow. For VCS, this is a very big market differentiator and offers our customers great value in comparison to our competition. (Marc Hirshfield, Director, Project Management Office, maxIT-VCS)

In the wake of integrating two companies under yet a third, the PMO will remain a cadre of experienced (and for the most part certified) project managers. This group is our "go to" band of PMs who can manage complex healthcare projects. This core also contributes to the project management methodology maintenance and training.

We currently are a Registered Education Provider with PMI.

We plan to develop additional internal training for all levels of project managers, as well as an internal company certification for attainment of specific skills in project management. (Heidi Wurtz, VP, Solution Center, maxIT-VCS)

**Indra**

Executive management [at Indra] is highly motivated to support project management development within the company. They regularly insist upon improving our training programs for project managers as well as focusing on the need that the best project management methods are in place.

Sometimes the success of a project constitutes a significant step in the development of a new technology, in the launch in a new market, or for the establishment of a new partnership, and, in those cases, the managing directorate usually plays an especially active role as sponsors during the project or program execution. They participate with the customer in steering committees for the project or the program, and help in the decision making or risk management processes.

For a similar reason, due to the significance of a specific project but at a lower level, it is not uncommon to see middle level management carefully watching its execution and providing, for instance, additional support to negotiate with the customer the resolution of a particular issue.

Getting middle level management support has been accomplished using the same set of corporate tools for project management at all levels and for all kind of projects in the company. No project is recognized if it is not in the corporate system, and to do that, the line managers must follow the same basic rules and methods, no matter if it is a recurring, a non-recurring effort, or other type of project. A well developed WBS, a complete foreseen schedule, a risk management plan and a tailored set of earned value methods, may be applied to any kind of projects. (Enrique Sevilla Molina, PMP, formerly Corporate PMO Director, Indra.)

## 7.10 GETTING LINE MANAGEMENT SUPPORT

Management support is not restricted exclusively to senior management, as was shown in the previous section. Line management support is equally crucial for project management to work effectively. Line managers are usually more resistant to project management and often demand proof that project management provides value to the organization before they support the new processes. This problem was identified previously in the journey to excellence in project management. It also appeared at Motorola. According to a spokesperson at Motorola[2]:

> This (getting line management support) was tough at first. It took years of having PMs provide value to the organization.

When organizations become mature in project management, sponsorship at the executive levels and at middle management levels becomes minimal and integrated project teams are formed where the integrated or core team is empowered to manage the project with minimal sponsorship other than for critical decisions. These integrated or core teams may or may not include line management. The concept of core teams became a best practice at Motorola[3]:

> Most project decisions and authority resides in the project core team. The core team is made up of middle- to low-level managers for the different functional areas (marketing, software, electrical, mechanical, manufacturing, system test, program management, quality, etc.) and has the project ownership responsibility. This core team is responsible for reviewing and approving product requirements and committing resources and schedule dates. It also acts as the project change control board and can approve or reject project scope change requests. However, any ship acceptance date changes must be approved by senior management.

## 7.11 DTE ENERGY

Jason Schulist, Manager, Continuous Improvement, Operating Systems Strategy Group at DTE Energy, comments on project sponsorship:

> The champions of continuous improvement projects that use the 4-Gate/9-Step methodology [Previously discussed in Chapter 4] are primarily Vice Presidents and Directors of DTE Energy. These champions sign off at every gate to approve the results of the process. In some business units, a Master Black Belt also signs off at every gate to ensure that rigorous analysis and the DTE Energy Operating System methodology is followed.

The importance of this statement is that senior management should be the champions of the methodology in addition to becoming project champions. In one auto supplier, senior management undertook the role of champion for the development of a project management methodology. After the methodology was developed, the executive took on a different role in the company. Three years later, the company realized that there were no continuous improvement efforts on the methodology because there was no executive champion in place. The executive then returned to championing the methodology and continuous improvement efforts occurred.

---

2. H. Kerzner, *Project Management Best Practices: Achieving Global Excellence*, Wiley, 2006, p. 307.
3. Ibid., p. 308.

## 7.12 INITIATION CHAMPIONS AND EXIT CHAMPIONS _____

As project management evolved, so did the role of the executive in project management. Today, there are three roles that the executive plays:

- The project sponsor
- The project (initiation) champion
- The exit champion

The role of the executive in project management as a project sponsor has become reasonably mature. Most textbooks on project management have sections that are dedicated to the role of the project sponsor.[4] The role of the project champion however, is just coming of age. Stephen Markham defines the role of the champion:

> Champions are informal leaders who emerge in a somewhat erratic fashion. Championing is a voluntary act by an individual to promote a particular project. In the act of championing, individuals rarely refer to themselves as champions; rather, they describe themselves as trying to do the right thing for the right company. A champion rarely makes a single decision to champion a project. Instead, he or she begins in a simple fashion and develops increasing enthusiasm for the project. A champion becomes passionate about a project and ultimately engages others based upon personal conviction that the project is the right thing for the entire organization. The champion affects the way other people think of the project by spreading positive information across the organization. Without official power or responsibility, a champion contributes to new product development by moving projects forward. Thus, champions are informal leaders who (1) adopt projects as their own in a personal way, (2) take on risk by promoting the projects beyond what is expected of people in their position, and (3) promote the project by getting other individuals to support it.[5]

With regard to new product development (NPD) projects, champions are needed to overcome the obstacles in the "valley of death," as seen in Figure 7–6.[6] The valley of death is the area in NPD where recognition of the idea/invention and efforts to commercialize the product come together. In this area, good projects often fall by the wayside and projects with less value often get added into the portfolio of projects. According to Markham[7],

> Many reasons exist for the Valley of Death. Technical personnel (left side of Figure 7–6) often do not understand the concerns of commercialization personnel (right side) and vice versa. The cultural gap between these two types of personnel manifests itself in the results prized by one side and devalued by the other. Networking and contact management may be important to sales people but seen as shallow and self-aggrandizing by technical

---

4. H. Kerzner, *Project Management: A Systems Approach to Planning, Scheduling and Controlling*, 11th ed., Hoboken, NJ: Wiley, 2013, Chapter 10. Also, H. Kerzner and F. Saladis, *What Executives Need to Know About Project Management*, Hoboken, NJ: Wiley and New York: IIL, 2009.
5. S. K. Markham, "Product Champions: Crossing the Valley of Death," in P. Belliveau, A. Griffin, and S. Somermeyer (Eds.), *The PDMA Toolbook for New Product Development*, Hoboken, NJ: Wiley, 2002, p. 119.
6. Ibid., p. 120.
7. Ibid., p. 120.

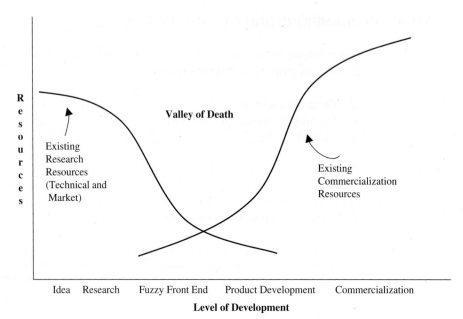

**Figure 7–6.** Valley of death.

people. Technical people find value in discovery and pushing the frontiers of knowledge. Commercialization people need a product that will sell in the market and often consider the value of discovery as merely theoretical and therefore useless. Both technical and commercialization people need help translating research findings into superior product offerings.

As seen in Figure 7–6, the valley of death seems to originate somewhere near the fuzzy front end (FFE).

> [FFE is] the messy "getting started" period of product development, which comes before the formal and well-structured product development process, when the product concept is still very fuzzy. It generally consists of the first three tasks (strategic planning, concept generation, and, especially, pretechnical evaluation) of the product development process. These activities are often chaotic, unpredictable, and unstructured. In comparison, the new product development process is typically structured and formal, with a prescribed set of activities, questions to be answered, and decisions to be made.[8]

Project champions are usually neither project managers nor project sponsors. The role of the champion is to sell the idea or concept until it finally becomes a project. The champion may not even understand project management and may not have the necessary skills to manage a project. Champions may reside much higher up in the organizational hierarchy than the project manager.

---

8. P. Belliveau, A. Griffin, and S. Somermeyer, *The PDMA Toolbook for New Product Development*, Hoboken, NJ: Wiley, 2002, p. 444.

Allowing the project champion to function as the project sponsor can be equally as bad as allowing them to function as the project manager. When the project champion and the project sponsor are the same person, projects never get canceled. There is a tendency to prolong the pain of continuing on with a project that should have been canceled.

Some projects, especially very long-term projects where the champion is actively involved, often mandate that a collective belief exist. The collective belief is a fervent, and perhaps blind, desire to achieve that can permeate the entire team, the project sponsor, and even the most senior levels of management. The collective belief can make a rational organization act in an irrational manner. This is particularly true if the project sponsor spearheads the collective belief.

When a collective belief exists, people are selected based upon their support for the collective belief. Champions may prevent talented employees from working on the project unless they possess the same fervent belief as the champion does. Nonbelievers are pressured into supporting the collective belief and team members are not allowed to challenge the results. As the collective belief grows, both advocates and nonbelievers are trampled. The pressure of the collective belief can outweigh the reality of the results.

There are several characteristics of the collective belief, which is why some large, high-technology projects are often difficult to kill:

- Inability or refusing to recognize failure
- Refusing to see the warning signs
- Seeing only what you want to see
- Fearful of exposing mistakes
- Viewing bad news as a personal failure
- Viewing failure as a sign of weakness
- Viewing failure as damage to one's career
- Viewing failure as damage to one's reputation

Project sponsors and project champions do everything possible to make their project successful. But what if the project champion, as well as the project team and sponsor, have blind faith in the success of the project? What happens if the strongly held convictions and the collective belief disregard the early warning signs of imminent danger? What happens if the collective belief drowns out dissent?

In such cases, an exit champion must be assigned. The exit champion sometimes needs to have some direct involvement in the project in order to have credibility, but direct involvement is not always a necessity. Exit champions must be willing to put their reputation on the line and possibly face the likelihood of being cast out from the project team. According to Isabelle Royer[9]:

> Sometimes it takes an individual, rather than growing evidence, to shake the collective belief of a project team. If the problem with unbridled enthusiasm starts as an unintended consequence of the legitimate work of a project champion, then what may be needed is

---

9. I. Royer, "Why Bad Projects are So Hard to Kill," *Harvard Business Review*, February 2003, p. 11. Copyright

a countervailing force—an exit champion. These people are more than devil's advocates. Instead of simply raising questions about a project, they seek objective evidence showing that problems in fact exist. This allows them to challenge—or, given the ambiguity of existing data, conceivably even to confirm—the viability of a project. They then take action based on the data.

The larger the project and the greater the financial risk to the firm, the higher up the exit champion should reside. If the project champion just happens to be the CEO, then someone on the board of directors or even the entire board of directors should assume the role of the exit champion. Unfortunately, there are situations where the collective belief permeates the entire board of directors. In this case, the collective belief can force the board of directors to shirk their responsibility for oversight.

Large projects incur large cost overruns and schedule slippages. Making the decision to cancel such a project, once it has started, is very difficult, according to David Davis[10]:

> The difficulty of abandoning a project after several million dollars have been committed to it tends to prevent objective review and recosting. For this reason, ideally an independent management team—one not involved in the projects development—should do the recosting and, if possible, the entire review. . . . If the numbers do not holdup in the review and recosting, the company should abandon the project. The number of bad projects that make it to the operational stage serves as proof that their supporters often balk at this decision.
>
> . . . Senior managers need to create an environment that rewards honesty and courage and provides for more decision making on the part of project managers. Companies must have an atmosphere that encourages projects to succeed, but executives must allow them to fail.

The longer the project, the greater the necessity for the exit champions and project sponsors to make sure that the business plan has "exit ramps" such that the project can be terminated before massive resources are committed and consumed. Unfortunately, when a collective belief exists, exit ramps are purposefully omitted from the project and business plans. Another reason for having exit champions is so that the project closure process can occur as quickly as possible. As projects approach their completion, team members often have apprehension about their next assignment and try to stretch out the existing project until they are ready to leave. In this case, the role of the exit champion is to accelerate the closure process without impacting the integrity of the project.

Some organizations use members of a portfolio review board to function as exit champions. Portfolio review boards have the final say in project selection. They also have the final say as to whether or not a project should be terminated. Usually, one member of the board functions as the exit champion and makes the final presentation to the remainder of the board.

---

10. D. Davis, "New Projects: Beware of False Economics," *Harvard Business Review*, March–April 1985, pp. 100–101. Copyright © 1985 by the President and Fellows of Harvard College. All rights reserved.

# 8 Training and Education

## 8.0 INTRODUCTION

Establishing project management training programs is one of the greatest challenges facing training directors because project management involves numerous complex and interrelated skills (qualitative/behavioral, organizational, and quantitative). In the early days of project management, project managers learned by their own mistakes rather than from the experience of others. Today, companies excellent in project management are offering a corporate curriculum in project management. Effective training supports project management as a profession.

Some large corporations offer more internal courses related to project management than most colleges and universities do. Such companies treat education almost as a religion. Smaller companies have more modest internal training programs and usually send their people to publicly offered training programs.

This chapter discusses processes for identifying the need for training, selecting the students who need training, designing and conducting the training, and measuring training's return on dollars invested.

## 8.1 TRAINING FOR MODERN PROJECT MANAGEMENT

During the early days of project management, in the late 1950s and throughout the 1960s, training courses concentrated on the advantages and disadvantages of various

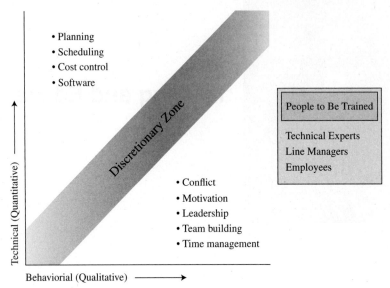

**Figure 8–1.** Types of project management training. *Source:* Reprinted from H. Kerzner, *In Search of Excellence in Project Management*, Hoboken, NJ: Wiley, 1998, p. 174.

organizational forms (e.g., matrix, traditional, projectized, and functional). Executives learned quickly, however, that any organizational structure could be made to work effectively and efficiently when basic project management is applied. Project management skills based in trust, teamwork, cooperation, and communication can solve the worst structural problems.

Starting in the 1970s, emphasis turned away from organizational structures for project management. The old training programs were replaced with two basic programs:

- Basic project management, which stresses behavioral topics such as multiple reporting relationships, time management, leadership, conflict resolution, negotiation, team building, motivation, and basic management areas such as planning and controlling.
- Advanced project management, which stresses scheduling techniques and software packages used for planning and controlling projects.

Today's project management training programs include courses on behavioral as well as quantitative subjects. The most important problem facing training managers is how to achieve a workable balance between the two parts of the coursework—behavioral and quantitative (see Figure 8–1). For publicly sponsored training programs, the seminar leaders determine their own comfort levels in the "discretionary zone" between technical and behavioral subject matter. For in-house trainers, however, the balance must be preestablished by the training director on the basis of factors such as which students will be assigned to manage projects, types of projects, and average lengths of projects (see Table 8–1).

**TABLE 8–1.   EMPHASES IN VARIOUS TRAINING PROGRAMS**

| Type of Person Assigned for PM Training (PM Source) | Training Program Emphasis | |
| --- | --- | --- |
| | Quantitative/Technology Skills | Behavior Skills |
| **Training Needed to Function as a Project Manager** | | |
| Technical expert on short-term projects | High | Low |
| Technical expert on long-term projects | High | High |
| Line manager acting as a part-time project manager | High | Low |
| Line manager acting as a full-time project manager | High | Average to high |
| Employees experienced in cooperative operations | High | Average to high |
| Employees inexperienced in cooperative operations | High | Average to high |
| **Training Needed for General Knowledge** | | |
| Any employees or managers | Average | Average |

*Source:* Reprinted from H. Kerzner, *In Search of Excellence in Project Management*, Hoboken, NJ: Wiley, 1998, p. 175.

## 8.2  NEED FOR BUSINESS EDUCATION

In the previous section, we discussed the importance of determining the right balance between quantitative skills and behavioral skills. That balance is now changing because of how we view the role of a project manager. Today, we have a new breed of project manager. Years ago, virtually all project managers were engineers with advanced degrees. These people had a command of technology rather than merely an understanding of technology. If the line manager believed that the project manager did in fact possess a command of technology, then the line manager would allow the assigned functional employees to take direction from the project manager. The result was that project managers were expected to manage people and provide technical direction.

Most project managers today have an understanding of technology rather than a command of technology. As a result, the accountability for the success of the project is now viewed as shared accountability between the project manager and all affected line managers. With shared accountability, the line managers must now have a good understanding of project management, which is why more line managers are now becoming PMP®s. Project managers are now expected to manage deliverables rather than people. Management of the assigned resources is more often than not a line function.

Another important fact is that project managers are treated as though they are managing part of a business rather than simply a project, and therefore they are expected to make sound business decisions as well as project decisions. Project managers must understand business principles. In the future, project managers may be expected to become externally certified by PMI® and internally certified by their company on the organization's business processes.

Now, when designing training courses, we determine the correct balance between quantitative skills, behavioral skills, and business skills. Soft skills and business acumen are crucial elements for a flawless project execution, says Benny Nyberg, formerly group assistant VP with responsibility for PM Methodologies and Talent Development at ABB:

> After implementing the ABB PM Process as a common high level process throughout the company's project sales organizations as well as several product development orga-

nizations, one thing was very clear. In a technical company employing large numbers of highly skilled engineers, some of which are promoted to project management, the technical aspects of project management such as planning, scheduling and cost control are the least difficult to implement. Junior employees do need training in this area but the real challenge for reaching operational excellence in project management, a flawless project execution, desirable projects result and a high level of customer satisfaction lies in identifying and developing project managers with the right business acumen. Project management is a management position requiring excellent commercial, communications and leadership skills. A project manager must be very business minded, be able to communicate effectively with a variety of different stake holders and possess the ability to lead and motivate people. For delivery projects, a precise understanding of the contract, i.e. terms and conditions, scope and any promises made is crucial for being able to deliver just that, meet customer expectations and as such assure customer satisfaction and project success. Contract understanding is further a requisite for maximizing financial outcome and recognizing up-selling opportunities as they occur.

The role of a project manager, especially for big contracts that take many months or several years to complete, is very close to the role of a key account manager. The following skills/abilities are among the most important for success; business mind, communication, negotiation, leadership, risk management, salesmanship.

In order to indentify and address training and other development activities required for the wide variety of competencies and skills, ABB have implemented a competency model. The model includes a definition of required competencies, questionnaires for self-assessment, interview questionnaires plus development guides and last but not least a number of selectable training modules leading to appropriate level of certification.

## 8.3 SAP: IMPORTANCE OF A PROJECT MANAGEMENT CAREER PATH[1]

In SAP Services we have established a clear project management development path for anyone supporting our customer delivery projects (or internal transformation projects). The project management career path spans from entry-level position of PM Associate through various levels of Project Managers all the way to the role of Program Manager or Delivery Executive. (See Figure 8–2)

Each role has a clearly defined skillset profile, including precise definition of required job skills and expected proficiency level for each skill that is required in the job. Each profile is also associated with the category of project that the person on this career level is expected to manage—the dimensions of this assignment includes project size, complexity, risk exposure, revenue, etc.

Each profile is tightly linked with HR job profile and respective personal development path that clearly specifies recommended training classes for each skill in the profile. The training offerings for the PM practice specify target roles for the training and the training learning objectives are tied to specific skills from the profile. The training catalogue covers

1. ©2013 by SAP. All rights reserved. Reproduced by permission. Material in this section has been provided by Jan Musil, Global Lead of Project Management Practice, SAP Field Services, SAP America, Inc.

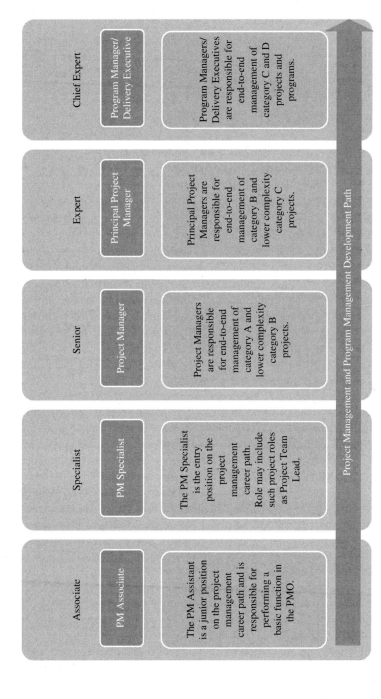

**Figure 8–2.** Project management and program management development path.

broad range of skills from core project management skills based on PMI standards like PMI *PMBOK® Guide* project management fundamentals; SAP specific knowledge (like SAP solution knowledge, SAP implementation methodology, Agile delivery, delivery and internal PM tools, etc.) to leadership and interpersonal skills.

Through structured career path the project managers grow their knowledge, expertise and experiences, which allows them to move along the PM career path. The PM career path is interlinked with other job profiles in SAP so that project managers can make career choices that allow them to move to different career paths like management, sales roles or remain in the PM career.

## 8.4 INTERNATIONAL INSTITUTE FOR LEARNING

Given the importance of project management now and in the future, there will be a continuing need for project management education. E. LaVerne Johnson (founder, president and CEO of the International Institute for Learning) comments on the growth of project management training. (For more information about IIL, please visit www.iil.com.)

In IIL's more than twenty-year history, we've worked with thousands of organizations around the world, designing and delivering training in project, program, and portfolio management. Our clients range across all industries and include large global companies as well as smaller organizations trying to gain an edge through effective programs. With IIL serving our clients' day-to-day needs as well as those just emerging on the horizon, we've been in a unique position to participate in and observe project management's growth into a full-fledged profession. We have been "on the scene" as project management has transitioned from an area of interest to an organizational imperative.

From our perspective, courses that were sufficient just a few years ago now fall short—and that's a real sign of progress. The global market and expanding importance of project management has dictated whole new families of courses, richer content, and a flexible range of delivery methods that allow students to learn when and where they need to—in live and virtual classrooms, alone at their desks whether at home or in the office, self-paced or instructor-led. It seems that learning and technology have come together to offer the project management profession extremely useful tools to keep on building the future. IIL takes pride in its Many Methods of Learning™ brand, which ensures that the education it provides serves a diversity of needs, styles, and interests.

**Evolutionary Years: Learning Trends** Training courses during the 1980s were mostly geared to advancing the project manager's skills. The focus of training was on the basics: the fundamental methodology and the know-how required to pass PMI®'s Project Management Professional (PMP®) Certification exam. In response, IIL launched training courses in project management fundamentals and established a comprehensive certification program that allowed individuals to prepare for and successfully pass PMI's PMP® exam. A small variety of books, classroom courses, and software products were made available to those individuals responsible for managing projects in their companies.

**Revolutionary Years: Marketplace Trends**

In recent years, a far greater variety of companies and industries have recognized the importance of managing projects more effectively and analyzing the ways in which projects meet overall corporate goals. Compared to previous years, a revolution is emerging. This has become evident in a number of trends.

- The volume of projects is increasing as more and more companies run their businesses via projects. Indeed some leading organizations undertake hundreds of thousands of individual projects each year—some small and simple, others huge and complex.
- The ability to effectively manage projects has become critically important to business, and good project management skills have become a competitive advantage for leading companies.
- As a result of this revolutionary growth, the status and value of the project manager has grown in importance—having this know-how allows a company to complete projects faster, at lower cost, with greater customer satisfaction, and with more desirable project outcomes.
- Knowledge that was once deemed "nice to have" is now considered mandatory. A company's economic success and survival depend on its ability to determine which projects support its overall strategic objectives, and to enable it to sequence them in ways to achieve that success.
- Today, employees with project management skills expand beyond project managers. Team members and middle and top management are developing expertise in the subject.
- The complexity and scope of project management methodologies has grown to include new skills and applications.
- Process development and improvement through direct, hands-on support or knowledge management solutions have become a requirement for economic survival in the most challenging of times. IIL's Unified Project Management® Methodology (UPMM™) Software Suite was developed to support consistency and quality in project, program, and portfolio management implementation.
- A large number of project management-related software programs have been developed to help manage projects (such as Microsoft® Project and Project Server, SharePoint, Primavera, and dozens of others including a wide range of open-source software).
- Project management certification has become an even more valuable asset to an individual's career path. As a result, there were more than 520,000 active PMPs and more than 411,000 PMI members as of early 2013 (the latter, an 8.8% increase over the previous year).
- Heretofore, the project manager's skill set has remained mostly technical. But today we are seeing project managers embrace additional knowledge areas such as leadership and interpersonal skills.
- Strategic planning for project management has taken on importance. Organizations are now seeking systematic ways to better align project management with business objectives.
- More and more companies are establishing project and portfolio management offices.

- Approaches to project management within an organization remain relatively varied and nonstandardized. Companies need to work toward a more matured and common methodology for more repeatable and predictable success.
- Companies and their project management offices are placing stronger emphasis on quality services, quality products, and improved processes using management tools such as Lean Six Sigma, Earned Valued, and Business Analysis.

**Revolutionary Years: Learning Trends**    In response to these trends, a greater variety of courses are available to a broadening number of industries. New methods of learning have been introduced to meet the growing diversity of client needs. Here are some examples of how IIL has responded to these needs and established best practices in training and education for project managers:

- IIL has dozens of different course titles in advanced areas of knowledge to increase the scope, application, and sophistication of the project manager. Such courses include advanced concepts in risk management, complex project management, requirements management, the design and development of a project office, and Agile project management.
- Courses addressing the "softer" side of project management are designed to hone facilitation skills, interpersonal skills, leadership skills, and other nontechnical areas.
- As organizations are increasing in their levels of project management maturity, there is a need for training in the effective use of enterprise project management software.
- More and more universities are recognizing project management as a part of their degree programs. IIL has partnered with the New York University School of Continuing and Professional Studies (NYU-SCPS) and the University of Southern California's Marshall School of Business to offer project management certificate programs.
- The way we learn is changing. Employees have less time to devote to classroom study. As a result, in addition to traditional classroom training, IIL offers delivery methods such as on-demand, virtual, mobile apps, and video.

**A Look into the Crystal Ball: Trends and Learning Responses**    It's always a challenge to try and predict the future, but there are some emerging trends that allow us to take a reasonable stab at this. For each of these trends, there will be the need to develop the appropriate learning responses.

- A key competitive factor in companies will be their ability to undertake and effectively manage multiple projects—projects to develop new products and services, get to market faster, reduce costs, improve customer satisfaction, increase sales, and so on. The more well-chosen projects are, the more competitive a company will be in the marketplace (particularly regarding projects to improve products, processes, and customer satisfaction).

- We anticipate a blending of project management methodologies with other proven business strategies (such as Six Sigma, Agile, quality management, risk management, and business analysis). Training in these subjects will similarly become blended.

- Project, program, and portfolio management will continue to grow in importance and become a strategic differentiating factor for organizations to remain competitive. Portfolio management software is now a requirement.

- Senior management will become more knowledgeable and involved in project management efforts. This will require project management training that meets the unique needs of executives.

- Strategic planning for project management will become a way of life for leading organizations. The role of the project office/portfolio office will increase and become commonplace and vital in companies. Membership will include the highest levels of executive management. Senior management will take leadership of the company's project portfolio management efforts.

- IIL will continue to partner with companies to offer dual certification for project managers. Project managers will be required to be certified in project management as well as in their internal business operations and processes.

- Executives will be increasingly involved in activities such as capacity planning, portfolio management, prioritization, process improvement, supply chain management, and strategic planning specifically for project management. In fact, more and more executives are earning project management certifications.

- Participation of upper managers will be obstructed by their limited experience and training in project, program, and portfolio management. An essential element will be to provide these managers with experience and training in how to manage project activities within their companies. This training must be tailored to the unique needs and responsibilities of upper management.

- The company's reward and recognition systems will change to stimulate and reinforce project management goals and objectives.

- Training in project management will be expanded to include all levels of the company hierarchy, including the non–project manager. Training will become responsive to an array of job functions, levels, and responsibilities.

- The status of the certified project manager will grow significantly and the project manager's skills will be both technical and managerial.

- We will witness the establishment of a corporate-level project management executive (chief project management officer).

- Project benchmarking and continuous project improvement will become essential for leading organizations. The project management maturity models will play an important role in this regard, as they will help companies identify their strengths, weaknesses, and specific opportunities for improvement.

- The expanding importance of project, program, and portfolio management will require more individuals who are trained in project management. This in turn will necessitate the development of new and improved methods of delivery. Online training will play an increasingly important role.

- We will see an order-of-magnitude increase in the number of organizations reaching the higher levels of project management maturity.
- More colleges and universities will offer degree programs in project management and seek to align their courses with international standards and best practices.
- Project management will focus on providing the knowledge and best practices to support sustainable initiatives. Project sustainability in the global economy through values, leadership, and professional responsibility will be the mandate of all project, program, and portfolio managers and sponsors.

## 8.5 IDENTIFYING THE NEED FOR TRAINING

Identifying the need for training requires that line managers and senior managers recognize two critical factors: first, that training is one of the fastest ways to build project management knowledge in a company and, second, that training should be conducted for the benefit of the corporate bottom line through enhanced efficiency and effectiveness.

Identifying the need for training has become somewhat easier in the past 10 years because of published case studies on the benefits of project management training. The benefits can be classified according to quantitative and qualitative benefits. The quantitative results include:

- Shorter product development time
- Faster, higher-quality decisions
- Lower costs
- Higher profit margins
- Fewer people needed
- Reduction in paperwork
- Improved quality and reliability
- Lower turnover of personnel
- Quicker "best practices" implementation

Qualitative results include:

- Better visibility and focus on results
- Better coordination
- Higher morale
- Accelerated development of managers
- Better control
- Better customer relations
- Better functional support
- Fewer conflicts requiring senior management involvement

Companies are finally realizing that the speed at which the benefits of project management can be achieved is accelerated through proper training.

## 8.6 SELECTING STUDENTS

Selecting the people to be trained is critical. As we have already seen in a number of case studies, it is usually a mistake to train only the project managers. A thorough understanding of project management and project management skills is needed throughout the organization if project management is to be successful. For example, one automobile subcontractor invested months in training its project managers. Six months later, projects were still coming in late and over budget. The executive vice president finally realized that project management was a team effort rather than an individual responsibility. After that revelation, training was provided for all of the employees who had anything to do with the projects. Virtually overnight, project results improved.

Dave Kandt, retired group vice president, quality, program management and continuous improvement at Johnson Controls, explained how his company's training plan was laid out to achieve excellence in project management:

> We began with our executive office, and once we had explained the principles and philosophies of project management to these people, we moved to the managers of plants, engineering managers, cost analysts, purchasing people, and, of course, project managers. Only once the foundation was laid did we proceed with actual project management and with defining the roles and responsibilities so that the entire company would understand its role in project management once these people began to work. Just the understanding allowed us to move to a matrix organization and eventually to a stand-alone project management department.

## 8.7 FUNDAMENTALS OF PROJECT MANAGEMENT EDUCATION

Twenty years ago, we were somewhat limited as to availability of project management training and education. Emphasis surrounded on-the-job training in hopes that fewer mistakes would be made. Today, we have other types of programs, including:

- University courses
- University seminars
- In-house seminars
- In-house curriculums
- Distance teaming (e-learning)
- Computer-based training (CBT)

With the quantity of literature available today, we have numerous ways to deliver the knowledge. Typical delivery systems include:

- Lectures
- Lectures with discussion
- Exams
- Case studies on external companies
- Case studies on internal projects
- Simulation and role playing

Training managers are currently experimenting with "when to train." The most common choices include:

- *Just-in-time training:* This includes training employees immediately prior to assigning them to projects.
- *Exposure training:* This includes training employees on the core principles just to give them enough knowledge so that they will understand what is happening in project management within the firm.
- *Continuous learning:* This is training first on basic, then on advanced, topics so that people will continue to grow and mature in project management.
- *Self-confidence training:* This is similar to continuous learning but on current state-of-the-art knowledge. This is to reinforce employees' belief that their skills are comparable to those in companies with excellent reputations for project management.

## 8.8 SOME CHANGES IN PROJECT MANAGEMENT EDUCATION

In the early years of project management, almost all project managers came from the engineering disciplines. The project managers were expected to possess a command of technology rather than just an understanding of technology. When earned value management (EVM) principles were developed, project management coursework emphasized cost and schedule control. Seminars appeared in the marketplace entitled "Project Management," but the content for these two- or three-day courses was almost entirely PERT and EVM. In all but a few industries, project management was viewed as part-time work, and an add-on to one's primary job. The need to fully understand the skills and competencies to be effective as a project management was not considered to be that important.

Today, project management is a career path position in almost all companies. Colleges and universities are now offering master's and doctorate degrees in project management. Typical coursework is such programs include:

*Core coursework:*
- Principles of Project Management
- Project Scheduling Techniques
- Project Estimating Techniques
- Project Financing Techniques
- Creativity and Brainstorming
- Problem Solving and Decision Making
- Global Project Management
- Managing Multiple Projects
- Project Management Leadership
- Managing Virtual Teams
- Project Portfolio Management

*Electives:*
- Advanced Project Management
- Project Quality Management
- Project Procurement and Contracting
- Project Ethics and the Code of Professional Conduct
- Project Monitoring and Control Techniques
- Project Reporting Practices
- Stakeholder Relations Management
- Conducting Project Health Checks
- Managing Troubled Projects
- Capturing Best Practices
- Managing Cultural Differences

Some educational institutions also offer the students specialized training for certification in various project management areas. The knowledge needed to pass the certification exams can come from specialized coursework or from the traditional core requirements and electives. A partial list of some certification programs related to project management might include:

- Project Management
- Program Management
- Project Risk Management
- Agile Project Management
- Business Analyst
- Managing Complex Projects
- Other Project Management Certifications
- Specialized or Customized Certifications

The greatest change in project management education appears to be in the softer skills requirements. This is understandable since projects require people working together. Emphasis in the behavioral areas is now being placed upon:

- Problem-Solving Skills
- Decision-Making Skills
- Conceptualization Skills
- Creativity/Brainstorming Skills
- Process Skills
- Coping with Stress/Pressure
- Leadership without Authority
- Multiple Boss Reporting
- Counseling and Facilitation
- Mentorship Skills
- Negotiating Skills
- Conflict Resolution Skills
- Presentation Skills

In the future, it is possible that social networking skills may be added to the list.

## 8.9 DESIGNING COURSES AND CONDUCTING TRAINING

Many companies have come to realize that on-the-job training may be less effective than more formal training. On-the-job training virtually forces people to make mistakes as a learning experience, but what are they learning? How to make mistakes? It seems much more efficient to train people to do their jobs the right way from the start.

Project management has become a career path. More and more companies today allow or even require that their employees get project management certification. One company informed its employees that project management certification would be treated the same as a master's degree in the salary and career path structure. The cost of the training behind the certification process is only 5 or 10 percent of the cost of a typical master's degree in a business administration program. And certification promises a quicker return on investment (ROI) for the company. Project management certification can also be useful for employees without college degrees; it gives them the opportunity for a second career path within the company.

Linda Zaval, formerly a trainer with the International Institute for Learning, explained what type of project management training worked best in her experience during the early 1990s:

> In our experience, we have found that training ahead of time is definitely the better route to go. We have done it the other way with people learning on the job, and that has been a rather terrifying situation at times. When we talk about training, we are not just talking about training. We want our project managers to be certified through the Project Management Institute. We have given our people two years to certify. To that end there is quite a bit of personal study required. I do believe that training from the formal training end is great, and then you can modify that to whatever the need is in-house.

There is also the question of which are better: internally based or publicly held training programs. The answer depends on the nature of the individual company and how many employees need to be trained, how big the training budget is, and how deep the company's internal knowledge base is. If only a few employees at a time need training, it might be effective to send them to a publicly sponsored training course, but if large numbers of employees need training on an ongoing basis, designing and conducting a customized internal training program might be the way to go.

In general, custom-designed courses are the most effective. In excellent companies, course content surveys are conducted at all levels of management. For example, the R&D group of Babcock and Wilcox in Alliance, Ohio, needed a project management training program for 200 engineers. The head of the training department knew that she was not qualified to select core content, so she sent questionnaires out to executive managers, line managers, and professionals in the organization. The information from the questionnaires was used to develop three separate courses for the audience. At Ford Motor Company, training was broken down into a two-hour session for executives, a three-day program for project personnel, and a half-day session for overhead personnel.

For internal training courses, choosing the right trainers and speakers is crucial. A company can use trainers currently on staff if they have a solid knowledge of project management, or the trainers can be trained by outside consultants who offer train-the-trainer programs. Either way, trainers from within the company must have the expertise the company needs. Some problems with using internal trainers include the following:

- Internal trainers may not be experienced in all areas of project management.
- Internal trainers may not have up-to-date knowledge of the project management techniques practiced by other companies.
- Internal trainers may have other responsibilities in the company and so may not have adequate time for preparation.
- Internal trainers may not be as dedicated to project management or as skillful as external trainers.

But the knowledge base of internal trainers can be augmented by outside trainers as necessary. In fact, most companies use external speakers and trainers for their internal educational offerings. The best way to select speakers is to seek out recommendations from training directors in other companies and teachers of university-level courses in project management. Another method is contacting speakers' bureaus, but the quality of the speaker's program may not be as high as needed. The most common method for finding speakers is reviewing the brochures of publicly sponsored seminars. Of course, the brochures were created as sales materials, and so the best way to evaluate the seminars is to attend them.

After a potential speaker has been selected, the next step is to check his or her recommendations. Table 8–2 outlines many of the pitfalls involved in choosing speakers for internal training programs and how you can avoid them.

**TABLE 8–2.   COMMON PITFALLS IN HIRING EXTERNAL TRAINERS AND SPEAKERS**

| Warning Sign | Preventive Step |
|---|---|
| Speaker professes to be an expert in several different areas. | Verify speaker's credentials. Very few people are experts in several areas. Talk to other companies that have used the speaker. |
| Speaker's résumé identifies several well-known and highly regarded client organizations. | See whether the speaker has done consulting for any of these companies more than once. Sometimes a speaker does a good job selling himself or herself the first time, but the company refuses to rehire him or her after viewing the first presentation. |
| Speaker makes a very dramatic first impression and sells himself or herself well. Brief classroom observation confirms your impression. | Being a dynamic speaker does not guarantee that quality information will be presented. Some speakers are so dynamic that the trainees do not realize until too late that "The guy was nice but the information was marginal." |
| Speaker's résumé shows 10–20 years or more experience as a project manager. | Ten to 20 years of experience in a specific industry or company does not mean that the speaker's knowledge is transferable to your company's specific needs or industry. Ask the speaker what types of projects he or she has managed. |
| Marketing personnel from the speaker's company aggressively show the quality of their company, rather than the quality of the speaker. The client list presented is the company's client list. | You are hiring the speaker, not the marketing representative. Ask to speak or meet with the speaker personally and look at the speaker's client list rather than the parent company's client list. |
| Speaker promises to custom design his or her materials to your company's needs. | Demand to see the speaker's custom-designed material at least two weeks before the training program. Also verify the quality and professionalism of view graphs and other materials. |

The final step is to evaluate the training materials and presentation the external trainer will use in the classes. The following questions can serve as a checklist:

- Does the speaker use a lot of slides in his or her presentation? Slides can be a problem when students do not have enough light to take notes.
- Does the instructor use transparencies? Have they been prepared professionally? Will the students be given copies of the transparencies?
- Does the speaker make heavy use of chalkboards? Too much chalkboard work usually means too much note taking for the trainees and not enough audiovisual preparation from the speaker.
- Does the speaker use case studies? If he or she does, are the case studies factual? It is best for the company to develop its own case studies and ask the speaker to use those so that the cases will have relevance to the company's business.
- Are role playing and laboratory experiences planned? They can be valuable aids to learning, but they can also limit class size.
- Are homework and required reading a part of the class? If so, can they be completed before the seminar?

## 8.10 MEASURING THE RETURN ON INVESTMENT ON EDUCATION

The last area of project management training is the determination of the value earned on the dollars invested in training. It is crucial to remember that training should not be performed unless there is a continuous return on dollars for the company. Keep in mind, also, that the speaker's fee is only part of the cost of training. The cost to the company of having employees away from their work during training must be included in the calculation. Some excellent companies hire outside consultants to determine ROI. The consultants base their evaluations on personal interviews, on-the-job assessments, and written surveys.

One company tests trainees before and after training to learn how much knowledge the trainees really gained. Another company hires outside consultants to prepare and interpret post-training surveys on the value of the specific training received.

The amount of training needed at any one company depends on two factors: whether the company is project-driven and whether it has practiced project management long enough to have developed a mature project management system. Figure 8–3 shows the amount of project management training offered (including refresher courses) against the number of years in project management. Project-driven organizations offer the most project management training courses, and organizations that have just started implementing project management offer the fewest. That's no surprise. Companies with more than 15 years of experience in applying project management principles show the most variance.

## 8.11 PROJECT MANAGEMENT IS NOW A PROFESSION

For several years, project management was viewed as a part-time occupation and, therefore, all training was designed for one's primary job description, whatever that may

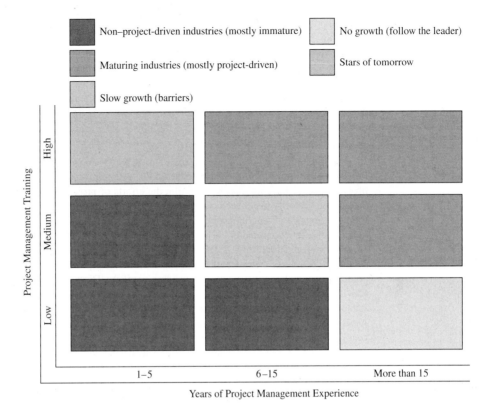

**Figure 8–3.** Amount of training by type of industry and year of project management experience. *Source:* Reprinted from H. Kerzner, *In Search of Excellence in Project Management*, Hoboken, NJ: Wiley, 1998, p. 185.

be, rather than project management. Because of this, there was no need to develop job descriptions for project and program managers. Today, these job descriptions exist, project management is viewed as a profession, and training programs are provided based upon the job descriptions. When asked if AT&T had job descriptions, a spokesperson for AT&T responded "yes" to both project and program management:

**Project Manager**

Provides end-to-end project management throughout the lifecycle of a project by directing the efforts of project team(s) using dotted-line authority to deliver a completed product and/or service. Has full accountability for managing larger low complexity to high complexity projects, or projects within programs which may span multiple regions and/or multiple functions; multiple concurrent projects may be managed. Includes estimating, scheduling, coordinating, assigning resources, ensuring that project funding is secured, and assisting in recommending business solutions/alternatives for projects. Assesses, plans for, and manages project risks, issues, jeopardies, escalations and problem resolutions. Manages project scope, project budgeting and cost reporting, and ensures completion of projects while meeting quality, schedule and cost objectives using the organization's standard processes. Acts as project liaison between IT partners, client organizations and IT leadership. May assist in supplier management of existing vendors. May direct Associate Project Managers to provide support with project communications and tracking project progress. Does not include the management of extremely large and complex programs, with multiple sub-programs, requiring senior level oversight and extensive executive communications. Must spend 80% or more of time performing the project management duties described above.

### Program Manager

Provides end-to-end project management and/or program management throughout the lifecycle of a project/program by directing the efforts of project/program team(s) using dotted-line authority to deliver a completed project and/or service. Has full accountability for managing concurrent high complexity projects and/or programs which may span multiple regions, functions and/or business units. Responsible for detailed planning including program/project structure & staffing, estimating, resource allocation and assignment, detailed scheduling, critical path analysis, consolidating project plans into an overall program plan and negotiating any sequencing conflicts. Directs project and/or program activities utilizing the organization's standard processes to ensure the timely delivery of stated business benefits, comparing actuals to plans and adjusting plans as necessary. Assesses, plans for, and manages project/program risks including mitigation & contingency plans; manages issues, jeopardies, escalations and problem resolutions. Defines project/program scope and ensures changes to scope and deliverables are managed using the change control process. Manages large program or project budgets and cost reporting. Acts as liaison with client and IT leadership, providing communication and status regarding the progress of the project/program. May assist with RFP development, evaluation, and supplier selection, as well as ongoing relationships with suppliers or consultants. Utilizes knowledge of business, industry and technology to incorporate business process improvements into the organization and/or to develop business strategies and functional/business/technical architectures. May direct the efforts of project managers when they manage a project or sub-program over which the Senior Project/Program Manager has authority. May include the management of extremely large and complex programs, with multiple sub-programs, requiring senior level oversight and extensive executive communications. Must spend 80% or more of time performing the project management duties described above.

The recognition of project management as a profession has spread worldwide. According to Enrique Sevilla Molina, formerly PMO director, Indra:

> Project management is considered the result of a specific blend of knowledge and experience, gained through dedication to achieve success in the projects under the project manager responsibility.
>
> We have a set of management roles associated to the different levels of responsibilities and expertise to manage projects, programs and portfolios, and to develop business opportunities (i.e. Project Managers, Program Directors, etc). For each role, a specific set of skills in a certain degree is defined, so performance and achievement may be assessed. The yearly evaluation of personal performance is done based on the job descriptions, the role maturity achieved so far, the expected performance for the role and the actual performance, so the evolution in personal development may also be assessed.

## 8.12 COMPETENCY MODELS

Twenty years ago, companies prepared job descriptions for project managers to explain roles and responsibilities. Unfortunately, the job descriptions were usually abbreviated and provided very little guidance on what was required for promotion or salary increases. Ten years ago we still emphasized the job description, but it was now supported by coursework, which was often mandatory. By the late 1990s, companies began emphasizing core competency models, which clearly depicted the skill levels needed to be effective as a project manager. Training programs were instituted to support the core competency models. Unfortunately, establishing a core competency model and the accompanying training is no easy task.

Eli Lilly has perhaps one of the most comprehensive and effective competency models in industry today. Martin D. Hynes, III, director, pharmaceutical projects management (PPM), was the key sponsor of the initiative to develop the competency model. Thomas J. Konechnik, operations manager, pharmaceutical projects management, was responsible for the implementation and integration of the competency model with other processes within the PPM group. The basis for the competency model is described here.

Lilly Research Laboratories project management competencies are classified under three major areas:

*Scientific/Technical Expertise*
- Knows the Business: Brings an understanding of the drug development process and organizational realities to bear on decisions.
- Initiates Action: Takes proactive steps to address needs or problems before the situation requires it.
- Thinks Critically: Seeks facts, data, or expert opinion to guide a decision or course of action.
- Manages Risks: Anticipates and allows for changes in priorities, schedules, and resources and changes due to scientific/technical issues.

*Process Skills*
- Communicates Clearly: Listens well and provides information that is easily understood and useful to others.
- Attention to Details: Maintains complete and detailed records of plans, meeting minutes, agreements.
- Structures the Process: Constructs, adapts, or follows a logical process to ensure achievement of objectives and goals.

*Leadership*
- Focuses on Results: Continually focuses own and others' attention on realistic milestones and deliverables.
- Builds a Team: Creates an environment of cooperation and mutual accountability within and across functions to achieve common objectives.
- Manages Complexity: Organizes, plans, and monitors multiple activities, people, and resources.

- Makes Tough Decisions: Demonstrates assurance in own abilities, judgments, and capabilities; assumes accountability for actions.
- Builds Strategic Support: Gets the support and level of effort needed from senior management and others to keep project on track.

We examine each of these competencies in more detail below.

1. Knows the Business: Brings an understanding of the drug development process and organizational realities to bear on decisions.

Project managers/associates who demonstrate this competency will:

- Recognize how other functions in Eli Lilly impact the success of a development effort.
- Use knowledge of what activities are taking place in the project as a whole to establish credibility.
- Know when team members in own and other functions will need additional support to complete an assignment/activity.
- Generate questions based on understanding of nonobvious interactions of different parts of the project.
- Focus attention on the issues and assumptions that have the greatest impact on the success of a particular project activity or task.
- Understand/recognize political issues/structures of the organization.
- Use understanding of competing functional and business priorities to reality test project plans, assumptions, time estimates, and commitments from the functions.
- Pinpoint consequences to the project of decisions and events in other parts of the organization.
- Recognize and respond to the different perspectives and operating realities of different parts of the organization.
- Consider the long-term implications (pro and con) of decisions.
- Understand the financial implications of different choices.

Project managers/associates who do not demonstrate this competency will:

- Rely on resource and time estimates from those responsible for an activity or task.
- Make decisions based on what ideally should happen.
- Build plans and timelines by rolling up individual timelines and so on.
- Perceive delays as conscious acts on the part of other parts of the organization.
- Assume that team members understand how their activities impact other parts of the project.
- Focus attention on providing accurate accounts of what has happened.
- Avoid changing plans until forced to do so.
- Wait for team members to ask for assistance.

Selected consequences for projects/business of not demonstrating this competency are:

- Project manager or associate may rely on senior management to resolve issues and obtain resources.

- Proposed project timelines may be significantly reworked to meet current guidelines.
- Attention may be focused on secondary issues rather than central business or technical issues.
- Current commitments, suppliers, and so on, may be continued regardless of reliability and value.
- Project deliverables may be compromised by changes in other parts of Lilly.
- Project plans may have adverse impact on other parts of the organization.

2. Initiates Action: Takes proactive steps to address needs or problems before the situation requires it.

Project managers/associates who demonstrate this competency will:

- Follow up immediately when unanticipated events occur.
- Push for immediate action to resolve issues and make choices.
- Frame decisions and options for project team, not simply facilitate discussions.
- Take on responsibility for dealing with issues for which no one else is taking responsibility.
- Formulate proposals and action plans when a need or gap is identified.
- Quickly surface and raise issues with project team and others.
- Let others know early on when issues have major implications for project.
- Take action to ensure that relevant players are included by others in critical processes or discussions.

Project managers/associates who do not demonstrate this competency will:

- Focus efforts on ensuring that all sides of issues are explored.
- Ask others to formulate initial responses or plans to issues or emerging events.
- Let functional areas resolve resource issues on their own.
- Raise difficult issues or potential problems after their impact is fully understood.
- Avoid interfering or intervening in areas outside own area of expertise.
- Assume team members and others will respond as soon as they can.
- Defer to more experienced team members on how to handle an issue.

Selected consequences for projects/business of not demonstrating this competency are:

- Senior management may be surprised by project-related events.
- Project activities may be delayed due to "miscommunications" or to waiting for functions to respond.
- Effort and resources may be wasted or underutilized.
- Multiple approaches may be pursued in parallel.
- Difficult issues may be left unresolved.

3. Thinks Critically: Seeks facts, data, or expert opinion to guide a decision or course of action.

Project managers/associates who demonstrate this competency will:

- Seek input from people with expertise or first-hand knowledge of issues and so on.

- Ask tough, incisive questions to clarify time estimates or to challenge assumptions and be able to understand the answers.
- Immerse self in project information to quickly gain a thorough understanding of a project's status and key issues.
- Focus attention on key assumptions and root causes when problems or issues arise.
- Quickly and succinctly summarize lengthy discussions.
- Gather data on past projects, and so on, to help determine best future options for a project.
- Push to get sufficient facts and data in order to make a sound judgment.
- Assimilate large volumes of information from many different sources.
- Use formal decision tools when appropriate to evaluate alternatives and identify risks and issues.

Project managers/associates who do not demonstrate this competency will:

- Accept traditional assumptions regarding resource requirements and time estimates.
- Rely on team members to provide information needed.
- Push for a new milestone without determining the reason previous milestone was missed.
- Summarize details of discussions and arguments without drawing conclusions.
- Limit inquiries to standard sources of information.
- Use procedures and tools that are readily available.
- Define role narrowly as facilitating and documenting team members' discussions.

Selected consequences for projects/business of not demonstrating this competency are:

- Commitments may be made to unrealistic or untested dates.
- High-risk approaches may be adopted without explicit acknowledgment.
- Projects may take longer to complete than necessary.
- New findings and results may be incorporated slowly only into current Lilly practices.
- Major problems may arise unexpectedly.
- Same issues may be revisited.
- Project plan may remain unchanged despite major shifts in resources, people, and priorities.

4. **Manages Risks:** Anticipates and allows for changes in priorities, schedules, resources, and changes due to scientific/technical issues.

   Project managers/associates who demonstrate this competency will:

   - Double-check validity of key data and assumptions before making controversial or potentially risky decisions.
   - Create a contingency plan when pursuing options that have clear risks associated with them.
   - Maintain ongoing direct contact with "risky" or critical path activities to understand progress.
   - Push team members to identify all the assumptions implicit in their estimates and commitments.
   - Stay in regular contact with those whose decisions impact the project.

- Let management and others know early on the risks associated with a particular plan of action.
- Argue for level of resources and time estimates that allow for predictable "unexpected" events.
- Pinpoint major sources of scientific risks.

Project managers/associates who do not demonstrate this competency will:

- Remain optimistic regardless of progress.
- Agree to project timelines despite serious reservations.
- Value innovation and new ideas despite attendant risks.
- Accept less experienced team members in key areas.
- Give individuals freedom to explore different options.
- Accept estimates and assessments with minimal discussion.

Selected consequences for projects/business of not demonstrating this competency are:

- Projects may take longer to complete than necessary.
- Project may have difficulty responding to shifts in organizational priorities.
- Major delays could occur if proposed innovative approach proves inappropriate.
- Known problem areas may remain sources of difficulties.
- Project plans may be subject to dramatic revisions.

5. Communicates Clearly: Listens well and provides information that is easily understood and useful to others.

Project managers/associates who demonstrate this competency will:

- Present technical and other complex issues in a concise, clear, and compelling manner.
- Target or position communication to address needs or level of understanding of recipient(s) (e.g., medical, senior management).
- Filter data to provide the most relevant information (e.g., does not go over all details but knows when and how to provide an overall view).
- Keep others informed in a timely manner about decision or issues that may impact them.
- Facilitate and encourage open communication among team members.
- Set up mechanisms for regular communications with team members in remote locations.
- Accurately capture key points of complex or extended discussions.
- Spend the time necessary to prepare presentations for management.
- Effectively communicate and represent technical arguments outside own area of expertise.

Project managers/associates who do not demonstrate this competency will:

- Provide all the available details.
- See multiple reminders or messages as inefficient.
- Expect team members to understand technical terms of each other's specialties.
- Reuse communication and briefing materials with different audiences.

- Limit communications to periodic updates.
- Invite to meetings only those who (are presumed to) need to be there or who have something to contribute.
- Rely on technical experts to provide briefings in specialized, technical areas.

Selected consequences for projects/business of not demonstrating this competency are:

- Individuals outside of the immediate team may have little understanding of the project.
- Other projects may be disrupted by "fire drills" or last-minute changes in plan.
- Key decisions and discussions may be inadequately documented.
- Management briefings may be experienced as ordeals by team and management.
- Resources/effort may be wasted or misapplied.

6. Pays Attention to Details: Systematically documents, tracks, and organizes project details.

   Project managers/associates who demonstrate this competency will:

- Remind individuals of due dates and other requirements.
- Ensure that all relevant parties are informed of meetings and decisions.
- Prepare timely, accurate, and complete minutes of meetings.
- Continually update or adjust project documents to reflect decisions and changes.
- Check the validity of key assumptions in building the plan.
- Follow up to ensure that commitments are understood.

   Project managers/associates who do not demonstrate this competency will:

- Assume that others are tracking the details.
- See formal reviews as intrusions and waste of time.
- Choose procedures that are least demanding in terms of tracking details.
- Only sporadically review and update or adjust project documents to reflect decisions and other changes.
- Limit project documentation to those formally required.
- Rely on meeting notes as adequate documentation of meetings.

   Selected consequences for projects/business of not demonstrating this competency are:

- Coordination with other parts of the organization may be lacking.
- Documentation may be incomplete or difficult to use to review project issues.
- Disagreements may arise as to what was committed to.
- Project may be excessively dependent on the physical presence of manager or associate.

7. Structures the Process: Constructs, adapts, or follows a logical process to ensure achievement of objectives and goals.

   Project managers/associates who demonstrate this competency will:

- Choose milestones that the team can use for assessing progress.
- Structure meetings to ensure agenda items are covered.
- Identify sequence of steps needed to execute project management process.

- Maintain up-to-date documentation that maps expectations for individual team members.
- Use available planning tools to standardize procedures and structure activities.
- Create simple tools to help team members track, organize, and communicate information.
- Build a process that efficiently uses team members' time, while allowing them to participate in project decision; all team members should not attend all meetings.
- Review implications of discussion or decisions for the project plan as mechanism for summarizing and clarifying discussions.
- Keep discussions moving by noting disagreements rather than trying to resolve them there and then.
- Create and use a process to ensure priorities are established and project strategy is defined.

Project managers/associates who do not demonstrate this competency will:

- Trust that experienced team members know what they are doing.
- Treat complex sequences of activities as a whole.
- Share responsibility for running meetings, formulating agendas, and so on.
- Create plans and documents that are as complete and detailed as possible.
- Provide written documentation only when asked for.
- Allow team members to have their say.

Selected consequences for projects/business of not demonstrating this competency are:

- Projects may receive significantly different levels of attention.
- Project may lack a single direction or focus.
- Planning documents may be incomplete or out of date.
- Presentations and briefings may require large amounts of additional work.
- Meetings may be seen as unproductive.
- Key issues may be left unresolved.
- Other parts of the organization may be unclear about what is expected and when.

8. Focuses on Results: Continually focuses own and others' attention on realistic project milestones and deliverables.

Project managers/associates who demonstrate this competency will:

- Stress need to keep project-related activities moving forward.
- Continually focus on ultimate deliverables (e.g., product to market, affirm/disconfirm merits of compound, value of product/program to Lilly) (manager).
- Choose actions in terms of what needs to be accomplished rather than seeking optimal solutions or answers.
- Remind project team members of key project milestones and schedules.
- Keep key milestones visible to the team.
- Use fundamental objective of project as means of evaluating option driving decisions in a timely fashion.
- Push team members to make explicit and public commitments to deliverables.
- Terminate projects or low-value activities in timely fashion.

Project managers/associates who do not demonstrate this competency will:

- Assume that team members have a clear understanding of project deliverables and milestones.
- Approach tasks and issues only when they become absolutely critical.
- Downplay or overlook negative results or outcomes.
- Keep pushing to meet original objectives in spite of new data/major changes.
- Pursue activities unrelated to original project requirements.
- Trust that definite plans will be agreed to once team members are involved in the project.
- Allow unqualified individuals to remain on tasks.
- Make attendance at project planning meetings discretionary.

Selected consequences for projects/business of not demonstrating this competency are:

- Milestones may be missed without adequate explanation.
- Functional areas may be surprised at demand for key resources.
- Commitments may be made to unreasonable or unrealistic goals or schedules.
- Projects may take longer to complete than necessary.
- Objectives and priorities may differ significantly from one team member to another.

9. Builds a Team: Creates an environment of cooperation and mutual accountability within and across functions to achieve common objectives.

Project managers/associates who demonstrate this competency will:

- Openly acknowledge different viewpoints and disagreements.
- Actively encourage all team members to participate regardless of their functional background or level in the organization.
- Devote time and resources explicitly to building a team identity and a set of shared objectives.
- Maintain objectivity; avoid personalizing issues and disagreements.
- Establish one-on-one relationship with team members.
- Encourage team members to contribute input in areas outside functional areas.
- Involve team members in the planning process from beginning to end.
- Recognize and tap into the experience and expertise that each team member possesses.
- Solicit input and involvement from different functions prior to their major involvement.
- Once a decision is made, insist that team accept it until additional data become available.
- Push for explicit commitment from team members when resolving controversial issues.

Project managers/associates who do not demonstrate this competency will:

- State what can and cannot be done.
- Assume that mature professionals need little support or team recognition.
- Limit contacts with team members to formal meetings and discussions.
- Treat issues that impact a team member's performance as the responsibility of functional line management.

- Help others only when explicitly asked to do so.
- Be openly critical about other team members' contributions or attitudes.
- Revisit decisions when team members resurface issues.

Selected consequences for projects/business of not demonstrating this competency are:

- Team members may be unclear as to their responsibilities.
- Key individuals may move onto other projects.
- Obstacles and setbacks may undermine overall effort.
- Conflicts over priorities within project team may get escalated to senior management.
- Responsibility for project may get diffused.
- Team members may be reluctant to provide each other with support or accommodate special requests.

10. Manages Complexity: Organizes, plans, and monitors multiple activities, people, and resources.

Project managers/associates who demonstrate this competency will:

- Remain calm when under personal attack or extreme pressure.
- Monitor progress on frequent and consistent basis.
- Focus personal efforts on most critical tasks: apply 80–20 rule.
- Carefully document commitments and responsibilities.
- Define tasks and activities to all for monitoring and a sense of progress.
- Break activities and assignments into components that appear doable.
- Balance and optimize workloads among different groups and individuals.
- Quickly pull together special teams or use outside experts in order to address emergencies or unusual circumstances.
- Debrief to capture "best practices" and "lessons learned."

Project managers/associates who do not demonstrate this competency will:

- Limit the number of reviews to maximize time available to team members.
- Stay on top of all the details.
- Depend on team members to keep track of their own progress.
- Let others know how they feel about an issue or individual.
- Rely on the team to address issues.
- Assume individuals recognize and learn from their own mistakes.

Selected consequences for projects/business of not demonstrating this competency are:

- Projects may receive significantly different levels of attention.
- Projects may take on a life of their own with no clear direction or attainable outcome.
- Responsibility for decisions may be diffused among team members.
- Exact status of projects may be difficult to determine.
- Major issues can become unmanageable.
- Activities of different parts of the business may be uncoordinated.
- Conflicts may continually surface between project leadership and other parts of Lilly.

11. Makes Tough Decisions: Demonstrates assurance in own abilities, judgments, and capabilities; assumes accountability for actions.

    Project managers/associates who demonstrate this competency will:

    - Challenge the way things are done and make decisions about how things will get done.
    - Force others to deal with the unpleasant realities of a situation.
    - Push for reassessment of controversial decisions by management when new information/data become available.
    - Bring issues with significant impact to the attention of others.
    - Consciously use past experience and historical data to persuade others.
    - Confront individuals who are not meeting their commitments.
    - Push line management to replace individuals who fail to meet expectations.
    - Challenge continued investment in a project if data suggest it will not succeed.
    - Pursue or adopt innovative procedures that offer significant potential benefits even where limited prior experience is available.

    Project managers/associates who do not demonstrate this competency will:

    - Defer to the ideas of more experienced team members.
    - Give others the benefit of the doubt around missed commitments.
    - Hold off making decisions until the last possible moment.
    - Pursue multiple options rather than halt work on alternative approaches.
    - Wait for explicit support from others before raising difficult issues.
    - Accept senior managers' decisions as "nonnegotiable."
    - Rely on the team to make controversial decisions.
    - Provide problematic performers with additional resources and time.

    Selected consequences for projects/business of not demonstrating this competency are:

    - Projects may take longer to complete than necessary.
    - Failing projects may be allowed to linger.
    - Decisions may be delegated upward.
    - Morale of team may be undermined by nonperformance of certain team members.
    - "Bad news" may not be communicated until the last minute.
    - Key individuals may "burn out" in effort to play catch-up.

12. Builds Strategic Support: Gets the support and level of effort needed from senior management and others to keep projects on track.

    Project managers/associates who demonstrate this competency will:

    - Assume responsibility for championing the projects while demonstrating a balance between passion and objectivity.
    - Tailor arguments and presentations to address key concerns of influential decision makers.
    - Familiarize self with operational and business concerns of major functions within Lilly.
    - Use network of contacts to determine best way to surface an issue or make a proposal.
    - Push for active involvement of individuals with the experience and influence needed to make things happen.

- Pinpoint the distribution of influence in conflict situations.
- Presell controversial ideas or information.
- Select presenter to ensure appropriate message is sent.
- Ask senior management to help position issues with other senior managers.

Project managers/associates who do not demonstrate this competency will:

- Meet senior management and project sponsors only in formal settings.
- Propose major shifts in direction in group meetings.
- Make contact with key decision makers when faced with obstacles or problems.
- Limit number of face-to-face contacts with "global" partners.
- Treat individuals as equally important.
- Avoid the appearance of "politicking."
- Depend on other team members to communicate to senior managers in unfamiliar parts of Lilly.

Selected consequences for projects/business of not demonstrating this competency are:

- Viable projects may be killed without clear articulation of benefits.
- "Cultural differences" may limit success of global projects.
- Decisions may be made without the input of key individuals.
- Resistance to changes in project scope or direction may become entrenched before merits of proposals are understood.
- Key individuals/organizations may never buy in to a project's direction or scope.
- Minor conflicts may escalate and drag on.

## 8.13 HARRIS CORPORATION

All too often, people attend seminars and courses leading to certification as a PMP and are overwhelmed with the knowledge presented in the course. They wonder how any why any corporation would perform all of the activities covered in the information presented in the course and why they must learn all of this information.

While it is true that many companies do not need or perform all of the activities that are covered, aerospace and defense contractors are required to perform all of these activities. Aerospace and defense contractors survive on how well they perform the project management processes. When companies like Harris Corporation become exceptionally good in project management and undergo continuous improvement in project management, they outperform their competitors. Harris Corporation has a history of success in project management. The remaining material in this section was provided by Alex Sadowski at Harris Corporation.[2] Perhaps after reading the remainder of this section, the reader will have a

---

2. Alex Sadowski, PMP, is a program manager with the Government Communications Systems Division of Harris Corporation. He has over 30 years of experience in a large, diverse customer base including various civilian government and military organizations. His project management activities are concentrated in the Aerospace and Defense Contractor business arena. He is also involved with initiatives for the Harris Division Process Group and the Division Training Steering Committee.

better understanding and appreciation of why this material is being taught and the complexities of performing in an industry whose life blood and survival rest upon maintaining superior project management capability.

1. The basic philosophy for Project Management applies to all projects no matter what size and what industry is involved. However, each industry has its own unique environment and culture and the application of Project Management philosophy can take different forms based upon this uniqueness. The Aerospace and Defense Contractor industry's unique environment can be characterized as follows:

   ● The environment is dynamic
   ● The schedule is most often aggressive
   ● Changing mission scenarios result in requirements changes
   ● Effective change management is absolutely necessary
   ● This is the first time a project of this type is being done
   ● In most cases the technology envelope is stretched
   ● To complete the project, the development of new technologies is often required.
   ● New methodologies have to be developed
   ● Risk management is of paramount importance
   ● Proprietary and/or classified information is involved
   ● This hampers open communication
   ● Security considerations affect all aspects of the project
   ● Being overly focused on new and challenging technologies can cause problems with managing the entire project
   ● The demands of applying new and developing technologies can be very much overwhelming
   ● The trick is to meet the technological challenge without losing control of cost and schedule

2. As an Aerospace and Defense Contractor, Harris Corporation, has over the years developed detailed processes and procedures that cover all phases of project management from the identification of an opportunity all the way through to final completion, sell-off, and closure.

3. These processes and procedures are documented in great detail in the command media, and are regularly promulgated via seminars, courses, and general meetings as appropriate.

4. This process involves many gates (e.g., Pursue/No-Pursue Reviews, Bid/No-Bid Meetings, Proposal Red Teams, Pricing Reviews, Job Start-up Reviews, Regular Program Reviews, Systems Requirements Reviews, Preliminary Design Reviews, Critical Design Reviews, Peer Reviews, Production Readiness Reviews, Test Readiness Reviews, Final Customer Reviews, Contract Closure, etc.)

5. The processes and methodologies are quite extensive, but there are a few key aspects of this methodology that make everything fall into place and facilitate project success.

6. Through much experience, sometimes quite painful, we have found that detailed up-front planning is the best approach to achieving project success.

7. All too often there is the danger that the overexuberance of the project team and the impatience of the customer results in the short circuiting of upfront planning. *This is a recipe for disaster!*

8. Proper planning has to be based upon understanding what the customer has contracted for, the price they have agreed to pay, and the date they expect to see completion.

9. A comprehensive Project Plan has to be developed before the full project team is engaged.

10. Effective planning depends on the following (see Figure 8–4):

- Work Breakdown Structure (WBS)
- Integrated Project Schedule (IPS)
- Earned Value Management System (EVMS)
- Change Management Planning
- Risk Management Planning

11. Both the Statement of Work (SOW) and the Systems Requirements Document (SRD) are used to develop the Work Breakdown Structure (WBS).

- The SOW identifies what is to be done, any constraints, any special customer gates, the deliverables and the timeframe for the project.
- The SRD is the official requirements definition upon which the technical aspects of the project are based. This is used to make certain that the required functionality is achieved.

12. The WBS documents the details of what needs to be accomplished.

- It breaks down the project into its essential parts.
- It provides the basis for assigning tasks and responsibilities.
- It is a hierarchical structure that shows how each of the individual tasks contribute to the major task.
- These individual tasks are essential in defining the schedule and in setting up the EVMS for the project.
- A WBS dictionary is developed that defines each task, and who is responsible for that task.

13. Change Management

- Change is inevitable in any project.
- Some changes have no overall impact on the project.
- Some changes can have a dramatic effect on either the cost, schedule, or technical integrity of the effort. There are times when all three of these will be affected.
- A plan has to developed that allows

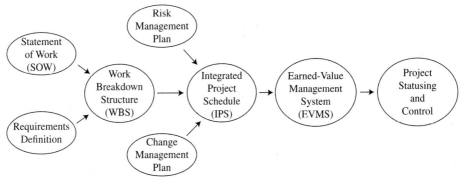

**Figure 8–4.** Effective planning steps.

- for the identification of change
- the impact of the change
- how such changes should be addressed (accepted or rejected)
- how changes are negotiated with the customer
- how changes to the contract are made and by whom

14. Risk Management

- All projects have risks
- A risk is anything that can affect the successful outcome of a program
  - A risk can impact
  - Cost
  - Schedule
  - Technical integrity (e.g., Functionality, Reliability, etc.)
- If a risk is not resolved, it then becomes a problem.
- At the very initial planning stages of a project the risks have to be identified.
- A plan has to be developed that addresses
  - Identification of risks
  - Definition of their severity
  - Definition of the probability of occurrence
  - Definition of the impact to project success (i.e., cost, schedule, etc.)
  - Risk ranking
  - Risk mitigation
  - Risk retirement

15. The Integrated Project Schedule

- Details the schedule of activities of the project from beginning to end
- Shows the start and stop times of the discrete individual activities
- Shows the interrelationships of all the individual activities
- Identifies the critical path
- Identifies all critical milestones

16. Earned Value Management System

- Facilitates the tracking of the progress of individual activities with respect to schedule and cost adherence
- Objectively shows which tasks are on schedule and whether they are within budget
- Provides a means of continually assessing whether or not a project is on schedule and within budget

17. Project Statusing & Control

- For any project to succeed, continual statusing and control is essential.
- Generally, reports from EVMS should be generated and reviewed every month. For problem projects or those of high risk, weekly statusing via EVMS would be most beneficial.
- EVMS provides a convenient, accurate and ongoing methodology for determining the progress of a project.
- The better the effort on the WBS and the IPS, the more accurate the information obtained from EVMS.

- Realistic and meaningful milestones, sufficiently covering all tasks, are key to the successful use of EVMS.
- Milestones must be tracked regularly (i.e., usually monthly, but as often as weekly if necessary).
- Once milestone accomplishments are accounted for then schedule and cost statusing can be accomplished.
- Generation of the Schedule Performance Index (SPI) provides an objective measure of whether or not the project is on schedule.
- Generation of the Cost Performance Index (CPI) provides an objective measure of whether or not the project is within budget.
- A new Estimate to Complete (ETC) should be generated regularly (i.e., at monthly statusing) and compared with the project's Budgeted-cost at Completion (BAC).
- Discrepancies between the ETC and BAC must be analyzed and reconciled. A result should be implementing a plan to get the program back on track.

A well-implemented EVMS is a most powerful management tool for the project manager.

- By considering the value of the SPI, the Project Manager knows if the project will meet schedule.
- If the value is less than 1, then the PM knows that the project is behind schedule, and must determine what tasks are falling behind and why, so that appropriate corrective action can be taken.
- If the value is greater than 1, then it indicates that the project is ahead of schedule. Too much joy should not be taken at this time since this could be only a temporary situation. If the value is much greater than 1, then that could indicate a problem in the original estimation and planning. This requires detailed analysis.
- By considering the value of the CPI, the Project Manager knows if the project will stay within budget.
- If the value is less than 1, then the PM knows that the project is over-running the budget, and must determine which tasks are exceeding cost and why, so that appropriate corrective action can be taken.
- If the value is greater than 1, then it means that the project is under budget. Caution must be taken here since this can be only a temporary situation. If the value is much greater than 1, then there is likely a problem with the original estimation and planning, and a detailed analysis must be done to determine why the discrepancy.
- When the project is planned, a budget is established. The budgeted cost of the project is what is expected to be expended at the conclusion of the project and is the BAC. As the project proceeds, actual costs are incurred. By reviewing the actual costs, and considering what still needs to be done, the PM can then arrive at an Estimate to Complete (ETC).
- By comparing the ETC and BAC, the Variance at Completion (VAC) can be calculated. The VAC indicates whether the project will over-run or under-run.
- If the VAC indicates an overrun, then an investigation and analysis must be accomplished to identify the problem. Then a course of action can be taken to mitigate the problem.

18. In the demanding and dynamic environment of Aerospace and Government projects, the use of a well defined and fully implemented Earned Value Management System can provide the necessary methodology to keep the project on schedule and within budget.

## 8.14 ALCATEL-LUCENT: RECOGNIZING THE VALUE OF A PMP

All too often, companies fail to capitalize on the intellectual property retained by PMPs and likewise cannot recognize the contribution they can make to the company. Some companies treat the PMP credentials as just something to put at the end of one's name on a business card or something to put into a proposal as part of competitive bidding efforts. Some companies even provide funding for training programs to obtain the credentials for fear that the workers will seek employment in other firms that provide employees with more opportunities for project management education.

But when companies like Alcatel-Lucent recognize the value that a PMP can make to the company, there can be a huge return-on-investment (ROI) on the PMPs. The ROI can be seen in the form of mentorship for project managers who aspire to become PMPs or wish to recertify their credentials, an advisory council or hotline for projects that are in trouble, or simply by establishing a global project management information network so that all employees worldwide understand current company developments and best practices with regard to project management.

Good things happen to companies like Alcatel-Lucent that recognizes the value of a PMP. According to Rich Maltzman, PMP, senior manager, Alcatel-Lucent Global PMO:

> Alcatel-Lucent has sought to increase PM maturity by establishing and deploying an integrated PM Development Framework including a dedicated career path where job descriptions are interlocked with a resource and skills management system connected directly to development options such as dedicated PM courseware. The company has also provided its PMs with a sophisticated Delivery Framework to help them implement and govern projects, measure and manage risk and to manage project financials. These frameworks provide the fundamental platform for achieving PM maturity. In particular, one focus group of PMPs has taken on a dynamic role to support soon-to-be PMPs and existing PMPs to facilitate the credentialing of PMs and to help existing PMPs not only to retain their PMP but to do so in a highly productive manner for their peers.
>
> The team, which gave itself the name PM-CERT for PM-Continuing Education and Recertification Team, meets weekly and focuses on several efforts to meet the goals spelled out in its title. Although an ad-hoc team, it works in concert with the Global PMO and Alcatel-Lucent University (a Registered Education Provider of the PMI), both of which are represented on the team. Here are some of PM-CERT's main activities.

**PMP Study Groups**

> Since the inception of this initiative, about 30 PMP Study groups have been established. These are time-zone based groups, each with about 10 members, who study collaboratively for the PMP exam under the tutelage of an instructor. This instructor is typically a recently credentialed PMP. Meeting at a frequency established by the team, and working in the language in which the majority of members expect to take the test, the Study Groups use the same materials and drill together on subject matter per a model syllabus established by the PM-CERT team, but adapted for the needs of the particular group. A special PMP Study Hall website has been established specifically for these teams to ask questions, share findings and, of course, post the fact that they have passed the exam! Students receive 30 hours of contact time which contributes the majority of what is needed for the exam application, and the instructor is able to claim valuable PDUs by virtue of their service as

an instructor. Instructors and students have both reported that this has been an enriching experience, stating, for example, that the Study Groups, "allowed me to meet and learn experiences from other PMs with a background in different areas. This helped me become a stronger PM by using the knowledge gained from these various backgrounds."

## Newsletter

For several years, the PM-CERT team has published a newsletter periodically, but at least quarterly, with featured stories about project management, news of upcoming PDU opportunities, profiles of projects, and announcements of PM-related events around the world. It had build up about 500 subscribers. Now the PM-CERT team uses a corporate blog based on Alcatel-Lucent's *Engage* corporate social media platform to distribute news and information.

## International PM Day Symposium

Alcatel-Lucent has decided to celebrate International PM Day by running an International PM Day Symposium. This is a web-based conference with connectivity by telephone and web-based video. It has been held since 2007, with attendance around 1,000 people at each. A theme is given to the Symposium and presenters submit proposals from which a PM-CERT committee selects for use at the session. Past themes have included "Project Management that Matters," and "Project Management—a World of Difference." Presenters from about 6–8 different countries share lessons learned or tackle a project management knowledge area such as Risk Management. In addition, guest speakers, such as noted authors Dr. Harold Kerzner and Jean Binder (winner of PMI's David Cleland Award), Dr. Charlie Pellerin from NASA, or officials of PMI are featured. The Symposium has been registered with Alcatel-Lucent University, and students collect 6 PDUs for their verified attendance. These sessions are archived and are continuously available for replay through Alcatel-Lucent University.

## Making PDUs Available—Learning Center

Again working with Alcatel-Lucent University, the PM-CERT has helped established a PM Learning Center—a resource-rich website with tips on how to apply for the PMP exam, how to keep your credential once earned, suggested readings and courses, and links to the appropriate PM Community *Engage* groups and to the established curricula of PM courseware.

## Roundtables

PM-CERT sponsors bimonthly "Discussion Roundtables," 1-hour lunchtime telephone conversations focused on a particular topic and led by a subject-matter expert from inside or outside of Alcatel-Lucent. The sessions are casual and provide the student with 1 PDU.

## Inspiration from PMP-CERT

The PM-CERT team has also inspired other PM activities. Based partially on the success of PM-CERT, Alcatel-Lucent's PM Community *Engage* Group is very active, covering general-interest PM topics, and has risen in the ranks to become one of the most popular corporate *Engage* groups. Based on this success, the company is now starting to use this as a model for other disciplines and job profiles in the company as a way to promote community learning and knowledge sharing.

## 8.15 INTEGRATED PROJECT MANAGEMENT AT TECH MAHINDRA LIMITED

The term "integration" has several meanings whether it is used with regard to IT, engineering, sociology, or economics. Integration is a process of combining various elements and also includes management and control principles as well to assemble all of the components. Tech Mahindra Limited has found an interesting use of integration for assembling all of the necessary elements to train and educate project managers. The remaining material in this section has been provided by Hirdesh Singhal, Tech Mahindra Limited Learning World, Tech Mahindra Limited.

Tech Mahindra Limited manages IT service projects for customers around the globe. To manage the rapid growth and fill the demand–supply gap for skilled human resources in the organization, it was critical that Tech Mahindra Limited rapidly grow project managers internally who could manage projects effectively and profitably meeting the expectations of various stakeholders.

A Project Management Competency Development Framework was available in Tech Mahindra Limited that focused on making each of role holders in the delivery effective and preparing them to take up the next role. This was aligned to the Project Manager Competency Development Framework from PMI, USA.

With this background, the **Integrated Project Management (IPM)** practice was developed for senior Project Leads, aligned with Project Manager Competency Development Framework so that they are quickly prepared to take up the role of PM.

**The practice was developed with following goals:**

- Create project managers faster than competition
- Save cost by reducing external hiring of PMs
- Leverage the knowledge and skills of existing project leads to create PMs
- Reduce the cycle time to get a PM on board
- Make new PMs effective from day one in their job
- Prepare existing associates to take higher level responsibilities
- Motivate associates to stay with the company for longer duration

**Some of the differentiators of Integrated Project Management practice include:**

- **Holistic learning** to aspiring PMs which not only includes project management processes and tools but also, management and behavior, culture and languages and leadership also
- **Certification from the User Community**—Delivery Units leaders, Delivery Integrators/Program Managers are involved through the cycle and assess and certify participants on their readiness to take up PM role

- **Action Learning Projects** during the course provides deep understanding of delivery issues
- **Daily Learning Log Book leading** to Personal Development Plan after the completion of the program

The design process is shown in Figure 8–5

**Selection of Participants:**

- Having more than 8 years of project experience
- Must have managed a team of at least 10 people
- Indication of abilities to take higher level responsibilities—a feedback from performance management system

We had ascertained that Project Management learning experience needs to be blended with behavioral, management and leadership learning experience to produce world class project managers. Participants could clearly understand the processes they need to follow, tools they need to use and various management techniques that can help them manage the project better. To improve the behavior, they did role plays and simulations to practice and demonstrate the change in behavior during the duration of the program. The practice leveraged various training programs and content that existed within the corporate university.

Tech Mahindra Limited integrated the data in Tables 8–3 and 8–4.

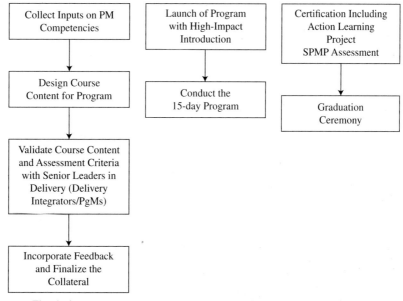

**Figure 8–5.**  The design process.

**TABLE 8–3.   PROJECT MANAGEMENT TRAINING PROGRAMS**

| S No | Name of the Course | Duration | Delivery Mode |
|------|--------------------|----------|---------------|
| 1 | Tech Mahindra Limited Project Management Practices | 3 Days | ILT |
| 2 | PM Processes & Tools | 2 Days | ILT |
| 3 | Software Estimations using Function Points | 2 Days | ILT |
| 4 | Risk Management | 5.5 hours | V-learning |
| 5 | Basic Concepts & Techniques of Estimation | 6.5 hours | V-learning |
| 6 | Software Project Measurement & Metrics | 4 hours | V-learning |
| 7 | Managing Project Full Lifecycle Businesses | 1 Day | ILT |

**TABLE 8–4.   MANAGEMENT, BEHAVIORAL, AND LEADERSHIP PROGRAMS**

| S No | Name of the course | Duration | Delivery Mode |
|------|--------------------|----------|---------------|
| 1 | Business Challenge Simulation | 2 days | ILT |
| 2 | Executive Presence | 1 day | ILT |
| 3 | Conflict management Tools | 1 day | ILT |
| 4 | Client Facing Skills | 1 day | ILT |

One of the important facets of delivery excellence is to make the project managers productive from day one. They were not only learning during the day, but at the end of the day they were also filling their Learning Log Books, indicating what they could learn well and what they need to work on or practice more. This also helped them develop their long term development plan.

The structure of the company is based on one key premise of collaboration. There was strong collaboration between the different functions of the company. See Figure 8–6.

- **Various schools of the corporate university** came together to design and deliver the program covering Project Management, General Management, Behavioral and Leadership components.
- **Real Time Leadership Center** helped with insights on measures and attributes that a PM needs to focus on while managing projects.
- The **Quality function** provided insights on the PM processes and tools in the organization.
- The **Resource Fulfillment Team** deployed the participants by matching the demand and supply of PMs throughout the company.

Collaboration of Tech Mahindra Limited Learning World with Other Functions of the Organization:

**Key lessons:**

1. Success of the practice is largely dependent on the selection of right participants for the program. Involvement of users and business units is crucial to right outcome.
2. This program helped participants learn all the aspects required to perform the role of PM.

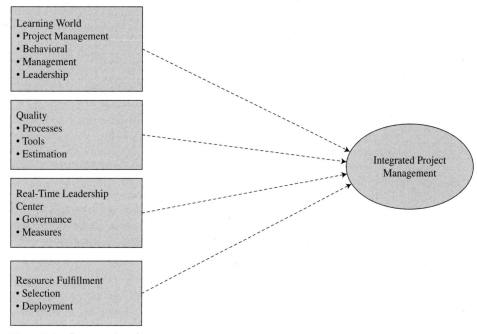

**Figure 8–6.** Integrated project management.

3. Create a holistic program that prepares for all-around stakeholder management. Thus, include management, behavioral and leadership aspects.
4. Validation of the design and content of the program by the users helped us make the program effective from day one.
5. Action Learning Projects focused on ground-level issues in the delivery of the projects. Issues that were not significant were removed. These topics were identified through a survey with the senior delivery leaders.
6. Strong support and sponsorship from senior executives for the practice is one of the key elements for the success.
7. Having prerequisite programs before participants come for the program increases the effectiveness of the program and provides more time for the participants to practice together in the classroom and learn for each others.

## 8.16 HEWLETT-PACKARD

The quality of the project management training and education a company's employees receive is, along with executive buy-in, one of the most important factors in achieving success and ultimately excellence in project management. The training could be for both the employees of the company as well as for its suppliers who must interface with the customer's project management methodology. Let's look at some case examples of effective training programs.

Hewlett-Packard is clearly committed to program and project management development. According to Jim Crotty, Americas Program Management Profession Leader:

### PM Development

HP Services has a comprehensive Program Management Development Program (PMDP) with courses that cover all aspects of program and project management training. A standard curriculum with over 100 courses is implemented throughout the world covering program/project leadership, management, communication, risk management, contracting, managing business performance, scheduling and cost control, and quality. The courses are based on PMI's Project Management Body of Knowledge (*PMBOK® Guide*). The curriculum also encompasses specialized courses on key HP internal topics, such as the Program Methodology, as well as essential business and financial management aspects of projects.

All courses taught in HP's PMDP curriculum are registered in PMI's Registered Education Provider (REP) Program to ensure a consistent basis and oversight. PMDP won an Excellence in Practice award in career development and organizational learning from the American Society for Training & Development (ASTD), the world's leading resource on workplace learning and performance issues.

Even the most experienced HP PMs continue development activities to strengthen their knowledge and skills. HP Services sponsors the Project Management University (PMU) Program. PMU consists of five day symposiums in each major geography (Americas, Asia/Pacific and Europe) and one day events held in HP offices. These events provide project managers with an opportunity to devote concentrated time to study and to exchange knowledge and ideas with other HP project managers from around the world and in their local geography. PMU has been recognized for excellence by both ASTD and PMI.

### PMP Certification

HP has a well-established program to encourage and support our project managers to achieve certification. HP Services has over 5000 individuals who have earned the PMP® (Project Management Professional) certification from PMI®.

### PMI Support

HP actively supports the Project Management Institute (PMI®), a nonprofit organization with more than 265,000 members. PMI® has set standards for Project Management excellence that are recognized by the industry and our customers worldwide.

HP employees participate on a number of PMI® boards and committees, including the Global Corporate Council. Global Accreditation Center, Research Program Membership Advisory Group, development of the Certified Associate in Project Management (CAPM), development of Certificates of Added Qualification (CAQ) in IT Systems, IT Networks, and Project Management Office, *PMBOK® Guide* review, and *PMBOK® Guide* Update Teams. Many HP employees hold leadership positions in PMI® Chapters and SIGs throughout the world.

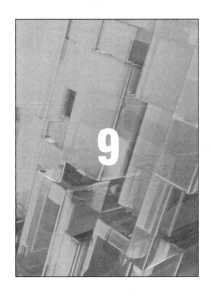

# 9 Informal Project Management

## 9.0 INTRODUCTION

Over the past 25 years, one of the most significant changes in project management has been the idea that informal project management does work. In the 1950s and 1960s, the aerospace, defense, and large construction industries were the primary users of project management tools and techniques. Because project management was a relatively new management process, customers of the contractors and subcontractors wanted evidence that the system worked. Documentation of the policies and procedures to be used became part of the written proposal. Formal project management, supported by hundreds of policies, procedures, and forms, became the norm. After all, why would a potential customer be willing to sign a $10 million contract for a project to be managed informally?

This chapter clarifies the difference between informal and formal project management, then discusses the four critical elements of informal project management.

## 9.1 INFORMAL VERSUS FORMAL PROJECT MANAGEMENT

Formal project management has always been expensive. In the early years, the time and resources spent on preparing written policies and procedures had a purpose: They placated the customer. As project management became established, formal documentation was created mostly for the customer. Contractors began managing more informally, while the customer was still paying for formal project management documentation. Table 9–1 shows the major differences between formal and informal project management. As you can see, the most relevant difference is the amount of paperwork.

**TABLE 9–1.  FORMAL VERSUS INFORMAL PROJECT MANAGEMENT**

| Factor | Formal Project Management | Informal Project Management |
|---|---|---|
| Project manager's level | High | Low to middle |
| Project manager's authority | Documented | Implied |
| Paperwork | Exorbitant | Minimal |

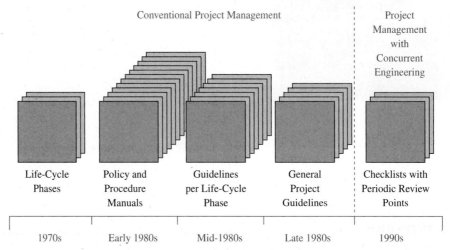

**Figure 9–1.**  Evolution of policies, procedures, and guidelines.  *Source:* Reprinted from H. Kerzner, *In Search of Excellence in Project Management*, Hoboken, NJ: Wiley, 1998, p. 196.

Paperwork is expensive. Even a routine handout for a team meeting can cost $500–$2,000 per page to prepare. Executives in excellent companies know that paperwork is expensive. They encourage project teams to communicate without excessive amounts of paper. However, some people are still operating under the mistaken belief that ISO 9000 certification requires massive paperwork.

Figure 9–1 shows the changes in paperwork requirements in project management. The early 1980s marked the heyday for lovers of paper documentation. At that time, the average policies and procedures manual probably cost between $3 million and $5 million to prepare initially and $1 million to $2 million to update yearly over the lifetime of the development project. Project managers were buried in forms to complete to the extent that they had very little time left for actually managing the projects. Customers began to complain about the high cost of subcontracting, and the paperwork boom started to fade.

Real cost savings did not materialize until the early 1990s with the growth of concurrent engineering. Concurrent engineering shortened product development times by taking activities that had been done in series and performing them in parallel instead. This change increased the level of risk in each project, which required that project management back away from some of its previous practices. Formal guidelines were replaced by less detailed and more generic checklists.

Policies and procedures represent formality. Checklists represent informality. But informality does not eliminate project paperwork altogether. It reduces paperwork requirements to

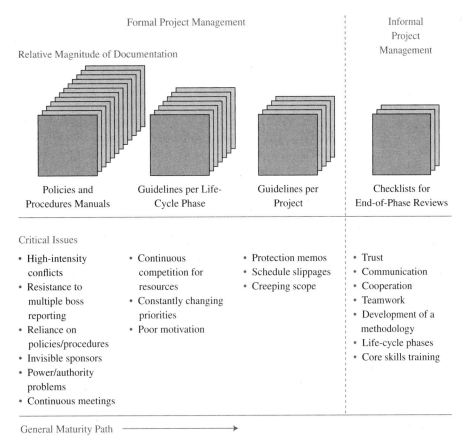

**Figure 9–2.** Evolution of paperwork and change of formality levels. *Source:* Reprinted from H. Kerzner, *In Search of Excellence in Project Management*, Hoboken, NJ. Wiley, 1998, p. 198.

minimally acceptable levels. To move from formality to informality demands a change in organizational culture (see Figure 9–2). The four basic elements of an informal culture are these:

- Trust
- Communication
- Cooperation
- Teamwork

Large companies quite often cannot manage projects on an informal basis although they want to. The larger the company, the greater the tendency for formal project management to take hold. A former vice president of IOC sales operations and customer service at Nortel Networks, believes that[1]:

The introduction of enterprise-wide project process and tools standards in Nortel Networks and the use of pipeline metrics (customer-defined, industry standard measures) provides a

---

1. H. Kerzner, *Project Management Best Practices: Achieving Global Excellence*, Hoboken, NJ: Wiley, 2006, p. 329.

framework for formal project management. This is necessary given the complexity of tele-com projects we undertake and the need for an integrated solution in a short time frame. The Nortel Networks project manager crosses many organizational boundaries to achieve the results demanded by customers in a dynamic environment.

Most companies manage either formally or informally. However, if your company is project-driven and has a very strong culture for project management, you may have to manage either formally or informally based upon the needs of your customers.

## 9.2 TRUST

Trusting everyone involved in executing a project is critical. You wake up in the morning, get dressed, and climb into your car to go to work. On a typical morning, you operate the foot pedal for your brakes maybe 50 times. You have never met the people who designed the brakes, manufactured the brakes, or installed the brakes. Yet you still give no thought to whether the brakes will work when you need them. No one broadsides you on the way to work. You do not run over anyone. Then you arrive at work and push the button for the elevator. You have never met the people who designed the elevator, manufactured it, installed it, or inspected it. But again you feel perfectly comfortable riding the elevator up to your floor. By the time you get to your office at 8 A.M., you have trusted your life to uncounted numbers of people whom you have never even met. Still, you sit down in your office and refuse to trust the person in the next office to make a $50 decision.

Trust is the key to the successful implementation of informal project management. Without it, project managers and project sponsors would need all that paperwork just to make sure that everyone working on their projects was doing the work just as he or she had been instructed. Trust is also key in building a successful relationship between the contractor/subcontractor and the client. Let's look at an example.

Perhaps the best application of informal project management that I have seen occurred several years ago in the Heavy Vehicle Systems Group of Bendix Corporation. Bendix hired a consultant to conduct a three-day training program. The program was custom designed, and during the design phase the consultant asked the vice president and general manager of the division whether he wanted to be trained in formal or informal project management. The vice president opted for informal project management. What was the reason for his decision? The culture of the division was already based on trust. Line managers were not hired solely based on technical expertise. Hiring and promotions were based on how well the new manager would communicate and cooperate with the other line managers and project managers in making decisions in the best interests of both the company and the project.

When the relationship between a customer and a contractor is based on trust, numerous benefits accrue to both parties. The benefits are apparent in companies such as Hewlett-Packard, Computer Associates, and various automobile subcontractors. Table 9–2 shows the benefits.

**TABLE 9–2.  BENEFITS OF TRUST IN CUSTOMER–CONTRACTOR WORKING RELATIONSHIPS**

| Without Trust | With Trust |
| --- | --- |
| Continuous competitive bidding | Long-term contracts, repeat business, and sole-source contracts |
| Massive documentation | Minimal documentation |
| Excessive customer–contractor team meetings | Minimal number of team meetings |
| Team meetings with documentation | Team meetings without documentation |
| Sponsorship at executive levels | Sponsorship at middle-management levels |

## 9.3 COMMUNICATION

In traditional, formal organizations, employees usually claim that communication is poor. Senior managers, however, usually think that communication in their company is just fine. Why the disparity? In most companies, executives are inundated with information communicated to them through frequent meetings and dozens of weekly status reports coming from every functional area of the business. The quality and frequency of information moving down the organizational chart are less consistent, especially in more formal companies. But whether it is a problem with the information flowing up to the executive level or down to the staff, the problem usually originates somewhere upstairs. Senior managers are the usual suspects when it comes to requiring reports and meetings. And many of those reports and meetings are unnecessary and redundant.

Most project managers prefer to communicate verbally and informally. The cost of formal communication can be high. Project communication includes dispensing information on decisions made, work authorizations, negotiations, and project reports. Project managers in excellent companies believe that they spend as much as 90 percent of their time on internal interpersonal communication with their teams. Figure 9–3 illustrates the communication channels used by a typical project manager. In project-driven organizations, project managers may spend most of their time communicating externally to customers and regulatory agencies.

Good project management methodologies promote not only informal project management but also effective communications laterally as well as vertically. The methodology itself functions as a channel of communication. A senior executive at a large financial institution commented on his organization's project management methodology, called Project Management Standards (PMS):

> The PMS guides the project manager through every step of the project. The PMS not only controls the reporting structure but also sets the guidelines for who should be involved in the project itself and the various levels of review. This creates an excellent communication flow between the right people. The communication of a project is one of the most important factors for success. A great plan can only go so far if it is not communicated well.

Most companies believe that a good project management methodology will lead to effective communications, which will allow the firm to manage more informally than formally. The question, of course, is how long it will take to achieve effective communications. With all employees housed under a single roof, the time frame can be short. For

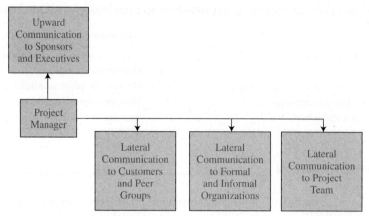

**Figure 9–3.** Internal and external communication channels for project management. *Source:* Reprinted from H. Kerzner, *In Search of Excellence in Project Management*, Hoboken, NJ: Wiley, 1998, p. 200.

global projects, geographical dispersion and cultural differences may mandate decades before effective communication will occur. Even then, there is no guarantee that global projects will ever be managed informally.

Suzanne Zale, Hewlett-Packard operations director, emphasized[2]:

> With any global project, communications becomes more complex. It will require much more planning up front. All constituents for buy-in need to be identified early on. In order to leverage existing subject matter, experts conversant with local culture, and suppliers, the need for virtual teams becomes more obvious. This increases the difficulty for effective communications.
>
> The mechanism for communication may also change drastically. Face-to-face conversations or meetings will become more difficult. We tend to rely heavily on electronic communications, such as video and telephone conferencing and electronic mail. The format for communications needs to be standardized and understood up front so that information can be sent out quickly. Communications will also take longer and require more effort because of cultural and time differences.

One of the implied assumptions for informal project management to exist is that the employees understand their organizational structure and their roles and responsibilities within both the organizational and project structure. Forms such as the linear responsibility chart and the responsibility assignment matrix are helpful. Communication tools are not used today with the same frequency as in the 1970s and 1980s.

For multinational projects, the organizational structure, roles, and responsibilities must be clearly delineated. Effective communications is of paramount importance and probably must be accomplished more formally than informally.

Suzanne Zale, Hewlett-Packard operations director stated:

> For any global project, the organizational structure must be clearly defined to minimize any potential misunderstandings. It is best to have a clearcut definition of the organizational

chart and roles and responsibilities. Any motivation incentives must also contemplate cultural differences. The drivers and values for different cultures can vary substantially.

The two major communication obstacles that must be overcome when a company truly wants to cultivate an informal culture are what I like to call hernia reports and forensic meetings. Hernia reports result from senior management's belief that that which has not been written has not been said. Although there is some truth to such a belief, the written word comes with a high price tag. We need to consider more than just the time consumed in the preparation of reports and formal memos. There is all the time that recipients spend reading them as well as all the support time taken up in processing, copying, distributing, and filing them.

Status reports written for management are too long if they need a staple or a paper clip. Project reports greater than 5 or 10 pages often are not even read. In companies excellent in project management, internal project reports answer these three questions as simply as possible:

- Where are we today?
- Where will we end up?
- Are there any problems that require management's involvement?

All of these questions can be answered on one sheet of paper.

The second obstacle is the forensic team meeting. A forensic team meeting is a meeting scheduled to last 30 minutes that actually lasts for more than 3 hours. Forensic meetings are created when senior managers meddle in routine work activities. Even project managers fall into this trap when they present information to management that management should not be dealing with. Such situations are an invitation to disaster.

## 9.4 COOPERATION

Cooperation is the willingness of individuals to work with others for the benefit of all. It includes the voluntary actions of a team working together toward a favorable result. In companies excellent in project management, cooperation is the norm and takes place without the formal intervention of authority. The team members know the right thing to do, and they do it.

In the average company (or the average group of any kind, for that matter), people learn to cooperate as they get to know each other. That takes time, something usually in short supply for project teams. But companies such as Ericsson Telecom AB, the General Motors Powertrain Group, and Hewlett-Packard create cultures that promote cooperation to the benefit of everyone.

## 9.5 TEAMWORK

Teamwork is the work performed by people acting together with a spirit of cooperation under the limits of coordination. Some people confuse teamwork with morale, but morale has more to do with attitudes toward work than it has to do with the work itself. Obviously, however, good morale is beneficial to teamwork.

In excellent companies, teamwork has these characteristics:

- Employees and managers share ideas with each other and establish high levels of innovation and creativity in work groups.
- Employees and managers trust each other and are loyal to each other and the company.
- Employees and managers are committed to the work they do and the promises they make.
- Employees and managers share information freely.
- Employees and managers are consistently open and honest with each other.

Making people feel that they are part of a team does not necessarily require a great deal of effort. Consider the situation at the Engineering and Construction Services Division of Dow Chemical Corporation several years ago. Dow Chemical had requested a trainer to develop a project management training course. The trainer interviewed several of the seminar participants before the training program to identify potential problem areas. The biggest problem appeared to be a lack of teamwork. This shortcoming was particularly evident in the drafting department. The drafting department personnel complained that too many changes were being made to the drawings. They simply could not understand the reasons behind all the changes.

The second problem identified, and perhaps the more critical one, was that project managers did not communicate with the drafting department once the drawings were complete. The drafting people had no idea of the status of the projects they were working on, and they did not feel as though they were part of the project team.

During the training program, one of the project managers, who was responsible for constructing a large chemical plant, was asked to explain why so many changes were being made to the drawings on his project. He said, "There are three reasons for the changes. First, the customers don't always know what they want up front. Second, once we have the preliminary drawings to work with, we build a plastic model of the plant. The model often shows us that equipment needs to be moved for maintenance or safety reasons. Third, sometimes we have to rush into construction well before we have final approval from the Environmental Protection Agency. When the agency finally gives its approval, that approval is often made contingent on making major structural changes to the work already complete." One veteran employee at Dow commented that in his 15 years with the company no one had ever before explained the reasons behind drafting changes.

The solution to the problem of insufficient communication was also easy to repair once it was out in the open. The project managers promised to take monthly snapshots of the progress on building projects and share them with the drafting department. The drafting personnel were delighted and felt more like a part of the project team.

## 9.6 COLOR-CODED STATUS REPORTING

The use of colors for status reporting, whether it be for printed reports or intranet-based visual presentations, has grown significantly. Color-coded reports encourage informal project management to take place. Colors can reduce risks by alerting management quickly that a potential problem exists. One company prepared complex status reports but color coded the right-hand margins of each page designed for specific audiences and levels of management. One executive commented that he now reads only those pages that are color-coded for him specifically rather than having to search through the entire report. In another company, senior management discovered that color-coded intranet status reporting allowed senior management to review more information in a timely manner just by focusing on those colors that indicated potential problems. Colors can be used to indicate:

- Status has not been addressed.
- Status is addressed, but no problems exist.
- Project is on course.
- A potential problem might exist in the future.
- A problem definitely exists and is critical.
- No action is to be taken on this problem.
- Activity has been completed.
- Activity is still active and completion date has passed.

## 9.7 CRISIS DASHBOARDS

Over the past several years, dashboards have become commonplace for presenting project status information to the project team, clients, and stakeholders. The purpose of a dashboard is to convert raw data into meaningful information that can easily be understood and used for informed decision making. The dashboard provides the viewer with "situational awareness" of what the information means now and what it might mean in the future if the existing trends continue. Dashboards function as communication tools that allow us to go to paperless project management, fewer meetings, and eliminate waste.

Projects in today's environment are significantly more complex than many of the projects managed in the past. With today's projects, governance is performed by a governance committee rather than just a single project sponsor. Each stakeholder or member of the governance committee may very well require different metrics and KPIs. If each stakeholder wishes to view twenty to thirty metrics, the costs of metric measurement and reporting can be significant and defeat the purpose of going to paperless project management.

The solution to effective communications with stakeholders and governance groups is to show them that they can most likely get all of the critical data they need for informed decision making with six to ten metrics or KPIs that can be displayed on one computer screen. This is not always the case and drill down to other screens may be necessary. But in general, one computer screen shot should be sufficient.

If an out of tolerance condition or crisis situation exists with any of the metrics or KPIs on the dashboard screen, then the situation should be readily apparent to the viewer. But what if the crisis occurs due to metrics that do not appear on the screen? In this case, the viewer will be immediately directed to a crisis dashboard which shows all of the metrics that are out of tolerance. The out-of-tolerance metrics will remain on the crisis dashboard until such time when the crisis or out-of-tolerance conditions are corrected. Each stakeholder will now see the regular screen shot and then be instructed to look at the crisis screen shot.

**Defining a Crisis**  A crisis can be defined as any event, whether or not expected, that can lead to an unstable or dangerous situation affecting the outcome of the project. Crises imply negative consequences that can harm the organization, its stakeholders, and the general public. The crisis can result in changes to the firm's business strategy, how it interfaces with the enterprise environmental factors, how the firm's social consciousness is exhibited, and the way it maintains customer satisfaction. A crisis does not necessarily mean that the project will fail nor does it mean that the project should be terminated. The crisis could simply be that the project's outcome will not occur as expected.

Some crises may appear gradually and can be preceded by early warning signs. These can be referred to as smoldering crises. The intent of metrics and dashboards is to identify trends that could indicate that a crisis may be approaching and provide the project manager with sufficient time to develop contingency plans. The earlier you know about the impending crisis, the more options you may have available as a remedy.

How do we determine whether the out of tolerance condition is just a problem or a crisis? The answer is in the potential damage that can occur. If any of the following can occur, then the situation would most likely be treated as a crisis:

- There is a significant threat to the outcome of the project
- There is a significant threat to the organization as a whole, its stakeholders and possibly the general public
- There is a significant threat to the firm's business model and strategy
- There is a significant threat to worker health and safety
- There is a significant threat to consumers, such as with product tampering
- There is a possibility for loss of life
- There is the possibility of work delays because systems are being redesigned
- There is the possibility of work delays due to necessary organizational changes
- There is a significant chance that the firm's image or reputation will be damaged
- There is a significant chance that the deterioration in customer satisfaction could result in present and future loss of significant revenue

It is important to understand the difference between risk management and crisis management. According to Wikipedia, the free encyclopedia,

> In contrast to risk management, which involves assessing potential threats and finding the best ways to avoid those threats, crisis management involves dealing with threats before, during, and after they have occurred. That is, crisis management is proactive, not merely reactive. It is a discipline within the broader context of management consisting of skills and techniques required to identify, assess, understand, and cope with a serious situation, especially from the moment it first occurs to the point that recovery procedures start.

Crises often require that immediate decisions be made. Effective decision making requires information. If one metric appear to be in a crisis mode and shows up on the crisis dashboard, the viewers may find it necessary to look at several other metrics, which may not be in a crisis mode and may not appear on the crisis dashboard but are possible causes of the crisis. Looking at metrics on dashboards is a lot easier than reading reports.

The differentiation between a problem and a crisis is like beauty; it is in the eyes of the beholder. What one stakeholder sees as a problem, another stakeholder may see it as a crisis. Table 9–3 shows how difficult it is to make the differentiation.

We can now draw the following conclusions about crisis dashboards:

- The definition of what is or is not a crisis is not always clear to the viewers
- Not all problems are crises
- Sometimes unfavorable trends are treated as a crisis and appear on the crisis dashboards
- The crisis dashboard may contain a mixture of crisis metrics and metrics that are treated as just problems

**TABLE 9–3.  DIFFERENTIATING BETWEEN A PROBLEM AND A CRISIS**

| Metric/KPI | Problem | Crisis |
|---|---|---|
| Time | The project will be late but still acceptable to the client. | The project will be late and client is considering cancellation. |
| Cost | Costs are being overrun but client can provide additional funding. | Costs are being overrun and no additional funding is available. Cancellation is highly probable. |
| Quality | The customer is unhappy with the quality but can live with it. | Quality of the deliverables is unacceptable, personal injury is possible, the client may cancel the contract and no further work may come from this client. |
| Resources | The project is either understaffed or the resources assigned have marginal skills to do the job. A schedule delay is probably. | The quality or lack of resources will cause a serious delay in the schedule and the quality of workmanship may be unacceptable such that the project may be cancelled. |
| Scope | There are numerous scope changes, which cause changes to the baselines. Delays and cost overruns are happening but are acceptable to the client for now. | The number of scope-changes has led the client to believe that the planning is not correct and more scope changes will occur. The benefits of the project no longer outweigh the cost and project termination is likely. |
| Action Items | The client is unhappy with the amount of time taken to close out action items, but the impact on the project is small. | The client is unhappy with the amount of time taken to close out action items and the impact on the project is significant. Governance decisions are being delayed because of the open action items and the impact on the project may be severe. |
| Risks | Significant risk levels exist, but the team may be able to mitigate some of the risks. | The potential damage that can occur because of the severity of the risks is unacceptable to the client. |
| Assumptions and Constraints | New assumptions and constraints have appeared and may adversely affect the project. | New assumptions and constraints have appeared such that significant project replanning will be necessary. The value of the project may no longer be there. |
| Enterprise Environmental Factors | The enterprise environmental factors have changes and may adversely affect the project. | The new enterprise environmental factors will greatly reduce the value and expected benefits of the project. |

- The metrics that appear on a traditional dashboard reporting system may have to be redrawn when placed on a crisis dashboard to make sure that the metrics are easily understood
- Crisis metrics generally imply that either this situation must be monitored closely or that some decisions must be made.

## 9.8 INFORMAL PROJECT MANAGEMENT AT WORK

Let's review two case studies that illustrate informal project management in action.

**Polk Lighting**

Polk Lighting (a pseudonym) is a $35 million company located in Jacksonville, Florida. The company manufactures lamps, flashlights, and a variety of other lighting instruments. Its business is entirely based in products and services, and the company does not take on contract projects from outside customers. The majority of the company's stock is publicly traded. The president of Polk Lighting has held his position since the company's startup in 1985.

In 1994, activities at Polk centered on the R&D group, which the president oversaw personally, refusing to hire an R&D director. The president believed in informal management for all aspects of the business, but he had a hidden agenda for wanting to use informal project management. Most companies use informal project management to keep costs down as far as possible, but the president of Polk favored informal project management so that he could maintain control of the R&D group. However, if the company were to grow, the president would need to add more management structure, establish tight project budgets, and possibly make project management more formal than it had been. Also, the president would probably be forced to hire an R&D director.

Pressure from the company's stockholders eventually forced the president to allow the company to grow. When growth made it necessary for the president to take on heavier administrative duties, he finally hired a vice president of R&D.

Within a few years, the company's sales doubled, but informal project management was still in place. Although budgets and schedules were established as the company grew, the actual management of the projects and the way teams worked together remained informal.

**Boeing Aerospace (1970s)**

Several decades ago, Boeing was the prime contractor for the U.S. Air Force's new short-range attack missile (SRAM) and awarded a subcontract for developing the missile's propulsion system to the Thiokol Corporation.

It is generally assumed that communication between large customers and contractors must be formal because of the potential for distrust when contracts are complex and involve billions of dollars. The use of on-site representatives, however, can change a potentially contentious relationship into one of trust and cooperation when informality is introduced into the relationship.

Two employees from Boeing were carefully chosen to be on-site representatives at the Thiokol Corporation to supervise the development of the SRAM's propulsion system. The working relationship between Thiokol's project management office and Boeing's on-site representatives quickly developed into shared trust. Team meetings were held without the exchange of excessive documentation. And each party agreed to cooperate with the other. The Thiokol project manager trusted Boeing's representatives well enough to give them raw data from test results even before Thiokol's engineers could formulate their own opinions on the data. Boeing's representatives in turn promised that they would not relay the raw data to Boeing until Thiokol's engineers were ready to share their results with their own executive sponsors.

The Thiokol–Boeing relationship on this project clearly indicates that informal project management can work between customers and contractors. Large construction contractors have had the same positive results in using informal project management and on-site representatives to rebuild trust and cooperation. Informality is not a replacement for formal project management activities. Rather, it simply means that some activities can be done more informally rather than formally. Formal and informal communications can exist simultaneously.

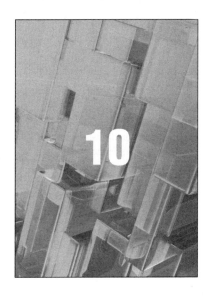

# 10 Behavioral Excellence

## 10.0 INTRODUCTION

Previously, we saw that companies excellent in project management strongly emphasize training for behavioral skills. In the past it was thought that project failures were due primarily to poor planning, inaccurate estimating, inefficient scheduling, and lack of cost control. Today, excellent companies realize that project failures have more to do with behavioral shortcomings—poor employee morale, negative human relations, low productivity, and lack of commitment.

This chapter discusses these human factors in the context of situational leadership and conflict resolution. It also provides information on staffing issues in project management. Finally, the chapter offers advice on how to achieve behavioral excellence.

## 10.1 SITUATIONAL LEADERSHIP

As project management has begun to emphasize behavioral management over technical management, situational leadership has also received more attention. The average size of projects has grown, and so has the size of project teams. Process integration and effective interpersonal relations have also taken on more importance as project teams have gotten larger. Project managers now need to be able to talk with many different functions and departments. There is a contemporary project management proverb that goes something like this: "When researcher talks to researcher, there is 100 percent understanding. When researcher talks to manufacturing, there is 50 percent understanding. When researcher talks to sales, there is zero percent understanding. But the project manager talks to all of them."

Randy Coleman, former senior vice president of the Federal Reserve Bank of Cleveland, emphasizes the importance of tolerance:

The single most important characteristic necessary in successful project management is tolerance: tolerance of external events and tolerance of people's personalities. Generally, there are two groups here at the Fed—lifers and drifters. You have to handle the two groups differently, but at the same time you have to treat them similarly. You have to bend somewhat for the independents (younger drifters) who have good creative ideas and whom you want to keep, particularly those who take risks. You have to acknowledge that you have some trade-offs to deal with.

A senior project manager in an international accounting firm states how his own leadership style has changed from a traditional to a situational leadership style since becoming a project manager:

I used to think that there was a certain approach that was best for leadership, but experience has taught me that leadership and personality go together. What works for one person won't work for others. So you must understand enough about the structure of projects and people and then adopt a leadership style that suits your personality so that it comes across as being natural and genuine. It's a blending of a person's experience and personality with his or her style of leadership.

Many companies start applying project management without understanding the fundamental behavioral differences between project managers and line managers. If we assume that the line manager is not also functioning as the project manager, here are the behavioral differences:

- Project managers have to deal with multiple reporting relationships. Line managers report up a single chain of command.
- Project managers have very little real authority. Line managers hold a great deal of authority by virtue of their titles.
- Project managers often provide no input into employee performance reviews. Line managers provide formal input into the performance reviews of their direct reports.
- Project managers are not always on the management compensation ladder. Line managers always are.
- The project manager's position may be temporary. The line manager's position is permanent.
- Project managers sometimes are a lower grade level than the project team members. Line managers usually are paid at a higher grade level than their subordinates.

Several years ago, when Ohio Bell was still a subsidiary of American Telephone and Telegraph, a trainer was hired to conduct a three-day course on project management. During the customization process, the trainer was asked to emphasize planning, scheduling, and controlling and not to bother with the behavioral aspects of project management. At that time, AT&T offered a course on how to become a line supervisor that all of the seminar participants had already taken. In the discussion that followed between the trainer and the course content designers, it became apparent that leadership, motivation, and conflict resolution were being taught from a superior-to-subordinate point of view in AT&T's course. When the course content designers realized from the discussion that project managers provide leadership, motivation, and conflict resolution to employees who do not report directly to them, the trainer was allowed to include project management–related behavioral topics in the seminar.

Organizations must recognize the importance of behavioral factors in working relationships. When they do, they come to understand that project managers should be hired for their overall project management competency, not for their technical knowledge alone. Brian Vannoni, formerly site training manager and principal process engineer at GE Plastics, described his organization's approach to selecting project managers:

> The selection process for getting people involved as project managers is based primarily on their behavioral skills and their skills and abilities as leaders with regard to the other aspects of project management. Some of the professional and full-time project managers have taken senior engineers under their wing, coached and mentored them, so that they learn and pick up the other aspects of project management. But the primary skills that we are looking for are, in fact, the leadership skills.

Project managers who have strong behavioral skills are more likely to involve their teams in decision making, and shared decision making is one of the hallmarks of successful project management. Today, project managers are more managers of people than they are managers of technology. According to Robert Hershock, former vice president at 3M:

> The trust, respect, and especially the communications are very, very important. But I think one thing that we have to keep in mind is that a team leader isn't managing technology; he or she is managing people. If you manage the people correctly, the people will manage the technology.

In addition, behaviorally oriented project managers are more likely to delegate responsibility to team members than technically strong project managers. In 1996, Frank Jackson, formerly a senior manager at MCI, said that:

> Team leaders need to have a focus and a commitment to an ultimate objective. You definitely have to have accountability for your team and the outcome of your team. You've got to be able to share the decision-making. You can't single out yourself as the exclusive holder of the right to make decisions. You have got to be able to share that. And lastly again, just to harp on it one more time, is communications. Clear and concise communication throughout the team and both up and down a chain of command is very, very important.

Some organizations prefer to have a project manager with strong behavioral skills acting as the project manager, with technical expertise residing with the project engineer. Other organizations have found the reverse to be effective. Rose Russett, formerly the program management process manager for General Motors Powertrain, stated:

> We usually appoint an individual with a technical background as the program manager and an individual with a business and/or systems background as the program administrator. This combination of skills seems to complement one another. The various line managers

are ultimately responsible for the technical portions of the program, while the key responsibility of the program manager is to provide the integration of all functional deliverables to achieve the objectives of the program. With that in mind, it helps for the program manager to understand the technical issues, but they add their value not by solving specific technical problems but by leading the team through a process that will result in the best solutions for the overall program, not just for the specific functional area. The program administrator, with input from all team members, develops the program plans, identifies the critical path, and regularly communicates this information to the team throughout the life of the program. This information is used to assist with problem solving, decision-making, and risk management.

## 10.2 CONFLICT RESOLUTION

Opponents of project management claim that the primary reason why some companies avoid changing over to a project management culture is that they fear the conflicts that inevitably accompany change. Conflicts are a way of life in companies with project management cultures. Conflict can occur on any level of the organization, and conflict is usually the result of conflicting objectives. The project manager is a conflict manager. In many organizations, the project managers continually fight fires and handle crises arising from interpersonal and interdepartmental conflicts. They are so busy handling conflicts that they delegate the day-to-day responsibility for running their projects to the project teams. Although this arrangement is not the most effective, it is sometimes necessary, especially after organizational restructuring or after a new project demanding new resources has been initiated.

The ability to handle conflicts requires an understanding of why conflicts occur. We can ask four questions, the answers to which are usually helpful in handling, and possibly preventing, conflicts in a project management environment:

- Do the project's objectives conflict with the objectives of other projects currently in development?
- Why do conflicts occur?
- How can we resolve conflicts?
- Is there anything we can do to anticipate and resolve conflicts before they become serious?

Although conflicts are inevitable, they can be planned for. For example, conflicts can easily develop in a team in which the members do not understand each other's roles and responsibilities. Responsibility charts can be drawn to map out graphically who is responsible for doing what on the project. With the ambiguity of roles and responsibilities gone, the conflict is resolved or future conflict averted.

Resolution means collaboration, and collaboration means that people are willing to rely on each other. Without collaboration, mistrust prevails and progress documentation increases.

The most common types of conflict involve the following:

- Manpower resources
- Equipment and facilities

- Capital expenditures
- Costs
- Technical opinions and trade-offs
- Priorities
- Administrative procedures
- Schedules
- Responsibilities
- Personality clashes

Each of these types of conflict can vary in intensity over the life of the project. The relative intensity can vary as a function of:

- Getting closer to project constraints
- Having met only two constraints instead of three (e.g., time and performance but not cost)
- The project life cycle itself
- The individuals who are in conflict

Conflict can be meaningful in that it results in beneficial outcomes. These meaningful conflicts should be allowed to continue as long as project constraints are not violated and beneficial results accrue. An example of a meaningful conflict might be two technical specialists arguing that each has a better way of solving a problem. The beneficial result would be that each tries to find additional information to support his or her hypothesis.

Some conflicts are inevitable and occur over and over again. For example, consider a raw material and finished goods inventory. Manufacturing wants the largest possible inventory of raw materials on hand to avoid possible production shutdowns. Sales and marketing wants the largest finished goods inventory so that the books look favorable and no cash flow problems are possible.

Consider five methods that project managers can use to resolve conflicts:

- Confrontation
- Compromise
- Facilitation (or smoothing)
- Force (or forcing)
- Withdrawal

Confrontation is probably the most common method used by project managers to resolve conflict. Using confrontation, the project manager faces the conflict directly. With the help of the project manager, the parties in disagreement attempt to persuade one another that their solution to the problem is the most appropriate.

When confrontation does not work, the next approach project managers usually try is compromise. In compromise, each of the parties in conflict agrees to trade-offs or makes concessions until a solution is arrived at that everyone involved can live with. This give-and-take-approach can easily lead to a win–win solution to the conflict.

The third approach to conflict resolution is facilitation. Using facilitation skills, the project manager emphasizes areas of agreement and deemphasizes areas of disagreement.

For example, suppose that a project manager said, "We've been arguing about five points, and so far we've reached agreement on the first three. There's no reason why we can't agree on the last two points, is there?" Facilitation of a disagreement does not resolve the conflict. Facilitation downplays the emotional context in which conflicts occur.

Force is also a method of conflict resolution. A project manager uses force when he or she tries to resolve a disagreement by exerting his or her own opinion at the expense of the other people involved. Often, forcing a solution onto the parties in conflict results in a win–lose outcome. Calling in the project sponsor to resolve a conflict is another form of force project managers sometimes use.

The least used and least effective mode of conflict resolution is withdrawal. A project director can simply withdraw from the conflict and leave the situation unresolved. When this method is used, the conflict does not go away and is likely to recur later. Personality conflicts might well be the most difficult conflicts to resolve. Personality conflicts can occur at any time, with anyone, and over anything. Furthermore, they can seem almost impossible to anticipate and plan for.

Let's look at how one company found a way to anticipate and avoid personality conflicts on one of its projects. Foster Defense Group (a pseudonym) was the government contract branch of a Fortune 500 company. The company understood the potentially detrimental effects of personality clashes on its project teams, but it did not like the idea of getting the whole team together to air its dirty laundry. The company found a better solution. The project manager put the names of the project team members on a list. Then he interviewed each of the team members one on one and asked each to identify the names on the list that he or she had had a personality conflict with in the past. The information remained confidential, and the project manager was able to avoid potential conflicts by separating clashing personalities.

If at all possible, the project manager should handle conflict resolution. When the project manager is unable to defuse the conflict, then and only then should the project sponsor be brought in to help solve the problem. Even then, the sponsor should not come in and force a resolution to the conflict. Instead, the sponsor should facilitate further discussion between the project managers and the team members in conflict.

## 10.3 STAFFING FOR EXCELLENCE

Project manager selection is always an executive-level decision. In excellent companies, however, executives go beyond simply selecting the project manager. They use the selection process to accomplish the following:

- Project managers are brought on board early in the life of the project to assist in outlining the project, setting its objectives, and even planning for marketing and sales. The project manager's role in customer relations becomes increasingly important.
- Executives assign project managers for the life of the project and project termination. Sponsorship can change over the life cycle of the project, but not the project manager.
- Project management is given its own career ladder.

- Project managers given a role in customer relations are also expected to help sell future project management services long before the current project is complete.
- Executives realize that project scope changes are inevitable. The project manager is viewed as a manager of change.

Companies excellent in project management are prepared for crises. Both the project manager and the line managers are encouraged to bring problems to the surface as quickly as possible so that there is time for contingency planning and problem solving. Replacing the project manager is no longer the first solution for problems on a project. Project managers are replaced only when they try to bury problems.

A defense contractor was behind schedule on a project, and the manufacturing team was asked to work extensive overtime to catch up. Two of the manufacturing people, both union employees, used the wrong lot of raw materials to produce a $65,000 piece of equipment needed for the project. The customer was unhappy because of the missed schedules and cost overruns that resulted from having to replace the useless equipment. An inquisition-like meeting was convened and attended by senior executives from both the customer and the contractor, the project manager, and the two manufacturing employees. When the customer's representative asked for an explanation of what had happened, the project manager stood up and said, "I take full responsibility for what happened. Expecting people to work extensive overtime leads to mistakes. I should have been more careful." The meeting was adjourned with no one being blamed. When word spread through the company about what the project manager did to protect the two union employees, everyone pitched in to get the project back on schedule, even working uncompensated overtime.

Human behavior is also a consideration in assigning staff to project teams. Team members should not be assigned to a project solely on the basis of technical knowledge. It has to be recognized that some people simply cannot work effectively in a team environment. For example, the director of research and development at a New England company had an employee, a 50-year-old engineer, who held two master's degrees in engineering disciplines. He had worked for the previous twenty years on one-person projects. The director reluctantly assigned the engineer to a project team. After years of working alone, the engineer trusted no one's results but his own. He refused to work cooperatively with the other members of the team. He even went so far as redoing all the calculations passed on to him from other engineers on the team.

To solve the problem, the director assigned the engineer to another project on which he supervised two other engineers with less experience. Again, the older engineer tried to do all of the work by himself, even if it meant overtime for him and no work for the others.

Ultimately, the director had to admit that some people are not able to work cooperatively on team projects. The director went back to assigning the engineer to one-person projects on which the engineer's technical abilities would be useful.

Robert Hershock, former vice president at 3M, once observed:

There are certain people whom you just don't want to put on teams. They are not team players, and they will be disruptive on teams. I think that we have to recognize that and make sure that those people are not part of a team or team members. If you need their expertise, you can bring them in as consultants to the team but you never, never put people like that on the team.

> I think the other thing is that I would never, ever eliminate the possibility of anybody being a team member no matter what the management level is. I think if they are properly trained, these people at any level can be participators in a team concept.

In 1996, Frank Jackson, formerly a senior manager at MCI, believed that it was possible to find a team where any individual can contribute:

> People should not be singled out as not being team players. Everyone has got the ability to be on a team and to contribute to a team based on the skills and the personal experiences that they have had. If you move into the team environment one other thing that is very important is that you not hinder communications. Communications is the key to the success of any team and any objective that a team tries to achieve.

One of the critical arguments still being waged in the project management community is whether an employee (even a project manager) should have the right to refuse an assignment. At Minnesota Power and Light, an open project manager position was posted, but nobody applied for the job. The company recognized that the employees probably did not understand what the position's responsibilities were. After more than 80 people were trained in the fundamentals of project management, there were numerous applications for the open position.

It's the kiss of death to assign someone to a project manager's job if that person is not dedicated to the project management process and the accountability it demands.

## 10.4 VIRTUAL PROJECT TEAMS

Historically, project management was a face-to-face environment where team meetings involved all players meeting together in one room. Today, because of the size and complexity of projects, it is impossible to find all team members located under one roof. Duarte and Snyder define seven types of virtual teams.[1] These are shown in Table 10–1.

Culture and technology can have a major impact on the performance of virtual teams. Duarte and Snyder have identified some of these relationships in Table 10–2.

The importance of culture cannot be understated. Duarte and Snyder identify four important points to remember concerning the impact of culture on virtual teams. The four points are[2]:

1. There are national cultures, organizational cultures, functional cultures, and team cultures. They can be sources of competitive advantages for virtual teams that know how to use cultural differences to create synergy. Team leaders and members who understand and are sensitive to cultural differences can create more robust outcomes

---

1. D. L. Duarte and N. Tennant Snyder, *Mastering Virtual Teams.* San Francisco: Jossey-Bass, 2001, p. 10. Reproduced by permission of John Wiley & Sons.

2. Ibid., p. 70.

**TABLE 10–1. TYPES OF VIRTUAL TEAMS**

| Type of Team | Description |
| --- | --- |
| Network | Team membership is diffuse and fluid; members come and go as needed. Team lacks clear boundaries within the organization. |
| Parallel | Team has clear boundaries and distinct membership. Team works in the short term to develop recommendations for an improvement in a process or system. |
| Project or product development | Team has fluid membership, clear boundaries, and a defined customer base, technical requirement, and output. Longer-term team task is nonroutine, and the team has decision-making authority. |
| Work or production | Team has distinct membership and clear boundaries. Members perform regular and outgoing work, usually in one functional area. |
| Service | Team has distinct membership and supports ongoing customer network activity. |
| Management | Team has distinct membership and works on a regular basis to lead corporate activities. |
| Action | Team deals with immediate action, usually in an emergency situation. Membership may be fluid or distinct. |

*Source*: D. L. Duarte and N. Tennant Snyder, *Mastering Virtual Teams.* San Francisco: Jossey-Bass, 2001, p. 10. Reproduced by permission of John Wiley & Sons.

**TABLE 10–2. TECHNOLOGY AND CULTURE**

| Cultural Factor | Technological Considerations |
| --- | --- |
| Power distance | Members from high-power-distance cultures may participate more freely with technologies that are asynchronous and allow anonymous input. These cultures sometimes use technology to indicate status differences between team members. |
| Uncertainty avoidance | People from cultures with high uncertainty avoidance may be slower adopters of technology. They may also prefer technology that is able to produce more permanent records of discussions and decisions. |
| Individualism–collectivism | Members from highly collectivistic cultures may prefer face-to-face interactions. |
| Masculinity–femininity | People from cultures with more "feminine" orientations are more prone to use technology in a nurturing way, especially during team startups. |
| Context | People from high-context cultures may prefer more information-rich technologies, as well as those that offer opportunities for the feeling of social presence. They may resist using technologies with low social presence to communicate with people they have never met. People from low-context cultures may prefer more asynchronous communications. |

*Source:* D. L Duarte and N. Tennant Snyder, *Mastering Virtual Teams,* San Francisco: Jossey-Bass, 2001, p. 60.

than can members of homogeneous teams with members who think and act alike. Cultural differences can create distinctive advantages for teams if they are understood and used in positive ways.

2. The most important aspect of understanding and working with cultural differences is to create a team culture in which problems can be surfaced and differences can be discussed in a productive, respectful manner.

3. It is essential to distinguish between problems that result from cultural differences and problems that are performance based.

4. Business practices and business ethics vary in different parts of the world. Virtual teams need to clearly articulate approaches to these that every member understands and abides by.

## 10.5 REWARDING PROJECT TEAMS

Today, most companies are using project teams. However, there still exist challenges in how to reward project teams for successful performance. The importance of how teams are rewarded is identified by Parker, McAdams, and Zielinski[3]:

> Some organizations are fond of saying, "We're all part of the team," but too often it is merely management-speak. This is especially common in conventional hierarchical organizations; they say the words but don't follow up with significant action. Their employees may read the articles and attend the conferences and come to believe that many companies have turned collaborative. Actually, though, few organizations today are genuinely team-based.
>
> Others who want to quibble point to how they reward or recognize teams with splashy bonuses or profit-sharing plans. But these do not by themselves represent a commitment to teams; they're more like a gift from a rich uncle. If top management believes that only money and a few recognition programs ("team of year" and that sort of thing) reinforce teamwork, they are wrong. These alone do not cause fundamental change in the way people and teams are managed.
>
> But in a few organizations, teaming is a key component of the corporate strategy, involvement with teams is second nature, and collaboration happens without great thought or fanfare. There are natural work groups (teams of people who do the same or similar work in the same location), permanent cross-functional teams, ad hoc project teams, process improvement teams, and real management teams. Involvement just happens.

Why is it so difficult to reward project teams? To answer this question, we must understand what a team is and is not[4]:

> Consider this statement: an organizational unit can act like a team, but a team is not necessarily an organizational unit, at least for describing reward plans. An organizational unit is just that, a group of employees organized into an identifiable business unit that appears on the organizational chart. They may behave in a spirit of teamwork, but for the purposes of developing reward plans they are not a "team." The organizational unit may be a whole company, a strategic business unit, a division, a department, or a work group.
>
> A "team" is a small group of people allied by a common project and sharing performance objectives. They generally have complementary skills or knowledge and an interdependence that requires that they work together to accomplish their project's objective. Team members hold themselves mutually accountable for their results. These teams are not found on an organization chart.

Incentives are difficult to apply because project teams may not appear on an organizational chart. Figure 10–1 shows the reinforcement model for employees.[5] For project teams, the emphasis is the three arrows on the right-hand side of Figure 10–1.

---

3. G. Parker, J. McAdams, and D. Zielinski, *Rewarding Teams*, San Francisco: Jossey-Bass, 2000, p. 17. Reproduced by permission of John Wiley & Sons.

4. Ibid., p. 17.

5. Ibid., p. 29.

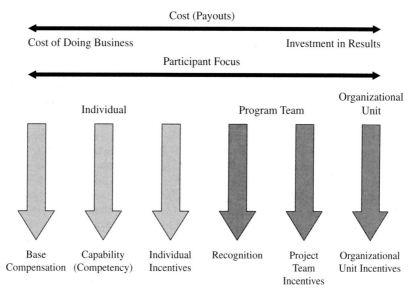

**Figure 10–1.** Reinforcement model.

Project team incentives are important because team members expect appropriate rewards and recognition[6]:

> Project teams are usually, but not always, formed by management to tackle specific projects or challenges with a defined time frame—reviewing processes for efficiency or cost-savings recommendations, launching a new software product, or implementing enterprise resource planning systems are just a few examples. In other cases, teams self-form around specific issues or as part of continuous improvement initiatives such as team-based suggestion systems.
>
> Project teams can have cross-functional membership or simply be a subset of an existing organizational unit. The person who sponsors the team—its "champion" typically creates an incentive plan with specific objective measures and an award schedule tied to achieving those measures. To qualify as an incentive, the plan must include preannounced goals, with a "do this, get that" guarantee for teams. The incentive usually varies with the value added by the project.

Project team incentive plans usually have some combination of these basic measures:

- *Project milestones:* Hit a milestone, on budget and on time, and all team members earn a defined amount. Although sound in theory, there are inherent problems in tying financial incentives to hitting milestones. Milestones often change for good reason (technological advances, market shifts, other developments) and you don't want the team and management to get into a negotiation on slipping dates to trigger the incentive. Unless milestones are set in stone and reaching them is simply a function of the team doing its

6. Ibid., pp. 38–39.

normal, everyday job, it's generally best to use recognition-after-the-fact celebration of reaching milestones—rather than tying financial incentives to it.

Rewards need not always be time based, such that when the team hits a milestone by a certain date it earns a reward. If, for example, a product development team debugs a new piece of software on time, that's not necessarily a reason to reward it. But if it discovers and solves an unsuspected problem or writes better code before a delivery date, rewards are due.

- *Project completion:* All team members earn a defined amount when they complete the project on budget and on time (or to the team champion's quality standards).
- *Value added:* This award is a function of the value added by a project, and depends largely on the ability of the organization to create and track objective measures. Examples include reduced turnaround time on customer requests, improved cycle times for product development, cost savings due to new process efficiencies, or incremental profit or market share created by the product or service developed or implemented by the project team.

One warning about project incentive plans: they can be very effective in helping teams stay focused, accomplish goals, and feel like they are rewarded for their hard work, but they tend to be exclusionary. Not everyone can be on a project team. Some employees (team members) will have an opportunity to earn an incentive that others (non-team members) do not. There is a lack of internal equity. One way to address this is to reward core team members with incentives for reaching team goals, and to recognize peripheral players who supported the team, either by offering advice, resources, or a pair of hands, or by covering for project team members back at their regular job.

Some projects are of such strategic importance that you can live with these internal equity problems and non–team members' grousing about exclusionary incentives. Bottom line, though, is this tool should be used cautiously.

Some organizations focus only on cash awards. However, Parker et al. have concluded from their research that noncash awards can work equally well, if not better, than cash awards[7]:

Many of our case organizations use non-cash awards because of their staying power. Everyone loves money, but cash payments can lose their motivational impact over time.

However, non-cash awards carry trophy value that has great staying power because each time you look at that television set or plaque you are reminded of what you or your team did to earn it. Each of the plans encourages awards that are coveted by the recipients and, therefore, will be memorable.

If you ask employees what they want, they will invariably say cash. But providing it can be difficult if the budget is small or the targeted earnings in an incentive plan are modest. If you pay out more often than annually and take taxes out, the net amount may look pretty small, even cheap. Non-cash awards tend to be more dependent on their symbolic value than their financial value.

Non-cash awards come in all forms: a simple thank-you, a letter of congratulations, time off with pay, a trophy, company merchandise, a plaque, gift certificates, special

---

7. Ibid., pp. 190–191.

services, a dinner for two, a free lunch, a credit to a card issued by the company for purchases at local stores, specific items or merchandise, merchandise from an extensive catalogue, travel for business or a vacation with the family, and stock options. Only the creativity and imagination of the plan creators limit the choices.

## 10.6 KEYS TO BEHAVIORAL EXCELLENCE

There are some distinguishing actions that project managers can take to ensure the successful completion of their projects. These include:

- Insisting on the right to select key project team
- Negotiating for key team members with proven track records in their fields
- Developing commitment and a sense of mission from the outset
- Seeking sufficient authority from the sponsor
- Coordinating and maintaining a good relationship with the client, parent, and team
- Seeking to enhance the public's image of the project
- Having key team members assist in decision-making and problem solving
- Developing realistic cost, schedule, and performance estimates and goals
- Maintaining backup strategies (contingency plans) in anticipation of potential problems
- Providing a team structure that is appropriate, yet flexible and flat
- Going beyond formal authority to maximize its influence over people and key decisions
- Employing a workable set of project planning and control tools
- Avoiding overreliance on one type of control tool
- Stressing the importance of meeting cost, schedule, and performance goals
- Giving priority to achieving the mission or function of the end item
- Keeping changes under control
- Seeking ways to assure job security for effective project team members

Earlier in this book, I claimed that a project cannot be successful unless it is recognized as a project and gains the support of top-level management. Top-level management must be willing to commit company resources and provide the necessary administrative support so that the project becomes part of the company's day-to-day routine of doing business. In addition, the parent organization must develop an atmosphere conducive to good working relationships among the project manager, parent organization, and client organization.

There are actions that top-level management should take to ensure that the organization as a whole supports individual projects and project teams as well as the overall project management system:

- Showing a willingness to coordinate efforts
- Demonstrating a willingness to maintain structural flexibility
- Showing a willingness to adapt to change
- Performing effective strategic planning

- Maintaining rapport
- Putting proper emphasis on past experience
- Providing external buffering
- Communicating promptly and accurately
- Exhibiting enthusiasm
- Recognizing that projects do, in fact, contribute to the capabilities of the whole company

Executive sponsors can take the following actions to make project success more likely:

- Selecting a project manager at an early point in the project who has a proven track record in behavioral skills and technical skills
- Developing clear and workable guidelines for the project manager
- Delegating sufficient authority to the project manager so that she or he can make decisions in conjunction with the project team members
- Demonstrating enthusiasm for and commitment to the project and the project team
- Developing and maintaining short and informal lines of communication
- Avoiding excessive pressure on the project manager to win contracts
- Avoiding arbitrarily slashing or ballooning the project team's cost estimate
- Avoiding "buy-ins"
- Developing close, not meddlesome, working relationships with the principal client contact and the project manager

The client organization can exert a great deal of influence on the behavioral aspects of a project by minimizing team meetings, rapidly responding to requests for information, and simply allowing the contractor to conduct business without interference. The positive actions of client organizations also include:

- Showing a willingness to coordinate efforts
- Maintaining rapport
- Establishing reasonable and specific goals and criteria for success
- Establishing procedures for making changes
- Communicating promptly and accurately
- Committing client resources as needed
- Minimizing red tape
- Providing sufficient authority to the client's representative, especially in decision making

With these actions as the basic foundation, it should be possible to achieve behavioral success, which includes:

- Encouraging openness and honesty from the start from all participants
- Creating an atmosphere that encourages healthy competition but not cutthroat situations or liar's contests
- Planning for adequate funding to complete the entire project
- Developing a clear understanding of the relative importance of cost, schedule, and technical performance goals

- Developing short and informal lines of communication and a flat organizational structure
- Delegating sufficient authority to the principal client contact and allowing prompt approval or rejection of important project decisions
- Rejecting buy-ins
- Making prompt decisions regarding contract okays or go-aheads
- Developing close working relationships with project participants
- Avoiding arm's-length relationships
- Avoiding excessive reporting schemes
- Making prompt decisions on changes

Companies that are excellent in project management have gone beyond the standard actions as listed previously. These additional actions for excellence include the following:

- The outstanding project manager has these demonstrable qualities:
  - Understands and demonstrates competency as a project manager
  - Works creatively and innovatively in a nontraditional sense only when necessary; does not look for trouble
  - Demonstrates high levels of self-motivation from the start
  - Has a high level of integrity; goes above and beyond politics and gamesmanship
  - Is dedicated to the company and not just the project; is never self-serving
  - Demonstrates humility in leadership
  - Demonstrates strong behavioral integration skills both internally and externally
  - Thinks proactively rather than reactively
  - Is willing to assume a great deal of risk and will spend the appropriate time needed to prepare contingency plans
  - Knows when to handle complexity and when to cut through it; demonstrates tenaciousness and perseverance
  - Is willing to help people realize their full potential; tries to bring out the best in people
  - Communicates in a timely manner and with confidence rather than despair
- The project manager maintains high standards of performance for self and team, as shown by these approaches:
  - Stresses managerial, operational, and product integrity
  - Conforms to moral codes and acts ethically in dealing with people internally and externally
  - Never withholds information
  - Is quality conscious and cost conscious
  - Discourages politics and gamesmanship; stresses justice and equity
  - Strives for continuous improvement but in a cost-conscious manner
- The outstanding project manager organizes and executes the project in a sound and efficient manner:
  - Informs employees at the project kickoff meeting how they will be evaluated
  - Prefers a flat project organizational structure over a bureaucratic one

- Develops a project process for handling crises and emergencies quickly and effectively
- Keeps the project team informed in a timely manner
- Does not require excessive reporting; creates an atmosphere of trust
- Defines roles, responsibilities, and accountabilities up front
- Establishes a change management process that involves the customer
- The outstanding project manager knows how to motivate:
  - Always uses two-way communication
  - Is empathetic with the team and a good listener
  - Involves team members in decision making; always seeks ideas and solutions; never judges an employee's idea hastily
  - Never dictates
  - Gives credit where credit is due
  - Provides constructive criticism rather than making personal attacks
  - Publicly acknowledges credit when credit is due but delivers criticism privately
  - Makes sure that team members know that they will be held accountable and responsible for their assignments
  - Always maintains an open-door policy; is readily accessible, even for employees with personal problems
  - Takes action quickly on employee grievances; is sensitive to employees' feelings and opinions
  - Allows employees to meet the customers
  - Tries to determine each team member's capabilities and aspirations; always looks for a good match; is concerned about what happens to the employees when the project is over
  - Tries to act as a buffer between the team and administrative/operational problems
- The project manager is ultimately responsible for turning the team into a cohesive and productive group for an open and creative environment. If the project manager succeeds, the team will exhibit the following behaviors:
  - Demonstrates innovation
  - Exchanges information freely
  - Is willing to accept risk and invest in new ideas
  - Is provided with the necessary tools and processes to execute the project
  - Dares to be different; is not satisfied with simply meeting the competition
  - Understands the business and the economics of the project
  - Tries to make sound business decisions rather than just sound project decisions

## 10.7 PROACTIVE VERSUS REACTIVE MANAGEMENT

Perhaps one of the biggest behavioral challenges facing a project manager, especially a new project manager, is learning how to be proactive rather than reactive. Kerry Wills, discusses this problem.

**Proactive Management Capacity Propensity[8]**

In today's world Project Managers often get tapped to manage several engagements at once. This usually results in them having just enough time to react to the problems of the day that each project is facing. What they are not doing is spending the time to look ahead on each project to plan for upcoming work, thus resulting in more fires that need to be put out. There used to be an arcade game called "whack-a-mole" where the participant had a mallet and would hit each mole with it when one would pop up. Each time a mole was hit, a new mole would pop up. The cycle of spending time putting out fires and ignoring problems that cause more fires can be thought of as "project whack-a-mole."

It is my experience that proactive management is one of the most effective tools that Project Managers can use to ensure the success of their projects. However, it is a difficult situation to manage several projects while still having enough time to look ahead. I call this ability to spend time looking ahead the "Proactive Management Capacity Propensity" (PMCP). This article will demonstrate the benefits of proactive management, define the PMCP, and propose ways of increasing the PMCP and thus the probability of success on the projects.

## Proactive Management

Overview Project Management involves a lot of planning up-front including work plans, budgets, resource allocations, and so on. The best statistics that I have seen on the accuracy of initial plans says there is a 30% positive or negative variance from the original plans at the end of a project. Therefore, once the plans have been made and the project has started, the Project Manager needs to constantly re-assess the project to understand the impact of the 60% unknowns that will occur.

The dictionary defines proactive as "acting in advance to deal with an expected difficulty." By "acting in advance" a Project Manager has some influence over the control of the unknowns. However, without acting in advance, the impacts of the unknowns will be greater as the Project Manager will be reacting to the problem once it has snowballed.

A Metaphor When I drive into work in the morning, I have a plan and schedule. I leave my house, take certain roads and get to work in forty minutes. If I were to treat driving to work as a Project (having a specific goal with a finite beginning and end), then I have two options to manage my commute [Figure 10–2]:

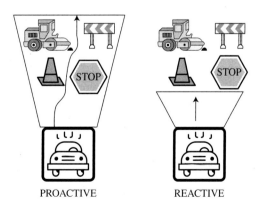

PROACTIVE                    REACTIVE

**Figure 10–2.** Driving metaphor.

8. The remainder of this section was provided by Kerry R. Wills, PMP®. Reproduced by permission of Kerry R. Wills.

By *proactively* managing my commute, I watch the news in the morning to see the weather and traffic. Although I had a plan, if there is construction on one of the roads that I normally take, then I can always change that plan and take a different route to ensure that my schedule gets met. If I know that there may be snow then I can leave earlier and give myself more time to get to work. As I am driving, I look ahead in the road to see what is coming up. There may be an accident or potholes that I will want to avoid and this gives me time to switch lanes.

A *reactive* approach to my commute could be assuming that my original plan will work fully. As I get on the highway, if there is construction then I have to sit in it because by the time I realize the impact, I have passed all of the exit ramps. This results in me missing my schedule goal. The same would happen if I walked outside and saw a foot of snow. I now have a chance to scope since I have the added activity of shoveling my driveway and car. Also, if I am a reactive driver, then I don't see the pothole until I have driven over it (which may lead to a budget variance since I now need new axles).

## Benefits

The metaphor above demonstrates that reactive management is detrimental to projects because by the time that you realize that there is a problem it usually has a schedule, scope or cost impact. There are several other benefits to proactive management:

- Proactively managing a plan allows the Project Manager to see what activities are coming up and start preparing for them. This could be something as minor as setting up conference rooms for meetings. I have seen situations were tasks were not completed on time because of something as minor as logistics.
- Understanding upcoming activities also allows for the proper resources to be in place. Oftentimes, projects require people from outside of the project team and lining them up are [*sic*] always a challenge. By preparing people in advance, there is a higher probability that they can be ready when needed.

## The Relationship

- The Project Manager should constantly be replanning. By looking at all upcoming activities as well as the current ones, it can give a gage of the probability of success which can be managed rather than waiting until the day before something is due to realize that the schedule cannot be met.
- Proactive management also allows time to focus on quality. Reactive management usually is characterized by rushing to fix whatever "mole" has popped up as quickly as possible. This usually means a patch rather than the appropriate fix. By planning for the work appropriately, it can be addressed properly which reduces the probability of rework.
- As previously unidentified work arises, it can be planned for rather than assuming that "we can just take it on."

Proactive management is extremely influential over the probability of success of a project because it allows for replanning and the ability to address problems well before they have a significant impact.

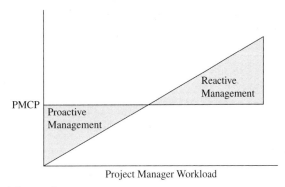

**Figure 10–3.** Proactivity graph.

Overview I have observed a relationship between the amount of work that a project manager has and their ability to manage proactively. As Project Managers get more work and more concurrent projects, their ability to manage proactively goes down.

The relationship between Project Manager workload and the ability to manage proactively is shown in Figure 10–3. . . . As Project Managers have increased work, they have less capacity to be proactive to and wind up becoming more reactive.

Not all Projects and Project Managers are equal. Some Project Managers can handle several projects well and some projects require more focus than others. I have therefore labeled this factor, the Project Management Capacity Propensity (PMCP). That is, the sum of those qualities that allow a Project Manager to proactively manage projects.

PMCP There are several factors that make up the PMCP that I have outlined below.

Project Manager skill sets have an impact on the PMCP. Having good Time Management and organization techniques can influence how much a PM can focus on looking ahead. A Project Manager who is efficient with their time has the ability to review more upcoming activities and plan for them.

Project Manager expertise of the project is also influential to the PMCP. If the PM is an expert in the business or the project, this may allow for quicker decisions since they will not need to seek out information or clarification (all of which takes away time).

The PMCP is also impacted by team composition. If the Project Manager is on a large project and has several team leads who manage plans, then they have an increased ability to focus on replanning and upcoming work. Also, having team members who are experts in their field will require less focus from the Project Manager.

Increasing the PMCP The good news about the PMCP is that it can be increased.

Project Managers can look for ways to increase their skillsets through training. There are several books and seminars on time management, prioritization, and organization. Attending these can build the effectiveness of the time spent by the PM on their activities.

The PM can also re-evaluate the team composition. By getting stronger team leads or different team members, the PM can offload some of their work and spend more time focusing on proactive management.

All of these items can increase the PMCP and result in an increased ability to manage proactively. The image . . . [in Figure 10–4] shows how a PMCP increase raises the bar and allows for more proactive management with the same workload.

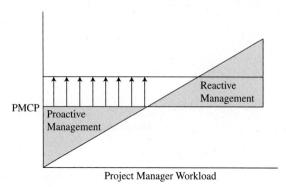

**Figure 10–4.** Increasing PMCP.

## Conclusion

To proactively manage a project is to increase your probability of being successful. There is a direct correlation between the workload that a PM has and their ability to look ahead. Project Managers do have control over certain aspects that can give them a greater ability to focus on proactive management. These items, the PMCP, can be increased through training and having the proper team.

Remember to keep your eyes on the road.

# Measuring Return on Investment on Project Management Training Dollars

## 11.0 INTRODUCTION

For almost three decades, the 1960s through the 1980s, the growth and acceptance of project management were restricted to the aerospace, defense, and heavy construction industries. In virtually all other industries, project management was nice to have but not a necessity. There were very few project management training programs offered in the public marketplace, and those that were offered covered the basics of project management with weak attempts to customize the material to a specific company. The concept of measuring the return on investment (ROI) on training, at least in project management courses, was nonexistent. Within the past ten years, there have been several studies on quantifying the benefits of project management with some pioneering work on the benefits of project management training.[1-4] There is still a great deal of effort needed, but at least we have recognized the need.

Today, our view of project management education has changed and so has our desire to evaluate ROI on project management training funds. There are several reasons for this:

● Executives realize that training is a basic necessity for companies to grow.
● Employees want training for professional growth and advancement opportunities.

1. W. Ibbs and J. Reginato, *Quantifying the Value of Project Management*, Newton Square, PA: Project Management Institute, 2002.
2. W. Ibbs and Y-H. Kwak, *The Benefits of Project Management*, Newton Square, PA: Project Management Institute, 1997.
3. W. Ibbs, "Measuring Project Management's Value: New Directions for Quantifying PM/ROI®," in *Proceedings of the PMI Research Conference*, June 21–24, 2000, Paris, France.
4. J. Knutson, A three-part series in PM *Network*, January, February, and July 1999.

- Project management is now viewed as a profession rather than a part-time occupation.
- The importance of becoming a PMP® has been increasing.
- There are numerous university programs available leading to MS, MBA, and PhD degrees in project management.
- There are certificate programs in various project management concepts such as risk management and program management.
- The pressure to maintain corporate profitability has increased, resulting in less money available for training. Yet more and more training funds are being requested by the workers who desire to become PMP®s and then must accumulate sixty professional development units (PDUs) every three years to remain certified.
- Management realizes that a significant portion of training budgets must be allocated to project management education, but it should be allocated for those courses that provide the company with the greatest ROI. The concept of educational ROI is now upon us.

## 11.1 PROJECT MANAGEMENT BENEFITS

In the early years of project management, primarily in the aerospace and defense industries, studies were done to determine the benefits of project management. In a study by Middleton, the benefits discovered were[5]:

- Better control of projects
- Better customer relations
- Shorter product development time
- Lower program costs
- Improved quality and reliability
- Higher profit margins
- Better control over program security
- Improved coordination among company divisions doing work on the project
- Higher morale and better mission orientation for employees working on the project
- Accelerated development of managers due to breadth of project responsibilities

These benefits were identified by Middleton through surveys and were subjective in nature. No attempt was made to quantify the benefits. At that time, there existed virtually no project management training programs. On-the-job training was the preferred method of learning project management, and most people learned from their own mistakes rather than from the mistakes of others.

Today the benefits identified by Middleton still apply, and we have added other benefits to the list:

- Accomplishing more work in less time and with few resources
- More efficient and more effective performance

5. C. J. Middleton, "How to Set Up a Project Organization," *Harvard Business Review*, March–April 1967, pp. 73–82.

- Increase in business due to customer satisfaction
- A desire for a long-term partnership relationship with customers
- Better control of scope changes

Executives wanted all of the benefits described here, and they wanted the benefits yesterday. It is true that these benefits could be obtained just by using on-the-job training efforts, but this assumed that time was a luxury rather than a constraint. Furthermore, executives wanted workers to learn from the mistakes of others rather than their own mistakes. Also executives wanted everyone in project management to look for continuous improvement efforts rather than just an occasional best practice.

Not every project management training program focuses on all of these benefits. Some courses focus on one particular benefit while others might focus on a group of benefits. Deciding which benefits you desire is essential in selecting a training course. And if the benefits can be quantified after training is completed, then executives can maximize their ROI on project management training dollars by selecting the proper training organizations.

## 11.2 GROWTH OF ROI MODELING

In the past several years, the global expansion of ROI modeling has taken hold. The American Society for Training and Development (ASTD) has performed studies on ROI modeling.[6] Throughout the world, professional associations are conducting seminars, workshops, and conferences dedicated to ROI on training.

According to the 2001 Training's annual report, more than $66 billion was spent on training in 2001. It is therefore little wonder that management treats training with a business mindset, thus justifying the use of ROI measurement. But despite all of the worldwide commitment and documented successes, there is still a very real fear in many companies preventing the use of ROI modeling. Typical arguments are: "It doesn't apply to us"; "We cannot evaluate the benefits quantitatively"; "We don't need it"; "The results are meaningless"; "It costs too much." These fears create barriers to the implementation of ROI techniques, but most barriers are myths that can be overcome.

In most companies, human resources development (HRD) maintains the lead role in overcoming these fears and performing the ROI studies. The cost of performing these studies on a continuous basis could be as much as 4–5 percent of the HRD budget. Some HRD organizations have trouble justifying this expense. And to make matters worse, the HRD personnel may have a poor understanding of project management.

The salvation in overcoming these fears and designing proper project management training programs could very well be the project management office (PMO). Since the PMO has become the guardian of all project management intellectual property as well as

---

6. J. J. Phillips, *Return on Investment in Training and Performance Improvement Programs*, 2nd ed., Burlington, MA: Butterworth-Heinemann, 2003, Chapter 1. In the opinion of this author, this is by far one of the best, if not the best, text on this subject.

designing project management training courses, the PMO will most likely take the lead role in calculating ROI on project management–related training courses. Members of the PMO might be required to become certified in educational ROI measurement the same way that they are certified as a PMP® or Six Sigma Black Belt.

Another reason for using the PMO is because of the enterprise project management (EPM) methodology. EPM is the integration of various processes such as total quality management, concurrent engineering, continuous improvement, risk management, and scope change control into one project management methodology that is utilized on a company-wide basis. Each of these processes has measurable output that previously may not have been tracked or reported. This has placed additional pressure on the PMO and project management education to develop metrics and measurement for success.

## 11.3 THE ROI MODEL

Any model used must provide a systematic approach to calculating ROI. It should be prepared on a life-cycle basis or step-by-step approach similar to an EPM methodology. Just like with EPM, there is an essential criterion that must exist for any model to work effectively.

Because criteria are considered essential, any ROI methodology should meet the vast majority of, if not all, criteria. The bad news is that most ROI processes do not generally meet all of these criteria. A typical model is shown in Figure 11–1. The definitions of the levels in Figure 11–1 are shown in Table 11–1.

**Figure 11–1.** The ROI model. *Source:* Adapted from J. J. Phillips, *Return on Investment in Training and Performance Improvement Programs*, 2nd ed., Burlington, MA: Butterworth-Heinemann, 2003 p. 37.

**TABLE 11–1. DEFINING LEVELS**

| Level | Description |
|---|---|
| 1: Reaction/satisfaction | Measures the participants' reaction to the program and possibly creates an action plan for implementation of the ideas |
| 2: Learning | Measures specific skills, knowledge, or attitude changes |
| 3: Application | Measures changes in work habit or on-the-job performance as well as application and implementation of knowledge learned |
| 4: Business impact | Measures the impact on the business as a result of implementation of changes |
| 5: Return on investment | Compares monetary benefits with the cost of the training and expressed as a percentage |

## 11.4 PLANNING LIFE-CYCLE PHASE

The first life-cycle phase in the ROI model is the development of evaluation plans and baseline data. The evaluation plan is similar to some of the PMBOK® Guide knowledge areas that require a plan as part of the first process step in each knowledge area. The evaluation plan should identify:

- The objective(s) of the program
- The way(s) the objective(s) will be validated
- The target audience
- Assumptions and constraints
- The timing of the program

Objectives for the training program must be clearly defined before ROI modeling can be completed. Table 11–2 identifies typical objectives. The objectives must be clearly defined for each of the five levels of the model. Column 3 in Table 11–2 would be representative of the objectives that a company might have when it registers a participant in a Project Management Certificate Program (PMCP) training course. In this example, the company funding the participant's training might expect the participant to become a PMP® and then assist the organization in developing an EPM methodology based upon the *PMBOK® Guide* with the expectation that this would lead to customer satisfaction and more business. Column 4 in Table 11–2 might be representative of a company that registers a participant in a course on best practices in project management. Some companies believe that if a seminar participant walks away from a training program with two good ideas for each day of the program and these ideas can be implemented reasonably fast, then the seminar would be considered a success. In this example, the objectives are to identify best practices in project management that other companies are doing and can be effectively implemented in the participant's company.

There can be differences in training objectives, as seen through the eyes of management. As an example, looking at columns 3 and 4 in Table 11–2, objectives might be:

- Learn skills that can be applied immediately to the job. In this case, ROI can be measured quickly. This might be representative of the PMCP course in column 3.
- Learn about techniques and advancements. In this case, additional money must be spent to achieve these benefits. ROI measurement may not be meaningful until

**TABLE 11–2.  TYPICAL PROGRAM OBJECTIVES**

|       |                      | Objectives | |
|-------|----------------------|-------------------------------------|-------------------------------------|
| Level | Description          | Typical PMCP Training               | Typical Best Practices Training Course |
| 1     | Reaction/satisfaction | Understand principles of *PMBOK® Guide* | Understand that companies are documenting their best practices |
| 2     | Learning             | Demonstrate skills or knowledge in domain groups and knowledge areas | Demonstrate how best practices benefit an organization |
| 3     | Application          | Development of EPM processes based upon the *PMBOK® Guide* | Develop a best practices library or ways to capture best practices |
| 4     | Business impact      | Measurement of customer and user satisfaction with EPM | Determine the time and/or cost savings from a best practice |
| 5     | Return on investment | Amount of business or customer satisfaction generated from EPM | Measure ROI for each best practice implemented |

after the techniques have been implemented. This might be representative of the best practices course in column 4.

● A combination of the above.

## 11.5  DATA COLLECTION LIFE-CYCLE PHASE

In order to validate that each level's objectives for the training course were achieved, data must be collected and processed. Levels 1–4 in Figure 11–1 make up the data collection life-cycle phase.

To understand the data collection methods, we revisit the course on best practices in project management, which covers best practices implemented by various companies worldwide. The following assumptions will be made:

● Participants are attending the course to bring back to their company at least two ideas that can be implemented in their company within six months.
● Collecting PDUs is a secondary benefit.
● The course length is two days.[7]

Typical data collection approaches are shown in Table 11–3 and explained below for each level.

**Level 1: Reaction and Satisfaction**

Level 1 measures the participant's reaction to the program and possibly an action plan for implementation of the ideas. The measurement for level 1 is usually an end-of-course questionnaire where the participant rates the information presented, quality of instruction, instructional material, and other such topics on a scale of 1–7. All too often, the questionnaire is answered based upon the

---

7. Some companies have one-day, two-day, and even week-long courses on best practices in project management.

**TABLE 11–3. DATA COLLECTION**

| Level | Measures | Data Collection Methods and Instruments | Data Sources | Timing | Responsible Person |
|---|---|---|---|---|---|
| Reaction/ satisfaction | A 1–7 rating on end-of-course critique | Questionnaire | Participant (last day of program) | End of program | Instructor |
| Learning | Pretest, posttest, CD-ROM, and case studies | In-class tests and skill practice sets | Instructor | Each day of course | Instructor |
| Application | Classroom discussion | Follow-up session or questionnaire | Participant and/or PMO | Three months after program[a] | PMO |
| Business impact | Measurement of EPM continuous improvement efforts | Benefit–cost monitoring by the PMO | PMO records | Six months after program | PMO |
| Return on investment | Benefit–cost ratios | PMO studies | PMO records | Six months after program | PMO |

[a]Usually for in-house program only. For public seminars, this may be done by the PMO within a week after completion of training.

instructor's presentation skills rather than the quality of the information. While this method is most common and often serves as an indication of customer satisfaction hopefully leading to repeat business, it is not a guarantee that new skills or knowledge have been learned.

## Level 2: Learning

This level measures specific skills, knowledge, or attitude changes learned during the course. Instructors use a variety of techniques for training, including:

- Lectures
- Lectures/discussions
- Exams
- Case studies (external firms)
- Case studies (internal projects)
- Simulation/role playing
- Combinations

For each training technique, a measurement method must be established. Some trainers provide a pretest at the beginning of the course and a posttest at the end. The difference in scores is usually representative of the amount of learning that has taken place. This is usually accomplished for in-house training programs rather than public seminars. Care must be taken in the use of pretests and posttests. Sometimes, a posttest is made relatively easy for the purpose of making it appear that learning has taken place. Out-of-class testing can also be accomplished using take-home case studies and CD-ROM multiple-choice questions.

Testing is necessary to validate that learning has taken place and knowledge has been absorbed. However, simply because learning has taken place is no guarantee that the information learned on best practices can or will be transferred to the company. The learning might simply confirm that the company is doing well and keeping up with the competitors.

**Level 3: Application of Knowledge**

This level measures changes in work habits or on-the-job performance as well as implementation of knowledge learned. Measurement at this level is normally done through follow-up sessions or follow-up questionnaires. However, for publicly offered courses with a large number of participants, it is impossible for the instructor to follow up with all participants. In such cases, the responsibility falls on the shoulders of the PMO. Participants may be required to prepare a short one- or two-page report on what they learned in the course and what best practices are applicable to the company. The report is submitted to the PMO that might have the final decision on the implementation of the ideas. Based on the magnitude of the best practices ideas, the portfolio management of projects may be impacted. However, there is no guarantee at this point that there will be a positive impact on the business.

**Level 4: Business Impact**

This level measures the impact on the business as a result of implementation of the changes. Typical measurement areas are shown in Figure 11–2. The critical terms in Figure 11–2 are:

● *Critical Success Factor (CSF):* This measures changes in the output of the project resulting from implementation of best practices. Hopefully, this will lead to improvements in time, cost, quality, and scope.
● *Key Performance Indicator (KPI):* This measures changes in the use of the EPM system and support received from functional management and senior management.
● *Business Unit Impact:* This is measured by customer satisfaction as a result of the implementation of best practices and/or future business opportunities.

The measurement at level 4 is usually accomplished by the PMO. There are several reasons for this. First, the information may be company sensitive and not available to the instructor. Second, since there may be a long time span between training and the

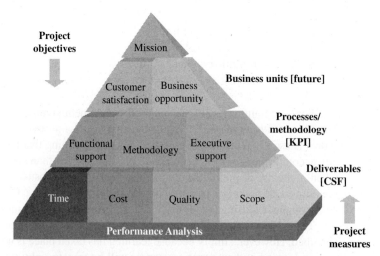

**Figure 11–2.** Postmortem pyramid. *Source:* From H. Kerzner, *Advanced Project Management: Best Practices in Implementation*, 2nd ed., Hoboken, NJ: Wiley, p. 302.

implementation of best practices, the instructor may not be available for support. Third, the company may not want anyone outside of the company talking to its customers about customer satisfaction. Although the implementation of best practices may have a favorable business impact, care must be taken that the implementation was cost effective.

As shown in Figure 11–1, an important input into level 4 is *isolate the effects of training.* It is often impossible to clearly identify the business impact that results directly from the training program. The problem is that people learn project management from multiple sources, including:

- Formal education
- Knowledge transfer from colleagues
- On-the-job experience
- Internal research on continuous improvements
- Benchmarking

Because of the difficulty in isolating the specific knowledge, this step is often overlooked.

## 11.6 DATA ANALYSIS LIFE-CYCLE PHASE

In order to calculate the ROI, the business impact data from level 4 must be converted to a monetary value. The information can come from interviews with employees and managers, databases, subject matter experts, and historical data. Very rarely will all of the information needed come from one source.

Another input required for data analysis is the cost of the training program. Typical costs that should be considered include:

- Cost of course design and development
- Cost of materials
- Cost of the facilitator(s)
- Cost of facilities and meals during training
- Costs of travel, meals, and lodgings for each participant
- Fully burdened salaries of participants
- Administrative or overhead cost related to the training course or approach of participants to attend training
- Possible cost (loss of income) of not having the participants available for other work during the time of training

Not all benefits can be converted to monetary values. This is the reason for the "identify intangible benefits" box in Figure 11–1. Some business impact benefits that are easily converted to monetary values include:

- Shorter product development time
- Faster, higher-quality decisions

- Lower costs
- Higher profit margins
- Fewer resources needed
- Reduction in paperwork
- Improved quality and reliability
- Lower turnover of personnel
- Quicker implementation of best practices

Typical benefits that are intangible and cannot readily be converted to monetary value include:

- Better visibility and focus on results
- Better coordination
- Higher morale
- Accelerated development of managers
- Better project control
- Better customer relations
- Better functional support
- Fewer conflicts requiring some management support

Despite the fact that these benefits may be intangible, every attempt should be made to assign monetary values of these benefits.

**Level 5: Return on Investment**   Two formulas are required for completion of level 5. The first formula is the BCR, which can be formulated as

$$BCR = \frac{Program\ benefits}{program\ costs}$$

The second formula is the ROI expressed as a percent. The formula is based upon "net" program benefits, which are the benefits minus the cost. Mathematically, we can describe it as

$$ROI = \frac{net\ program\ benefits}{program\ costs} \times 100$$

To illustrate the usefulness of this level, we consider three examples all based upon the same training course. You attend a two-day seminar on best practices in project management. Your company's cost for attending the course is:

| | |
|---|---|
| Registration fee | $ 475 |
| Release time (16 hr at $100/hr) | 1,600 |
| Travel expenses | 800 |
| | $2,875 |

When the seminar is completed, you come away with three best practices to recommend to your company. Your company likes all three ideas and assigns you as the project manager to implement all three best practices. Additional funds must be spent to achieve the benefits desired.

---

## Example 1

During the seminar, you discover that many companies have adopted the concept of paperless project management by implementing a "traffic light" status-reporting system. Your company already has a web-based EPM system but you have been preparing paper reports for status review meetings. Now, every status review meeting will be conducted without paper and with an LCD projector displaying the web-based methodology with a traffic light display beside each work package in the work breakdown structure.

The cost of developing the traffic light system is:

| | |
|---|---|
| Systems programming (240 hr at $100/hr) | $24,000 |
| Project management (150 hr at $100/hr) | 15,000 |
| | $39,000 |

The benefits expressed by monetary terms are:

- Executive time in project review meeting (20 hr per project to 10 hr per project × 15 projects × 5 executives per meeting × $250/hr): $187,500
- Paperwork preparation time reduction (60 hr/project × 15 projects × $100/hr): $90,000
- Total additional benefit is therefore $275,500:

$$\text{BCR} = \frac{\$275,000 - \$39,000}{\$2875} = 82$$

$$\text{ROI} = \frac{\$275,000 - \$39,000 - \$2875}{\$2875} = 8109$$

This means that for every dollar invested in the training program, there was a return of $8,109 in net benefits! In this example, it was assumed that workers were fully burdened at $100/hr and executives at $250/hr. The benefits were one-year measurements and the cost of developing the traffic light system was not amortized but expensed against the yearly benefits.

Not all training programs generate benefits of this magnitude. Lear in Dearborn, Michigan, has a project management traffic light reporting system as part of its web-based EPM system. Lear has shown that in the same amount of time that it would review the status of one project using paper, it now reviewed the status of all projects using traffic light reporting.

## Example 2

During the training program, you discover that other companies are using templates for project approval and initiation. The templates are provided to you during the training program and it takes a very quick effort to make the templates part of the EPM system and inform everyone about the update. The new templates will eliminate at least one meeting per week at a savings of $550:

$$\text{Benefit} = (\$500/\text{meeting}) \times (1\ \text{meeting/week}) \times 50\ \text{weeks} = \$27{,}500$$

$$\text{BCR} = \frac{\$27{,}500}{\$2875} = 9.56$$

$$\text{ROI} = \frac{\$27{,}500 - \$2875}{\$2875} = 8.56$$

In this example, for each $1 invested in the best practices program, a net benefit of $8.56 was recognized.

## Example 3

During the training program, you learn that companies are extending their EPM systems to become more compatible with systems utilized by their customers. This should foster better customer satisfaction. The cost of updating your EPM system to account for diversified customer report generators will be about $100,000.

After the report generator is installed, one of your customers with whom you have four projects per year informs you they are so pleased with this change that they will now give you sole-source procurement contracts. This will result in a significant savings in procurement costs. Your company typically spends $30,000 preparing proposals:

$$\text{BCR} = \frac{(4\ \text{projects} \times \$30{,}000) - \$100{,}000}{\$2875} = 6.96$$

$$\text{ROI(\%)} = \frac{(4 \times \$30{,}000) - \$100{,}000 - \$2875}{\$2875} = 5.96$$

In this case, for every dollar invested in the best practices program there was a net benefit of $5.96 received.

To date, there have been very few attempts to measure ROI specifically on project management education other than work done by Phillips. However, there have been some successes. In an insurance company, a $100 million project was undertaken. All employees were required to undergo project management training prior to working on the project. The project was completed 3 percent below budget.

Unsure of whether the $3 million savings was due to better project management education or poor initial estimating, the company performed a study on all projects where the employees were trained on project management prior to working on project teams. The result was an astounding 700 percent return on training dollars.

In another organization, the HRD people worked with project management to develop a computer-based project management training program. The initial results indicated a 900 percent ROI. The workers took the course on their own time rather than company time.

Perhaps this is an indication of the benefits of e-learning programs. The e-learning programs may produce a much higher ROI than traditional courses because the cost of the course is significantly reduced with the elimination of the cost of release time.

## 11.7 REPORTING LIFE-CYCLE PHASE

The final life-cycle phase in Figure 11–1 is reporting. The acceptance of the results could very well be based upon how the report is prepared. The report must be self-explanatory to all target groups. If assumptions are made concerning costs or benefits, then they must be justified. If the ROI numbers are inflated to make a training program look better than it was, then people may be skeptical and refuse to accept the results of future ROI studies. All results should be factual and supported by realistic data.

## 11.8 CONCLUSIONS

Because of the quantity and depth of available project management training programs, the concept of measuring ROI on training dollars can be expected to grow. Executives will recognize the benefits of this approach and its application to project management the same way it is applied to other training programs. Project management training organizations will be required to demonstrate expertise in ROI analysis. Eventually, PMI might even establish a special investigation group on ROI measurement.

# 12  The Project Office

## 12.0 INTRODUCTION

As companies begin to recognize the favorable effect that project management has on profitability, emphasis is placed upon achieving professionalism in project management using the project office (PO) concept. The concept of a PO or project management office (PMO) could very well be the most important project management activity in this decade. With this recognition of importance comes strategic planning for both project management and the project office. Maturity and excellence in project management do *not* occur simply by using project management over a prolonged period of time. Rather, it comes through strategic planning for both project management and the PO.

General strategic planning involves the determination of where you wish to be in the future and then how you plan to get there. For PO strategic planning, it is often easier to decide which activities should be under the control of the PO than determining how or when to do it. For each activity placed under the auspices of the PO, there may appear pockets of resistance that initially view removing this activity from its functional area as a threat to its power and authority. Typical activities assigned to a PO include:

- Standardization in estimating
- Standardization in planning
- Standardization in scheduling
- Standardization in control
- Standardization in reporting
- Clarification of project management roles and responsibilities
- Preparation of job descriptions for project managers

- Preparation of archive data on lessons learned
- Benchmarking continuously
- Developing project management templates
- Developing a project management methodology
- Recommending and implementing changes and improvements to the existing methodology
- Identifying project standards
- Identifying best practices
- Performing strategic planning for project management
- Establishing a project management problem-solving hotline
- Coordinating and/or conducting project management training programs
- Transferring knowledge through coaching and mentorship
- Developing a corporate resource capacity/utilization plan
- Supporting portfolio management activities
- Assessing risks
- Planning for disaster recovery
- Auditing the use of the project management methodology
- Auditing the use of best practices

In the first decade of the twenty-first century, the PO became commonplace in the corporate hierarchy. Although the majority of activities assigned to the PO had not changed, there was now a new mission for the PO:

- The PO now has the responsibility for maintaining all intellectual property related to project management and to actively support corporate strategic planning.

The PO was now servicing the corporation, especially the strategic planning activities for project management, rather than focusing on a specific customer. The PO was transformed into a corporate center for control of project management intellectual property. This was a necessity as the magnitude of project management information grew almost exponentially throughout the organization.

During the past 15 years, the benefits to executive levels of management of using a PO have become apparent. They include:

- Standardization of operations
- Company rather than silo decisionmaking
- Better capacity planning (i.e., resource allocations)
- Quicker access to higher-quality information
- Elimination or reduction of company silos
- More efficient and effective operations
- Less need for restructuring
- Fewer meetings that rob executives of valuable time
- More realistic prioritization of work
- Development of future general managers

All of the above benefits are either directly or indirectly related to project management intellectual property. To maintain the project management intellectual property, the PO must maintain the vehicles for capturing the data and then disseminating the data to the various stakeholders. These vehicles include the company project management intranet, project websites, project databases, and project management information systems. Since much of this information is necessary for both project management and corporate strategic planning, then there must exist strategic planning for the PO.

The recognition of the importance of the PMO has now spread worldwide. Enrique Sevilla Molina, formerly Corporate PMO director for Indra, states:

> We have a PMO at corporate level and local PMOs at different levels throughout the company, performing a variety of functions. The PMO at corporate level provides directions on different project management issues, methodology clarifications, and tool use to local PMOs.
>
> Besides supporting the local PMOs and Project Managers as requested, the main functions of the corporate PMO include acting on the following areas:
>
> - Maintenance and development of the overall project management methodology, including the extensions for Program and Portfolio levels
> - Definition of the training material and processes for the PMs
> - Management of the PMP certification process and candidates training and preparation
> - Definition of the requirements for the corporate PM tools

The Corporate PMO reports to the financial managing director.

A typical PMO does not have profit and loss responsibility on projects, nor does a typical PMO manage projects for external clients. According to Jim Triompo, group senior vice president at ABB:

> The project office does not deliver projects. The projects managed by the project management office are limited to process/tools development, implementation, and training. The project management office is sometimes requested to perform reviews, participate in division-level risk reviews, and operational reviews in various countries.

Most PMOs are viewed as indirect labor and therefore subject to downsizing or elimination when a corporation is under financial stress. To minimize the risk, the PMO should set up metrics to show that the PMO is adding value to the company. Typical metrics include:

- Tangible measurements include:
  - Customer satisfaction
  - Projects at risk
  - Projects in trouble
  - The number of red lights that need recovery, and by how much added effort

- Intangible elements may also exist and these may not be able to be measured.
  - Early-on identification of problems
  - Quality and timing of information

## 12.1 BOEING[1]

Not all companies use the term Project Management Office. In some companies it is also call a Community of Excellence or a Community of Practice. Every company has their own unique goals and objectives for a PMO. As such, the responsibilities of the PMO can vary from company to company. The following information on Boeing was provided by Kerry Braaflat – Boeing Executive Sponsor for Project Management; Stephen Markgraf – Leader Boeing Enterprise Project Management Office; Scott Sears – Leader Boeing Project Management Community of Excellence; Christine Baker – Prior Leader Boeing Project Management Community of Excellence; and Sherry Kytonen –Boeing Senior Portfolio Project Manager:

The Boeing Enterprise Project Management Office sponsors a Project Management Community of Excellence (PjMCoE) that exemplifies Project Management best practices and promotes the Project Management disciple across The Boeing Company.

A Community of Excellence (CoE) within Boeing is a formally chartered group that functionally aligns with at least one business organization, has enterprise representation, and is committed to business engagement, knowledge sharing and application across the entire Boeing Company.

The PjMCoE operates as a voluntary interest group with more than 5,400 active Boeing members and is one of the largest interest groups within Boeing. The PjMCoE started out in 1997 as an informal Project Management interest group that included only 75 members. The purpose of the PjMCoE can be described as having the responsibility to lead and manage a Boeing-wide forum for increased awareness of the skills, discipline and profession of project management. The PjMCoE is foundational to project management success at Boeing and provides the following services to both its members and the business:

- Provide a Forum of Networking, Collaboration & Support
- Provide Project Management Mentoring, Coaching & Training
- Host Annual PMP® Study Group
- Assist Hiring Managers
- Provide Project Managers and Project Teams with available Resources

The PjMCoE has its own Steering Team with defined responsibilities that support CoE products and services and is representative across all Boeing operating groups and sites. The Steering Team is foundational to the achievement of the CoE charter, facilitation of regular meetings, oversees operation of an effective management information system, ensures all engagement requests receive a response, and notifies aligned organizations when contacted by project teams, programs or functions with requests for support.

In support of the continued development of the project management skill, the PjMCoE provides support to its members by offering the following:

- Libraries containing Project Management–related books, periodicals, methodologies, software, periodicals and presentations
- SharePoint and web site with regional chapter information and news

---

- Knowledge centers providing mentoring and coaching services and information on Project Management certifications and degrees
- An annual PMP® study group hosting 16 live and recorded training sessions and lessons learned to assist in preparing for the PMP® exam. The first PMP® study group started in 2000 and has had a 95% pass rate for those who actually took the PMP® exam. By 2013 the PMP study group kick-off meeting had more than 400 employees in attendance
- Assistance for managers seeking to hire or promote Project Managers
- Bi-monthly WebEx meetings hosting guest speakers and offering Professional Development Units used toward accreditation
- Annual Project Management Conference and Expo attracting Boeing employees and contractors from around the world

PjMCoE maintains a strong ongoing relationship with PMI and is a registered PMI Education Provider. As such, the PjMCoE provides bi-weekly training events to its members, provides instructors to support the delivery of Boeing specific project management classes and offers other ad-hoc training offerings that provide Boeing employees with an opportunity to obtain Professional Development Units towards the re-certification on their PMP® certifications. Training modules in MS Project, Milestones Professional and Risk, Issue and Opportunity Management, Leadership, Communication and Virtual Team Management are also available. In 2012 alone, the PjMCoE provided training to more than 10,000 employees by offering more than 15,000 hours of training opportunities.

Many employees take advantage of the Mentoring and/or mentee and consultations services to assist them in career development and project management skills. Mentoring also provides visibility of opportunities for both managers and employees.

In 2013 the PjMCoE will be partnering with Boeing's Global Corporate Citizenship team to identify volunteer opportunities for Project Managers to support community service projects.

## 12.2 PHILIPS HEALTHCARE SOFTWARE CUSTOMER SERVICES[2]

**Project Management Excellence with a Global Project Management Office (PMO)**

Michael Bauer, head of the Global PMO at Philips Healthcare Software Customer Services, describes how the PMO supports a global operating clinical informatics solutions business to reach Project Management Excellence.

### Philips Healthcare and Software Customer Services

Philips is a strong, diversified health and well-being company, operating in all global geographies. About 118,000 employees work in the three businesses Healthcare, Consumer Lifestyle and Lighting.

---

2. ©2013 by Philips. Reproduced by permission. All rights reserved.

Philips Healthcare[3] provides meaningful innovations that improve the quality of care, enhance patients' lives and enable the delivery of better outcomes at lower cost. Philips Healthcare operates in five main business areas: Imaging Systems; Patient Care and Clinical Informatics (PCCI); Home Healthcare Solutions; Global Customer Services; and Healthcare Transformation Services.

Philips is a world leader in cardiology solutions, and has a strong presence in cardio-pulmonary, oncology and women's health. Philips helps clinicians diagnose, treat and manage today's most prevalent diseases such as congestive heart failure, breast and other cancers, respiratory and other coronary artery diseases as quickly, effectively and efficiently as possible. The focus is on understanding the complete cycle of care—from disease prevention to screening and diagnosis through to treatment, monitoring and health management—and choosing to participate in the areas where Philips Healthcare can add significant value. Philips Healthcare believes clinical excellence and continuous innova-tion around the patient experience can fundamentally change healthcare as we know it.

Patient Care and Clinical Informatics (PCCI) provides clinical informatics and patient care solutions that simplify clinician workflow, improve financial outcomes, and help improve and save lives.

Philips Healthcare Software Customer Services (SCS) implements and supports clinical information solution projects in hospital organizations. Philips Healthcare SCS provides project management, implementation services and solution customer care to ensure superior customer experience across the entire solution life cycle. SCS is a global organization operating worldwide with a single focus—the successful implementation and customer care for Philips Healthcare Clinical Informatics solutions.

The Clinical Informatics solutions (based on highly configurable software products) offered are:

- Radiology PACS (picture archiving and communication system) and RIS (radiol-ogy information system) systems
- Cardiology PACS systems
- Cardiovascular information systems
- Intensive care, anesthesia and OB (obstetrics) care clinical information systems

The breadth of complexity in Clinical Informatics projects spans across:

- Standalone solutions in group practice or small departments
- Multiple applications across multiple departments
- Multi-hospital deployment
- "Greenfield" implementations across all modalities[4] and applications

Implementing Clinical Informatics projects is a local activity performed in each hos-pital organization in each country, often in the local language. SCS operates with both local and centralized resources. This global/local organizational design often leads to a

---

3. See as well http://www.healthcare.philips.com/main/ for more information.
4. E.g., imaging modalities, like XR, MRI, CT.

virtual working environment with specific requirements to efficiently run the projects. The requirements and maturity levels in each country/market and from each hospital customer greatly varies. Each project in a hospital is unique and varies in duration (from weeks to years), in size (up to multi-million euros/ dollars) and in complexity (from stand-alone solution for one clinician to regional distributed solution for thousands of users).

SCS has an awareness that each organization leaves an imprint with the customer, an experience made up of rational and emotional aspects which determines what Clinical Informatics customers associate with the Philips brand, what Philips means to them. This is especially pronounced in a services business. Customer experience[5] is at the heart of a relationship that translates into whether customers repeatedly rely upon the organization's capabilities and embrace them as a trusted advisor.

This fundamental understanding of the vital importance of customer experience—the way it is created and delivered—in forming the basis for continued business success and differentiation in the market has lead SCS to taking a holistic approach to customer experience management. SCS has a vision and strategy based on customer experience and has shaped culture and operations to support the realization of this vision.

SCS applies this customer experience focused approach across the entire solution life cycle from the point in time the customers share their vision and mandate Philips with the realization, through solution implementation and post-go-live customer care. In this context, **Project Management Excellence** is a key strategic ingredient in ensuring SCS reliably and repeatedly delivers the desired customer experience. Hence, building and sustaining Project Management Excellence and reaching a high level of project management maturity with Clinical Informatics solution implementation projects is an adamant ambition of vital importance for both, the customer and Philips.

According to the Project Management Institute's 2012 Pulse of the Profession survey[6] only 73% of projects at organizations with high project management maturity meet their original business goals and intent. For Philips Healthcare Software Customer Services the ambitions for successful project implementation are high and require high project management maturity for a Clinical Informatics business. This was the key strategic driver to implement a Global PMO. SCS uses the term **Project Management Excellence** with the focus on the following important aspects:

- **People**: Well educated, certified, skilled (hard, soft) and continuously trained project managers (and project teams) with a professional mindset, appearance and behavior. This also includes recruiting the best project manager talents.[7]
- **Processes**: Highly efficient, standardized, repeatable and well documented processes, which are continuously improved.
- **Tools**: Highly integrated and efficient tools, templates and applications from the project acquisition until the end of the project.

---

5. Sources for customer experience concepts: http://www.cxpa.org/, http://www.temkingroup.com/, http://www.beyondphilosophy.com/

6. Source: PMI's *Pulse of the Profession* March 2012, page 6.

7. See as well for importance of Project Management talent management: PMI's *Pulse of the Profession In Depth Study: Talent Management*, March 2013.

Project Management Excellence is not seen as a static goal; the ambition is to continuously raise the bar for project management maturity as well as oversee overall project management (PM) competencies and project delivery capacity.

## Project Management Office

The establishment of a global Project Management Office (PMO) was a clear strategic decision and has full support from top management to drive project management strategically.[8] The PMO enables the ambition of **Project Management Excellence**, where the following aspects need to be highlighted:

- Project Management Excellence matters to SCS – key aspect to value and improving skills, processes and tools.
- Change Management – identify, drive and implement improvements and changes in the organization.
- Standardization – enable standardized practices and processes across product domains and regions.[9]
- Continuous learning – train, review and mentor as required.
- Facilitation of a PM Community of Practice – key aspect to enable to Share, Learn, Leverage, Network and Communicate.[10]

In building of the global Project Management Office (PMO) the following conditions were considered:

1. PMO positioning: Senior experts in Project Management that guide the organization through changes towards Project Management Excellence with strong Change Management.
2. PMO focus: The PMO itself does not implement customer-facing projects in hospitals themselves but focuses on supporting the local project manager/project teams.
3. Project Management Consultant: Implementation of a dedicated senior subject matter expert role to provide consultancy for the project managers in all the countries. The aim of the Project Management Consultant position is to advance the project management processes, tools and methodologies to truly reflect the ambitions of Project Management Excellence.
4. Project Support: Implementation of a dedicated operational project management process role to support operations, maintain the various platforms for information and knowledge sharing, train on operational aspects.
5. PMO integration in operations and geographic regions: The Global PMO is highly linked to the project implementation organizations in the respective countries to enable close alignment. PMO Team members are located in various locations to allow for a truly global reach.

---

8. See as well for importance of strategic support of Project Management: PMI's *Pulse of the Profession*, March 2013, page 3.

9. "High-performing organizations are almost three times more likely than low-performing organizations (36 percent vs. 13 percent) to use standardized practices throughout the organization, and have better project outcomes as a result." Source: PMI's *Pulse of the Profession*, March 2013, page 10.

10. See as well http://wenger-trayner.com/Intro-to-CoPs/ for more information about Community of Practice (CoP).

*Project Management Office Services Offering*

All PMO services are positioned around the Project Managers. The Project Managers are central to all initiative areas and projects of the PMO, as shown in Figure 12–1.

- The initiative area **Skills and Competencies** focuses on all aspects to train, educate and certify the project managers (and project teams). This includes the following services:

  - Designing and maintaining the SCS Project Manager curriculum
  - Organization of PMI PMP/CAPM[11] and Prince2 Foundation and Practitioner certification preparation training
  - Organization of SCS specific training, conferences and webinars
  - Mentoring of new project managers
  - Creation of trainings around internal tooling/process/methodology
  - Facilitation of best practices, lessons learned, knowledge sharing and management
  - Talent management

**Figure 12–1.** Project excellence pyramid diagram.

---

11. PMI (Project Management Institute) Project Management Professional (PMP)/Certified Associate in Project Management (CAPM)

- The initiative area **Tools, Processes, Methodology** focuses on all aspects to improve tools, processes and methodology. This includes the following services:

  - Standardization of Project Management practices and processes
  - Design of the SCS Project Management framework SUCCESS 2.0
  - Improvement of the IT system landscape related to project management
  - Creation of templates (e.g., SOW, WBS, risk register) adapted to used methodologies and related business
  - Support of all improvement by strong change management

- The initiative area **Operations** focuses on all aspects to ease the life of the project managers (and their managers) from an operational perspective:

  - Support specific projects in trouble, project recovery, project health checks
  - Support project operations on system landscape
  - Maintain the platforms for information and knowledge sharing
  - Train on operational aspects
  - KPIs and reporting
  - Standardized communications.

### Project Manager Community of Practice

As mentioned the project managers are central to the PMO services, in particular the Community of Practice (CoP). A CoP has the following characteristics:[12]

1. Shared domain of interest (project management)
2. Members are practitioners (project managers)
3. Engagement of the members in joint activities

The Global PMO supports and facilitates the CoP in close teamwork with a CoP core team. The latter consists of volunteering project manager practitioners from different geographic regions. The goal of the SCS PM CoP is to Share, Learn, Leverage, Network and Communicate among the project managers. Engagement in the CoP varies from very active contributing members to passive participation. In addition, the PM CoP is one way to reach out to the project managers to share information.

The exchange in the CoP itself helps build individual and group competencies, resolve problems and avoid re-inventing the wheel. The CoP itself also provides a feedback mechanism to the PMO to help enhance and develop its services and its future direction via various tools (e.g., with using polling technology, the CoP can advise on preferences and priorities). From the beginning, the project managers are involved actively and their voice is taken into account.

Activities include regular online meetings to exchange information on the key topics, support in the preparation of face to face training and conferences, as well as build new artifacts in creation of new content for a specific product domain or project management processes. Online CoP Meetings are a repeated success. It's the best way to get the instant feedback from the community for the community.

---

12. Source: http://wenger-trayner.com/Intro-to-CoPs/

*Processes*

Within Software Customer Services standard processes and methodologies are used throughout the lifecycle of a project. Figure 12–2 gives an overview:

It shows how Philips Healthcare processes are structured, how PMI process groups are mapped and how the process frameworks overlap for smooth handoff between areas of responsibilities within SCS.

- **Order Acquisition**: Tight teamwork between sales and project management is highly important in Clinical Informatics solution projects following a defined Sales Bid Management process.
- **Order Realization**: SUCCESS 2.0 is the SCS project management framework supporting the right milestones, templates, acceptance, reporting, etc.—all consistent with PMI and Prince2 combined with best practices from clinical informatics. SUCCESS 2.0 is closely related to the Sales Bid Management Process and enables a smooth handoff to the Customer Care area.
- **Customer Care**: ITIL[13] is an industry best practice to set up a highly efficient, state-of-the art IT services concept.
- **Customer Experience**: To make current customer experience visible at different project milestones and later in the customer care phase customers are invited at key touch points in the Software Customer Lifecycle to share their feedback with the SCS organization. The survey process continues beyond implementations and upgrades, with ongoing support and maintenance. Feedback received is evaluated

**Figure 12–2.** Overview of standard processes and methodology.

---

13. ITIL (IT Infrastructure Library) developed by the Office of Government in Commerce (OGC) in the United Kingdom.

under the aspect of congruence with the desired customer experience, identifying room for improvement and also learning about strengths with regards to Project Management Excellence and later on Customer Care. This closed loop process is a highly effective approach and key characteristic of a continuously learning and improving organization.

An important takeaway is that the different processes are all inter-related and thus break the traditional silo approach. Communication and teamwork are some of the key aspects that prevailed during the definitions of these processes. Especially in a global organization like Philips Healthcare SCS it is important to implement, train and improve harmonized and standardized processes. It is also important that everyone speaks the same Project Management language and uses the same terms. This is one of the reasons why every SCS Project Manager needs to be PMP certified (for the UK & Ireland it is Prince2 Foundation & Practitioner according to market needs), and is trained on the SUCCESS 2.0 Project Management framework and is at least ITIL Foundation certified.

*Key Takeaways for Global Project Management Excellence with a Global PMO*

The key takeaways for achieving **Project Management Excellence** with a Global PMO could be summarized as follows:

- **Project Management Excellence**, even though it is not an absolute objective per se, it is considered a **proactive way** to anticipate and fulfill the **needs of the Clinical Informatics Industry**.
- Project Management Excellence is not a static objective. It requires **continuous improvement** around **People, Processes and Tools**.
- A **PMO is a key enabler** for **Project Management Excellence** where specific roles focus and setup is very important for the success.
- A **Community of Practice** of the Project Managers is a state-of-the art approach and a key and recognized best-practice to Share, Learn, Leverage, Network and Communicate together.

**Process Harmonization and Standardization** is highly important for the success of an organization operating globally. Tight integration in the upstream processes (e.g., sales, bid management) and downstream processes (e.g., customer care) are very important too. This has to be supported by solid change management and training activities.

## 12.3 maxIT-VCS[14]

maxIT-VCS has a PMO, which serves as the "home" for all of our project management consultants. The PMO reports directly to our senior vice president, who reports to our CEO. The group is matrixed into our functional or practice specific divisions. The PMO

---

14. The material on maxIT-VCS was provided by Marc Hirshfield, PMP, Director, Project Management Office at maxIT-VCS.

serves the company in two primary ways (internally and externally). Internally, we offer employees a career path into project management. This includes access to methodology, as well as standards and educational opportunities (i.e., PMI's PMP® credential) that are solely focused on project management. Externally, we offer our customers services with respect to healthcare information technology project management. This includes certified project managers with expertise in the industry (average at VCS is 14 years). We also offer our customers support with methodology and the tools to build the foundation of a PMO via our PMO JumpStart solution. maxIT-VCS has provided project and program management support services to our customer base for approximately seven years.

maxIT-VCS decided to invest and create the PMO to meet both internal and external market demands. Internally, the PMO allows maxIT-VCS to offer a home and path for employees who want to focus solely on the career path of project management. It provided opportunities across different software vendor suites and offers a formal methodology and documented approach to running a healthcare IT initiative (project). Externally, it met the needs of our customer base who were looking for qualified project managers to help support their healthcare IT initiatives.

The biggest challenge with a PMO is following one methodology or standard across multiple initiatives which vary in size, scope and complexity. The key to success is making sure your PMO follows a consistent format, provides the tools needed to get the job done, but allows for scalability based on the individual PM's needs. A "one size fits all" mentality just does not work when running a PMO.

When we started the PMO at maxIT-VCS, we had full executive (principal) support. Everyone was on-board with the value in the practice and this fact alone lead to our success. As we all know, having executive buy-in and support is not only critical to a successful project, but also integral in building a PMO.

Not only is the role of a PM at maxIT-VCS considered a career path for our employees, but it is one of our solutions. Most healthcare organizations we provide services to are still learning the true value and ROI of a project manager. Some early adopter organizations that have invested and supported a PMO properly are seeing a large return on their investment. We have job descriptions for our project managers, which include levels one through three (three being our senior PM).

Heidi Wurtz, VP, Solution Center, maxIT-VCS states that:

maxIT-VCS, an SAIC company, continues to support a PMO as we work to integrate our combined management and strategic consulting services. The PMO resides within our Solutions Group, the management consulting arm of our integrated organization. The Solutions Group has defined a Portfolio Management initiative with a vision to support maxIT-VCS's effort to improve its portfolio and program management capabilities. It seeks to capture and integrate key project data (status, schedules, resource assignments, etc.) in a data warehouse and present needed information to managers with specific needs. This will provide a foundation for business intelligence across the entire organization, while eliminating or streamlining manual processes. The Portfolio Management Opportunity Summary includes:

- Optimize leadership and management time spent on internal program and project communications
- Provide a business intelligence environment for all levels of the organization to help manage programs and projects
- Standardized and focused key documentation for maxIT-VCS PM and PD engagements

- Leverage data entered into a project management database
- Leverages intellectual property—archives documents and data for future use
- Support and standardize expectations of PMs, PDs and analyst roles across practices
- Identify those projects requiring management attention based on business criteria (standardizing Project Review process)
- Client quality review score will be included, supporting appropriate interactions with clients, and obtaining critical data for ongoing improvement, proposals, and sales and marketing

## 12.4 AVIVA[15]

**Structure**

### Market Context

Aviva-Canada is one of the leading Property and Casualty insurance groups in Canada providing home, automobile, recreational vehicle, group and business insurance to more than three million customers. The Company is a wholly owned subsidiary of UK-based Aviva plc and has more than 3,000 employees, 25 locations and 1,700 independent broker partners. Aviva Canada and its employees invest in positive change, including through the Aviva Community Fund and Eva's Initiatives,[16] its partner in Aviva's global Street to School program to help homeless and other at-risk youth reach their potential. (http://Avivacanada.com/content/member-companies.)

### Industry Context

Business megatrends[17] have started to shape, and impact many industries. Customer service need for speed, increased transparency, and a personalized ability to track products and services is our present reality. As a result, the business megatrends translate to insurance megatrends (to all: General Insurance, Life, Health Care, etc.) where all are pressured to become more agile (increase productivity—do more with less), digitize the value chain (both vertically, and horizontally), and human capital management is even more important than ever (up skills and invest in the skills of the not so distant future). See Figure 12–3.

We describe Aviva Canada's EPMO as market-driven type, operating in a hybrid-federated model. Our structure and processes allow the function to better respond to, and support an organization that focuses on profitable growth, and being continually more nimble and efficient. Allowing for a dynamic approval of initiatives enable agility and nimbleness. As a result initiatives are submitted as the environment justifies the investment (legal, regulatory, response to a competitor or cost pressure, etc.) Our approach is to manage within

---

15. ©2013 by Aviva. Reproduced by permission. All rights reserved. Information provided by Ian Laliberté, Vice-President, IT Strategy and Change

16. Eva is the name of the youth center Aviva Canada supports

17. *Service 2020: Megatrends for the Decade Ahead*, a BDO Report, written by the Economist Intelligence Unit, Summer 2011, 37 pp

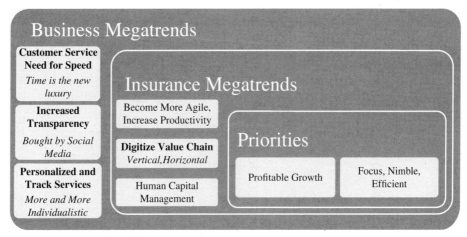

**Figure 12–3.** Business and insurance megatrends shaping priorities.

the total portfolio investment budget for the year (we set it at a % of revenue target) and this is supported by a strong expand and contract resourcing model providing flexibility in human capital. Traditional EPMOs mostly operate where initiatives are submitted once a year (and bottom-up) during the budgeting and planning season, where tremendous effort is spent on selection, prioritization and forecasting demand and supply over the next 12–18 months. We do not believe in this approach; if the environment demands the business to respond then it must be presented for evaluation and decision—let's hear it! Our job is to then make it happen if approved—this is why we are here for: leading and facilitating enterprise-wide changes.

## Leadership

A culture of leadership is a critical ingredient to a market-driven type EPMO. To continually strive for speed, efficiency, adaptability and flexibility each and every single person (and not just management) must be able to challenge status quo, bust bureaucracy as required, and make their approach fit for purpose within the directional governance framework. Leading from the edge as opposed to *within*, or *top down* is how we partner and engage.

### BHAG, 4 Pillars

To achieve an element of thrust and connection with everyone, a clear big, hairy, and audacious goal (BHAG) is set—a transformative statement that continually reminds everyone where we come from, what we are aiming for and by doing what. To achieve the envisioned future and bring focus into how we will do this, four pillars support the BHAG. A simple statement that connects with full time employees and contractors. This is shown in Figure 12–4.

This might be seen as difficult to maintain in a market-driven model, but is quite the opposite. This is core to our decision-making for hiring (interviewees are asked about their initial reaction and comment on the spur of the moment), for continuous improvement initiatives we should have (process and operations), for how we must behave internally and externally, and most importantly how we align with our stakeholders and customers. Each FTE is assigned to one of the pillars to embody and drive the change required.

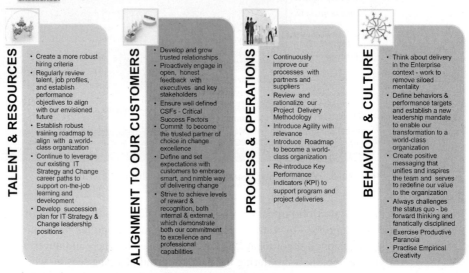

**Figure 12–4.**   IT strategy and change vision, BHAG and pillars.

### Industry Leadership and Get-Together

To have employees and contractors alike to strive to be thought leaders, versatile (do whatever it takes to get the job done even if not in the job description),we also recognized the team must be highly engaged and visible externally in various project management communities. To unlock the power of thought leadership, connecting, sharing and discussing challenges, few employees get together at least twice a year with our functional peers, from various industries. These half-day sessions are worth their value in gold for every organization, as cross-pollination of knowledge tried and true approach enable all of us to quickly disseminate key learning in each of our organization, and expand everyone's network. This also enables us to share our vision and approach, so that as staff and contractors change organizations part as their career, we try to shape what the future can look like. As of 2012, the most senior leaders connect with other EPMO leaders, then the same take place with our Enterprise Business Analysis practice, and Change Delivery and BPM practice.

Perfect example is also taking this opportunity in Kerzner's latest release. This is another platform to raise the thought leadership required to elevate such an important, core function in many organizations.

## Human Capital

### Solid Talent Management Review

Aviva as an organization focus is on its talent, and has a good talent management process and framework. In the EPMO function, we make sure every one person has his or her yearly objectives set, a personal development plan in place, and do have their monthly 1:1 with their respective line managers to discuss progress on their objectives and development. At the mid-year, the management team comes together and discusses all their staff and how they have been performing over time, and how their agility has been developing. This is a great way for the EPMO SLT to hear about how their delivery staff is being perceived by the rest of the organization, in fact, critical feedback for the development of our staff. The line manager then provides that feedback back to their staff in the following monthly 1:1, where honest feedback is provided, and an action plan is put in place. An important process, as it also empowers the employee to rectify its trajectory between the mid year and end of year performance review, for where the same process repeats but this time to discuss year end performance for merit and bonus increase.

### Clear Career Path

Career path is important, and every role does have a clear position profile that describes its responsibilities, but also demonstrate how one person can move from a junior, to intermediate, to senior role, and even laterally in other project or portfolio management roles and enterprise business analysis roles. All position profiles up to the vice president are made available to the staff.

### Visible Training Plan

The training plan is a bottom-up view of all staff personal development plan, and top down view of what the management team also envision what some of the individuals or teams will require in the present year based on the direction and focus of the organization as a whole. No one is ever left behind, and we make the training plan visible, as we are serious when we say we invest in everyone. Our admin staff also goes through productivity tools and project management training, this is important for everyone to speak the same language.

### Rewards and Recognitions

Rewards and recognition is also core and instrumental in making our team a success; in fact, a lot of what we do is learned and applied in other area of the organization. Celebration as part of project successes take place, rewards are given to individual and teams, and sometime simple "thank you cards," movie passes or coffee cards go a long way when it is one of your peer that is just simply saying thank you. Contractors are considered as much as part of the team as our full time employees; team unity and uniformity is core for us to become a world-class organization, and as such, everyone must feel they are the trusted partners of choice.

### Project Assignment Supporting Career Growth

To support the last three points above on career pathing, training and reward and recognition, our tiered project governance model also ensures more junior staff can take full accountability and responsibility of small to medium size type initiatives that fit their seniority. More will be discussed around this topic around this in the Functions and Operations sections.

### Hire for Success

A lot is being written in this section, but that is because if there are no people, you simply do not exist as a function within an organization. Therefore, people are the nucleus of success. To sustain this, we ensure our hiring managers look at both "the what, and the how" of a candidate. To ensure focus around these two areas, our people managers agree on who will focus on what before the interview takes place; remember, to become world-class there are no compromises on the quality of individuals we bring in; full time or contractors we follow the same process. Professional designations are a must (PMI, Prince2, CBAP) – if years of experience justify the hiring, we make sure the designation gap is closed within 12 months of hiring. Delivery or management type roles (up to the VP level), designations are a must. To be clear, we constantly re-enforce the message that operating in this space requires great energy and drive; individual and team contribution are important, and we do recognize not everyone can adapt to such an environment—we understand and make sure that we exit underperformers from the organization.

### Role-based Onboarding Program

Once a candidate is selected, most organization would welcome the new team members on their start day and give them a tour around, meet their new key stakeholders, etc. Not us. Onboarding is serious business; nimbleness and speed start day 1. We have developed role-based onboarding profiles where it could take 1–2 weeks of learning based on the role and seniority. Everything is covered; BHAG and our four pillars, tools, process, governance, facilities/logistics, approval and decision-making process, organizational charts, etc. Investing 1–2 weeks upfront save you weeks of running around and inefficiencies, and worst, minimize potential costly mistakes as a result of just not knowing what is and what isn't. As simple as this might sound, we are always surprised to hear how much this is nonexistent in other organizations from new staff we bring onboard, but also ecstatic to always hear how great our onboarding program is.

### Embedding and Cross-pollination

Finally, this section wouldn't be complete if we didn't close the loop on our hybrid-federated model. Federated because our tentacles are in every function of the organization through our project delivery, and when we do have large, complex, multi year programs in execution, we do want our people to change reporting lines, and in others, we maintain the internal capacity but offer a different service; we are hybrid and adaptive. In the federated sense, we welcome secondment opportunities (input) and letting our best people go (output) because the in and out of staff further pushes the management and delivery across the organization, and in turn they coach and mentors others in this operating space; again, constantly driving toward enabling focus, nimbleness and efficiencies in enterprise-wide delivery. Hybrid because it isn't a true federated model, but also because in some instances we developed a program team that only support a specific function (our Personal Lines)—a great opportunity for PMs and BAs to learn about the insurance business operations, and even consider a career change in the function. Again, we never consider this a lost, but a tremendous win-win for both functions, and the organization as a whole.

**Function**

## Functional Teams—Our Strategic Partnership Ecosystem

We have found the right mix of roles and functions required allowing to quickly expand and contract based on the market dynamics. As its core, our team (including management) is made of about 30 FTs, and can easily be double in size based on portfolio load. The strength of our operation and

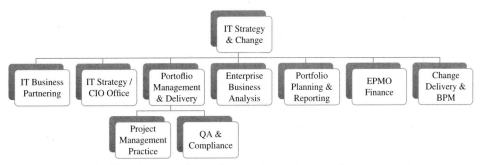

**Figure 12–5.** Aviva Canada's IT strategy and change functional overview.

what enables us to be seen more and more as the strategic partner in change excellence is because of the breath of our unit that goes beyond traditional project, program and portfolio delivery. EPMO operates within the IT Strategy & Change function, where the following teams could be seen as leading from the edges, leading from the core, and a mix of the two—our true enterprise delivery system. See Figure 12–5 for IT Strategy & Change functional overview.

*Leading from the Edges:*
- **IT Business Partnering:** At two levels. At the first level business partnering we have senior managers where their role is to be a 'triangular' escalation point between project / program managers, functional leaders from the area they represent (can be project or BAU related), or IT specific that might / might not include a 3rd party. At the second level we have the most senior IT leaders appointed as a partner to the other largest functional areas (like Sales, Finance, Underwriting, etc.). Their role is to be the bridge between other EC members and their SLT to the CIO office. The most senior leader in IT strategy and Change is appointed to Broker Distribution (Sales).
- **IT Strategy / CIO Office:** This section directly supports the CIO Office in managing all requests for information, Executive Committee and Board presentations, as well as ensuring a solid information and technology roadmap is in place that supports directly the strategic direction of the organization as well as answering the tactical priorities, mainly through its portfolio delivery.

*Leading from the Core*
- **Portfolio management and delivery:** This is the project management delivery arm of Aviva's portfolio; this is where PCs, PMs and PgMgrs report into. This is also where our Project Management practice resides.
- **Enterprise Business Analysis:** Being an enterprise PMO requires a wide business speak transcending many functional areas – all BAs supporting the portfolio delivery report in this function
- **Portfolio Planning and Reporting:** This team is the application owner of the project and portfolio management package, where the portfolio demand and supply

information reside. This team produces all the business intelligence information required to all various stakeholders; our portfolio performance, health and status is made visible to all 3,500 employees across the organization.

- **EPMO Finance:** We have a team of sr. project financial analysts (SPFAs), where each of them have a mix of projects to support; no PMs or program managers are left alone. We manage our portfolio like a mutual fund; project fluctuates up and down every month and we make ensure we do not burden the delivery management team with maintaining proper Cash and P&L views. We have professionals that do this for a living (just like PMs or BAs), and this ensures that proper finance standards are followed with our investment and reporting. This team also supports the development and initial financial details of any business case, and works closely with respective functional finance business partners as required. Also it tracks all statement of work, and invoices as they come in to ensure they match the initial agreed commitment.

*Leading in Both Spaces*

- **Project Management Practice Lead:** Leading from the core and because of our expand and contract model, contractors do need a lot more coaching on how to navigate and work within our governance model. This is also where our project rescue function takes place—when things go wrong, we have someone who can quickly step in and work in partnership with the owners and delivery team to help bring things on track. In addition to leading from the edges, this area also owns the PMCoP (Community of Practice) where practitioners and functional leaders performing in a project manager role come together and discuss specific topics of interest and issues, and fosters an environment of networking with peers across the organization.

- **Change Delivery and BPM:** This area leads from the core where one needs to holistically understand all project interdependencies (schedules, resources) and in using this viewpoint, actively defines initiatives to continuously optimize the delivery channel and also have business process re-engineering work based on project assignments (mostly when it involve transformational type work in nature). Leading from the edge where external interdependencies must be understood (we all know about BAU type initiatives) during the initiate and plan phase, so that a sound schedule is put in place. Business process re-engineering also take place outside of the portfolio of projects, mostly supporting strategic and IT business partners as required when inefficiencies are encountered in various areas of the organization.

- **QA and Compliance:** This isn't a policing function, but an enablement function to expedite and provide simple directions to what is next from a governance perspective, or whom should be contacted based on where you are in the lifecycle. This function is also accountable to maintain all lifecycles (PMLC, SDLC, Agile, Iterative, Waterfall, and Hybrid). This team support specific projects, and will partner with various areas of the organization where functional leaders might need help to navigate the technology delivery landscape.

The strength of our model is that we are part of the decision-making process, we support all direct reports to respective EC members through our partnership model, we are a

**Figure 12–6.** Enterprise delivery model and interactions.

key link between various functions and portfolio delivery teams, we also help shape and support the strategic development of future investments, and leverage every single asset existing in Aviva's value chain to ensure delivery success. This is what we define as our enterprise delivery engine (See Figure 12–6):

Operating within this model is quite unique; we get to be exposed to the breadth and depth of the organization, and have a wide network of relationships and connections that break silo thinking. The challenge of any EPMOs is to feel they are one team. The nature of the work we do is to constantly dispatch the work request across the entire organization; when an initiative is approved, you then get assigned to work with the sponsor and owner of a function, bring people together and get the job done. Unless it is a really large project or program, you would get to work with a few key members, but otherwise, only small pods of the EPMO get to work together at any point in time. As a result, it is important to constantly perform team-building activities, and even though there is a functional structure in place, it wouldn't be complete if we didn't talk about the activities that are the glue to this function:

- **Monthly PM/PC meetings:** This is when all project managers and coordinators get together and discuss their initiatives and struggles; it's amazing to see how much dependency can exist between teams (resources, system promotion windows, vendor requests), and how others are willing to support and help each other. Creating an awareness of what else is in the pipeline through a simple roundtable achieve a lot.
- **Quarterly Departmental Meetings:** it is important to bring the department together once a quarter to discuss the latest top down news, decisions, org changes,

rewards and recognitions, and discuss any hot topics. These updates and discussions are critical because these teams must continually navigate through the organizational landscape.

- **Social Committee/Get Together:** In the summer nothing better than inviting everyone for a late afternoon refreshments—it allow everyone to disconnect and get to know each other outside the corporate walls
- **Corporate Social Responsibility:** We like to mix fun, team get together, and impacting our community at all at once. Aviva allow each employee to give 15-hr/year in helping the communities we live and work in. We take this seriously, and make it a part of our personal objectives. Every year we raise a few thousand dollars and do work (local youth organization, tree planting, etc.).
- **Continuous Optimization:** Every year we deliver between 10 and 20 continuous improvements initiatives for our area alone. Adaptation is key, and our EPMO is in constant flux to always better respond to the need of the organization. Therefore, we make sure to include this in everyone's objectives and ensure that all contribute to make this a better, flexible place to work in.
- **Quarterly Roundtable with PMs/PCs:** When many projects are on the go, at various stages of the lifecycle, it is important our project managers come together and share with their peers about their initiative and struggles. This cross-pollination of information sharing elevates the organizational awareness of all change initiatives, but more importantly, help identify interdependencies between projects.

**Operations**

### Governance: A Partnership Model

Our model is about enabling the organization to be focused, nimble, [and] efficient and provide the speed to market required. Therefore, we understand that our governance cannot be about control and policing, but rather being fit for purpose and delivering agility. To achieve such, our lifecycle is: business speak (proposal, initiate, plan, execute, closeout), nimble and self-served (we have 3-tier governance), adaptive (waterfall, iterative, agile, hybrid), agile and effective (5 phases, 5 gates, 3 approvals). See Figure 12–7.

### Business Speak

This looks a lot like PMI's proposed lifecycle at the macro level and this is true. Aviva Canada being in North America coupled with our expand-contract resource operating model and hiring practices, we want to make sure our project management professionals (PMI or Prince2) can quickly grasp our organizational blueprint in a language they are all familiar with. This also applies to any new senior management joining the organization who becomes either a Project Owner or Executive Sponsor.

### Self-Served and Tiered Governance

Not one size fits all. Speed to market, ensures two things: the organization is enabled to lead and manage their own small to medium-size initiatives and allows those initiatives to move quickly through the organization by requiring less bureaucracy and

**Figure 12–7.**  Aviva Canada's adaptive, flexible and market-driven lifecycle.

approvals regardless of who is leading it. This is supported by a self served governance assessment (as EPMO mostly lead and manage the large, complex, transformative type initiatives). Our self-served governance assessment is completed by check boxes, where we ask simple questions around performance (expected duration, team size, value) and complexity (strategic value, business risk, business and customer impact, technical risk, technical impact). The output will indicate if the project should navigate the organization in a light, medium or full governance approach. Now if we tie this back to our Human Capital section around career paths and development, this ensures our most senior delivery personnel are assigned to the most complex and large type initiatives, and also gives the opportunity to our junior staff to lead and manage small to medium-size initiatives, having P&L accountability relevant to their career experience and maturity. See Figure 12–8.

### Self-Served and Adaptive

No one size fits all, and no one delivery approach fits all. We ensure the delivery methodology applied to the initiative is also fit for purpose, to ensure maximum delivery throughput to achieve the desired outcomes and realize benefits as soon as possible. As such, a self-serve model is also used here. We use ten simple questions to help determine if the delivery methodology should be waterfall, iterative, agile or hybrid. Respective process flow maps are integrated within the PMLC; therefore, once the delivery methodology is identified, one simply needs to pull the chart and follow the process flow steps. See Figure 12–9 for an example of our self-served project methodology selection and see Figure 12–10 for an example of how we map our iterative workflow within our planning stage of our lifecycle.

### Agile and Effective—Portfolio Management

Having a great tiered governance approach, supported by multiple fit for purpose delivery models will only get you so far until the senior management has full transparency and

**AVIVA** GOVERNANCE ASSESSMENT

| INITIATIVE | <Project Name and ID> |
| DATE | <Information complete as of YYYY-MM-DD> |

Check 1 box in the appropriate column of each of the performance and complexity indicator rows below to calculate your governance level.

**PERFORMANCE INDICATORS**

| | | | | | |
|---|---|---|---|---|---|
| DURATION | < 6 weeks | < 6 months | < 12 months | < 24 months | > 24 months |
| TEAM SIZE | ≤ 5 resources | ≤10 resources | ≤ 15 resources | ≤ 25 resources | > 25 resources |
| COST | ≤ $50,000 | ≤ $250,000 | ≤ $1 Million | ≤ $5 Million | > $5 Million |

**COMPLEXITY INDICATORS**

| | | | | | |
|---|---|---|---|---|---|
| STRATEGIC VALUE | Minimal Focus is internal operations (i.e., maintenance) | Limited Business related objectives | Visible Influences cost, operations, revenue or customer loyalty | Distinct Visible impact on industry and market position. | Critical Major impact and visibility on business outlook |
| BUSINESS RISK | No Appreciable Business Risk | Some General Business Risk (<$1 million) | Some Key Business Risk ($1-$10 million) | Increased Key Business Risk ($1-$10 million); Mandatory; Regulatory | Extensive Critical Business Risk (>$10 million) |
| BUSINESS AND CUSTOMER IMPACT | Single Internal Business Unit; Non-customer Facing Employees | Multiple Internal Business Units; Some Customer Facing Employees | Multiple Internal Business Units; One or More Vendors; Customer Facing Employees | Multiple Internal Business Units or Vendors at Multiple Locations; Some Policy Holders and Brokers | Multiple Internal Business Units or Vendors at International Locations; Policy holders or Brokers |
| TECHNICAL RISK | Existing Aviva System; Off the Shelf; Industry Standard | New to Aviva; Off the Shelf; Industry Standard | New to Aviva; Customized; Industry Advanced | Industry New; Custom Built (bespoke) | New Leading Edge; Custom Built (bespoke) |
| TECHNICAL IMPACT | No Change to Infrastructure; No Change to Applications, Interfaces, or Data Model | No Change to Infrastructure; Minor changes to an Application, Interface or Data Model | Minor Change to Infrastructure; Changes to One or More Applications, Interfaces or Data Models | Major Change to Infrastructure; Major Changes to Multiple Applications or Interfaces; Changes to Shared Data Model | Multiple Major Changes to Infrastructure; New Application Family, Multiple New Interfaces or Multiple New Data Models |

| RELEASE MANAGEMENT LEVEL | | GOVERNANCE | |

| GOVERNANCE LEVELS | Light | Medium | Full |
|---|---|---|---|
| | Minimal technical risk or impact; limited business risk or customer impact | Moderate strategic value, technical impact, and/or business impact | High strategic value; major technical risk and impact; and/or business risk or customer impact |

**POTENTIAL DATA QUALITY ISSUES**

Please review the Data Issues Log and list those Data Issues that this project plans to address/remediate, or those that could impact this project if they are not remediated. Indicate any other concerns that may exist with the quality of the data thtat is involved with this project.

**NOTES (ATTACH OR REFERENCE SUPPORTING DOCUMENTS WITH VERSION AND DATE)**

SEND COMPLETED ASSESSMENT TO THE PG AND COMPLIANCE MANAGER

**Figure 12–8.** Self-served governance assessment.

visibility into the portfolio. Strategic directions and decisions are made at the top; therefore, the knowing-doing translation starts right at the Executive Committee table. It is critical that right at that table we have a simple definition of a project (to remove the BAU vs. project stigma) and all agree on new change investments required, and are clear on present progress, status and health of the portfolio.

Back in 2010, our CEO endorsed and published from the Executive Office what is the definition of a project.

**Figure 12–9.** Self-served project methodology selection.

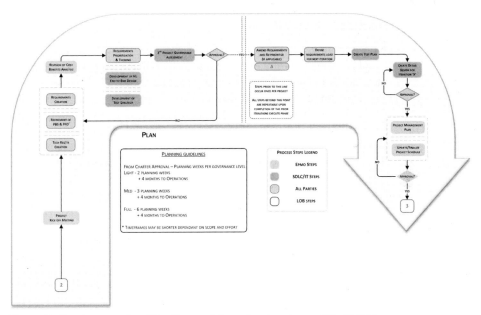

**Figure 12–10.** An example of iterative workflow planning stage within PMLC.

A simple definition all employees understand. To support our relentless focus on execution and alignment, part of this published definition is that no work should be done on projects that are not listed in the portfolio. If an employee is requested to work on an initiative not listed on the portfolio, the employee needs to escalate with their respective line manager (we publish our portfolio on the intranet, every month). See Figure 12–11 for Aviva Canada's project definition.

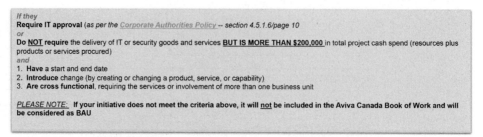

*If they*
**Require IT approval** (*as per the Corporate Authorities Policy -- section 4.5.1.6/page 10*
*or*
**Do NOT require** the delivery of IT or security goods and services **BUT IS MORE THAN $200,000** in total project cash spend (resources plus products or services procured)
*and*
1. **Have** a start and end date
2. **Introduce** change (by creating or changing a product, service, or capability)
3. **Are cross functional**, requiring the services or involvement of more than one business unit

*PLEASE NOTE:* **If your initiative does not meet the criteria above, it will <u>not</u> be included in the Aviva Canada Book of Work and will be considered as BAU**

**Figure 12–11.** Aviva Canada's project definition.

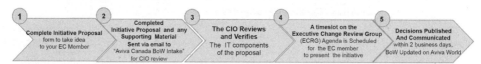

**Figure 12–12.** Adding a project to Aviva Canada's portfolio—initiative proposal and approval workflow.

## No Hidden Iceberg

As per our lifecycle, anyone that has an idea can submit it, by using our Initiative Proposal template. Simple information is asked to help shape the idea and bring consistency to how ideas are presented for funding approval. See Figure 12–12 to see the simple steps.

This simple process ensure that all projects are first vetted with their respective Executive, so that it can be validated right away if this is considered a priority or not. From there and if their Executive approve, it is submitted to our CIO to ensure high level IT estimates (order of magnitude) include all the components required. Once validated by the CIO, we submit for the next Executive Committee meeting where the respective EC member will present and gain the EC approval. During that meeting, decisions are made on classification (run transform, grow), priority level (P1, P2 or P3 bucket), funding type (Corporate, Business), and who should be leading this (EPMO, Business). We then communicate the decision.

## ARM—Assess Resource and Methodology

Every month we have a 30-min timeslot scheduled at the EC meeting for any Initiative Proposal approvals (if any) and to review the portfolio's health and financial status. We learned from experience that once an initiative is approved, everyone wants to start right away; but yet assignment of resources by various line managers has not yet taken place. To expedite from the approval to resourcing so we can start our Initiate stage and choose the right delivery methodology, we have scheduled ARM meetings one business day following the EC meeting so that line managers come together to review the approved proposal and then identify their appropriate lead assignment (who's the best fit and based on supply available). The immediate effect is that all respective managers are in synch about what are the initiatives and know who and by when they need to provide resources. This setup has provided us with a faster launch approach.

**Figure 12–13.** Example of our portfolio financial view.

## One Version of the Truth, Status and Health of the Portfolio

The second piece we review at the EC meeting is the financial health of our portfolio. All projects are classified by Run, Transform or Grow. This enables us to keep a check on priorities alignment and where we need to bring focus or re-balance when required. In addition, every functional leader at the top know exactly how their change investment portfolio is progressing. See Figure 12–13.

## Company-Wide Visibility and Transparency

Once we close the month end status reporting and financials, our portfolio team produces relevant monthly dashboards that get published on our intranet, and a link is sent to all Aviva Canada's management team (Sr. Manager and above). On our reporting landing page we update and publish our portfolio status including: projects by stage and gates (everyone is aware of what is coming, what is happening, and what is closing), how our projects and resources aligned against the strategic priorities of the organization, our overall risks and issues status, and resource demand and supply view by project, resource type and portfolio. See Figure 12–14 for an example of our resource demand and supply dashboard.

## Customer-Centricity

The term "Customer" in a service department can have different meanings causing confusion or overlooking the true business customer. As strategic partners to the business we have expanded our understanding from our internal Business customer, to our external customer. This allows us to speak the same language as the business as it relates to a new product or service placement within a customer segment. For Aviva our Customer is our Brokers and in order to enable a foundational knowledge, we use a "Broker-On-A-Page" tool. The Broker-On-A-Page tools provides project teams and executives with a single snapshot of a specific customer segment, including critical success factors, desired interaction model and verbatims directly from the segment. This defines the segment distinctly to enable focused delivery. The Broker-On-A-Page tool is used as a vital decision-making tool to manage scope-creep and keep what is important top of mind. See Figure 12–15.

## Active Listening for Future Planning

Active listening requires comprehending key messages being sent to us by our stakeholders, peers, customers and market, and then acting upon them. Active listening is critical to

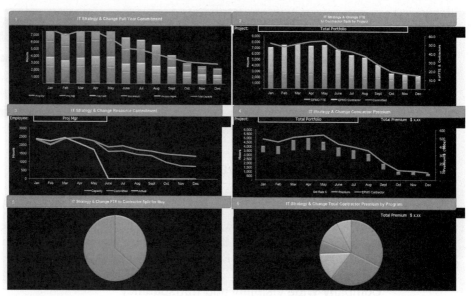

**Figure 12–14.** Example of one of the six dashboards: Resource demand and supply allocation.

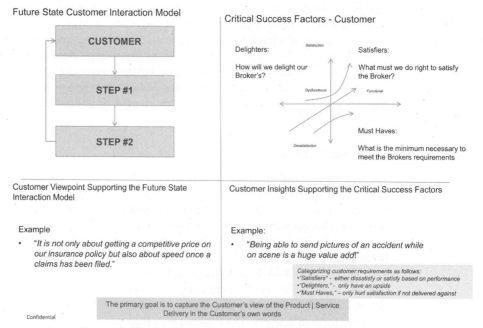

**Figure 12–15.** Example of Broker-On-A-Page tool (or Customer-On-A-Page).

maintain a world-class EPMO which allows us to focus on "what matters" today and proactively plan and optimize for the future. This core competency has delivered several ideas which are at various stages of development for potential future deployment. For example, we are investigating the implications of a modernized relevant method of strategic thinking based on scientific research of living systems. This self-enabled thinking method is planned to support asking better strategic questions, further strengthening our decision making. Listening, understanding, sharing and responding to core messages from broad set of sources, internally and externally, powers the team to focus on what is important today, and while remaining forward-thinking.

## 12.5 CHURCHILL DOWNS INCORPORATED (CDI): ESTABLISHING A PMO

Deciding to implement a PMO is easy. Being able to do it requires that certain obstacles be overcome. Chuck Millhollan, director of program management at CDI, discusses the chronology of events his organization went through and some of the obstacles that had to be overcome:

> One of our primary barriers to implementing structured project, program and portfolio management processes was "familiarity." The Churchill Downs Incorporated PMO, chartered in April 2007, is the first in the thoroughbred horseracing industry. Our senior-most leadership understood the need for a structured, standardized approach to requesting, approving, and managing projects and maintaining the project portfolio; however, many of the organizational resources had never been exposed to formal project management concepts.
>
> Our executives took an active role in the implementation process.
>
> I would say this is one of the primary factors influencing the early success enjoyed by the Churchill Downs Incorporated PMO. We chartered our PMO with clearly defined vision and mission statements and business objectives. Our CEO signed the charter, granting authority to the PMO to expend organizational resources as related to managing capital projects.

> i.  Our PMO was chartered in April 2007.
> ii. We developed a three-fold mission focused on the need identified by our senior leadership:
>
>   1. Establish, facilitate, and manage the project portfolio selection and funding process.
>   2. Create a foundation for consistent project success throughout the organization through development of a strong and pervasive Project Management discipline within CDI's project teams.
>   3. Guide key projects to a successful conclusion by providing project management leadership, while improving the quality and repeatability of related processes.
>
> iii. We defined the PMO's business objectives and linked progress to the PMO Director's compensation plan. Objectives included:
>
>   1. Develop and Implement Standards for Project Selection.
>   2. Develop and Implement a Standardized Project Management Methodology.
>   3. Build Project Management Professionalism among CDI Staff.
>   4. Manage the CDI Project Portfolio.
>   5. Direct Project Management for Key Strategic Initiatives.
>   6. Ensure Processes for Benefit Realization.
>
> iv. We also conducted training classes on project management, team building, critical thinking, etc. . .. to not only share our knowledge, but also to build relationships with project team members and other stakeholders.

     v. The PMO has facilitated a book club (also chartered with clearly defined objectives) for the last year. This process received recognition throughout the organization and has directly contributed to developing relationships between different departments. Our Book Club membership includes representatives from 9 different departments, ranging from Vice President level members to individual contributors.

       1. Objective 1: Personal growth through completing chosen books and active involvement in discussions.
       2. Objective 2: Explore creative ideas and ways of addressing real-world business issues through practical application of concepts and shared learning as related to Churchill Downs and respective teams.
       3. Objective 3: Promote interaction among different functional areas within the Churchill team by active participation in book club discussions and sharing opportunities for addressing real-world work-related issues in a safe, confidential environment.
       4. Objective 4: Share learning within respective teams through intradepartmental discussion and implementation of learning related concepts.

     vi. The primary driving factors behind Churchill Downs Incorporated's decision to staff a PMO were challenges with defining and managing the scope of projects, effectively allocating resources amongst multiple projects, and brining projects to a defined closure.

## 12.6 CHURCHILL DOWNS INCORPORATED (CDI): MANAGING SCOPE CHANGES

Mature PMOs either participate directly in the scope changes above a certain dollar level or set up processes for controlling scope changes. Chuck Millhollan, director of program management at CDI, identifies six steps necessary for scope definition and change control[18]:

**Step 1: Be Lean**

Trying to introduce any type of structure or control in an organization or environment that has been absent of controls can present a significant challenge. Before a project management organization can address scope change control, they must implement a process to define scope. Getting organizational decision makers to accept the project management precepts is not overly difficult, but changing organizational behavior to leverage these principles is another matter altogether. The more change we attempt to introduce into an environment, the more difficulty that environment has in adapting to, accepting, and embracing that change. To avoid the natural resistance to excessive change, a logical approach is to limit the scope of change and focus on immediate needs. Focus on the foundation and basics. Why have a complex, highly mature process if you are not consistently performing the basics well?

**Step 2: Define Preliminary Scope**

The immediate need for an organization without processes for capturing the business objectives associated with project requests is to define a structured approach for documenting, evaluating, and approving the prelimi-

---

18. C. Millhollan, "Scope Change Control: Control Your Projects or Your Project Will Control You!." Copyright ©2008 by Chuck Millhollan. Reproduced by permission of Chuck Millhollan.

nary scope of work. Note that *approving* the scope of work involves more than shaking heads, shaking hands, or a causal agreement on broad, subjective criteria. Approvals, in project management, imply documented endorsements More simply, signatures that provide evidence of agreement and a foundation to build upon. It is important to emphasize to stakeholders and sponsors unfamiliar with our profession's structured approach to managing projects that accepting a preliminary scope of work does not mean that you are locked in for the remainder of execution. Nothing could be farther from the truth. Instead, you are protecting their interests by beginning to set boundaries upon which effective planning can begin. In other words, you are increasing the probability (remember the research) that the project will be successful.

## Step 3: Develop Understanding of What Final Acceptance Means to Project Sponsor or Sponsors

How do we know when we have arrived at a destination? When traveling, we know our trip is complete when we reach our intended destination. Likewise, we know that a project is complete when we have delivered on the business objectives identified in the project charter, right? Well, yes . . . and then some. The "and them some" is the focus of scope change control. How does your organization define final sponsor acceptance? The recommended approach is to define sponsor acceptance for stakeholders using plain language. Sponsor acceptance is the formal recognition that the objectives defined in the original agreed upon scope of work have been met, plus the objectives agreed upon in all of the formally approved change requests. This plain definition helps to avoid the differing perceptions around what was wanted versus what was documented.

## Step 4: Define, Document, and Communicate a Structured Approach to Requesting, Evaluating, and Approving Change Requests

What is a change request? Some schools of thought suggest that changes are limited to requests for additional features, deliverables, or work. While this paper is focused on these types of change requests, or scope change requests, it is important to note that any change that has the potential to impact expectations should follow a formalized change request, approval, and communication process. Remember, aggressively managing expectations is our best opportunity to influence our stakeholder's perception of value. Scope, budget, schedules, and risks are typically interdependent and directly influence our stakeholder perceptions. Also, remember that the most effective change control processes include risk assessments that evaluate the potential risks of either approving or disapproving a change request.

Keep in mind that too much bureaucracy, too much analysis, or too much unnecessary paperwork will give stakeholders an incentive to circumvent your process. If you want your stakeholders to avoid, ignore or completely by-pass your process, include a great deal of administrivia. Administrivia is the new word for "trivial administrative process." (As the author, I reserve the right to add to the English language.) Remember, our profession's focus is on delivery and business results, not just adherence to a predefined process. Taking a lean approach to scope change request documentation can help influence acceptance of this sometimes painful, but vital, process for capturing change.

Process tip: Determine early (either as an enterprise standard or for your specific project) what the tripwires and associated levels of authority are for approving a requested change. What level of change can be approved internally? For example, a change with an impact of less than one week schedule delay or budget impact of under $10,000 may be approved by the project manager. What needs to be escalated to the project sponsor, what needs to be reviewed by a change control board or governance council? Determining these decision points in advance can remove a great deal of the mystique around how to manage change.

Ensure that everyone understands the difference between the natural decomposition process and identifying new work that must be accomplished to deliver on a previously agreed upon business objective and work associated with new or modified deliverables. Remember that omissions and errors in planning may lead to schedule and budget changes, but are usually not scope changes.

**Step 5: Document and Validate Full Scope of Work (Create Work Breakdown Structure)**

A great approach for defining all of the work required to complete a project is to start with the desired end state and associated expected benefits. What work is required to provide those benefits? What work is required to reach the approved end state goals (or business objectives)? Plan to the level of detail necessary to effectively manage the work. Decomposing work packages beyond the level required for effective management is considered Administrivia. Note that defining and communicating the processes for final sponsor acceptance and requesting changes both come before traditional decomposition. Why? Terrific question! The natural planning processes that we follow in breaking down business objectives into definable work packages can be a catalyst for change requests. We want to communicate up front that change is not free and that additional requests will need to be formally requested, documented, agreed upon, and approved before being included in the project scope of work.

**Step 6: Manage Change**

Your foundation is laid, you have documented the preliminary scope, you have defined processes for sponsor acceptance, you have defined and documented scope change request processes, and you have developed your WBS, now the only thing left is to manage according to your policies and plan. Almost forgot . . . you have to manage the change requests that are guaranteed to come too! Scope change control protects the project manager, and the performing organization, from scope creep and contributes to managing stakeholder expectations.

A question that frequently comes up among practitioners is "what do I do when my leadership does not allow me to define, document, and manage change?" This is a real, practical question that deserves a response. The instinctive approach is to communicate the necessity for a structured approach to documenting and managing scope. As our peers will confess, this is not always sufficient to get the support we need to set organizational policy. We can attempt to implement these processes without formalization, or just "do it anyway." This can be an effective approach for demonstrating the value, but can also be perceived as a self-protective measure instead of a process used to increase the likelihood of project success. People can be leery of someone else documenting requests, justifications, etc. . . for their needs. Ensure that you share the information and provide an explanation as to why this approach is designed to ensure you are managing to their expectations. In general, people have difficulty not accepting altruistic approaches to meeting their needs.

**Learn from Other's Lessons: A Real-World Application**

Leveraging experience, best practices, and lessons learned, the Churchill Program Management Office began with the basics; they chartered their PMO. The three-fold mission of the newly founded PMO was to establish, facilitate, and manage the project portfolio selection and funding process; create a foundation for consistent project success throughout the organization through development of a strong and pervasive Project Management discipline; and to guide key projects to a successful conclusion by providing project management leadership, while improving the quality and repeatability of related processes. Sounds fairly standard, right? The mission was then broken down into specific objectives and successful completion of these objectives was tied to the PMO Director's compensation.

**PMO Objectives Included**

1. Develop and implement standard processes for project requests, evaluation, and funding to ensure that approved projects were aligned with Churchill Downs Incorporated's business goals and objectives.
2. Develop and implement a standardized project management methodology, to include policy, standards, guidelines, procedures, tools and templates.
3. Build project management professionalism by providing mentorship, training, and guidance to project teams as they learn and adopt project management processes and best practices.
4. Manage the Churchill Downs Incorporated project portfolio by ensuring required documentation is in place and that stakeholders are properly informed about the ongoing progress of the project portfolio through effective reporting of key performance indicators.
5. Direct project management for key strategic initiatives.
6. Ensure benefit realization by using processes for clearly defining business cases and the associated metrics for measuring project success. Facilitate post-implementation benefit measurement and reporting.

As related to change control, we wanted to ensure that the process was lean, that our stakeholders understood the importance of the process, and finally . . . arguably most important . . . communicated in a way that our stakeholders understood and could follow the change request processes. Here is a thought provoking question for our practitioners: "Why do we expect our stakeholders to learn and understand our vernacular?" To aid in understanding and training, we developed visual tools documenting our overall project management processes in a language that they understood. For example, the project "race track" (see Figure 4–10) demonstrated to our leadership and project team members what we, in our profession, take for granted as universally understood; that projects have a defined start, a defined finish, and require certain documentation throughout the planning and execution processes to ensure everyone understands expectations and that we will realize the intended benefit from the investment.

For Churchill Downs Incorporated, scope change control begins with the foundation of a completed Investment Request Worksheet (or business case) and an agreed to scope of work as outlined in a signed charter. The work is then decomposed to a level of detail required to control the effort and complete the work necessary to deliver on the requested and approved objectives as detailed in that charter and approved scope change requests. A scope change request consists of a simple to understand, fill in the blank template, and the process is facilitated by the project manager. More important, the scope change request form is used to document the business objectives for a change request, the metrics needed to ensure the change's benefits are realized, the impacts on schedule and costs, the funding source, and the necessary approvals required for including the request in the overall scope of work.

Some of the benefits that Churchill Downs Incorporated has realized to date from this structured approach to documenting and controlling scope include:

1. Retroactively documenting scope for legacy projects, which resulted in canceling projects that were plagued with uncontrolled change to the point that the final product would no longer deliver the benefits presented in the business case.
2. Denying scope change requests based on factual return on investment and impact analysis.

3. Ensuring that requested scope changes would contribute to the business objectives approved by the investment council.
4. Empowering project team members to say "no" to informal change requests that may or may not provide a quantifiable benefit.
5. Demonstrating that seemingly great ideas might not stand up to a structured impact analysis.

## 12.7 TYPES OF PROJECT OFFICES

There exist three types of POs commonly used in companies.

- *Functional PO:* This type of PO is utilized in one functional area or division of an organization such as information systems. The major responsibility of this type of PO is to manage a critical resource pool, that is, resource management. Many companies maintain an IT PMO, which may or may not have the responsibility for actually managing projects.
- *Customer group PO:* This type of PO is for better customer management and customer communications. Common customers or projects are clustered together for better management and customer relations. Multiple customer group POs can exist at the same time and may end up functioning as a temporary organization. In effect, this acts like a company within a company and has the responsibility for managing projects.
- *Corporate (or Strategic) PO:* This type of PO services the entire company and focuses on corporate and strategic issues rather than functional issues. If this PMO does manage projects, it is usually projects involving cost reduction efforts.

As will be discussed later, it is not uncommon for more than one type of PMO to exist at the same time. For example, American Greetings maintained a functional PMO in IT and a corporate PMO at the same time. As another example, consider the following comments provided by a spokesperson for AT&T:

Client Program Management Office (CPMO) represents an organization (e.g., business unit, segment) managing an assigned set of Portfolio Projects and interfaces with:

- Client Sponsors and Client Project Managers for their assigned projects
- Their assigned DPMO (see DPMO Roles and Responsibilities for contacts)
- Their assigned Portfolio Administration Office (PAO) representative
- CPO-Resource Alignment (RA) organization Factories

The Department Portfolio Management Office (DPMO) supports their client organization's Executive Officer, representing their entire department Portfolio. It serves as the primary point of contact between the assigned CPMOs within their Client organization and the PAO for management of the overall departmental Portfolio in the following areas:

- Annual Portfolio Planning
- Capital and Expense Funding within portfolio capital and expense targets
- In Plan List Change Management and Business Case Addendums
- Departmental Portfolio Project Prioritization

The PMO is led by an Executive Director that is a peer to the line Project Management Executive Directors. All Executive Directors report to the Vice President—Project Management.

The functions of the PMO include: Define, document, implement, and continually improve project management processes, tools, management information, and training requirements to ensure excellence in the customer experience. The PMO establishes and maintains:

- Effective and efficient project management processes and procedures across the project portfolio.
- Systems and tools focused on improving efficiency of project manager's daily activities while meeting external and internal customer needs.
- Management of information that measures customer experience, project performance, and organizational performance.
- Training/certification curriculum supporting organizational goals.

## 12.8 DELL INC.[19]

**Increasing Efficiencies and Improving Project Delivery Quality through Standardization and Governance**

How Dell Services standardized on project management, and an overarching PMO Framework, to support consistency in execution with reliable, predictable, high-quality results for complex projects and programs, thus providing unified seamless delivery, enabling our customers to do more, through integrated, holistic, multi-service business solutions.

### About Dell Services

Texas-based Dell Inc. is the world's No. 3 supplier of desktop and notebook PCs. Its technology products—including network servers, printers, data storage systems, and projectors—reach a global customer base. It also offers third-party software and hardware, as well as IT services. With more than 100,000 employees, Dell reported $62 billion in revenue for 2012.

A key business unit, Dell Services has 52,000 employees in more than 90 countries offering:

- Application services
- Managed services
- Cloud services
- IT support services
- Configuration services
- Deployment services
- OEM services
- Information security services
- End-user computing

---

19. ©2013 by Dell Inc. Reproduced by permission.

- Training services
- Enterprise consulting for data centers and for workloads
- Business process outsourcing

Created largely through the acquisition of Perot Systems in 2009, Dell Services supports 95 percent of global Fortune 500 companies, 10 million small businesses, 400,000 classrooms, every G20 government, 200,000 physicians, 75,000 channel partners, and 60,000 retail locations

### Dell Services PM3 Framework Evolution

For more than 28 years, Dell has empowered countries, communities, customers and people everywhere to use technology to realize their dreams. Dell is committed to delivering the services, solutions and products customers need to drive their business goals and suit their lifestyles.

The Dell acquisition of Perot Systems in 2009 combined two iconic IT brands. The expanded Dell became better positioned for immediate and long-term growth and efficiency driven by: providing a broader range of IT services and solutions and optimizing how they're delivered; extending the reach of Perot Systems' capabilities, including in the most dynamic customer segments, around the world; and supplying Dell computer systems to more Perot Systems customers.

In 2010, the Dell Services Business Unit President sponsored the launch of the Enterprise Project Management Standardization (EPMS) Program. In order to drive efficiencies and improve the overall customer experience in offering integrated, end-to-end, multiservice solutions for Dell's customers, it is imperative that the project teams demonstrate unified, seamless, high-quality delivery. Over the past 3 years, the EPMS Program has been successful in achieving the ultimate goal, to establish a minimum standard for project management practices to increase project management expertise, efficiency, and effectiveness across Dell Services, ultimately increasing the success of global project delivery over time.

The EPMS Program team started with the best-in-existence project management components already contributing to operational success across the organization. Through collaboration with representatives from all Dell Services' segments and delivery teams across the globe, the Enterprise Project Management Standardization (EPMS) Program worked to establish one standard project management framework called the "PM3" (**P**roject **M**anagement, **P**rogram **M**anagement, and **P**ortfolio **M**anagement). See Figure 12–16.

The patented PM3[20] is the Dell Services Global Project Delivery Framework, which encompasses the Project and Program Management Framework, the Project Management Office (PMO) Framework, and the internal Project Delivery Governance standards & processes.

### The PM3 Project/Program Management Framework:

a. Addresses the people, process and tool aspects of project, program and portfolio management.
b. Is flexible, scalable and applicable to any type of project.

---

20. US Patent 8,407,078 B1: Method of and System for Managing Projects, Programs and Portfolios Throughout the Project Lifecycle.

The Dell Services Global Project Delivery Framework

**Figure 12–16.** The Dell service global project delivery framework.

   c. Is completely aligned to industry-recognized best practices from the Project Management Institute's A Guide to the Project Management Body of Knowledge (*PMBOK® Guide*), as well as PMI's® Standard for Portfolio Management.

   d. Uniquely packages and integrates the methodology assets online, via SharePoint to help PM team members quickly and easily navigate the processes, templates and supporting toolkit, through a variety of user-friendly views.

*The PM3 PMO Framework (See Figure 12–17):*

- Focuses on portfolio-level standards, processes, tools and templates, addressing six standard PMO functions. These tools and processes, when properly applied, increase project delivery quality and effectiveness through better governance, overall support and leverage of best practices and lessons learned.
- Provides the standard processes, tools and templates to execute a group of standard functions that support business success through:

  - Project Portfolio Governance
  - Project Portfolio Management
  - Portfolio Resource Management
  - Portfolio Quality Management
  - PMO Support for Project & Program Management
  - PMO Organizational Planning, Design and Management

- Represents the portfolio management component of the PM3; It works within the PM3 Framework to provide a deeper organizational insight into all aspects of project execution both at the tactical and strategic levels. The PMO also helps to efficiently allocate the resources among the ongoing initiatives and governs the projects from a strategic viewpoint, helping to increase alignment of project delivery with the customer's strategic goals.

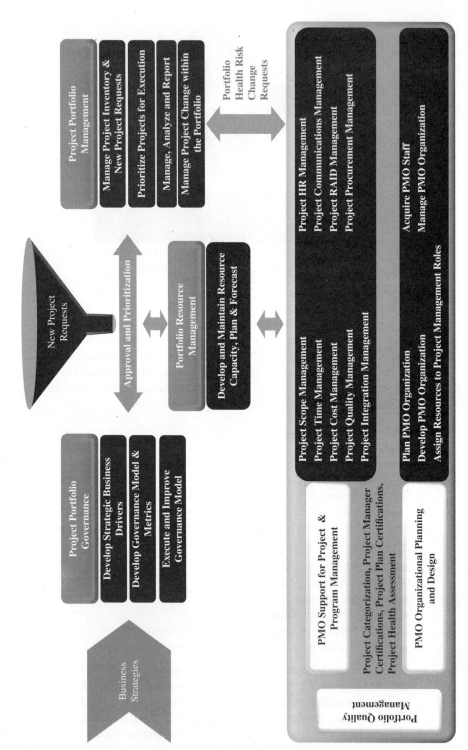

Figure 12–17.  PMO process flow.

- The PMO . . .

  - Streamlines Project, Program and Portfolio Management effort and increases project delivery quality and effectiveness.
  - Provides data that can be applied to driving up project throughput, thus maximizing a customer's strategic gain for their investment.
  - Is designed to accelerate organizational change management.
  - Provides clear and efficient lines of project leadership and authority for issue escalation and resolution.

- Overall the Dell Services PM3 Framework contributes to operational excellence for both Dell and our customers by:
  - Demonstrating unified, seamless delivery to our customers, standardizing project management across integrated, end-to-end, multi-service solutions.
  - Increasing likelihood of successful delivery, with proven, repeatable industry-aligned best practices.
  - Providing continuous project performance monitoring & reporting, using quantitative early warning metrics; proactively identifying & minimizing negative impacts which ultimately contributes to customer loyalty.
  - Enabling project managers to grow and thrive through flexible, cost-effective, industry-aligned training & certification, providing on-going professional development and more clearly defined career paths.

## Assuring Quality through Project Delivery Governance

Project Delivery Governance is the process and associated accountability framework to oversee, monitor and control global project delivery performance, as well as the compliance to the PM3 standards for project management.

*The PM3 Project Delivery Governance process:*
- Is applicable to both internal and external projects and programs.
- Is scaled appropriately based upon standardized Project Complexity Categorization (PCC) –which scores each project based on a common questionnaire of size, complexity and risk factors.
- Includes responsibilities of the Account PMO, Business Line/Delivery Team PMO and/or Governance Leads, as well as the Services BU PMO. This standard also defines the role of the operations management in the governance of project delivery and how we all work together.
- Is accomplished through routine governance, reporting, quality assurance and quality control activities.

This governance process improves visibility of project performance at the Dell Services Business Line and Business Unit (BU) levels, as well as defines standards for performance measurements and reporting. Information reported as part of the governance process provides early warning and is used to trigger quality control, as well as intervention and remediation processes for underperforming projects.

*The scope of control for this Dell Services BU-wide governance process is:*
- Global Project Delivery Performance
- Compliance to Project Management Standards

*PM3 Project Delivery Governance Benefits include:*
- Improves visibility of project performance across Dell Services.
- Provides continuous monitoring, early warning, identification and inspection of underperforming projects through timely, objective, quantitative performance reporting.
- Facilitates rapid response, proactive intervention and remediation for underperforming projects and programs.
- Defines a consistent, common standard for performance metrics and reporting.
- Ultimately contributes to reducing the negative impacts to the company's financials and customer satisfaction.
- Increases our ability to replicate excellence in all of our projects.
- Enables consistency in execution, with reliable, predictable results.
- Ensures standardized project management methodology and toolkit across Dell Services supporting scalability and growth targets.

### Enabling Success through Project Manager Development, Certification and Leadership Skills

### The "Art" and "Science" of Project Management

The patented PM3 methodology and supporting documents require qualified project management team members to interpret and apply the methodology, standards, and tools appropriately, based on the specific needs of the project or program. Project Management cannot be executed successfully, nor can the value and benefits be fully realized by following a checklist, or step-by-step procedure. Standards, process, and tools are only half of the equation. Successful project management relies on strong leadership, decision-making, and expert judgment, as well. To reap the maximum benefits, the use of the PM3 Project and Program Management Framework, Project Managers must strike a balance between the science of disciplined execution and the art of using sound judgment in leading the effort. The value comes when the processes and tools are applied properly and most efficiently for both our customers and Dell Services, balancing risk with the degree of rigor applied.

The PM3 processes and supporting tools and templates are designed to mitigate risk and deliver predictable and repeatable results–these processes and tools are what we refer to as "the Science" of project management. Project Managers must focus not only on the processes they need to follow but also on the intent of the processes and the results, or outcomes, these processes and standards are designed to produce.

The "Art" of project management is the judicious and cost-effective application of the science to the business problem and environment. The methodology and tools are flexible and require experienced, qualified project managers to apply them appropriately. Although our methodology provides guidelines for scaling based on the project size, complexity and risk, every customer engagement is different, and this scaling requires judgment and experience on the part of the project manager to decide where to customize, and where to flex.

The PM3 framework is a means to an end. There can be several paths that lead the PM practitioner to the critical outcomes that are necessary for project success. The strong project manager will balance the "art" and "science" to ensure the critical outcomes are achieved.

### Developing and Maintaining Proficiency through the Internal PM3 Project Management Certification Program

*Dell Services Internal PM3 Project Management Certification Program*

Is the standard for aligning the right project manager, based on their skills & experience, to the right project, based on size, complexity and risk factors (or project complexity categorization—PCC Level)

- Enables better allocation of project management resources to project complexity levels, to reduce risk and increase probability of project success, in the most cost-effective manner.
- Establishes a consistent standard across Dell Services to enable career mobility and promote career growth, through a more clearly defined development path.
- Provides a consistent level of qualification and quality assurance of the PM population, which is critical in a leveraged resource model.
- Is a multi-tiered project management certification program that certifies project management resources according to skills, training and most importantly, successful project experience.
- Is a key differentiator and value-add to the customer, demonstrating quality assurance and professional development for our project management team members, which increases customer confidence level with our project management resources.
- Is measured and driven by organizational priority and need.
- Is complimentary to Dell's people strategy and PMI® project management certifications
- Formally recognizes project manager success, experience and knowledge.
- Encourages and fosters a culture of mentoring.
- Promotes proficiency and consistency in project management standards and best practices across Dell Services.
- Ensures assignment of a well-rounded project manager, who has demonstrated effective application of the standards and proven success with like-sized/complexity projects and programs.
- Aligns with and compliments the PMI® project management certifications.
- Provides more visibility into career opportunities for project managers.

### Investing in Project Managers' Development to Drive Quality and Success

In addition to technical skills related to the specific project management processes and tools, or what is sometimes referred to as the "science" of project management, the Dell Services Project Management Learning System (PMLS) emphasizes the importance of human performance skills, or the "art" part of project management. Excellent leadership skills and good judgment are critical to the success of any project manager.

The PM3 Project Management Certification Program and associated PM3 Training are critical to long-term sustainability of the standards. The PM3 includes a comprehensive curriculum that provides project management team members the opportunity to build project management skills and understand the Dell Services approach. These experiential courses are based on the PM3 and associated toolkit, with additional real-world examples and case studies built to address common project management challenges. The project management curriculum has been approved by the Project Management Institute (PMI®), contributing to Dell Services PMI® Registered Education Provider (REP®) status, which qualifies participants to earn Professional Development Units (PDUs) toward their PMI® Project Management Professional (PMP), or other PMI® project management certification, or recertification.

All of the PM3 training is web-based, offered online to provide easy access and navigation, in the most convenient and cost-effective manner for Dell Services project management team members. A training homepage is available on the intranet to assist the user in navigating through all the available training resources, organized by:

> Latest highlights and shortcuts to newly released courses
> Important links to other related sites, team-specific training
> A direct link to the PM3 repository and PM3 Project Management (PM) Certification
>> Program
> Training categorized by Learning Paths or Programs, aligned to:

- Role and PM3 PM Certification Level
- Leadership Competencies
- Technical & Functional Competencies

### Driving Long-Term Adoption and Sustainability through Organization Change Management Techniques

A major critical success factor to the success of the Enterprise Project Management Standardization (EPMS) program has been in the approach and techniques used to drive and sustain major change over time, across the business unit.

The team found it most effective to leverage various Organizational Change Management (OCM) techniques, beginning with:

Committed executive-level sponsors established from the beginning and maintained throughout, including as the program transitioned to steady-state governance operations.

*Top-Down Communication from Sponsors*
- Reinforcing priority and alignment to overall organizational strategy
- On a regular basis

  - minimum twice/year with a long-running program and especially after leadership changes and significant re-organizations;
  - In conjunction with and after the strategy is communicated.

A "Change Readiness Assessment"—a standard element of OCM, leverages interviews and solicits input from a sample group of impacted stakeholders from all levels of the organization. This input was critical to the PM3 implementation, training and communication plans. In addition, it supports buy-in because the program team took the time to include and value stakeholder involvement and input right from the start of the program.

*Ensure involvement from all impacted stakeholder groups and teams*
- Global, Services Business Unit–Wide involvement and collaboration with subject matter expert (SME) representatives from all teams with project delivery.
- Solicit and incorporate feedback through formal deliverables and routine weekly/bi-weekly meetings.
- Establish a "leadership steering committee" for the program, with business unit-wide representation at the level of leadership with authority to enforce standards, to remove obstacles and escalate/resolve issues.
- Be collaborative versus dictatorial; flexing to adapt to the business situation and level of maturity of the organization, without sacrificing quality or the intent of the standard. Focus on the intended outcome.

*Establish and maintain credibility of the PMO and program team, across the business unit*
- Lead by example; demonstrate the value through practical application versus theory.
- Communicate alignment with industry-recognized best practices.
- Ensure implementation and/or improvement plans demonstrate value to the business early, and regularly, in the rollout to maintain interest, momentum and visibility to the value—with "quick wins," "low-hanging fruit" aligned to hottest business "pain points."
- Be supportive and flexible, balancing risk with rigor, scaling and streamlining where appropriate without sacrificing quality;
- The PMO exists to serve the business, our customers and the PM community; be supportive and accommodating, not intrusive, nor disruptive.
- Listen; continue to seek to understand the business and be sensitive to customer priorities.
- Get involved with outside organizations, take advantage of opportunities to showcase your best practices to solicit outside industry-recognition and validation; use these accomplishments, awards and recognition in communications to establish and lend credibility to the standards and the PMO team driving the change.
- Communicate, Communicate, Communicate! Sustains visibility and momentum.

## Contributors

PM3 contributions by subject matter experts from across Dell Services, with specific authorship for this publication by the Services Business Unit PMO Team:
**Michele A. Caputo, PMP®**
Dell Services Business Unit Project Management Office (BU PMO) Leader & Enterprise Project Management Standardization (EPMS) Program Director
**Herm Barringhaus, PMP®, CSM**
**Allison Bass, PMP®, ITIL, 6σGreenBelt**
**Bonnie Flynn, PMP®**

Tracy F. Grimes, PMP®
Brad Horner, PMP®
Tama S. Kinsey, PMP®
David H. Meyer, PMP®, CSM
Steve Sheppard, PMP®, PRINCE2®

## 12.9 COMPUTER SCIENCES CORPORATION (CSC)

Consider for a moment the following situation. Your company wants to start up a PMO to function as the guardian of project management intellectual property for your company. You hire a person who has done this for a previous company with some success. What this person brings to your company are the successes and failures that he or she encountered at a previous company. This individual will make decisions based upon his or her experiences. This approach can work, and can work well. But there may be a better way to accomplish the goal of implementation of a PMO.

More companies today appear to be hiring consulting companies to help them implement a PMO. Consulting companies maintain files of successes and failures among all of their clients and the consultant(s) you hire can draw upon the experiences provided within the consulting company's knowledge repository. If the consultant has issues where help is needed, the consultant can call upon other company consultants for advice and guidance. Simply stated, the consultant brings to the table perhaps hundreds of man-years of documented experiences and best practices whereas the person you hired above may bring to the table only five or ten years of personal experience.

Another problem occurs if the person you hire is unfamiliar with your industry. Consulting companies often maintain files by industry classification. This can significantly benefit the on-site consultant. With consulting companies, it is often easier to find industry-specific consultants. When starting up a PMO in a health-care environment, this is essential.

To illustrate how hiring a consulting company can be beneficial, consider CSC[21]:

> CSC's purpose is to deliver innovative business and technology solutions that help our commercial and government customers worldwide achieve what they want most—results. Focus is placed upon the uniqueness of each customer around the globe. CSC has over 91,000 employees in 80 countries worldwide and has been in business for over 49 years. CSC focuses upon a wide variety of industries and provides consulting, implementation, outsourcing and integration services at all levels.

**Best Practices/Culture/ Governance Structure/ Methodology**

In the above description of CSC, it is important to understand how critical the words "uniqueness of each customer" are. When starting up a PMO in a health-care environment, where improving and maintain-

---

21. The material on CSC in this section was provided by Nani Sadowski-Alvarez, PMP, CPHIMS Senior Management Consultant, Enterprise Program Management & Architecture, CSC.

ing health are of prime concern, setting up a PMO based upon an engineering, construction, or manufacturing company may not be advisable. Nani Sadowski-Alvarez continues:

> Best practices were utilized extensively during the incorporation of a PMO (Project Management Office) at a large 9,000 employee based Health Care System located in Baton Rouge, Louisiana entitled Franciscan Missionaries of our Lady. This implementation was conducted through a partnership with CSC over a duration of one year in order to formulate a streamlined, fully operational, optimized, and mature PMO team capable of facilitating and leading a large and balanced portfolio consisting of a blend of technical, clinical and business focused projects, programs and extended service requests. The Best Practices were established via a variety of experiences and lessons learned during prior PMO implementations at other client locations as well as through lessons learned shared via professional affiliations within the health care industry as well as a cross section of industries, project management based statistics, industry trends, and healthcare based trends and regulations. In addition, client surveys (both formal and informal) were conducted to determine and unveil:

- Clients' perception of areas within their organization that have proven to be operating in an efficient and effective manner
- Target areas that are recognized as being in need of improvement
- Areas and processes requiring focused attention that would enhance overall needs, mission/vision, desired organizational direction as well as provide desired emphasis upon the creation of positive and productive outcomes.
- What and who are the perceived drivers of this change.
- Overall organizational "pulse" and view surrounding the implementation of project methodology.
- Areas that require fine tuning and revisions.
- Internal employee views and feedback on the audits and assessments conducted by other consulting and audit agencies.
- The "understood" organizational strategic goals and initiatives.
- Overall perception and view of the change and the transformation under way.
- Environment and tone of interaction between organizational facilities and departments.

With the fact being that the implementation was driven/lead by consultants who closely partnered with the Health Care System, there are several additional factors to be taken into consideration with utilization of Best Practices as a guiding foundation. Following is a list of these *"Challenges and Opportunities as a Consultant Surrounding Incorporation/ Utilization of Best Practices"*:

- Accurate interpretation of the organizational culture
- Listening effectively to all signs provided by the organization (verbal and nonverbal) encircling their needs
- Customization of Best Practices to a level that provides the most beneficial results.
- Obtaining and encouraging ample feedback and input from the leaders and end users to ensure that the first impression of the Best Practices is as positive as possible.
- Delivery of effective communications that target the impact that the initiative/project and the incorporated Best Practices will have.
- Providing adequate answers to the Who, What, When, Where, Why and How surrounding the need for the transformation/change underway.

- Instilling a balanced and flexible approach to use the proven Best Practices as a foundation with the addition of adjustments that will ensure they blend with the organization acquiring and implementing them.
- Creating the proper balance to ensure silos are removed allowing for more transparent and fluid communication and utilization of agreed upon Best Practices.

**Best Practices Defined**—Overall Best Practices are processes, procedures, standards, methodologies, techniques, activities, etc. that have proven to be more effective and successful in providing measurable efficiencies and results than other comparable processes, procedures, standards, activities, and techniques.

Factors and questions considered in the evaluation of best practices are:

- What is it? A process, procedure, template, methodology, action?
- Are they repeatable?
- Is it something that results in efficiencies?
- Is there impact to clinical care? How and what regulations (Patient Safety Goals, JCAHO—The Joint Commission on Accreditation of Healthcare Organizations, HIPAA Compliance Regulations) are met or impacted by this?
- Is there an impact to Business Initiatives and Technological Advancements?
- What "trial and error" processes were underway to test a sample set of options and determine what the "best route" would be to proceed with?
- What evaluation parameters were implemented in order to assess and determine if a process/procedure/template/methodology/action were in fact meeting the needs of the impacted areas and were there efficiencies related to it?
- Was there clear and definable collaboration involved by stakeholders to ensure that all factors were considered?

**Samples of Best Practices with Health-Care PMOs**

- Structuring all projects in a consistent manner with a core methodology in place (that is required to be adhered to)
- Undergoing periodic random project audits from an external department (either Legal Services or The Quality and Audit Department) that are made available to all internal requesting resources
- Conducting formalized project initiation and closing presentations with all stakeholders being invited to participate as well as Core Team member's attendance being required. These events/milestones are then followed-up and communicated via internal marketing communications accessible to all employees.
- Ensuring all projects (clinical, technical, and business) receive input and scoring/rating from an experienced clinical executive representative.
- Validating that project scoring/prioritizing contains a section directly related to the organizations strategic initiatives as well as process flows and quality of care.
- Creating detailed project metrics that are available on a weekly basis for all stakeholders and employees within the organization.
- Creating transparency with all project related documentation in order to ensure that any interested parties are able to assess the current action items, risks, issues, change requests, etc. with the ability to then provide suggestions/opportunities for improvement if they feel so inclined.
- Instilling that the PMs are parties to the contract review and negotiation process and that they facilitate the completion of a contract checklist.

- All Project/Program and Extended Service Request goals must be compiled in alignment with the SMART Goal philosophy in order to validate that the goals are measurable.
- All Projects/Programs and Extended Service Requests must undergo formal Operational Turnover prior to the Project being flagged as "closed." This turnover occurs with participating of Core Team Members, impacted areas, and key stakeholders and requires sign off from Management at all areas receiving support responsibilities.
- Projects must undergo the formal resource request process in order to validate that the Resource Managers will be able to support the dedication required to make a project successful.

## Storage and Indication of Best Practices

Potential Best Practices are assessed and evaluated throughout the duration of the project life cycle. These are then documented in the Lessons Learned template and made accessible to all organizational employees/ Team Members. Project documentation is stored in SharePoint with view access provided to all Core Team Members, approving Committee Members, and Key Stakeholders as well as requesting parties.

Determined Best Practices are recorded as standards and often created into Policies/ Procedures with the need for tiered sign off approval in order to ensure that supporting and enforcing levels of authority have the opportunity to provide feedback and input.

## Methodology

The PMO was created at this Healthcare System (XYZ Health System) under the direction of the organization as well as the full-time support of the CIO/CMIO, Dr. Stephanie Mills and the Director of the PMO, Claudia Blackburn, PMP, CPHIMS. Nani Sadowski-Alvarez, PMP, CPHIMS was brought in from FCG/CSC to lead the organizational transformation Track focused upon the implementation. A blend of methodologies was utilized. The core foundation of methodology resided upon PMI based practices with a blend of ISO 9000 (for segments on the technology driven projects) lean project management principals from Six Sigma and best practices already established within the organization. With the foundation being PMI and *PMBOK® Guide* focused, the core phases of project management (Initiation, Planning, Executing/ Controlling, and Closing) were infiltrated throughout all project implementations. Supporting policies and procedures were developed along with standards and detailed process flows that targeted all stakeholders and explained the methodology step by step. Each of these tools provided input surrounding templates that would be utilized within the various phases as well as a listing of what the outputs and results would be. The key with all of these, was to keep the concepts and explanations direct and simple and to provide details surrounding the resource responsible for executing the particular steps in the process, anticipated turnaround times, outcomes, and reference to related and interdependent processes. All templates surrounding the methodology, processes, procedures, standards, workflows were standardized and available to anyone within the organization as well as requesting partner vendors. This creation of a transparent department enabled all stakeholders to have access to pertinent project details and methodology at any point in time.

## Governance

The transformation of the organization was divided into tracks with 1 Tracks main focus upon creation and Implementation of the PMO, another Tracks focus upon IT Governance, another tracks focus upon implementation of a PPM and internal IT Resource and Capacity Management tool, and the 4th track focused upon organizational governance. The key to success with this was that all tracks had to work collectively in order to ensure collaboration and cohesiveness of the undertakings of each track. The governance for the healthcare system was designed to ensure buy in from all key areas—Business, Clinical, and Technical. The PM (in partnership with the PMO and the Project Program Sponsor) was responsible for facilitation of the process and ensuring that the Business, Clinical, and Technical committees had the opportunity to review the Project/Program Charter (and provide feedback—suggestions) prior to delivery to the Steering Council. From there, the project may or may not go to the Executive Council (Pending project cost, size, impact to resources as indicated in Governance Policies) as well as be presented to the board. The parameters were all clearly defined and created with participation from key resources throughout the organization. This ensured that all areas had equal input and that there was a greater likelihood of rapid buy in and support for the governance structure.

**Project/Program Success**     Project success is measured based upon the delivery of the projects on time and within budget along with their ability to meet the project mission/vision as well as the SMART Goals. Now, with that being said, it is accurate to state that Change Management methodology is solidly in place and acceptable in re-aligning the project and its scope with adherence to the triple constraints and agreement that change requests must be signed by the Project/Program Sponsor. It is the PMs Responsibility to update project documentation accordingly and to adhere to the project change management Policy, procedure, standard, and process flow. Project/Program Status is available 24/7 to all requesting parties. It is encouraged that vendors participate in the status updates as well. Critical Success factors are clearly detailed within the project scope document (that is also signed by key stakeholders) and when changes are 25K or above they involve approval via the Governance Structure. PMs (Program and Project Managers) are required to conduct milestone updates and to report on them via the status report. PMs are also measured and evaluated based upon their ability to achieve the role based performance standards/indicators. These indicators align with their daily duties, project based duties as well as methodology perspectives and organizational initiatives.

Overall, project success is defined by the organization at large. The project charters contain full detail surrounding project mission/vision and the project goals. A balanced representation is present at Approving Council meetings (from the Financial/Business side, clinical and technological side of the house to include lead Executives, CFOs, Physicians, Technology Directors, Departmental Leads, SMEs, etc.) to ensure that the checks and balances are in place prior to voting a project to the approved state. Once it is approved the detailed scope is reviewed (containing critical success factors, known issues/risks, detailed process flows, communication matrix, etc.) and open for suggestion/revision prior to sign off. Once signoff has occurred, the formal Change Management methodology is adhered to. This methodology does contain a 'break the glass' procedure in the situation that an item/opportunity for change arises that is of urgent and immediate attention status. The execution of this process does require executive level approval due to the magnitude of individuals whom will be impacted.

Excellence is defined clearly by the $E^3$ theory of Excellence Exceeding Expectations. With any effort to exceed the initially set and approved expectations, there is the ability to achieve excellence where streamlining, enhancing, cost savings, patient and employee satisfaction are present.

**Questions to Ask Surrounding Program/Project Success**

- Have milestones been met?
- Has effective change management taken place?
- Were Risks efficiently mitigated?
- Were status reports provided per the policy and procedure guidelines throughout the duration of the project?
- Were there regularly scheduled Core Team Meetings?
- Were the Signed Project/Program Goals and Objectives achieved?
- Was the project methodology followed throughout the project life cycle?
- What are the Sponsor(s) and Key Stakeholder(s) view of the end results of the Project/Program?
- Have all participants had the ability to provide their view of project successes and opportunities for improvement?
- Is the project within Budget?
- Was the project delivered on time?
- What are the results of the project audit(s)?

NOTE: We have discovered and confirmed that templates need to be fine-tuned for all clients in order to ensure that they meet the needs and initiatives of that organization. The length of the templates can at points hinder the success of the project. As a result, the recommendation is for the PMO to continually review and discuss the utilization of the templates and to adjust them accordingly.

**Project Management Offices (PMO)**

The project management office that was created for the Health Care Facility consists of a Director PMO, 10 PMs (2 of whom are detailed as Senior PMs), 2 Extended Service Request Coordinators, and 2 technology engineers (who focus upon guidance with project technology assessments and evaluations as well as audits). This PMO reports to the CIO/CMIO as well as the Steering Council and other governing Councils. Programs (such as the implementation of a Heart Hospital) are also facilitated through the PMO office.

The vision of the PMO is to "provide consistent and standardized project management with focus upon excellence in a collaborative setting. All approved projects will contain measurable goals and objectives directly related to the goals and standards of the Health System with convergence towards achieving a comprehensive continuum of quality care."

There is a solid governance process in place as referenced in the governance section.

Support for the PMO trickles directly from the C-Level Executives on down at this Health Care organization as well as the consistent view point of the value that the PMO adds as a whole. Prior to the implementation of the PMO, an organizational audit occurred with the suggestion of more formalized methodology and standards as well as enhanced transparency surrounding project focused endeavors and initiatives.

Sponsors receive a one-on-one session with the Director of the PMO and the assigned PM in order to review a PowerPoint presentation and several hand outs that clearly define what the role of a project sponsor is. In this conversation it is also assessed if they feel prepared to take on the role of a Sponsor. If not, then this realization is brought to the Steering

Council for discussion and determination as to who an appropriate Sponsor(s) might be. During the initial session with the Sponsor, items such as Project Communication, frequency of status reporting, regular standing meetings between the PM and Sponsor, etc. are all covered and details are aligned. Project Sponsors learn to become supportive of their projects and to value the partnership that the PM and PMO form with them in order to ensure that their project is a success. A PMO marketing tri-fold is also presented to them that contains information surrounding the following:

- Process flow for the project approval process
- Definitions of key roles on the project team and what their responsibility is
- Critical success factors for all projects
- Definition of project versus program versus extended service request
- High-level communication expectations (to include their role in assisting with removal of barriers to the project's success)
- High-level review of project documentation that will consistently be provided
- Brief overview of project phases and methodology

Sponsors are held accountable for the success of their projects and are requested to be present in the initial presentation to the Steering Council requesting approval. They speak to the "need" of their project (for the organization), the ROI, the SMART Goals, overall added value, impact to resources, duration, scoring, etc. along with the PM and participate in the question/answer segment. As major changes (exceeding 25K in value or significant impact to Resources, or project schedule) arise, the Sponsor and PM are requested to attend the Steering Council meeting to provide the details as well as discuss the options for resolution with the Council and obtain a vote of approval for the newly targeted "direction" to follow.

**Management Support**     Sponsors are required to be management level and above. For Extended Service Requests (initiatives that are 500 hours or less in duration and require formal structure in order to execute) a Sponsor may reside at the Managerial level. For Projects and Programs, however, a Sponsor must at least reside at the Director level in order to ensure that they have the leverage and authority to assist with the barrier removal.

## 12.10 SLALOM CONSULTING: UNDERSTANDING THE NATURE OF A PMO

In the previous examples in this chapter, we saw companies that implemented a PMO after a well-thought-out process. But too often a company will rush into the implementation of a PMO well before it decides on the intended use of the PMO, what functions it will perform, who will be assigned, and potential internal political issues. Not every PMO is the same, even in companies that have multiple PMOs. Carl Manello, practice lead—delivery effectiveness, Slalom Consulting, provides us with his insights on PMOs:

What's in a name? More than most of us would consider. Far too often the three letter acronym "PMO" is bandied about with far too little consideration for what it refers to in its implementation. I have seen this carelessness occur with my clients as well as within consulting organizations that offer solutions in this arena. The over-simplification of project management organizations into a "PMO" has led to some disastrous results. What do "PMO's" do after all? Are they:

- Project Management Centers of Excellence (e.g., process owners)
- Governance Offices (i.e., overseeing compliance)
- Project manager resource pools
- Resource management centers charged with balancing supply and demand

I have seen of each of these implemented with the same name: PMO. "So what?", you may ask. "Why is it so important what we call the function?" A quick anecdote may illustrate the point well.

A CFO reads in the latest trade magazine about the much celebrated success of a peer company in their implementation of a PMO. "Fantastic," she thinks and proceeds to call the CIO. "Bill, we need to get your team to build a PMO. I know that they can make a difference, have a financial impact and deliver results. Go make it happen!" After hanging up the phone, Bill is not sure where to turn. Does the CFO want a Project Management Office, Program Management Office, Program Office, Enterprise Program Office or Governance Office?

As professional project managers we must help our companies distinguish the difference in functional role of each of these different organizations. We must also help to clarify the span of control differences—say, between a Program Office and an Enterprise Program Office. I coach my colleagues and clients that we should stay away from a "one name fits all" reference of the mythical "PMO" and instead talk about operational functions.

For example, Figure 12–17 illustrates an IT Governance Office which has a variety of accountabilities. However, it is focused within IT and does not reach into the "business" side of the company. Focused solely within IT, Governance takes on many roles. This "PMO" focuses on providing assistance to project managers, aligning and monitoring the assignment of project managers, assisting the Office of the CIO with strategic planning, monitoring IT financial/project spend and overall IT reporting for leadership. By contrast, this "PMO" does not contain a pool of project managers to be assigned to projects throughout the information technology organization. Neither does this governance organization "PMO" have the same span of control as a "PMO" (Program Management Office) operating within the confines of a large business initiative.

To illustrate the difference in span of control, let's look at the following models. While not perfect solutions for every implementation, these figures show the range of possibilities and begin to break-down the view of a one-size-fits-all PMO. Figure 12–18 (I'm sure you'll align figure numbers later, just wanted to be sure you're pointing the right ones) provides one of the simplest models: The Project Office.[22] The Project Office provides oversight, monitoring and reporting and possibly even some governance functions over non-related projects. By contrast, the Program Management Office in Figure 12–6 is contained within a single large-scale business initiative. This PMO has accountability for driving a business program, keeping track of multiple associated projects and providing monitoring and reporting to the key business executives. In some implementations, the PMO is not just the project governing body of the program, but it is the home of the key executives running the initiative. At Sears, during the integration with K-Mart, and at Donnelley for the Year

---

22. The symbol in the upper right-hand corner of Figure 12–19 illustrates that each of the successive PMO structures in Figures 12–20 through 12–22 leverages the same project management capabilities in PM*f* that was discussed in Figure 4–20.

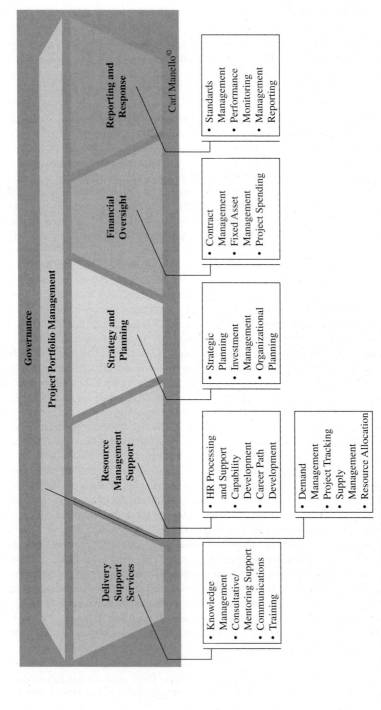

**Figure 12–18.** Example functions of a PMO.

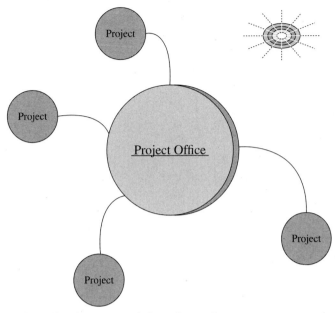

**Figure 12–19.** Project office that oversees independent projects.

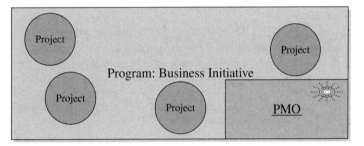

**Figure 12–20.** Program management office internal to program.

2000 efforts, the PMO was the executive leadership team of the business focused effort. There were of course supporting teams of program and project managers on-board, but when the chairman referred to the "PMO" he was talking about leadership. Figure 12–7 illustrates the next wider governance organization: the Program Office. Much like the Project Office, the Program Office oversees non-related projects with the added responsibility for governance over programs. By nature of its oversight, this organization is usually set up with accountabilities across large functions (e.g., across business units). Program Office members work with their counter-parts in Program Management Offices to collect metrics, status and key updates, however, usually focused at a higher level. While the PMO works at a detailed level collecting, reporting and managing a single program, the Program Office summarizes cross-initiative information for consolidation at the executive level. The last variant is the Enterprise Program Office in Figure 12–18, which is typically implemented in very large organizations. Its role is to further roll-up, consolidate and coordinate initiatives across the enterprise. In addition to governance responsibilities, this group may

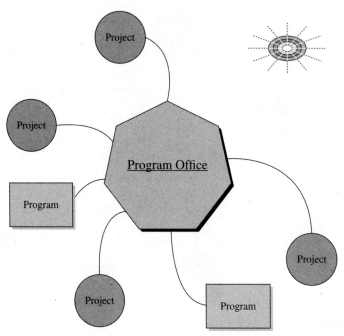

**Figure 12–21.**   Program office that oversees programs and projects.

be charged with portfolio management activities. With its view of the entire enterprise, the EPO is well suited to operate with the goals of the whole organization in mind, as opposed to siding with one business unit over another. As part of the first EPO at R. R. Donnelley & Sons in 1997, we worked on portfolio management for the whole enterprise, master schedule development, creation of project management methods and processes and we provided project support. We worked across corporate functions, Research & Development, all the business units (e.g., Book and Magazine) and were involved in several key business programs. Given that there are such a variety of project management organization incarnations, with varying spans of control, it is no surprise that there are also a wide variety of reporting accountabilities; a project organization does not always report to the same function in every company. Typically, I have seen what I identify as *Project Offices* put in place within the Information Technology organization. Many if not all IT initiatives report into or have a link to the IT Project Office, however, initiatives that do not involve IT as a major component are typically not tied to this governing body. My R. R. Donnelley & Sons *EPO* reported into the head of the Research & Development organization (at that time, Donnelley did not have a full-time CIO controlling the IT organization). My *IT Governance Office* at Zurich North America reported into the Vice President of IT while the Sears implementation of a *Program Management Office* was accountable to the Office of the Chairman. The Sears' PMO was a special reporting relationship and was due to the importance and the visibility of the businesses program we were leading. However, most other Program Management Offices usually report directly into the executive champion in charge of the business initiative.

While the illustrations I have provided are certainly not exhaustive, I think it clear that there are a number of different types of organization. There are also a variety of span of control or reporting accountabilities. Given this variability of implementation, it should

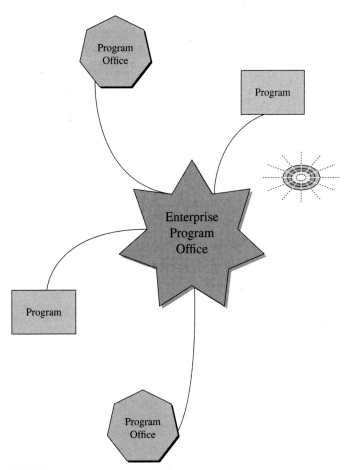

**Figure 12–22.** EPO that spans programs and POs.

be clear to the reader that a generic label of "PMO," is wholly inadequate and may in fact be short-sighted.

As mentioned in an earlier chapter, customers are now *requiring* their contractors to assess their organization and determine their level of maturity in project management. This assessment process, whether customer-driven or internally driven, most often requires PMO involvement. Carl Manello continues:

With so many different types of "PMOs" (as discussed above), what is the right one to build? Put another way, how can an organization assess where it is at in the project management maturity continuum, in order to determine which type of "PMO" to implement? The answer is a bit more complex than one might first assume. One needs first to understand the environment into which the project management organization will be housed. For to plan a project management organization that is neither linked to the organization's current state (i.e., competencies and capabilities), nor tied to a vision for growth and development is a recipe for a sub-optimal solution.

There are scores of models for assessing project management maturity. Each organization must carefully consider the implications of the model they choose and understand where the implementation based on the model will lead them. One maturity model is illustrated in the diagrams below.

Much like the CMMI provides a stepped maturity model for software development, the model in Figure 12–23 . . . creates a maturity model for project management organizations. The various steps are loosely aligned as follows:

- Building Basics—establishing the Project Management Functions (PM*f*)
- Deliverable Focus—implementing a Project Office to coordinate/oversee discrete projects
- Business Focus—putting a Program Management Office in an accountable role within key business initiatives
- Integrated View—the Program Office, bringing together disparate projects and programs into a holistic view for the organization
- Value Proposition—realization of the Enterprise Program Office

While there is no absolute rule about progressing through maturity one step at a time, I think any organization that attempts to realize the value proposition of implementing an *EPO* before it has any experience with the less mature organizations, is looking for trouble. Just as one cannot jump maturity levels in CMMI, that same view should be considered when developing project management organizations. Another parallel to CMMI comes from the assessed performance level of an organization. To be "CMMI" in the truest sense, an organization must be assessed by a third-party against an international standard. While there is no such international standard to assess the maturity level of PM organizations, companies need to be brutally honest about their capabilities and competencies.

Too many of the organizations that I have worked for have had an over-optimistic view of where they lie on the CMMI scale (many companies thinks that they are at Level 3 or *higher*!). Similarly, companies that jump levels by over-optimistically assessing their PM maturity may actually diminish the value proposition of the "PMO" and in fact negatively impact the enterprise.

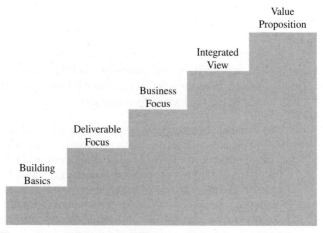

**Figure 12–23.** The project value maturity model.

Figure 12–24 is a related view of the maturity progression, focusing on six project management organizational challenges, aligned with the six maturity levels. Note that this particular model was developed to illustrate the developmental challenges of an IT project management organization. Depending on the type of "PMO," the model will be different.

- Business Alignment—the movement of the IT organization from separate business unit to service provider to the rest of the organization
- Project Focus—the growth from discrete IT projects to aligned initiatives focused on the business
- Governance—moving from no oversight to a metrics and measures based management organization
- Standards—evolving from the "wild west" of project management to a organization run with standard practices
- Planning Scope—evolving from independent operating projects to ones planned in a standard repeatable way and integrated as change initiatives for the business
- Community of Expertise Sphere of Influence—parallels the growth of "PMO" organizations outlined above, from individual oversight to enterprise-wide influence

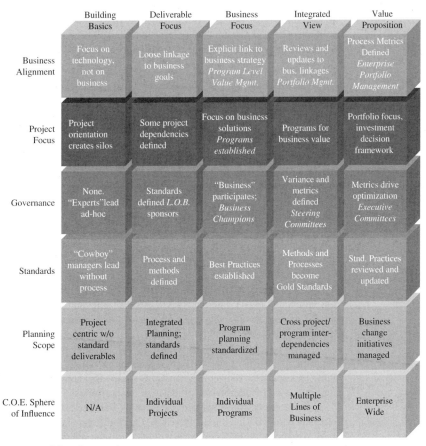

| | Building Basics | Deliverable Focus | Business Focus | Integrated View | Value Proposition |
|---|---|---|---|---|---|
| Business Alignment | Focus on technology, not on business | Loose linkage to business goals | Explicit link to business strategy *Program Level Value Mgnt.* | Reviews and updates to bus. linkages *Portfolio Mgnt.* | Process Metrics Defined *Enterprise Portfolio Management* |
| Project Focus | Project orientation creates silos | Some project dependencies defined | Focus on business solutions *Programs established* | Programs for business value | Portfolio focus, investment decision framework |
| Governance | None. "Experts"lead ad-hoc | Standards defined *L.O.B.* sponsors | "Business" participates; *Business Champions* | Variance and metrics defined *Steering Committees* | Metrics drive optimization *Executive Committees* |
| Standards | "Cowboy" managers lead without process | Process and methods defined | Best Practices established | Methods and Processes become Gold Standards | Stnd. Practices reviewed and updated |
| Planning Scope | Project centric w/o standard deliverables | Integrated Planning; standards defined | Program planning standardized | Cross project/ program inter-dependencies managed | Business change initiatives managed |
| C.O.E. Sphere of Influence | N/A | Individual Projects | Individual Programs | Multiple Lines of Business | Enterprise Wide |

**Figure 12–24.** Maturity matrix.

## 12.11 DTE ENERGY

Although functional POs can be developed anywhere in an organization, they are most common in an information systems environment. DTE Energy maintains an information systems PMO. According to Tim Menke, PMP®, Senior Continuous Improvement Expert, DO Performance Management:

> At DTE Energy the Customer Service Project Management Office (PMO) is a function with the Customer Service Program Management (CSPM) group reporting to the Vice President of Customer Service. The Customer Service Program Management group consists of the PMO, the process management team, and the customer contact channel management team.
>
> The PMO provides project management support for the continuous improvement initiatives within the department. These initiatives are designed to achieve the Customer Service Strategy.
>
> Specific functions of the PMO include:
>
> - Developing and maintaining the project management methodology
> - Maintaining the portfolio of Customer Service continuous improvement projects
> - Collecting and disseminating project data and metrics
> - Providing project management tools and templates
>
> The Customer Service portfolio contains projects from the five departments within Customer Service (Consumer Affairs, Customer Care, Billing, Data Acquisition, and Customer Service Program Management). The projects improve customer service, increase operational efficiency, and/or achieve savings from operations.
>
> The PMO enables and facilitates the application of the project management process. The employees in the PMO have extensive backgrounds in project management and act as consultants, liaisons, and coaches for their respective projects.

As stated previously, the PMO can also participate in the portfolio management of projects. This is common in companies that wish to make maximum use of the talent in their PMO. Tim Menke explains:

> We select projects at the enterprise level based on various indicators including Return On Investment (ROI), Internal Rate of Return (IRR) and Net Present Value (NPV). This annual process involves the highest levels of organizational leadership and is integral to the prioritization and budgeting process.
>
> Our Project Management Office (PMO) engages with the project manager on "approved" projects. Our PMO aggregates projects into portfolios aligned by business unit. This approach allows us to analyze "trade-offs" between projects within a business unit in an effort to elevate performance of the portfolio.
>
> As successes from this approach mount, our interest in performing portfolio management across business units increases. Our future focus includes a greater emphasis on resource allocation in accordance with enterprise strategies as opposed to business unit strategies.

## 12.12 CHUBB[23]

Chubb has a federated IT structure with a divisional PMO in each major IT unit in our U.S. zone, including Commercial Lines, Personal Lines, Claims, Corporate, and Infrastructure. We also have a PMO in our international zones including Canada, Europe, Asia Pacific, and Latin America. An increase in shared services and efficiency goals have driven the need for enterprise standards and consistency. To that end, leaders of the divisional PMOs work together with Enterprise PMO (EPMO) to develop practices largely in connection with our PPM tool and standard SDLC. Our drive toward standards requires local resources and management attention that compete with business priorities and local preferences, sometimes requiring negotiation and balance. Following are key accomplishments and challenges:

*Clarity is our project and portfolio management tool and includes:*

- Time Tracking to calculate project cost in hours. We also map tasks to ten standard categories that roll up to discretionary, non-discretionary and operations & administration. Chubb IT's Finance Manager has assumed ownership of this content and is in the process of developing targets for each spend category. Reporting accuracy was an initial challenge, but continues to improve as management increases use of time tracking results.
- Demand Management to standardize our work request process, This is particularly important for an organization that's growing shared services, but was challenging to implement given that each business unit already had local tools tailored to fit with business governance processes.
- Resource Management to improve capacity planning. This requires vertical "resource managers" to allocate employee time to projects and update the allocations weekly. Maintaining the necessary level of rigor for accuracy is developing. We ultimately plan to use resource allocations as the basis for project cost plans.
- Financial Management to calculate actual project cost by multiplying standard labor rates by role to work hours from time tracking. We also add contractor invoices, but do not currently include business resources needed for total cost.
- Project Management in 2014 to integrate time tracking with project schedules. This will eliminate double entry between time tracking and MS Project schedules, but requires proficiency in effort based scheduling which is a developing skill for some of our project managers.
- Portfolio Management in 2015 to analyze projects across all BU's against selected criteria (e.g., risk vs. return).

*Project Path is our standard SDLC*

- Each Discipline (PMO, BA, architecture, development, and QA) own their respective artifacts, but EPMO is responsible for overall consistency, integration, and repository via our home page. Each artifact is supported by a guideline, template,

---

23. Material on Chubb has been provided by Patrick Gerrity, Vice President, Enterprise PMO, on behalf of the employees in Chubb's PMO Discipline; reproduced by permission of Chubb.

and example. Gaining agreement on the definition of best practices across BU's has not been difficult.

- Use of Project Path is governed by a corporate policy and compliance is measured for projects over 1000 hours. PMs have flexibility to decide which attributes in an artifact are applicable to the project. Whole artifact's may also be excluded with approval from the PMO manager.
- Our big improvement initiative for 2013 is to improve our estimating model. This will include a

  - lower level of detail in the existing model (e.g., technology specific versions)
  - standard calculation of contingency based on risks identified. To that end, a new risk profile will be implemented to evaluate risks more comprehensively.
  - description of how estimates interact with our existing business case, project schedule, time tracking and status reporting requirements.

All people that participate in project teams will be trained.

*A PM Assurance function in the EPMO that:*
- conducts health checks of our top projects (>10k hrs) to ensure important risks are transparent and PM's are complying with enterprise standards. Over the past two years the focus of these health checks has evolved from artifact compliance to surfacing material risks.
- reports the status of our top projects in a monthly dashboard. EPMO decides the status color (vs. the project manager). This report is reviewed each month in a meeting that includes the worldwide CIO, divisional CIOs, divisional PMO leads, and EPMO leads.

Longer term I'd like to rotate the program managers who perform health checks with those that are managing programs. This is needed to maintain a fresh perspective and dispel the notion that the "auditors" are not practitioners.

## 12.13 HEWLETT-PACKARD

Another company that has recognized the importance of global project management is Hewlett-Packard. According to Sameh Boutros, PMP, formerly director of program and project management Practice at Hewlett-Packard:

For large, global companies the need for project management standardization is essential in order to deliver higher-value services at competitive costs. At Hewlett-Packard in the HP-EDS business group, there is a network of Program and Project Management (PPM) Practices in the Americas, Europe, and Asia Pacific regions. The mission of these practices are:

To provide PM Services to HP Clients through Account PMO Leaders and PMs that lead IT services projects. The PPM Practice achieves its objectives when PMs consistently deliver projects on time, on budget, and to the client satisfaction, using disciplined and mature best practices. The PPM Practice supports the business objectives of efficient use of resources, profitability, growth, and customer satisfaction. It also provides profession leadership to ensure that the PMs are prepared to meet the needs of the business and have the opportunity to develop and grow their careers.

Project management development involves formal training and certification as well as informal development. Project management is a core skill and competency for HP Services. The award winning Project Management Development Program is organized by core project management courses, advanced project management topics, courses specific to HP Services practices, and professional skills training. Other activities that support project management development include:

- Driving project management certification programs
- Updating and managing the formal training curriculum in coordination with workforce development
- Driving and participating in major events like PMI congresses and regional project management training/networking events
- Encouraging informal communication and mentoring
- Providing mentorship to field project managers

The Global Method for Program Management provides project managers with methodologies and a standardized approach using industry best practices and incorporating the added value of HP's experience. This is shown in Figure 12–25.

Doug Bolzman, Consultant Architect, PMP®, ITIL Expert at HP, discusses the PMO approach:

> Most organizations have a PMO established and this was generated from the view that their individual projects required oversight. This is a significant jump for many organizations that 15 years ago, did not see value in project managers and are now funding a PMO. But most of them are paying the price to staff the PMO, but still do not see the value, they see it as a necessary evil. In other words, things would probably be worse if we did not staff the PMO.

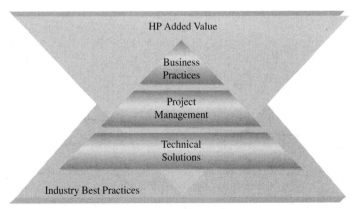

**Figure 12–25.** Global method, program methodology: standardized approach using industry best practices with company added value.

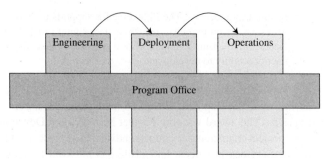

**Figure 12–26.** Using the PMO for facilitation.

| Planning Stage | Integration Stage | Deployment Stage | Operations Stage |
|---|---|---|---|
| Program Office | Program Office | Program Office | Program Office |
|  | Deployment | | Deployment |
|  | Operations | Deployment |  |
| Deployment |  | | Operations |
|  | Engineering | Operations |  |
| Operations |  |  | |
|  |  | Engineering |  |
| Engineering |  |  | Engineering |

**Figure 12–27.** Mapping the PMO to functionality.

Major functions include project oversight, status reporting and project conformance. Since release frameworks were not in place, companies had the situation where their main supplier organizations simply through the solution over the fence to the next supplier. The Program Management Office was created to facilitate these transactions. See Figure 12–26.

The problem with the implementation of this approach is that there never was a single model developed for this type of framework and the PMO would add additional constraints, bureaucracy or workloads. PMO was looked at to plan the direction of the company though the implementation of individual projects.

Instead, another model was developed to have all of the suppliers contribute to every stage of a release, which shares the accountability of planning and designing, while providing the PMO the proper level of functionality. See Figure 12–27.

## 12.14 STAR ALLIANCE[24]

Star Alliance is the world's first and largest airline alliance with 27 carriers and in the process of integrating one more (EVA). Current membership can always be found by going to

24. Information on Star Alliance has been provided by John Donohoe, PMP, Director, Project Management Office, Star Alliance Services GmbH.

Staralliance.com. Overall, the Star Alliance network offers more than 21,900 daily flights to 1,329 destinations in 194 countries. Its members carried a total of 670.5 million passengers with a turnover of US$181 billion in 2012. Each member of Star Alliance has a PMO.

- Adria Airways
- Aegean Airlines
- Air Canada
- Air China
- Air New Zealand
- ANA
- Asiana Airlines
- Austrian
- Avianca, TACA Airlines
- Brussels Airlines
- Copa Airlines
- Croatia Airlines
- EGYPTAIR
- Ethiopian Airlines
- LOT Polish Airlines
- Lufthansa
- Scandinavian Airlines
- Shenzhen Airlines
- Singapore Airlines
- South African Airways
- SWISS
- TAM Airlines
- TAP Portugal
- THAI
- Turkish Airlines
- United
- US Airways

The Star PMO does not act as a "super-PMO" to member carriers PMO offices. The Star Alliance PMO provides project management services across the Star enterprise. The Star PMO carries out for the business units include such topics as information technology, marketing, sales, products, services, and frequent-flier programs, as well as common sourcing projects, which are projects that use the combined purchasing power of all carriers to jointly purchase common commodities (spare parts, in-flight services, economy class seats, etc.).

Star Alliance projects are projects are aimed at providing a common travel experience across all carriers or those that leverage our size to develop common IT apps, common networks, common lounges, check-in services, or seamless Frequent Flyer upgrades across carriers. Project team members are normally business experts from member carriers distributed world-wide. We need to be very good in cultural awareness and consensus building.

In 2011 Agile project management processes were implemented to supplement their traditional waterfall method for selected projects. An assessment is made at the beginning of the project to determine which delivery method would be more effective according to defined criteria. Additionally, the Star Alliance PMO will assist and coordinate several member carriers to a Star Alliance Common IT Platform. The Star Alliance Common IT Platform is a strategic programme, focused on the effort to better serve the customer, markedly lower IT costs and significantly increase the speed of delivering new products to market. Once implemented, it will enable participating member airlines to improve customer services and enhance operational capabilities. It is based on Amadeus' pioneering new-generation Customer Management Solution portfolio which consists of Altéa Reservation, Altéa Inventory and Altéa Departure Control solutions.

## 12.15 PROJECT AUDITS AND THE PMO

In recent years, the necessity for a structured independent review of various parts of a business, including projects, has taken on a more important role. Part of this can be attributed to the Sarbanes-Oxley Law compliance requirements. These audits are now part of the responsibility of the PMO.

These independent reviews are audits that focus on either discovery or decision-making. They also can focus on determining the health of a project. The audits can be scheduled or random, and can be performed by in-house personnel or external examiners.

There are several types of audits. Some common types include:

- *Performance audits:* These audits are used to appraise the progress and performance of a given project. The project manager, project sponsor, or an executive steering committee can conduct this audit.
- *Compliance audits:* These audits are usually performed by the PMO to validate that the project is using the project management methodology properly. Usually the PMO has the authority to perform the audit but may not have the authority to enforce compliance.
- *Quality audits:* These audits ensure that the planned project quality is being met and that all laws and regulations are being followed. The quality assurance group performs this audit.
- *Exit audits:* These audits are usually for projects that are in trouble and may need to be terminated. Personnel external to the project, such as an exit champion or an executive steering committee, conduct the audits.
- *Best practices audits:* These audits can be conducted at the end of each life-cycle phase or at the end of the project. Some companies have found that project managers may not be the best individuals to perform the audit. In such situations, the company may have professional facilitators trained in conducting best practices reviews.

Checklists and templates often are the best means of performing audits and health checks. Nani Sadowski-Alvarez, PMP, CPHIMS senior management consultant, enterprise program management & architecture at Computer Sciences Corporation (CSC), shares with us a template for auditing a project (see Table 12–1).

**TABLE 12–1.  TEMPLATE FOR AUDITING A PROJECTS**

| | | | |
|---|---|---|---|
| **Project Sponsor:**_____ | | **Project Manager:** _____ | |
| **Project Go Live Date:**_____ | | **Project Final Audit Date:**_____ | |

| Validation | Document/Item to be validated | Rating | Comments |
|---|---|---|---|
| **Yes** ☐ **No** ☐ | **Council Approvals** (i.e., meeting minutes, sign-off, etc. indicating that the project was approved and has been signed off on for execution and implementation). | ☐ | |
| **Yes** ☐ **No** ☐ | **Signed Project Scope** (with original signatures and/or faxed/ electronic signatures attached). | ☐ | |
| **Yes** ☐ **No** ☐ | **Project Kickoff Presentation and Kickoff Meeting Agenda** (to include date of kickoff). | ☐ | |
| **Yes** ☐ **No** ☐ | Current/up-to-date **Project Capital and Operating Expense Cost Sheet.** | ☐ | |
| **Yes** ☐ **No** ☐ | All project-specific **Change Requests** (with detail in relation to triple constraints—schedule/budget/resource), with all corresponding change request sign-off approvals attached. | ☐ | |
| **Yes** ☐ **No** ☐ | Signed (by project sponsor) **Project Acceptance—Closure Letter and Operational Turnover.** | ☐ | |
| **Yes** ☐ **No** ☐ | **Project Closure Presentation** along with agenda and date of closure meeting. | ☐ | |
| **Yes** ☐ **No** ☐ | **Vendor Contracts and Statements of Work (SOW)** signed [by all impacted parties—i.e., vendor, sponsor(s), legal, etc.] | ☐ | |
| **Yes** ☐ **No** ☐ | Completed **Project Closure Checklist** upon closure of the project. | ☐ | |

| Validation | Document/Item to be validated | Rating | Comments |
|---|---|---|---|
| | **Initiation** | | |
| **Yes** ☐ **No** ☐ | **Project Charter and Project Score** | ☐ | |
| **Yes** ☐ **No** ☐ | **Finalized Approved Project Budget** | ☐ | |
| **Yes** ☐ **No** ☐ | Early phase **Workflows** (where applicable) created to demonstrate project need, streamlining of work effort, etc. NOTE: If project is strictly surrounding hardware (HW)/equipment—diagram provided can be an equipment or netware (NW) diagram. | ☐ | |

| Validation | Document/Item to be validated | Rating | Comments |
|---|---|---|---|
| | **Planning** | | |
| **Yes** ☐ **No** ☐ | **Signed Contracts** from Vendor along with SOW. | ☐ | |
| **Yes** ☐ **No** ☐ | **Project Schedule** that has been created and maintained via Clarity and Work Bench. (NOTE: Initial signed copy should be included as an addendum to the project scope. PM will show current schedule to auditor in Clarity/WorkBench upon request). | ☐ | |

**TABLE 12–1.** *(Continued)*

| Validation | | Document/Item to be validated | Rating | Comments |
|---|---|---|---|---|
| | | **Planning** | | |
| Yes ☐ | No ☐ | **Signed Project Scope**—full document to include project organization chart, communication plan, high-level milestones, technical/business/ clinical assessments, as well as project financial materials (when detailed financials are incorporated into the roll-out of projects) etc. | ☐ | |
| Yes ☐ | No ☐ | **Technical Assessment and Relating Diagrams** | ☐ | |
| Yes ☐ | No ☐ | **Provisioning Assessment** (where applicable) to include cost estimates, forecasted equipment, etc. | ☐ | |
| Yes ☐ | No ☐ | **Security Assessment** as established by financial information security officer (FISO). | ☐ | |
| Yes ☐ | No ☐ | **Finalized Approved Project Budget** | ☐ | |
| **Validation** | | **Document/Item to be validated** | **Rating** | **Comments** |
| | | **Executing/Controlling** | | |
| Yes ☐ | No ☐ | **Design/Build**—Any applicable design/build documentation. | ☐ | |
| Yes ☐ | No ☐ | **Risks and Issues**—PM should document issues and risks via the issues/risk/change request tab in Clarity and upload the affiliated documents to the specific risk or issue. | ☐ | |
| Yes ☐ | No ☐ | **Change Requests and Affiliated Sign-Off's**—The PM should document change requests for their projects, via the issues/risk/ change request tab. Change requests will also be uploaded into SharePoint under the applicable project folder. | ☐ | |
| Yes ☐ | No ☐ | **Revised and Enhanced Process Flows and Procedures** | ☐ | |
| Yes ☐ | No ☐ | **Testing**—All project testing related documentation (testing scripts, testing plan, etc.). | ☐ | |
| Yes ☐ | No ☐ | **Training**—All project training related documentation (training plan/ schedule, course information, etc.). | ☐ | |
| Yes ☐ | No ☐ | Appropriate measures and forms in place surrounding **Provisioning** (to include any necessary processes for obtaining sign-off for security access). | ☐ | |
| Yes ☐ | No ☐ | **Purchase Order Details** (purchase requests, cost estimates, etc.). | ☐ | |
| Yes ☐ | No ☐ | **All Invoices** (these will also be tracked via Clarity with the review and guidance of the IS accountant). | ☐ | |
| Yes ☐ | No ☐ | **Pre-Go Live Security Assessment** as preformed by FISO. | ☐ | |
| Yes ☐ | No ☐ | **Pre-Go Live Process Flows** (where applicable) detailing current state and what the forecasted state is following the implementation of the project; if the project involves equipment. | ☐ | |
| Yes ☐ | No ☐ | **Overview Assessment with IS Accountant** | ☐ | |

| Validation | | Document/Item to be validated | Rating | Comments |
|---|---|---|---|---|
| | | **Executing/Controlling** | | |
| Yes ☐ | No ☐ | **Activation/Go Live Plan** to include all necessary details for the core team and other impacted parties to achieve a successful and efficient project go live. | ☐ | |

| Validation | | Document/Item to be validated | Rating | Comments |
|---|---|---|---|---|
| | | **Closing** | | |
| Yes ☐ | No ☐ | **Project Budget**—Approved budget details vs. final actual closing budget. | ☐ | |
| Yes ☐ | No ☐ | **Lessons Learned**—As compiled by the PM and the entire project core team. | ☐ | |
| Yes ☐ | No ☐ | **Operational Turnover** documentation. | ☐ | |
| Yes ☐ | No ☐ | **Scanned Project Acceptance (Closure) Letter with Sponsor(s) Signature** | ☐ | |

| Validation | | Document/Item to be validated | Rating | Comments |
|---|---|---|---|---|
| | | **Agendas/Minutes** | | |
| Yes ☐ | No ☐ | All project-related **Agendas** (saved with date format in the title to ensure ease of use for reference). | ☐ | |
| Yes ☐ | No ☐ | All project-related **Meeting Notes** (saved with date format in title to ensure ease of use for reference). | ☐ | |

| Validation | | Document/Item to be validated | Rating | Comments |
|---|---|---|---|---|
| | | **Presentations** | | |
| Yes ☐ | No ☐ | **Kickoff PowerPoint Presentation** | ☐ | |
| Yes ☐ | No ☐ | **Vendor-Related Demos/Presentations** | ☐ | |
| Yes ☐ | No ☐ | **Internal Presentations Utilized for Project Approval** (if applicable). | ☐ | |
| Yes ☐ | No ☐ | **Project Closure PowerPoint Presentation** | ☐ | |

## 12.16 PROJECT HEALTH CHECKS

Quite often, projects undergo health checks, but by the wrong people. The concept of performing a health check is sound practice provided the right people are performing the health check and the right information is being discussed. The purpose of the health check should be to provide constructive criticism and evaluate alternative approaches as necessary. Too often, the meetings end up being a personal attack on the project team. Executive-level reviews and likewise reviews by the PMO may not provide the project manager with the constructive information that the project manager desires. Eric Maurice of NXP and Mark Gray, formerly of NXP and now CEO of SigmaPM, have identified an innovative way of doing this. They title this approach: *If two heads are better than one, then why not use three or four?*[25]

In our drive to increase the probability of success for projects, one mechanism that is often used is peer-reviews whereby we get other experts to critically analyse our project and give their judgment and advice. The main problem with this approach is the fact that it is often a one-shot, very brief review of the project management documentation with little or no insight to the actual mechanics of the project.

Eric Maurice, a project manager in the R&D area of NXP Semiconductors has come up with a novel way of getting real value out of the approach—the *multi-brained project manager!* (This is sometimes referred to as a Hydra project manager.)

Triggered by the findings of an MBTI (Myers Briggs Type Indicator) exercise done at the team level, Eric realised that there is a significant danger of becoming too biased towards one perspective of the project with the possibility of overlooking what could be obvious issues. This is further exacerbated by the level of complexity of the projects and the number of (sometimes conflicting) data that the project manager has to consider at the start-up.

Eric then approached several project management peers (from across the organisation) via the local network and asked them to become a part of a neural network—sharing ideas, concepts and viewpoints in the context of this particular project. The reason for using this approach rather than a simple peer review was to overcome the constraint of just having a one-shot input while also showing the added value opportunity.

Of course, to make this work required some preparation and starting conditions:

- Given that the peers were from vastly different project backgrounds, quite a lot of preparation had to go into explaining the project context to the group.
- Ground rules for mutual trust, openness, honesty and constructive criticism were needed (although not formally stated). This was especially helpful in identifying potential weaknesses in the risk plan, and helping to face the (sometimes brutal) truth.
- Overcoming one's own barriers to showing weak points is never easy—this again relies on a good level of trust and cooperation in the group.
- The frequency needs to be reasonably regular—in this case it was once per month (on a 20 month project). This is in order to secure the shared view of the project is kept current.

From the experience we have the following observations and outcomes:

- A tangible outcome was a reduction in the risk level for the project—a review of the risk plan helped to secure the content and response planning.
- Strong ties were built between the peer project managers that in fact remained in operation outside of this project context. This also helped to reinforce the value of networking in the organisation.

---

25. The material in this section was provided by Mark Gray, formerly Senior Project Manager at NXP Semiconductors and now CEO of SigmaPM, and Eric Maurice, PMP, Project Manager at NXP Semiconductors.

- The feedback from the participants was also positive, with appreciation shown for the opportunity to share and to learn from each other.
- An element that was seen as adding to the success was the decision to focus on just one specific topic area for each session (planning, risk. . .). This setting of a "theme" allowed the peers to apply their knowledge (or learn from the others) on a specific focus area in the context of a real and understood project.
- The small but dynamic group (between 3 and 6 people was seen as the ideal size) also served as a real incubator for new ideas, as well as an excellent conduit for lessons learned to be transmitted between projects and across the organisation.

In conclusion we can safely say that the usage of the multi-brained project manager approach has a clear value-added in getting to project execution excellence, much more so than either formal project reviews or the normal peer-review "snapshots." This gain is not only for the project, but also for the participants and the organisation as a whole!

**Some Recommendations— and a Health Warning!**

Since the process of setting up and running Hydra sessions takes a non-trivial investment in time, some consideration should be given to when this would be appropriate. We have some suggestions for when (and when not) this may be an appropriate approach:

- This process would have a good return on investment where either the project has a very high level of perceived risk, or where the desire is to use it as an opportunity for mentoring (either the project manager is new on the job or the peers have an opportunity to learn from an "adept").
- We would not recommend using this on very short-term projects (a few months duration) as this reduces the possibility of traction, or on projects with a low-level of strategic importance as this will reduce the level of interest from the peers.
- It's not a good idea to have project managers involved in multiple Hydra sessions—not just from point of view of the time required but also this would dilute the focus too much.
- Putting a Hydra approach in place on a project should come from the project manager themselves; forcing it turns it into a chore, or worse still indicates a lack of trust in the project manager.

Some would say that this should be the domain of the PMO (where one exists) but here we would like to give a health warning: The PMO should of course be the person(s) that help put it in place, set it up, and support the capturing of results—but the true value comes from having the peers really involved in the project under scrutiny. In the authors opinion if the Hydra becomes the domain of the PMO it has the risk of becoming the monster of Greek legend . . .

This approach is not intended to become just another "monitoring and controlling" tool—the real benefit is the shared learning and the multiple perspectives on the day-to-day functioning of the project.

## 12.17 PMO OF THE YEAR AWARD

Some people contend that the most significant change in project management in the first decade of the twenty-first century has been the implementation of the PMO concept. As such, it is no big surprise that The Center for Business Practices initiated the "PMO of the Year" award.[26]

**Award Criteria**     The PMO of the Year Award is presented to the PMO that best illustrates—through an essay and other documentation—their project management improvement strategies, best practices, and lessons learned. Additional support documentation—such as charts, graphs, spreadsheets, brochures, etc.—could not exceed five documents. While providing additional documentation was encouraged, each eligible PMO clearly demonstrated its best practices and lessons learned in the awards essay. Judges reviewed the essays to consider how the applicant's PMO linked project management to their organization's business strategies and played a role in developing an organizational project management culture. The essays were judged on validity, merit, accuracy, and consistency in addition to the applicant PMO's contribution to project and organizational success.

Types of best practices judges looked for include:

- Practices for integrating PMO strategies to manage projects successfully
- Improvements in project management processes, methodologies, or practices leading to more efficient and/or effective delivery of the organization's projects
- Innovative approaches to improving the organization's project management capability
- Practices that are distinctive, innovative, or original in the application of project management
- Practices that promote an enterprisewide use of project management standards
- Practices that encourage the use of performance measurement results to aid decision making
- Practices that enhance the capability of project managers

Best practice outcomes included:

- Evidence of realized business benefits—customer satisfaction, productivity, budget performance, schedule performance, quality, ROI, employee satisfaction, portfolio performance, strategic alignment
- Effective use of resources
- Improved organizational project management maturity
- Executive commitment to a project management culture expressed in policies and other documentation
- A PMO that exhibits an organizational business results focus
- Effective use of project management knowledge and lessons learned
- Individual performance objectives and potential rewards linked to measurement of project success
- Project management functions applied consistently across the organization

The essay comprised three sections. Incomplete submissions were disqualified.

---

26. The material in this section has been provided by The Center for Business Practices, Rockwell Automation, and Alcatel-Lucent. For additional information on the Center for Business Practices and The PMO of The Year Award, visit their website: www.cbponline.com.

**Completing the Essay**

Section 1: Background of the PMO?In no more than 1000 words, the applicants described their PMO, including background information on its scope, vision and mission, and organizational structure. In addition, they described:

- How long the PMO has been in place
- Their role within the PMO
- How the PMO's operation is funded
- How the PMO is structured (staff, roles and responsibilities, enterprisewide, departmental, etc.)
- How the PMO uses project management standards to optimize its practices

Section 2: PMO Innovations and Best Practices?In no more than 1500 words, the applicants addressed the challenges their organization encountered prior to implementing the new PMO practices and how they overcame those challenges. They described clearly and concisely the practices implemented and their effect on project and organizational success.

Section 3: Impact of the PMO and Future Plans?In no more than 500 words, the applicants described the overall impact of the PMO over a sustained period (e.g., customer satisfaction, productivity, reduced cycle time, growth, building or changing organizational culture, etc.). If available, the applicants provided quantitative data to illustrate the areas in which the PMO had the greatest business impact. Finally, they briefly described their PMO's plans for 2009 and how those plans will potentially impact their organization.

Two of the companies discussed in this book competed for the award: Rockwell Automation, which won the award of the 2009 PMO of the Year, and Alcatel-Lucent, which was recognized as one of the finalists for the award. Both of their profiles are discussed below.

**Rockwell Automation: 2009 PMO of the Year Winner**

Software Program Management Office
Type of organization: manufacturing
Headquarters: Milwaukee, Wisconsin, USA
Number of full-time employees (FTEs): 21,000 +
PMO FTEs: 30
PMO annual operating budget: $3.2 million
James C. Brown, Formerly Director, A&S Program Management Office
Presenting challenge: Rolling out a consistent product and project management practice across 16 businesses
Business benefits: Increased predictability and productivity; faster pace of innovation; delivery of a major release comprising 20 + projects on time and under cost for the first time in company history.
Website: http://www.rockwellautomation.com/

**Rockwell Automation: From "Clean Slate" to Global Innovator in under Five Years**

Rockwell Automation was formed by bringing two major automation companies, Allen-Bradley and Reliance Electric, together in the late 1980s. Over the years, Rockwell Automation has continued to acquire other leading automation suppliers as a growth strategy and also as a way to bring new advanced automation technologies into the company. In 2005, as Rockwell Automation was planning the rollout of a new SAP business system, the company recognized the need for a new, common product development (CPD) process that would be based on company best practices combined with industry best practices for product development. This effort resulted in a CPD process that allowed for enterprisewide adoption. All 16 different product businesses ranging from high-volume component suppliers to complex continuous process control systems solution suppliers now use the same high-level process framework for their new product developments.

Since project managers are instrumental in the execution of a product development process, it was quickly realized that introducing an end-to-end process to a company built from many related but very different product businesses would require consistent application of project management across all the product lines. To complicate matters, each business segment was at a different maturity level relative to all aspects of product development. A formal project management organization, established in 2004, already existed and was capitalized upon to support this effort. Says PMO Director James C. Brown, "If consistency, transparency and risk mitigation are important to your business, and they are for us, then we believe that a formal well recognized and managed project management entity is paramount."

Brown, hired in 2004 to help implement the PMO, called the project management environment "a clean slate," other than those people who were already identified as project managers. The PMO is structured by function with program managers overseeing programs and project managers overseeing projects, assisted by an additional two resources supporting tools such as MSProject Server and Sharepoint.

Rockwell's new PMO ramped up quickly, getting everyone PMP certified within two months, which caught the attention of the senior VP for the division. From there they began establishing processes and methodologies, establishing scorecards and metrics, and deploying tools in support of new product development and services. They moved from a waterfall approach in driving projects to an agile approach, driving processes from 20+ pages to fewer than 5 pages and moving them from notebook binders to electronic media. As Brown says, "We moved from reporting on everything to reporting on exceptions only."

The PMO grew from 10 people to 30 over just four years, and its reach went from North American to global. The number of projects under its direction grew from 12–15 to over 50 concurrently. On the way, Rockwell Automation deployed a portfolio management process in their Architecture and Software Group. The goals and purpose of the process are to link investments to business strategy, maximize the value of the portfolio, achieve a desired balance (mix) of projects, and focus the organization's efforts. The portfolio management process links to related processes, such as idea management, strategy development, program and project management, and the recently deployed CPD process, and has become an integral part of the planning process. Brown focuses on the human side of project portfolio management (PPM)'s benefits: "It's about people reaching consensus using trusted data, and a common decision making framework."

Of course, company culture is hard to change, so governance is critical; and that requires management commitment. The driving force behind management's commitment to implement this new process was the vision of a common consistent methodology for new product development across the enterprise. This consistency was prioritized from the top (direct management involvement in stage gate reviews) down, in order to realize benefit as soon as possible.

All too often, businesses were forced to deny funding for strategic projects due to the never-ending incremental product improvements that just kept coming. By forcing business management to approve each project's passage from one phase to the next, the new processes pushed the visibility of every project, every resource, and every dollar up to the decision-makers who wrestled with trying to find dollars to fund the real game-changers.

This visibility also made it easier for the business owners to kill projects with questionable returns or to delay a project in order to free up critical resources. This contrasted with the old way of executing a project, where reviews were informal and haphazard. Teams were able to continue spending and even overspending without any real fear of cancellation. Under the new process, every dependent organization is represented at the appropriate review and given the chance to agree or disagree with the project manager that all deliverables are available. The intent is to have the go/no-go decision made by both the primary organization responsible for the deliverables during the previous phase and the primary organization responsible for the deliverables in the subsequent phase. Both these organizations are required at each stage gate review. In this way, the process helps Rockwell Automation avoid surprises during the later phases by ensuring transparency during the earlier phases.

All of this has been implemented with a light touch that has eased acceptance of the new processes. As Brown notes, "There is a fine line between rigor and burden, the trick is to push this line hard to insure rigorous implementation without slowing the progress of the project team down."

The PMO has been instrumental in Rockwell Automation's quest for increased predictability, productivity, and visibility. By delivering—for the first time in the company's history—a major release that contained in excess of four programs and 20+ projects, on time and under budget, the organization has proven its business value.

**Alcatel-Lucent: 2009 PMO**        Global Program Management Office
**of the Year Finalist**             Type of Organization: telecom
                                     Headquarters: Paris, France
Number of full-time employees (FTEs): 70,000
PMO FTEs: 10
PMO annual operating budget: $4.5 million
PMO senior manager: Rich Maltzman, PMP, Senior Manager, Learning and Professional
    Advancement
Presenting challenge: Combining the project management improvement efforts of two
    companies into one supercharged initiative
Business benefits: Improvements in a wide array of project metrics on projects that impact
    customer satisfaction
Website: http:www.alcatel-lucent.com

The Alcatel-Lucent Global Program Management Office (GPMO) combines the best project management practices of Alcatel and Lucent, both of which were already in the midst of major efforts to revitalize project management as a discipline at the time that Lucent merged with Alcatel in November 2006. Both organizations had already researched best practices in project management, and the discipline was given priority by the new company leadership. A core team was assigned to combining the project management efforts into one new initiative in late 2006. The initial focus of the enterprisewide GPMO was on the 2000+ customer-facing project managers who oversee the turnover of new solutions to customers. The GPMO focused on two major "frameworks"—a project delivery framework and a project management development framework.

The project delivery framework, dedicated to the methodologies and tools that project managers use across the company, brings a new level of project management maturity by offering practice consistency across business units and geographical regions. At its core is a gate-based methodology called the contract implementation process (CIP). The CIP points to a collection of tools that can be used as appropriate by the customer-facing project managers on a region-by-region and unit-by-unit basis. Each CIP methodology is traceable to a *PMBOK® Guide* process. The company is now in the process of integrating CIP into a large enterprise project management software system for its population.

The project manager development framework is a nine-piece integrated model that recognizes the interconnectedness between such key project manager development elements such as a competency model, a career path, project manager training, industry certification, internal accreditation and recognition, and project manager skills management. The GPMO has set stringent targets for PMP® certification for its project managers over the next two years. Alcatel-Lucent was featured in PMI's *Leadership in Project Management* annual for its work in this area. The company's depth of commitment to providing a supportive environment for project managers is illustrated by a number of programs, including:

- *Project Management Professional Accreditation.* Alcatel-Lucent has its own program of accreditation, above and beyond the PMP® certification, the General Project Manager Accreditation, an accreditation above and beyond the PMP® that honors excellence in real-world deployment of external customer projects. It requires the completion of an extensive case-based set of advanced project management courseware and is subject to extremely strict criteria, including a formal jury board. This certification helps guarantee that project managers have not only the general project management wisdom needed for their work but also the particular experience and background in actual projects in the telecom field to support customers. Started with the project manager population (2000 people), professional accreditation was so successful that it is now being rolled out to all of the contributors in the services organization—approximately 18,000 people.
- *Competency Model.* The heart of the project management development framework is a competency model that takes the best from the heritage of Alcatel and Lucent. This is a living model, updated every year to keep up with changes in the project management discipline as well as the fast-paced telecom business.

**Alcatel-Lucent: Two Best-Practice Telecoms Unite Their PM Strengths**

- *RSMS.* The resource and skills management system facilitates project managers' ability to monitor their own progress in development, using their job profile as a basis in identifying skills gaps and suggesting development options to fill those gaps.
- *Alcatel-Lucent University.* As a PMI Registered Education Provider (REP), Alcatel-Lucent University provides access to a wide array of Web-based and instructor-led

training, some of which is highly customized and case based to allow project managers to learn from real successful projects.

- *PMP Study Groups.* The GPMO, through the work of the PM-CERT team (see Chapter 8), has assembled PMP study groups, 8–12 individuals who pool their efforts in studying for the PMP exam. A PMP instructor guides the group, which meets at a frequency of their own choosing via teleconference, and completes the study using a recommended book and set of practice questions. The program also benefits the instructor, yielding valuable PDUs, as we have registered this program with PMI as part of Alcatel-Lucent University.
- *International Project Management Day Symposium.* The second annual International Project Management Day Symposium featured presentations from seven countries on project management topics, with speakers such as Dr. Harold Kerzner. This program is now in its seventh year. About 1000 project managers participate annually in the symposium, which consistently receives excellent feedback scores.

**Integration: A Best Practice in Itself**

According to Rich Maltzman, PMP, leader, learning and professional advancement—a role that focuses on the human side of project management, including the career path, training, internal recognition programs, and skills management, "We feel that the integrated nature of the PM Development Framework is a best practice. It forces the interaction between supporting elements of a successful Project Management career, and in turn a PM discipline that can best support customer projects and thus increase the company's financial position."

In addition, some of the primary best practice tools include:

- *Project Delivery Framework.* The CIP, the heart of the delivery framework, is the standardized process for managing the project life cycle in the company and is focused on the handoffs at key gates in that project life cycle. Recognizing that people make projects work, it provides responsibility matrices to show which role is responsible for which activities at each handoff point. The extensive number of tools and processes that it defines for project managers provides a uniform and effective means to manage projects. CIP in turn is now at the heart of a major program to add enterprise level software to the arsenal of PM tools at the company.
- *Project Management Community Building.* The GPMO Web page provides news, executive messages, links to PMP exam preparation resources, CIP, and an increasingly popular *Engage* group centered on the PM Community. The GPMO also encourages project managers to provide "live" lessons learned rather than to rely on a static repository.
- *Project Team of the Year Award.* A dedicated program was established in January 2008 to recognize the most outstanding project teams in each region and in the entire company. The program was designed with peer project managers. Nominations for 32 teams were received and judged by a panel of project managers and executives. During the year, feature stories based on the nominations were placed on the GPMO and other key corporate websites. Nine finalists were chosen, and from that, a single winner was selected. All of the finalists will receive a team dinner, and

the winning team sends a representative to Paris to receive a special award from the Alcatel-Lucent CEO. This program has continued at Alcatel-Lucent, and in 2013 remains one of its most popular recognition programs.

● *Maturity Assessment and Improvement.* In 2008, the GPMO began a deliberate measurement of project management maturity using a custom survey tool built from recommendations from the Software Engineering Institute (Carnegie Mellon University, Capability Maturity Measurement Integrated, and the Project Management Office Executive Council (part of the Corporate Executive Board). The number of questions was limited to 25, divided among areas such as governance, project performance, resource management, and financials. The response rate was very high (700 respondents) and the answers have yielded mathematical data as well as verbatim feedback that provide a way to map out improvements.

**Tracking the Benefits**   Enrollment in Resource and Skills Management System (RSMS) has increased in the months since it was introduced to the point where—despite significant turnover in the workforce—well over 90% have actively started managing their skills using the dedicated skills programs customized for the four project manager job profiles. Over 100 new PMPs have been certified thanks to the establishments of targets and the use of the PMP study groups. In addition, 36 new general project manager accreditations were awarded in 2008, an increase of almost 30%.

The GPMO spearheads the dissemination of project management thought leadership by Alcatel-Lucent via:

● Presentations at PMI world congresses
● Article in PMI's *Leadership in Project Management* magazine feature
● Alcatel-Lucent presentations at the PMO Summit (Florida) and the PMO Symposium (Texas)
● Operations committee membership for the fourth edition of *PMBOK® Guide*
● Contributions to the fifth edition of the *PMBOK® Guide*

Finally, in terms of statistical results, all of the following measures demonstrated improvements over end-of-year 2007 or even within 2008:

● Baseline of projects under financial margin control is up 160 percent.
● Percent of projects with (approved) upscope has nearly doubled.
● Percent of projects with underruns has more than doubled.
● Percent of projects covered by the enterprise project management system went up from 52 to 87 percent within the year 2008.
● Percent of CFPMs with formal project management development plans has gone from 52 to over 80 percent.

Clearly, a dual focus on people and processes has served the company well and assisted the projects under the umbrella of the GPMO to sail through the merger period—a significant achievement in itself.

# 13 Six Sigma and the Project Management Office

## 13.0 INTRODUCTION

In the previous chapter, we discussed the importance of the PMO for strategic planning and continuous improvements. In some companies, the PMO was established specifically for the supervision and management of Six Sigma projects. Six Sigma teams throughout the organization would gather data and make recommendations to the PMO for Six Sigma projects. The Six Sigma project manager, and possibly the team, would be permanently assigned to the PMO.

Unfortunately, not all companies have the luxury of maintaining a large PMO where the Six Sigma teams and other supporting personnel are permanently assigned to the PMO. It is the author's belief that the majority of the PMOs have no more than four or five people permanently assigned. Six Sigma teams, including the project manager, may end up reporting "dotted" to the PMO and administratively "solid" elsewhere in the organization. The PMO's responsibility within these organizations is primarily for the evaluation, acceptance, and prioritization of projects. The PMO may also be empowered to reject recommended solutions to Six Sigma projects.

For the remainder of this chapter, we will focus on organizations that maintain small PMO staffs. The people assigned to the PMO may possess a reasonable knowledge concerning Six Sigma but may be neither Green nor Black Belts in Six Sigma. These PMOs can and do still manage selected Six Sigma projects but perhaps not the traditional type of Six Sigma projects taught in the classroom.

## 13.1 PROJECT MANAGEMENT—SIX SIGMA RELATIONSHIP

Is there a relationship between project management and Six Sigma? The answer is definitely "yes." The problem is how to exchange the benefits such that the benefits of Six

Sigma can be integrated into project management and, likewise, the benefits of project management can be integrated into Six Sigma. Some companies, such as EDS, have already recognized this important relationship, especially the input of Six Sigma principles to project management. Doug Bolzman, consultant architect, PMP®, ITIL Expert at Hewlett-Packard, discusses this relationship:

> We have incorporated the Information Technology Information Library (ITIL) into the information technology enterprise management (ITEM) framework to design the operational model required to maintain and support the release. ITIL operations components are evaluated and included within each release that requires an operational focus. Six Sigma models have been generated to assist the organization in understanding the capabilities of each release and how to manage the requirements, standards, and data for each of the established capabilities.

Today, there is a common belief that the majority of traditional, manufacturing-oriented Six Sigma failures are because of the lack of project management; nobody is managing the Six Sigma projects as projects. Project management provides Six Sigma with structured processes as well as faster and better execution of improvements.

From a project management perspective, problems with Six Sigma Black Belts include:

- Inability to apply project management principles to planning Six Sigma projects
- Inability to apply project management principles to the execution of Six Sigma projects
- Heavy reliance on statistics and minimum reliance on business processes
- Inability to recognize that project management is value added

If these problem areas are not resolved, then Six Sigma failures can be expected as a result of:

- Everyone plans but very few execute improvements effectively.
- There are too many projects in the queue and poor prioritization efforts.
- Six Sigma stays in manufacturing and is not aligned with overall business goals.
- Black Belts do not realize that executing improvements are projects within a project.

Six Sigma people are project managers and, as such, must understand the principles of project management, including statements of work, scheduling techniques, and so on. The best Six Sigma people know project management and are good project managers; Black Belts are project managers.

A possible solution to some of the Six Sigma failures is to require Six Sigma personnel to use the enterprise project management methodology. Jason Schulist, manager—continuous improvement, Operating Strategy Group at DTE Energy, discusses this:

## Example

In 2002, DTE Energy developed an Operating System Framework to enable systemic thinking around continuous improvement. We blended a lean tool and Six Sigma implementation strategy to develop our current "Lean Sigma" systemic approach. This approach utilizes a 4-Gate/9-Step project management model.[1]

Members of the various business units submit ideas for projects. A review committee prioritizes the projects within each business unit using a project selection document. Once prioritized, each business unit allocates 1–2 percent of its organizational staff to full-time continuous improvement initiatives. Most of these resources are either Lean Sigma Black Belt certified or in training. These resources use the four-gate/nine-step project management model for all projects.

## The 4-Gate/9-Step Management Model

In Steps 1–3, the project lead scopes the project opportunity, forms a team, and analyzes the current reality using rigorous data analysis techniques. By using the $y = f(x)$ tool, teams are able to quantify their metrics and scope their projects to the appropriate level. All Black Belt (BB) certified projects must achieve at least $250,000 in savings or have a significant impact in safety or customer service. The team develops a project charter, which the Champion signs along with the Gate 1 review form.

In Steps 4–6, the team defines the ideal state design that most effectively improves the metrics agreed to in Gate 1. The team identifies gaps and develops countermeasures [using the failure mode and effect analysis (FMEA)] to migrate the initiative from the current state to the ideal state. The team develops a master plan for implementing the changes and commits to targets for each of the metrics. The team measures both input and output metrics and measures the success of the project with respect to improving these metrics. The Champion signs off Gate 2.

In Step 7, the team implements its plan and course corrects as necessary in order to achieve the metrics. At Gate 3 the team reviews its performance to plan, addresses gaps and plan countermeasures, discusses progress to project targets, and ensures that outcomes will be achieved.

In Steps 8 and 9, the team measures project progress, sustains the goals, acknowledges the team, reflects on the project, and communicates the results. The team performs an after-action review (AAR) to inculcate the learning from the project and reduce mistakes in future implementations. The team must achieve the project metrics from Gate 2 before the sponsor signs off at Gate 4. If the team does not reach its targets, the team most likely returns to Gate 2 to redefine the ideal state and confirm that the original targets are still feasible.

The DTE Energy Sarah Sheridan award recognizes many successful completed projects using the 4-Gate/9-Step project management model. Operating systems improvements using the project methodology have saved DTE Energy over $40 million in 2003 and over $100 million in 2004.

---

1. The model is shown in the DTE Energy Section in Chapter 4.

## 13.2 INVOLVING THE PMO

The traditional PMO exists for business process improvements and supports the entire organization, including Six Sigma Black Belts, through the use of the enterprise project management methodology. Project managers, including Black Belts, focus heavily upon customer value-added activities, whether it be an internal or external customer. The PMO focuses on corporate value-added activities.

The PMO can also assist with the alignment of Six Sigma projects with strategy. This includes the following:

- Continuous reprioritization may be detrimental. Important tasks may be sacrificed and motivation may suffer.
- Hedging priorities to appease everyone may result in significant work being prolonged or disbanded.
- A cultural change may be required during alignment.
- Projects and strategy may be working toward cross-purposes.
- Strategy starts at the top whereas projects originate at the middle of the organization.
- Employees can recognize projects but may not be able to articulate strategy. Selecting the proper mix of projects during portfolio management of projects cannot be accomplished effectively without knowing the strategy. This may result in misinterpretation.
- "Chunking" breaks a large project into smaller ones to better support strategy. This makes it easier for revitalization or rejection.

The PMO can also assist in solving some of the problems associated with capturing Six Sigma best practices, such as:

- Introducing a best practice can "raise the bar" too soon and pressure existing projects to possibly implement a best practice that may not be appropriate at that time.
- Employees and managers are unaware of the existence of the best practices and do not participate in their identification.
- Knowledge transfer across the organization is nonexistent and weak at best.
- Falling prey to the superstitious belief that most best practices come from failures rather than from successes.

Simply stated, the marriage of project management with Six Sigma allows us to manage better from a higher level.

## 13.3 TRADITIONAL VERSUS NONTRADITIONAL SIX SIGMA

In the traditional view of Six Sigma, projects fall into two categories: manufacturing and transactional. Each category of Six Sigma is multifaceted and includes a management strategy, metric, and process improvement methodology. This is shown in Figure 13–1. Manufacturing Six Sigma processes utilize machines to produce products, whereas

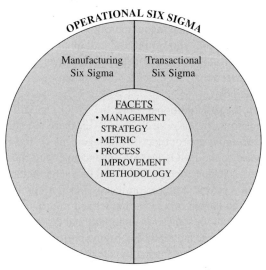

**Figure 13–1.** Six Sigma categories (traditional view).

transactional Six Sigma processes utilize people and/or computers to produce services. The process improvement methodology facet of Six Sigma addresses both categories. The only difference is what tools you will use. In manufacturing, where we utilize repetitive processes that make products, we are more likely to use advanced statistical tools. In transactional Six Sigma, we might focus more on graphical analysis and creative tools/techniques.

The traditional view of a Six Sigma project has a heavy focus on continuous improvement to a repetitive process or activity associated with manufacturing. This traditional view includes metrics, possibly advanced statistics, rigor, and a strong desire to reduce variability. Most of these Six Sigma projects fit better for implementation in manufacturing than in the PMO. Six Sigma teams manage these manufacturing-related projects.

Not all companies perform manufacturing and not all companies support the PMO concept. Companies without manufacturing needs might focus more on the transactional Six Sigma category. Companies without a PMO rely heavily upon the Six Sigma teams for the management of both categories of projects.

Those companies that do support a PMO must ask themselves the following three questions:

- Should the PMO be involved in Six Sigma projects?
- If so, what type of project is appropriate for the PMO to manage even if the organization has manufacturing capability?
- Do we have sufficient resources assigned to the PMO to become actively involved in Six Sigma project management?

PMOs that are actively involved in most of the activities described in Chapter 12 do not have the time or resources required to support all Six Sigma projects. In such a case, the PMO must be selective as to which projects to support. The projects selected are commonly referred to as nontraditional projects that focus more on project management–related activities than manufacturing.

Figure 13–2 shows the nontraditional view of Six Sigma. In this view, operational Six Sigma includes manufacturing activities and all other activities from Figure 13–1, and transactional Six Sigma now contains primarily those activities to support project management.

In the nontraditional view, the PMO can still manage both traditional and nontraditional Six Sigma projects. However, there are some nontraditional Six Sigma projects that are more appropriate for management by the PMO. Some of the projects currently assigned to the PMOs include enhancements to the enterprise project management methodology, enhancements to the PMO tool set, efficiency improvements, and cost avoidance/reduction efforts. Another project assigned to the PMO involves process improvements to reduce the launch of a new product and improving customer management. Experts in Six Sigma might view these as nontraditional types of projects. There is also some concern as to whether these are really Six Sigma projects or just a renaming of a continuous improvement project to be managed by a PMO. Since several companies now refer to these as Six Sigma projects, the author will continue this usage.

Strategic planning for Six Sigma project management is not accomplished merely once. Instead, like any other strategic planning function, it is a cycle of continuous improvements. The improvements can be small or large, measured quantitatively or qualitatively, and designed for either internal or external customers.

There almost always exists a multitude of ideas for continuous improvements. The biggest challenge lies in effective project selection and then assigning the right players. Both of these challenges can be overcome by assigning Six Sigma project management best practices to the project management office. It may even be beneficial having Six Sigma specialists with Green Belts or Black Belts assigned to the PMO.

**Figure 13–2.** Six Sigma categories (nontraditional view).

## 13.4 UNDERSTANDING SIX SIGMA

Six Sigma is not about manufacturing widgets. It is about a focus on processes. And since the PMO is the guardian of the project management processes, it is only fitting that the PMO have some involvement in Six Sigma. The PMO may be more actively involved in identifying the "root cause" of a problem than in managing the Six Sigma solution to the problem.

Some people contend that Six Sigma has fallen short of expectations and certainly does not apply to activities assigned to a PMO. These people argue that Six Sigma is simply a mystique that some believe can solve any problem. In truth, Six Sigma can succeed or fail but the intent and understanding must be there. Six Sigma gets you closer to the customer, improves productivity, and determines where you can get the biggest returns. Six Sigma is about process improvement, usually repetitive processes, and reducing the margin for human and/or machine error. Error can only be determined if you understand the critical requirements of either the internal or external customer.

There are a multitude of views and definitions of Six Sigma. Some people view Six Sigma as merely the renaming of total quality management (TQM) programs as Six Sigma. Others view Six Sigma as the implementation of rigorous application of advanced statistical tools throughout the organization. A third view combines the first two views by defining Six Sigma as the application of advanced statistical tools to TQM efforts.

These views are not necessarily incorrect but are incomplete. From a project management perspective, Six Sigma can be viewed as simply obtaining better customer satisfaction through continuous process improvement efforts. The customer could be external to the organization or internal. The word "satisfaction" can have a different meaning whether we are discussing external or internal customers. External customers expect products and services that are a high quality and reasonably priced. Internal customers may define satisfaction in financial terms, such as profit margins. Internal customers may also focus on such items as cycle time reduction, safety requirements, and environmental requirements. If these requirements are met in the most efficient way without any non–value-added costs (e.g., fines, rework, overtime), then profit margins will increase.

Disconnects can occur between the two definitions of satisfaction. Profits can always be increased by lowering quality. This could jeopardize future business with the client. Making improvements to the methodology to satisfy a particular customer may seem feasible but may have a detrimental effect on other customers.

The traditional view of Six Sigma focused heavily on manufacturing operations using quantitative measurements and metrics. Six Sigma tool sets were created specifically for this purpose. Six Sigma activities can be defined as operational Six Sigma and transactional Six Sigma. Operational Six Sigma would encompass the traditional view and focus on manufacturing and measurement. Operational Six Sigma focuses more on processes, such as the enterprise project management methodology, with emphasis on continuous improvements in the use of the accompanying forms, guidelines, checklists, and templates. Some people argue that transactional Six Sigma is merely a subset of operational Six Sigma. While this argument has merit, project management and specifically the PMO spend the majority of their time involved in transactional rather than operational Six Sigma.

The ultimate goal of Six Sigma is customer satisfaction, but the process by which the goal is achieved can differ whether we are discussing operational or transactional Six.

**TABLE 13–1.  GOALS OF SIX SIGMA**

| Goal[a] | Method of Achievement |
|---|---|
| Understand and meet customer requirements (do so through defect prevention and reduction instead of inspection) | Improvements to forms, guidelines, checklists, and templates for understanding customer requirements |
| Improve productivity | Improve efficiency in execution of the project management methodology |
| Generate higher net income by lowering operating costs | Generate higher net income by streamlining the project management methodology without sacrificing quality or performance |
| Reduce rework | Develop guidelines to better understand requirements and minimize scope changes |
| Create a predictable, consistent process | Continuous improvement on the processes |

[a]From *The Fundamentals of Six Sigma*, International Institute for Learning, New York, 2008, pp. 1–24.

**TABLE 13–2.  GOALS VERSUS FOCUS AREAS**

| Executive Goals | PMO Focus Areas |
|---|---|
| Provide effective status reporting | Identification of executive needs |
|  | Effective utilization of information |
|  | "Traffic light" status reporting |
| Reduce the time for planning projects | Sharing information between planning documents |
|  | Effective use of software |
|  | Use of templates, checklists, and forms |
|  | Templates for customer status reporting |
|  | Customer satisfaction surveys |
|  | Extensions of the enterprise project management methodology into the customer's organization |

Sigma. Table 13–1 identifies some common goals of Six Sigma. The left-hand column lists the traditional goals that fall more under operational Six Sigma, whereas the right-hand column indicates how the PMO plans on achieving the goals.

The goals for Six Sigma can be established at either the executive levels or the working levels. The goals may or may not be able to be completed with the execution of just one project. This is indicated in Table 13–2.

Six Sigma initiatives for project management are designed not to replace ongoing initiatives but to focus on those activities that may have a critical-to-quality and critical-to-customer-satisfaction impact in both the long and short terms.

Operational Six Sigma goals emphasize reducing the margin for human error. But transactional Six Sigma activities managed by the PMO may involve human issues such as aligning personal goals to project goals, developing an equitable reward system for project teams, and project career path opportunities. Fixing people problems is part of transactional Six Sigma but not necessarily of operational Six Sigma.

## 13.5 SIX SIGMA MYTHS[2]

Ten myths of Six Sigma are given in Table 13–3. These myths have been known for some time but have become quite evident when the PMO takes responsibility for project management transactional Six Sigma initiatives.

**Works Only in Manufacturing**   Much of the initial success in applying Six Sigma was based on manufacturing applications; however, recent publications have addressed other applications of Six Sigma. Breyfogle[3] includes many transactional/service applications. In GE's 1997 annual report, CEO Jack Welch proudly states that Six Sigma "focuses on moving every process that touches our customers—every product and *service* (emphasis added)—toward near-perfect quality."

**Ignores Customer in Search of Profits**   This statement is not myth but rather a misinterpretation. Projects worthy of Six Sigma investments should (1) be of primary concern to the customer and (2) have the potential for significantly improving the bottom line. Both criteria must be met. The customer is driving this boat. In today's competitive environment, there is no surer way of going out of business than to ignore the customer in a blind search for profits.

**Creates Parallel Organization**   An objective of Six Sigma is to eliminate every ounce of organizational waste that can be found and then reinvest a small percentage of those savings to continue priming the pump for improvements. With the large amount of downsizing that has taken place throughout the world during the past decade, there is no room or inclination to waste money through the duplication of functions. Many functions are understaffed as it is. Six Sigma is about nurturing any function that adds significant value to the customer while adding significant revenue to the bottom line.

#### TABLE 13–3.   TEN MYTHS OF SIX SIGMA

1. Works only in manufacturing
2. Ignores the customer in search of bottom-line benefits
3. Creates a parallel organization
4. Requires massive training
5. Is an add-on effort
6. Requires large teams
7. Creates bureaucracy
8. Is just another quality program
9. Requires complicated, difficult statistics
10. Is not cost effective

2. Adapted from F. W. Breyfogle III, J. M. Cupello, and B. Meadows, *Managing Six Sigma,* Hoboken, NJ: Wiley, 2001, pp. 6–8.

3. F. W. Breyfogle, III, *Implementing Six Sigma; Smarter Solutions Using Statistical Methods,* Hoboken, NJ: Wiley, 1999.

**Requires Massive Training**    Peter B. Vaill states:

> Valuable innovations are the positive result of this age (we live in), but the cost is likely to be continuing system disturbances owing to members' nonstop tinkering. Permanent white water conditions are regularly taking us all out of our comfort zones and asking things of us that we never imagined would be required. It is well for us to pause and think carefully about the idea of being continually catapulted back into the beginner mode, for that is the real meaning of being a continual learner. We do not need competency skills for this life. We need incompetency skills, the skills of being effective beginners.

**Is an Add-On Effort**    This is simply the myth "creates a parallel organization" in disguise. Same question, same response.

**Requires Large Teams**    There are many books and articles in the business literature declaring that teams have to be small if they are to be effective. If teams are too large, the thinking goes, a combinational explosion occurs in the number of possible communication channels between team members, and hence no one knows what the other person is doing.

**Creates Bureaucracy**    A dictionary definition of bureaucracy is "rigid adherence to administrative routine." The only thing rigid about wisely applied Six Sigma methodology is its relentless insistence that the customer needs to be addressed.

**Is Just Another Quality Program**    Based upon the poor performance of untold quality programs during the past three to five decades[4], an effective quality program would be welcome. More to the point[5], Six Sigma is "an entirely new way to manage an organization."

**Requires Complicated, Difficult Statistics**    There is no question that a number of advanced statistical tools are extremely valuable in identifying and solving process problems. We believe that practitioners need to possess an analytical background and understand the wise use of these tools but do not need to understand all the mathematics behind the statistical techniques. The wise application of statistical techniques can be accomplished through the use of statistical analysis software.

**Is Not Cost Effective**    If Six Sigma is implemented wisely, organizations can obtain a very high rate of return on their investment within the first year.

---

4. J. Micklethwait and A. Wooldridge, *The Witch Doctors of the Management Gurus,* New York: Random House, 1997.

5. T. Pyzdek, "Six Sigma Is Primarily a Management Program, *Quality Digest,* 1999, p. 26.

## 13.6 USE OF ASSESSMENTS

One of the responsibilities that can be assigned to a PMO is the portfolio management of projects. Ideas for potential projects can originate anywhere in the organization. However, ideas specifically designated as transactional Six Sigma projects may need to be searched out by the PMO.

One way to determine potential projects is through an assessment. An assessment is a set of guidelines or procedures that allows an organization to make decisions about improvements, resource allocations, and even priorities. Assessments are ways to:

- Examine, define, and possibly measure performance opportunities
- Identify knowledge and skills necessary for achieving organizational goals and objectives
- Examine and solve performance gap issues
- Track improvements for validation purposes

A gap is the difference between what currently exists and what it should be. The gaps can be in cost, time, quality, and performance, or efficiency. Assessments allow us to pinpoint the gap and determine the knowledge, skills, and abilities necessary to compress the gap. For project management gaps, the assessments can be heavily biased toward transactional rather than operational issues, and this could easily result in behavior modification projects.

There are several factors that must be considered prior to performing an assessment. These factors might include:

- Amount of executive-level support and sponsorship
- Amount of line management support
- Focus on broad-based applications
- Determining who to assess
- Bias of the participants
- Reality of the answers
- Willingness to accept the results
- Impact on internal politics

The purpose of the assessment is to identify ways to improve global business practices first and functional business practices second. Because the target audience is usually global, there must exist unified support and understanding for the assessment process and that it is for the best interest of the entire organization. Politics, power, and authority issues must be put aside for the betterment of the organization.

Assessments can take place at any level of the organization. These can be:

- Global organizational assessments
- Business unit organizational assessments
- Process assessments
- Individual or job assessments
- Customer feedback assessments (satisfaction and improvements)

There are several tools available for assessments. A typical list might include:

- Interviews
- Focus groups
- Observations
- Process maps

Assessments for Six Sigma project management should not be performed unless the organization believes that opportunities exist. The amount of time and effort expended can be significant, as shown in Figure 13–3.

The advantages of assessment can lead to significant improvements in customer satisfaction and profitability. However, there are disadvantages, such as:

- Costly process
- Labor intensive
- Difficulty in measuring which project management activities can benefit from assessments
- May not provide any meaningful benefits
- Cannot measure a return on investment from assessments

Assessments can have a life of their own. There are typical life-cycle phases for assessments. These life-cycle phases may not be aligned with the life-cycle phases of the enterprise project management methodology and may be accomplished more informally than formally. Typical assessment life-cycle phases include:

- Gap or problem recognition
- Development of the appropriate assessment tool set
- Conducting the assessment/investigation
- Data analyses
- Implementation of the changes necessary
- Review for possible inclusion in the best practices library

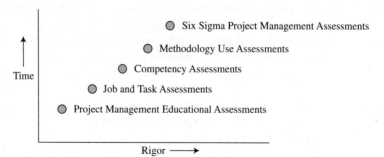

**Figure 13–3.** Time and effort expended.

**TABLE 13–4.  SCALES**

| Strongly agree | Under 20% |
|---|---|
| Agree | Between 20 and 40% |
| Undecided | Between 40 and 60% |
| Disagree | Between 60 and 80% |
| Strongly disagree | Over 80% |

Determining the tool set can be difficult. The most common element of a tool set is a focus on questions. Types of questions include:

- Open ended
  - Sequential segments
  - Length
  - Complexity
  - Time needed to respond
- Closed ended
  - Multiple choice
  - Forced choices (yes–no, true–false)
  - Scales

Table 13–4 illustrates how scales can be set up. The left-hand column solicits a qualitative response and may be subjective whereas the right-hand column would be a quantitative response and more subjective.

It is vitally important that the assessment instrument undergo pilot testing. The importance of pilot testing would be:

- Validate understanding of the instructions
- Ease of response
- Time to respond
- Space to respond
- Analysis of bad questions

## 13.7 PROJECT SELECTION

Six Sigma project management focuses on continuous improvements to the enterprise project management methodology. Identifying potential projects for the portfolio is significantly easier than getting them accomplished. There are two primary reasons for this:

- Typical PMOs may have no more than three or four employees. Based upon the activities assigned to the PMO, the employees may be limited as to how much time they can allocate to Six Sigma project management activities.
- If functional resources are required, then the resources may be assigned first to those activities that are mandatory for the ongoing business of the firm.

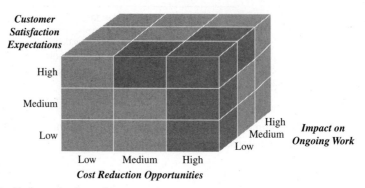

**Figure 13–4.** Project selection cube.

The conflict between ongoing business and continuous improvements occurs frequently. Figure 13–4 illustrates this point. The ideal Six Sigma project management activity would yield high customer satisfaction, high-cost reduction opportunities, and significant support for the ongoing business. Unfortunately, what is in the best interest of the PMO may not be in the best, near-term interest of the ongoing business.

All ideas, no matter how good or how bad, are stored in the "idea bank." The ideas can originate from anywhere in the organization, namely:

- Executives
- Corporate Six Sigma champions
- Project Six Sigma champions
- Master Black Belts
- Black Belts
- Green Belts
- Team members

If the PMO is actively involved in the portfolio management of projects, then the PMO must perform feasibility studies and cost–benefit analyses on projects together with prioritization recommendations. Typical opportunities can be determined using Figure 13–5. In this figure, $\Delta X$ represents the amount of money (or additional money) being spent. This is the input to the evaluation process. The output is the improvement, $\Delta Y$, which is the benefits received or cost savings. Consider the following example.

**Convex Corporation**　　　　Convex Corporation identified a possible Six Sigma project involving the streamlining of internal status reporting. The intent was to eliminate as much paper as possible from the bulky status reports and replace it with color-coded "traffic light" reporting using the company intranet. The PMO used the following data:

- Burdened hour at the executive level $240
- Typical number of project status review meetings per project 8

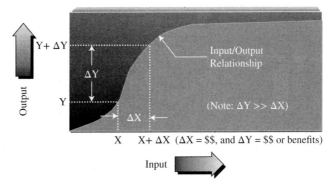

**Figure 13–5.** Six Sigma quantitative evaluation.

- Duration per meeting 2 hours
- Number of executives per meeting 5
- Number of projects requiring executive review 20

Using the above information, the PMO calculated the total cost of executives as:

(8 meetings) × (5 executives) × (2 hr/meeting) × ($240/hr) × (20 projects) = $384,000

Convex assigned one systems programmer (burdened at $100/hr) for four weeks. The cost for adding traffic light reporting to the intranet methodology was $16,000.

Six months after implementation, the number of meetings had been reduced to five per project for an average of 30 minutes in duration. The executives were now focusing on only those elements of the project that were color coded as a potential problem. On a yearly basis, the cost for the meetings on the 20 projects was now about $60,000. In the first year alone, the company identified a savings of $324,000 for one investment of $16,000.

## 13.8 TYPICAL PMO SIX SIGMA PROJECTS

Projects assigned to the PMO can be operational or transactional but mainly the latter. Typical projects might include:

- *Enhanced Status Reporting:* This project could utilize traffic light reporting designed to make it easier for customers to analyze performance. This could be intranet based. The intent is to achieve paperless project management. The colors could be assigned based upon problems, present or future risks, or title, level, and rank of the audience.
- *Use of Forms:* The forms should be user friendly and easy to complete. Minimal input by the user should be required and the data inputted into one form should service multiple forms if necessary. Nonessential data should be eliminated. The forms should be cross-listed to the best practices library.

- *Use of Checklists/Templates:* These documents should be comprehensive yet easy to understand. They should be user friendly and easy to update. The forms should be flexible such that they can be adapted to all situations.
- *Criteria of Success/Failure:* There must exist established criteria for what constitutes success or failure on a project. There must also exist a process that allows for continuous measurement against these criteria as well as a means by which success (or failure) can be redefined.
- *Team Empowerment:* This project looks at the use of integrated project teams, the selection of team members, and the criteria to be used for evaluating team performance. This project is designed to make it easier for senior management to empower teams.
- *Alignment of Goals:* Most people have personal goals that may not be aligned with goals of the business. This includes project versus company goals, project versus functional goals, project versus individual goals, project versus professional goals, and other such alignments. The greater the alignment between goals, the greater the opportunity for increased efficiency and effectiveness.
- *Measuring Team Performance:* This project focuses on ways to uniformly apply critical success factors and key performance indicators to team performance metrics. This also includes the alignment of performance with goals and rewards with goals. This project may interface with the wage and salary administration program by requiring two-way and three-way performance reviews.
- *Competency Models:* Project management job descriptions are being replaced with competency models. A competency criterion must be established, including goal alignment and measurement.
- *Financial Review Accuracy:* This type of project looks for ways of including the most accurate data into project financial reviews. This could include transferring data from various information systems such as earned-value measurement and cost accounting.
- *Test Failure Resolution:* Some PMOs maintain a failure-reporting information system that interfaces with FMEA. Unfortunately, failures are identified but there may be no resolution on the failure. This project attempts to alleviate this problem.
- *Preparing Transitional Checklists:* This type of project is designed to focus on transition or readiness of one functional area to accept responsibility. As an example, it may be possible to develop a checklist on evaluating the risks or readiness of transitioning the project from engineering to manufacturing. The ideal situation would be to develop one checklist for all projects.

This list is by no means comprehensive. However, the list does identify typical projects managed by the PMO. Some conclusions can be reached by analyzing this list. First, the projects can be both transactional and operational. Second, the majority of the projects focus on improvements to the methodology. Third, having people with Six Sigma experience (i.e., Green, Brown, or Black Belts) would be helpful.

When a PMO takes the initiative in Six Sigma project management, the PMO may develop a Six Sigma toolbox exclusively for the PMO. These tools most likely will not include the advanced statistics tools that are used by Black Belts in manufacturing but may be more process-oriented tools or assessment tools.

# 14 Project Portfolio Management

## 14.0 INTRODUCTION

Your company is currently working on several projects and has a waiting list of an additional 20 projects that they would like to complete. If available funding will support only a few more projects, how does a company decide which of the 20 projects to work on next? This is the project portfolio management process. It is important to understand the difference between project management and project portfolio management. Debra Stouffer and Sue Rachlin have made this distinction for IT projects[1]:

> An IT portfolio is composed of a set or collection of initiatives or projects. Project management is an ongoing process that focuses on the extent to which a specific initiative establishes, maintains, and achieves its intended objectives within cost, schedule, technical and performance baselines.
>
> Portfolio management focuses attention at a more aggregate level. Its primary objective is to identify, select, finance, monitor, and maintain the appropriate mix of projects and initiatives necessary to achieve organizational goals and objectives.
>
> Portfolio management involves the consideration of the aggregate costs, risks, and returns of all projects within the portfolio, as well as the various tradeoffs among them. Of course, the portfolio manager is also concerned about the "health" and well-being of each project that is included within the IT portfolio. After all, portfolio decisions, such as whether to fund a new project or continue to finance an ongoing one, are based on information provided at the project level.

---

1. D. Stouffer and S. Rachlin, "A Summary of First Practices and Lessons Learned in Information Technology Portfolio Management," prepared by the Chief Information Officer (CIO) Council, Washington, DC, March 2002, p. 7.

Portfolio management of projects helps determine the right mix of projects and the right investment level to make in each of them. The outcome is a better balance between ongoing and new strategic initiatives. Portfolio management is not a series of project-specific calculations such as ROI, NPV, IRR, payback period, and cash flow and then making the appropriate adjustment to account for risk. Instead, it is a decision-making process for what is in the best interest of the entire organization.

Portfolio management decisions are not made in a vacuum. The decision is usually related to other projects and several factors, such as available funding and resource allocations. In addition, the project must be a good fit with other projects within the portfolio and with the strategic plan.

The selection of projects could be based upon the completion of other projects that would release resources needed for the new projects. Also, the projects selected may be constrained by the completion date of other projects that require deliverables necessary to initiate new projects. In any event, some form of a project portfolio management process is needed.

## 14.1 WHY USE PORTFOLIO MANAGEMENT?

Not all companies assign the same degree of importance to portfolio management. In some companies, it is a manual process, while in others it mandates the use of sophisticated tools. Some companies believe that portfolio management is part of the strategic planning efforts while others see it as a support function for capacity planning. Carl Manello, PMP, practice director—delivery effectiveness, Slalom Consulting, believes that:

> Portfolio management is critical to ensure that companies are spending their limited resources in the best possible way. I have worked with companies at various levels to implement some type of portfolio management. In some cases, portfolio management is simply an extension of the annual budgetary process. In the best cases, it develops into a fully vetted assessment and on-going prioritization process.
>
> But there is no one best approach to portfolio management. The process of portfolio management has varying degrees to which it can be implemented. For example, one can choose to put [a] process in place with or without a sophisticated project portfolio management tool. Even the process by which prioritization is conducted may vary. Depending on the capability of the organization, the approach may be an inventory of current and proposed projects with a forced ranking created by the PMO. For the more sophisticated enterprise, a full blown process may engage senior and executive management into a negotiation to decide where to invest scarce corporate resources based on business unit, strategic objective and projected returns. Figure 14–1 illustrates a typical portfolio matrix. At one company, I worked to develop a relatively simplistic numerical algorithm for assessing projects. After criteria were established for *Business Importance* and *Complexity* (these were the two measures defined by the Enterprise Program Office and the Chief Financial Officer), each project was plotted on a simple two by two matrix. This visualization offered us the ability to compare and contrast dissimilar projects with very little effort.
>
> In 2012, Slalom has focused on leveraging this "proven practice" to larger and larger portfolios. At a global financial trading company, we implemented an automated model to support objective, quantifiable and repeatable assessment of a department's portfolio of approximately 100 initiatives. For a cellular carrier's enterprise focus, the model was used to represent almost 400 initiatives. We are currently getting ready to start an engagement

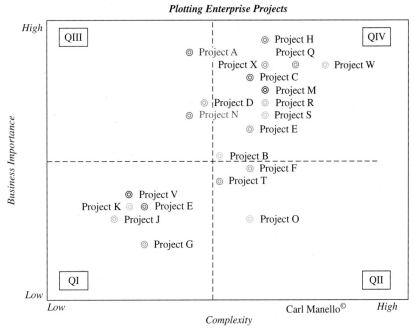

**Figure 14–1.** Typical portfolio matrix.

with a financial services company to implement the tool with visibility across the enterprise, inclusive of upward of 900 projects. Notwithstanding my well known bias as a tool bigot {Harold, can we reference page 170 paragraph on driving for process first, then the tool}, I recognize the value that tools bring to the complex world of portfolio management.

Even without committing to the market-leading tools for project portfolio management, companies can leverage the power of basic tools to help them through their analysis, ratification, prioritization and decision making. When the portfolio management exercise becomes one of creating visibility to critical decision-making data, not only are executives more like to be aligned to the process and participate, but decisions become one of *business decisions* and not the appeasement of the loudest department or senior vice president.

## 14.2 INVOLVEMENT OF SENIOR MANAGEMENT, STAKEHOLDERS, AND THE PMO ▬

The successful management of a project portfolio requires strong leadership by individuals who recognize the benefits that can be accrued from portfolio management. The commitment by senior management is critical. Stouffer and Rachlin comment on the role of senior management in an IT environment in government agencies[2]:

Portfolio management requires a business and an enterprise-wide perspective. However, IT investment decisions must be made both at the project level and the portfolio level. Senior

---

2. Ibid., p. 8.

government officials, portfolio and project managers, and other decision makers must routinely ask two sets of questions.

First, at the project level, is there sufficient confidence that new or ongoing activities that seek funding will achieve their intended objectives within reasonable and acceptable cost, schedule, technical, and performance parameters?

Second, at the portfolio level, given an acceptable response to the first question, is the investment in one project or a mix of projects desirable relative to another project or a mix of projects?

Having received answers to these questions, the organization's senior officials, portfolio managers, and other decision makers then must use the information to determine the size, scope, and composition of the IT investment portfolio. The conditions under which the portfolio can be changed must be clearly defined and communicated. Proposed changes to the portfolio should be reviewed and approved by an appropriate decision making authority, such as an investment review board, and considered from an organization-wide perspective.

Senior management is ultimately responsible for clearly defining and communicating the goals and objectives of the project portfolio as well as the criteria and conditions considered for the portfolio selection of projects. According to Stouffer and Rachlin, this includes[3]:

- Adequately define and broadly communicate the goals and objectives of the IT portfolio.
- Clearly articulate the organization's and management's expectations about the type of benefits being sought and the rates of returns to be achieved.
- Identify and define the type of risks that can affect the performance of the IT portfolio, what the organization is doing to avoid and address risk, and its tolerance for ongoing exposure.
- Establish, achieve consensus, and consistently apply a set of criteria that will be used among competing IT projects and initiatives.

Senior management must also collect and analyze data in order to assess the performance of the portfolio and determine whether or not adjustments are necessary. This must be done periodically such that critical resources are not being wasted on projects that should be canceled. Stouffer and Rachlin provide insight on this through their interviews[4]:

According to Gopal Kapur, President of the Center for Project Management, organizations should focus on their IT portfolio assessments and control meetings on critical project vital signs. Examples of these vital signs include the sponsor's commitment and time, status of the critical path, milestone hit rate, deliverables hit rate, actual cost versus estimated cost, actual resources versus planned resources, and high probability, high impact events. Using a red, yellow, or green report card approach, as well as defined metrics, an organization can establish a consistent method for determining if projects are having an adverse impact on the IT portfolio, are failing and need to be shut down.

Specific criteria and data to be collected and analyzed may include the following:

---

3. Ibid., p. 13.
4. Ibid., p. 18.

- Standard financial measures, such as return on investment, cost benefit analysis, earned value (focusing on actuals versus plan, where available), increased profitability, cost avoidance, or payback. Every organization participating in the interviews included one or more of these financial measures.
- Strategic alignment (def.ined as mission support), also included by almost all organizations.
- Client (customer) impact, as defined in performance measures.
- Technology impact (as measured by contribution to, or impact on, some form of defined architecture).
- Initial project and (in some cases) operations and schedules, as noted by almost all organizations.
- Risks, risk avoidance (and sometimes risk mitigation specifics), as noted by almost all participants.
- Basic project management techniques and measures.
- And finally, data sources and data collection mechanisms also are important. Many organizations interviewed prefer to extract information from existing systems; sources include accounting, financial, and project management systems.

One of the best practices identified by Stouffer and Rachlin for IT projects was careful consideration of both internal and external stakeholders[5]:

Expanding business involvement in portfolio management often includes the following:

- Recognizing that the business programs are critical stakeholders, and improving that relationship throughout the life cycle
- Establishing service level agreements that are tied to accountability (rewards and punishment)
- Shifting the responsibilities to the business programs and involving them on key decision making groups

In many organizations, mechanisms are in place to enable the creation, participation and "buy-in" of stakeholder coalitions. These mechanisms are essential to ensure the decision making process is more inclusive and representative. By getting stakeholder buy-in early in the portfolio management process, it is easier to ensure consistent practices and acceptance of decisions across an organization. Stakeholder participation and buy-in can also provide sustainability to portfolio management processes when there are changes in leadership.

Stakeholder coalitions have been built in many different ways depending on the organization, the process and the issue at hand. By including representatives from each major organizational component who are responsible for prioritizing the many competing initiatives being proposed across the organization, all perspectives are included. The approach, combined with the objectivity brought to the process by using predefined criteria and a decision support system, ensures that everyone has a stake in the process and the process is fair.

Similarly, the membership of the top decision making body is comprised of senior executives from across the enterprise. All major projects, or those requiring a funding source, must be voted upon and approved by this decision making body. The value of

---

5. Ibid., pp. 22–23.

getting stakeholder participation at this senior level is that this body works toward supporting the organization's overall mission and priorities rather than parochial interests.

More and more companies today are relying heavily upon the PMO for support with portfolio management. Typical support activities include capacity planning, resource utilization, business case analysis, and project prioritization. The role of the PMO in this regard is to support senior management, not to replace them. Portfolio management will almost always remain as a prime responsibility for senior management, but recommendations and support by the PMO can make the job of the executive a little easier. In this role, the PMO may function as more of a facilitator. Chuck Millhollan, director of program management at Churchill Downs Inc. (CDI), describes portfolio management in his organization:

Our PMO is responsible for the portfolio management process and facilitates portfolio reviews by our "Investment Council." We have purposefully separated the processes for requesting and evaluating projects (having projects approved in principle) and authorizing work (entry into the active portfolio).

When asked to describe the PMO's relationship to portfolio management, Chuck Millhollan commented:

Investment Council: The Investment Council is comprised of senior (voting) members (CEO, COO, CFO, EVPs) and representatives from each business unit. There are regularly scheduled monthly meetings, facilitated by the PMO, to review and approve new requests and review the active portfolio. The Investment Council's goals and objectives include:

1. Prioritize & allocate capital to projects.
2. Approve/disapprove requested projects based on the merit of the associated Business Case.
3. Act individually and collectively as vocal and visible project champions throughout their representative organizations.
4. As necessary, take an active role in approving project deliverables, helping resolve issues and policy decisions, and providing project related direction and guidance.

Request, Evaluation & Approval: We use an "Investment Request Worksheet" to standardize the format in which projects (called investment requests) are presented to the Investment Council. Elements include request description, success criteria and associated metrics, a description of the current and future state, alignment to strategic goals, preliminary risk assessment, identification of dependent projects, preliminary resource availability and constraint assessment and a payback analysis for ROI and Cost-Out initiatives.

Work Authorization: If projects were approved during the annual operational planning processes and are capital investments that generate ROI or result in a Cost-Out, come back to the Investment Council for work authorization and addition to the portfolio of active projects. This can be done concurrently with request, evaluation and approval for projects that are initiated mid-planning cycle.

Portfolio Maintenance: We use a bi-weekly project status reporting process and only include projects that the Investment Council has identified as requiring portfolio review and/or oversight. The portfolio reports are provided bi-weekly and presented monthly during the Investment Council meetings.

When the PMO supports or facilitates the portfolio management process, the PMO becomes an active player in the strategic planning process and supports senior management by making sure that the projects in the queue are aligned with strategic objectives. The role might be support or monitoring and control. Enrique Sevilla Molina, PMP, formerly corporate PMO director at Indra, discusses portfolio management in his organization:

> Portfolio management is strongly oriented to monitor and control the portfolio performance, and to review its alignment with the strategic planning. A careful analysis of trends and forecasts is also periodically performed, so the portfolio composition may be assessed and reoriented if required.
>
> Once the strategic targets for the portfolio have been defined and allocated through the different levels in the organization, the main loop of the process includes reporting, reviewing and taking actions on portfolio performance, problems, risks, forecasts and new contracts planning. A set of alerts, semaphores and indicators have been defined and automated in order to focus the attention on the main issues related with the portfolio management. Those projects or proposals marked as requiring specific attention are carefully followed by the management team, and a specific status reporting is provided for those.
>
> One of the key tools used for [the] portfolio management process is our Projects Monitor. It is a web based tool that provides a full view of the status of any predefined set of projects (or portfolio), including general data, performance data, indicators and semaphores. It has also the capability to produce different kinds of reports, at single project level, at portfolio level, or a specialized risk report for the selected portfolio.
>
> Besides the corporate PMO, major Business Units throughout the company use local PMOs in their portfolio management process. Some of them are in charge of risk status reporting for the major projects or programs in the portfolio. Others are in charge of an initial definition of the risk level for the projects and operations in order to provide an early detection of potential risk areas. And others play a significant role in providing the specific support to the portfolio managers when reporting the status to upper level management.
>
> Our corporate level PMO defines the portfolio management processes in order to be consistent with the project management level and, in consequence, the requirements for the implementation of those processes in the company tools and information systems.

Some companies perform portfolio management without involvement by the PMO. This is quite common when portfolio management might include a large amount of capital spending projects. According to a spokesperson at AT&T:

> Our PMO is not part of portfolio management. We maintain a Portfolio Administration Office (PAO) which approves major capital spending projects and programs through an annual planning process. The PAO utilizes change control for any modifications to the list of approved projects. Each Project Manager must track the details of their project and update information in the Portfolio Administration Tool (PAT). The Corporate Program Office uses data in PAT to monitor the health and well being of the projects. Individual projects are audited to ensure adherence to processes and reports are prepared to track progress and status.

## 14.3 PROJECT SELECTION OBSTACLES[6]

Portfolio management decision-makers frequently have much less information to evaluate candidate projects than they would wish. Uncertainties often surround the likelihood of success for a project, the ultimate market value of the project, and its total cost to completion. This lack of an adequate information base often leads to another difficulty: the lack of a systematic approach to project selection and evaluation. Consensus criteria and methods for assessing each candidate project against these criteria are essential for rational decision-making. Although most companies have established organizational goals and objectives, they are usually not detailed enough to be used as criteria for project portfolio management decision-making. However, they are an essential starting point.

Portfolio management decisions are often confounded by several behavioral and organizational factors. Departmental loyalties, conflicts in desires, differences in perspectives, and an unwillingness to openly share information can stymie the project selection, approval, and evaluation processes. Much project evaluation data and information is necessarily subjective in nature. Thus, the willingness of the parties to openly share and put trust in each other's opinions becomes an important factor.

The risk-taking climate or culture of an organization can also have a decisive bearing on the project selection process. If the climate is risk adverse, high-risk projects may never surface. Attitudes within the organization toward ideas and the volume of ideas being generated will influence the quality of the projects selected. In general, the greater the number of creative ideas generated, the greater the chances of selecting high-quality projects.

## 14.4 IDENTIFICATION OF PROJECTS

The overall project portfolio management process is a four-step approach, as shown in Figure 14–2. The first step is the identification of the ideas for projects and needs to help support the business. The identification can be done through brainstorming sessions, market research, customer research, supplier research, and literature searches. All ideas, regardless of merit, should be listed.

Because the number of potential ideas can be large, some sort of classification system is needed. There are three common methods of classification. The first method is to place the projects into two major categories, such as survival and growth. The sources and types of funds for these two categories can and will be different. The second method comes from typical R&D strategic planning models, as shown in Figure 14–3. Using this approach, projects to develop new products or services are classified as either offensive or defensive projects. Offensive projects are designed to capture new markets or expand market share within existing markets. Offensive projects mandate the continuous development of new products and services.

Defensive projects are designed to extend the life of existing products or services. This could include add-ons or enhancements geared toward keeping present customers or finding new customers for existing products or services. Defensive projects are usually easier to manage than offensive projects and have a higher probability of success.

---

6. W. Souder, *Project Selection and Economic Appraisal*, New York: Van Nostrand Reinhold, 1984, pp. 2–3.

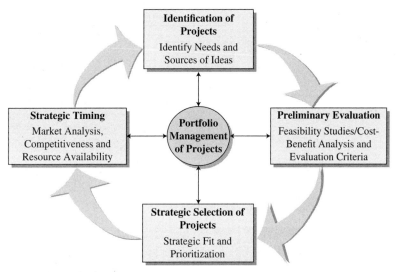

**Figure 14–2.** Project selection process.

**Figure 14–3.** R&D strategic planning process.

Another method for classifying projects would be:

● Radical technical breakthrough projects
● Next-generation projects
● New family members
● Add-ons and enhancement projects

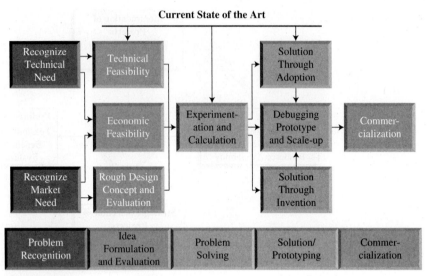

**Figure 14–4.** Modeling the innovation process.

Radical technological breakthrough projects are the most difficult to manage because of the need for innovation. Figure 14–4 shows a typical model for innovation. Innovation projects, if successful, can lead to profits that are many times larger than the original development costs. Unsuccessful innovation projects can lead to equally dramatic losses, which is one of the reasons why senior management must exercise due caution in approving innovation projects. Care must be taken to identify and screen out inferior candidate projects before committing significant resources to them.

There is no question that innovation projects are the most costly and difficult to manage. Some companies mistakenly believe that the solution is to minimize or limit the total number of ideas for new projects or to limit the number of ideas in each category. This could be a costly mistake.

In a study of the new-product activities of several hundred companies in all industries, Booz, Allen, and Hamilton[7] defined the new-product evolution process as the time it takes to bring a product to commercial existence. This process began with company objectives, which included fields of product interest, goals, and growth plans, and ended with, hopefully, a successful product. The more specifically these objectives were defined, the greater guidance would be given to the new-product program. This process was broken down into six manageable, fairly clear sequential stages:

- *Exploration:* The search for product ideas to meet company objectives.
- *Screening:* A quick analysis to determine which ideas were pertinent and merit more detailed study.
- *Business Analysis:* The expansion of the idea, through creative analysis, into a concrete business recommendation, including product features, financial analysis, risk analysis, market assessment, and a program for the product.

7. *Management of New Products*, Booz, Allen & Hamilton, McLean, VA, 1984, pp. 180–181.

- *Development:* Turning the idea-on-paper into a product-in-hand, demonstrable and producible. This stage focuses on R&D and the inventive capability of the firm. When unanticipated problems arise, new solutions and trade-offs are sought. In many instances, the obstacles are so great that a solution cannot be found, and work is terminated or deferred.
- *Testing:* The technical and commercial experiments necessary to verify earlier technical and business judgments.
- *Commercialization:* Launching the product in full-scale production and sale; committing the company's reputation and resources.

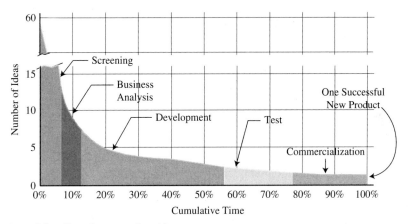

**Figure 14–5.**   Mortality of new product ideas.

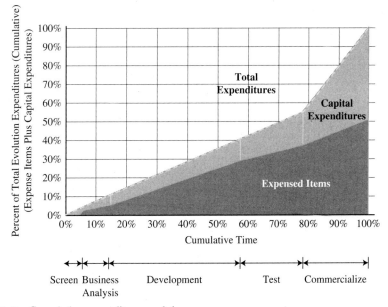

**Figure 14–6.**   Cumulative expenditures and time.

In the Booz, Allen & Hamilton study, the new-product process was characterized by a decay curve for ideas, as shown in Figure 14–5. This showed a progressive rejection of ideas or projects by stages in the product evolution process. Although the rate of rejection varied between industries and companies, the general shape of the decay curve is typical. It generally takes close to 60 ideas to yield just one successful new product.

The process of new-product evolution involves a series of management decisions. Each stage is progressively more expensive, as measured in expenditures of both time and money. Figure 14–6 shows the rate at which expense dollars are spent as time accumulates for the average project within a sample of leading companies. This information was based on an all-industry average and is, therefore, useful in understanding the typical industrial new-product process. It is significant to note that the majority of capital expenditures are concentrated in the last three stages of evolution. It is, therefore, very important to do a better job of screening for business and financial analysis. This will help eliminate ideas of limited potential before they reach the more expensive stages of evolution.

## 14.5 PRELIMINARY EVALUATION

As shown in Figure 14–2, the second step in project selection is preliminary evaluation. From a financial perspective, preliminary evaluation is basically a two-part process. First, the organization will conduct a feasibility study to determine whether the project can be done. The second part is to perform a benefit-to-cost analysis to see whether the company should do it (see Table 14–1).

The purpose of the feasibility study is to validate that the idea or project meets feasibility of cost, technological, safety, marketability, and ease of execution requirements. It is possible for the company to use outside consultants or subject matter experts (Smells) to assist in both feasibility studies and benefit-to-cost analyses. A project manager may not be assigned until after the feasibility study is completed because the project manager may not have sufficient business or technical knowledge to contribute prior to this point in time.

If the project is deemed feasible and a good fit with the strategic plan, the project is prioritized for development along with other approved projects. Once feasibility is

**TABLE 14–1. FEASIBILITY STUDENTS AND COST–BENEFITS ANALYSES**

|  | Feasibility Study | Cost–Benefit Analysis |
|---|---|---|
| Basic question | Can we do it? | Should we do it? |
| Life-cycle phase | Preconceptual | Conceptual |
| Project manager selected | Not yet | Perhaps |
| Analysis | **Qualitative** | **Quantitative** |
|  | Technical | NPV |
|  | Cost | Discounted cash flow (DCF) |
|  | Quality | IRR |
|  | Safety | ROI |
|  | Legal | Assumptions |
|  | Economical | Reality |
| Decision criteria | Strategic fit | Benefits > cost |

determined, a benefit-to-cost analysis is performed to validate that the project will, if executed correctly, provide the required financial and nonfinancial benefits. Benefit-to-cost analyses require significantly more information to be scrutinized than is usually available during a feasibility study. This can be an expensive proposition.

Estimating benefits and costs in a timely manner is very difficult. Benefits are often defined as:

- Tangible benefits, for which dollars may be reasonably quantified and measured
- Intangible benefits, which may be quantified in units other than dollars or may be identified and described subjectively

Costs are significantly more difficult to quantify, at least in a timely and inexpensive manner. The minimum costs that must be determined are those that are used specifically for comparison to the benefits. These include:

- The current operating costs or the cost of operating in today's circumstances.
- Future period costs that are expected and can be planned for.
- Intangible costs that may be difficult to quantify. These costs are often omitted if quantification would contribute little to the decision-making process.

There must be careful documentation of all known constraints and assumptions that were made in developing the costs and the benefits. Unrealistic or unrecognized assumptions are often the cause of unrealistic benefits. The go or no-go decision to continue with a project could very well rest upon the validity of the assumptions.

## 14.6 STRATEGIC SELECTION OF PROJECTS

From Figure 14–2, the third step in the project selection process is the strategic selection of projects, which includes the determination of a strategic fit and prioritization. It is at this point where senior management's involvement is critical because of the impact that the projects can have on the strategic plan.

Strategic planning and the strategic selection of projects are similar in that both deal with the future profits and growth of the organization. Without a continuous stream of new products or services, the company's strategic planning options may be limited. Today, advances in technology and growing competitive pressure are forcing companies to develop new and innovative products while the life cycle of existing products appears to be decreasing at an alarming rate. Yet, at the same time, executives may keep research groups in a vacuum and fail to take advantage of the potential profit contribution of R&D strategic planning and project selection.

There are three primary reasons that corporations work on internal projects:

- To produce new products or services for profitable growth
- To produce profitable improvements to existing products and services (i.e., cost reduction efforts)
- To produce scientific knowledge that assists in identifying new opportunities or in "fighting fires"

Successful project selection is targeted, but targeting requires a good information system, and this, unfortunately, is the weakest link in most companies. Information systems are needed for optimum targeting efforts, and this includes assessing customer and market needs, economic evaluation, and project selection.

Assessing customer and market needs involves opportunity-seeking and commercial intelligence functions. Most companies delegate these responsibilities to the marketing group, and this may result in a detrimental effort because marketing groups appear to be overwhelmed with today's products and near-term profitability. They simply do not have the time or resources to adequately analyze other activities that have long-term implications. Also, marketing groups may not have technically trained personnel who can communicate effectively with the R&D groups of the customers and suppliers.

Most organizations have established project selection criteria, which may be subjective, objective, quantitative, qualitative, or simply a seat-of-the-pants guess. The selection criteria are most often based upon suitability criteria, such as:

- Similar in technology
- Similar marketing methods used
- Similar distribution channels used
- Can be sold by current sales force
- Will be purchased by the same customers as current products
- Fits the company philosophy or image
- Uses existing know-how or expertise
- Fits current production facilities
- Both research and marketing personnel enthusiastic
- Fits the company long-range plan
- Fits current profit goals

In any event, there should be a valid reason for selecting the project. Executives responsible for selection and prioritization often seek input from other executives and

Total Weighted Score  = Σ (Criterion Score × Criterion Weight)
\* Scale: 10=Excellent;     1=Unacceptable

**Figure 14–7.**  Scoring model.

managers before moving forward. One way to seek input in a quick and reasonable manner is to transform the suitability criteria shown above into rating models. Typical rating models are shown in Figures 14–7, 14–8, and 14–9.[8] These models can be used for both strategic selection and prioritization.

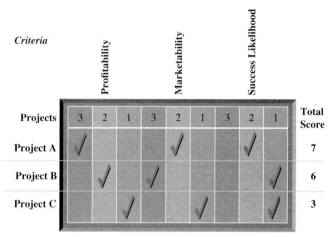

Figure 14–8. Checklist for three projects.

Figure 14–9. Scaling model for one project, project A.

---

8. W. Souder, *Project Selection and Economic Appraisal,* New York: Van Nostrand Reinhold, 1984, pp. 66–69.

Prioritization is a difficult process. Factors such as cash flow, near-term profitability, and stakeholder expectations must be considered. Also considered are a host of environmental forces, such as consumer needs, competitive behavior, existing or forecasted technology, and government policy.

Being highly conservative during project selection and prioritization could be a road map to disaster. Companies with highly sophisticated industrial products must pursue an aggressive approach to project selection or risk obsolescence. This also mandates the support of a strong technical base.

## 14.7 STRATEGIC TIMING

Many organizations make the fatal mistake of taking on too many projects without regard for the limited availability of resources. As a result, the highly skilled labor is assigned to more than one project, creating schedule slippages, lower productivity, less than anticipated profits, and never-ending project conflicts.

The selection and prioritization of projects must be made based upon the availability of qualified resources. Planning models are available to help with the strategic timing of resources. These models are often referred to as *aggregate planning models.*

Another issue with strategic timing is the determination of which projects require the best resources. Some companies use a risk–reward cube, where the resources are assigned based upon the relationship between risk and reward. The problem with this approach is that the time required to achieve the benefits (i.e., payback period) is not considered.

Aggregate planning models allow an organization to identify the overcommitment of resources. This could mean that high-priority projects may need to be shifted in time or possibly be eliminated from the queue because of the unavailability of qualified resources. It is a pity that companies also waste time considering projects for which they know that the organization lacks the appropriate talent.

Another key component of timing is the organization's tolerance level for risk. Here, the focus is on the risk level of the portfolio rather than the risk level of an individual project. Decision makers who understand risk management can then assign resources effectively such that the portfolio risk is mitigated or avoided.

## 14.8 ANALYZING THE PORTFOLIO

Companies that are project-driven organizations must be careful about the type and quantity of projects they work on because of the resources available. Because of critical timing, it is not always possible to hire new employees and have them trained in time or to hire subcontractors that may end up possessing questionable skills.

Figure 14–10 shows a typical project portfolio.[9] Each circle represents a project. The location of each circle represents the quality of resources and the life-cycle phase that

---

9. This type of portfolio was adapted from the life-cycle portfolio model commonly used for strategic planning activities.

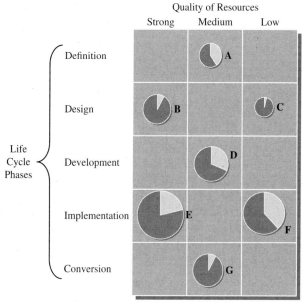

**Figure 14–10.** Typical project portfolio.

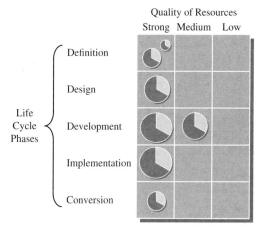

**Figure 14–11.** High-risk portfolio.

the project is in. The size of the circle represents the magnitude of the benefits relative to other projects, and the pie wedge represents the percentage of the project completed thus far.

In Figure 14–10, project A has relatively low benefits and uses medium-quality resources. Project A is in the definition phase. However, when project A moves into the design phase, the quality of resources may change to low-quality or strong quality. Therefore, this type of chart has to be updated frequently.

Figures 14–11, 14–12, and 14–13 show three types of portfolios. Figure 14–11 represents a high-risk project portfolio where strong resources are required on each project. This

**Figure 14–12.**   Profit portfolio.

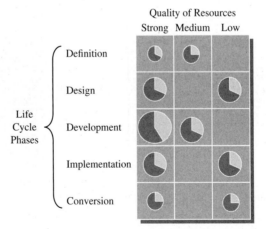

**Figure 14–13.**   Balanced portfolio.

may be representative of project-driven organizations that have been awarded large, highly profitable projects. This could also be a company in the computer field that competes in an industry that has short product life cycles and where product obsolescence occurs six months downstream.

Figure 14–12 represents a conservative, profit portfolio where an organization works on low-risk projects that require low-quality resources. This could be representative of a project portfolio selection process in a service organization or even a manufacturing firm that has projects designed mostly for product enhancement.

Figure 14–13 shows a balanced portfolio with projects in each life-cycle phase and where all levels of resources are being utilized, usually quite effectively. A very delicate juggling act is required to maintain this balance.

## 14.9 PROBLEMS WITH MEETING EXPECTATIONS

Why is it that, more often than not, the final results of either a project or an entire portfolio do not meet senior management's expectations? This problem plagues many corporations and the blame is ultimately (and often erroneously) rationalized as poor project management practices. As an example, a company approved a portfolio of 20 R&D projects for 2001. Each project was selected on its ability to be launched as a successful new product. The approvals were made following the completion of the feasibility studies. Budgets and timetables were then established such that the cash flows from the launch of the new products would support the dividends and the cash needed for ongoing operations.

Full-time project managers were assigned to each of the twenty projects and began with the development of detailed schedules and project plans. For eight of the projects, it quickly became apparent that the financial and scheduling constraints imposed by senior management were unrealistic. The project managers on these eight projects decided not to inform senior management of the potential problems but to wait awhile to see if contingency plans could be established. Hearing no bad news, senior management was left with the impression that all launch dates were realistic and would go as planned.

The eight trouble-plagued projects were having a difficult time. After exhausting all options and failing to see a miracle occur, the project managers then reluctantly informed senior management that their expectations would not be met. This occurred so late in the project life cycle that senior management became quite irate and several employees had their employment terminated, including some of the project sponsors.

Several lessons can be learned from this situation. First, unrealistic expectations occur when financial analysis is performed from "soft" data rather than "hard" data. In Table 14–1 we showed the differences between a feasibility study and a benefit-to-cost analysis. Generally speaking, feasibility studies are based upon soft data.

Therefore, critical financial decisions based upon feasibility study results may have significant errors. This can also be seen from Table 14–2, which illustrates the accuracy of typical estimates. Feasibility studies use top-down estimates that can contain significant error margins.

**TABLE 14–2.  COST/HOUR ESTIMATES**

| Estimating Method | Generic Type | WBS Relationship | Accuracy | Time to Prepare |
|---|---|---|---|---|
| Parametric | ROM[a] | Top down | 25% to + 75% | Days |
| Analogy | Budget | Top down | −10% to + 25% | Weeks |
| Engineering (grass roots) | Definitive | Bottom up | −5% to + 10% | Months |

[a]Rough order of magnitude.

Benefit-to-cost analyses should be conducted from detailed project plans using more definitive estimates. Benefit-to-cost analysis results should be used to validate that the financial targets established by senior management are realistic.

Even with the best project plans and comprehensive benefit-to-cost analyses, scope changes will occur. There must be a periodic reestimation of expectations performed on a timely basis. One way of doing this is by using the rolling wave concept shown in Figure 14–14. The rolling wave concept implies that as you get further along in the project, more knowledge is gained, which allows us to perform more detailed planning and estimating. This then provides additional information from which we can validate the original expectations.

Continuous reevaluation of expectations is critical. At the beginning of a project, it is impossible to ensure that the benefits expected by senior management will be realized at project completion. The length of the project is a critical factor. Based upon project length, scope changes may result in project redirection. The culprit is most often changing economic conditions, resulting in invalid original assumptions. Also, senior management must be made aware of events that can alter expectations. This information must be made known quickly. Senior management must be willing to hear bad news and have the courage to possibly cancel a project.

Since changes can alter expectations, project portfolio management must be integrated with the project's change management process. According to Mark Forman, the associate director for IT and e-government in the Office of Management and Budget[10]:

> Many agencies fail to transform their process for IT management using the portfolio management process because they don't have change management in place before starting. IT will not solve management problems—re-engineering processes will. Agencies have to train their people to address the cultural issues. They need to ask if their process is a simple process. A change management plan is needed. This is where senior management vision and direction is sorely needed in agencies.

Although the comments here are from government IT agencies, the problem is still of paramount importance in nongovernment organizations and across all industries.

**Figure 14–14.**   Rolling wave concept.

---

10. See note 1, p. 1.

## 14.10 PORTFOLIO MANAGEMENT AT ROCKWELL AUTOMATION[11] _____

Rockwell Automation has deployed a portfolio management process in their Architecture and Software Group. The goals and purpose of the process are to link investments to our business strategy, maximize the value of the portfolio, achieve a desired balance (mix) of projects, and focus the organization's efforts. The Portfolio Management process puts strategic focus on how we manage our investment dollars, by becoming an integral part of our planning process. It is about people reaching consensus using trusted data, and a common decision making framework. The Portfolio Management Process links to related processes, such as Idea Management, Strategy Development, Program and Project Management and our recently deployed Common Product Development Process. (See Figure 14–15.)

Investment Proposals are qualified through a common Concept Scorecard, which is a dynamic spreadsheet that quantifies and scores the attractiveness of a concept through an Investment Proposal, and if approved is utilized to quantify a project in the stage gate process, which includes funding events. The Data build for decision-making starts with less early and builds to more later as the accuracy and certainty of estimates improve. The sum of all nonfunded (proposals) and funded (projects) are managed through the Ranked Ordered Concept List, which is fed from the Concept Scorecard. (See Figure 14–16.)

**Figure 14–15.** Portfolio management and a common product development stage gate process.

11. This section on Rockwell Automation was provided by James C. Brown, PgMP, PMP, OPM3 AC, MPM, CIPM, CSP, CSSMBB, Director, A&S Enterprise Program Management Office; Karen Wojala, Manager, Business Planning; and Matt Stibora, Lean Enterprise Manager.

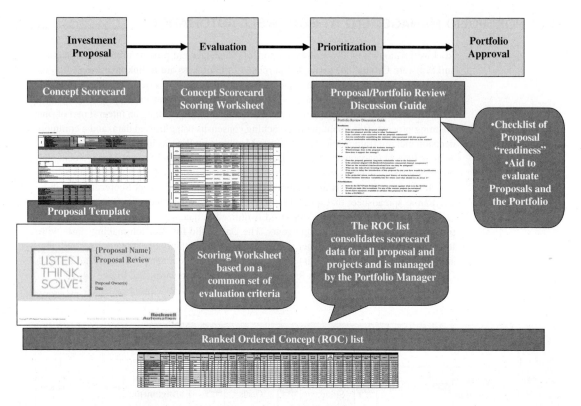

**Figure 14–16.**   Project management process and common template overview.

## 14.11 WORLD WILDLIFE FUND (WWF)[12]

**Portfolio Management:**
**Measuring Short- and**
**Long-Term Results in WWF**

WWF is one of the world's largest conservation organizations and is based on a network of country offices delivering a common Global Program Framework (GPF), an ambitious portfolio of biodiversity and footprint priorities to focus the WWF network's efforts until 2020. WWF is implementing a suite of global priority programs focusing on priority geographic areas (ecoregions), flagship species, ecological footprint[13] and drivers[14] in order to deliver on the GPF goals.

WWF International acts as a secretariat to coordinate the network and provide central management services and to set standards and best practices. Like all organizations, WWF

---

12. Any reproduction in full or in part of this article must mention the title and credit WWF as the copyright owner. Text 2013 © WWF-World Wide Fund for Nature (also known as World Wildlife Fund). All rights reserved. Material was written by Peter J. Stephenson, PhD, Director, and William Reidhead, MSc, Advisor, Design and Impact, Conservation Strategy and Performance Unit, WWF International.

13. Ecological footprint refers to the cropland, grazing land, forest and fishing grounds required to produce the food, fibre and timber consumed in a country, to absorb the wastes emitted in generating the energy it uses, and to provide space for its infrastructure.

14. A driver is defined as a social, economic or political factor that leads to a direct impact on the environment through a change in either the state of biodiversity and/or the ecological footprint.

needs to monitor its portfolio performance to maximize value for money, to manage risks, and to identify and share best practices.

However, as a nonprofit organization working in a very complex and ever-changing environment, WWF faces a number of unique challenges to portfolio management, in particular:

- A strongly decentralized organizational structure with independent management and approval systems, locally specific priorities, and nonstandard sets of performance measures
- Limited global financial and human resources, and hence a compelling need to prioritize programs that will maximize collective impact
- Continually evolving global context, heavily influenced by global economic and geopolitical trends
- Significant delays between intervention and measurable impact, and difficult attribution.

To address these challenges, WWF has developed a global portfolio monitoring and management system that empowers local management while informing global decision-making; that demonstrates short, medium, and long-term results; that detects emerging trends and opportunities; and that enables the most conservation-efficient allocation of resources.

The WWF portfolio management system, being rolled out in July 2013, will therefore provide programs with the information they need for adaptive management, will allow governance bodies to explore progress within and between programs, and will allow aggregation at a global level of enough data for meaningful analyses of overall delivery of the GPF and conservation impacts and trends.

*The portfolio management system includes three pillars:*
a.  Assessing program performance

     WWF recognizes that few measurable conservation results will be observable in the short-term, and in many cases programs will require 5 years or more to demonstrate change. Consequently, conservation programs place substantial emphasis on articulating and testing their theories of change, that is, the program's logic for how short-term interventions will roll up, over time, to increasingly large scale outcomes and impacts. A well-articulated theory of change will specify planned intermediate results as the basis for short-term work planning and the unit for regular program management. The first pillar of the WWF portfolio management system requires each program to self-assess annually, on a continuous scale of 1 to 7, its progress toward its annual planned intermediate results. To improve the objectivity of this process, each program is subject to independent peer review. An overall "conservation achievement KPI" is thus generated for each of WWF's priority programs; this KPI can be used at the portfolio level as a snapshot of performance and as an early warning system of under-performing programs or programs requiring additional support from the WWF network.
b.  Measuring outcomes and impacts

     The second pillar of the portfolio management system requires the organization to look beyond programs' short-term performance toward selected outcomes and impacts.

**Outcomes** are usually related to reducing threats to biodiversity and will be defined by a program's stated objectives. Outcome monitoring will answer questions such as: Were new protected areas established and effectively managed? Were fisheries catches improved and bycatch reduced? Were key policy changes agreed and implemented?

**Impact** is a measure of how well a program is delivering on its ultimate stated goals related directly to the biodiversity it is trying to conserve or the ecological footprint it is trying to reduce. It will answer questions such as: Did tiger numbers increase? Did forest cover in the Amazon remain stable? Did a cod fishery recover? Has energy consumption decreased? Are local communities better off as a result of the program?

The WWF Network agreed on a set of 20 "common indicators" intended to provide a picture of the results programs are attaining against a common set of comparable measures. The common indicators are articulated around state (habitat cover and connectivity; flagship species populations; ocean health; species diversity; environmental flows), pressures (habitat loss and degradation; river fragmentation; species offtake and over-exploitation, $CO_2$ emissions), responses (protected areas size and management effectiveness; sustainable production of commodities, energy and water; wildlife trade) and social impact (beneficiaries). Similar indicators from similar programs can be clustered to allow a global analysis of use to governance bodies in portfolio management.

c. Portfolio dashboards

The first and second pillars of the portfolio management system systematically provide program-level data on short-term performance (the conservation achievement KPI) and medium and long-term performance (the common outcome and impact indicators). These data can be presented together in a dashboard to provide a holistic view of how each program is performing, independent of any other. The data will be presented in the context of program goals to provide a relative measure of progress in delivering outcome and impacts. The dashboards will comprise the core of WWF's annual Global Conservation Program Report and will be accompanied by a narrative to put the results in the context of program actions and theories of change.

By using common short- and longer-term measures, the system also permits these data to be viewed across the WWF portfolio, allowing the comparison of performance across programs, and permitting the aggregate rollup of data to measure how WWF is performing as an organization against the global goals it set for itself in the Global Program Framework. Figure 14–17 provides a mock-up of how data might be aggregated and used at the portfolio level.

The portfolio dashboard will serve as a central management tool for use by a number of different audiences, including program management teams and governance bodies, WWF network oversight and governance bodies, as well as donors and other stakeholders. Key questions answered by the portfolio management system will include:

● Are WWF programs meeting their goals and objectives and having an impact on the state of biodiversity?

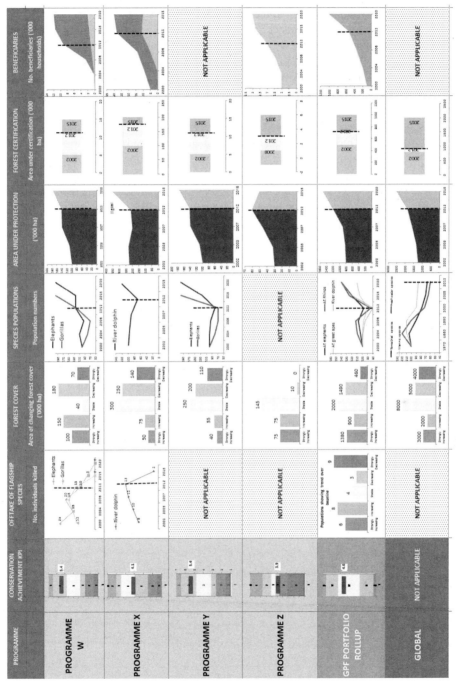

**Figure 14–17.** Sample consolidated dashboard for WWF forest programmes (all data hypothetical)

605

- Which are the programs where WWF's investment is having the greatest/least impact toward the goals of the Global Program Framework?
- What are common technical and operational factors that are influencing programs' performance?
- What are the best practices and lessons learned within and between programs?
- What is the WWF network's contribution to conservation globally? What can be attributed to WWF's work?
- What are emerging trends, challenges, and opportunities for conservation globally?

It is important to note that the effectiveness of the portfolio management system is closely linked to the quality of strategic plans and monitoring being practiced in each program. As a result, WWF programs are strongly encouraged, and supported, in following best practices in planning, monitoring and evaluation[15], in setting aside resources for data collection and analysis, and mainstreaming the resultant information into program-level management decision-making and learning.

---

15. WWF follows the "WWF Standards for Conservation Project and Programme Management" based on the "Open Standards for the Practice of Conservation," a set of best practices developed and promoted by the Conservation Measures Partnership (www.conservationmeasures.org).

# 15 Global Project Management Excellence

## 15.0 INTRODUCTION

In the previous chapters, we discussed excellence in project management, the use of project management methodologies, and the hexagon of excellence. Many companies previously described in the book have excelled in all of these areas. In this chapter we will focus on seven companies, namely IBM, Computer Associates, Microsoft, Deloitte, COMAU, Fluor, and Siemens, all of which have achieved specialized practices and characteristics related to in-depth globalized project management:

- They are multinational.
- They sell business solutions to their customers rather than just products or services.
- They recognize that, in order to be a successful solution provider, they must excel in project management rather than just being good at it.
- They recognize that that they must excel in all areas of project management rather than just one area.
- They recognize that a global project management approach must focus more on a framework, templates, checklists, forms, and guidelines, rather than rigid policies and procedures, and that the approach can be used equally well in all countries and for all clients.
- They recognize the importance of knowledge management, lessons learned, capturing best practices and continuous improvement.
- They understand the necessity of having project management tools to support their project management approach.
- They understand that, without continuous improvement in project management, they could lose clients and market share.
- They maintain a project management office or center of excellence.

- They perform strategic planning for project management.
- They regard project management as a strategic competency.

These characteristics can and do apply to all of the companies discussed previously, but they are of the highest importance to multinational companies.

## 15.1 IBM[1]

**Culture**

In his book, *Who Says Elephants Can't Dance*, former IBM Chairman, Lou Gerstner wrote:

- The issue, in essence, is: How do we deal with the IBM matrix? How can we execute our strategies in such a complex company?
- . . . IBM's unique value to customers is our ability to build integrated solutions. . .
- . . . our ability to integrate products, skills and insight is our *only* sustainable competitive advantage.
- . . . there is no long-term, sustainable competitive advantage in technology.
- The question then is: what kind of management system supports our fundamental strategy of integration? . . . The classic chain-of-command system is not only too slow for the pace of this industry, but it opposes our ability to work across organizations. Hierarchy, which seems to support integration, actually fights it. Hierarchy erects vertical lines and boundaries and fosters the infamous "silo" mentality.
- . . . a well-managed matrix is highly fluid and adaptable. Roles change often. Teams form and disband. Decisions about which business unit will lead in any particular situation are not codified. This puts a premium on the judgment of leaders at every level.[2]

In the mid-1990s, industry dynamics such as worldwide competition, resource pressures, rapid change in customer segments, and technology drove IBM to a different organizational structure than its traditional hierarchical management approach. Additionally, IBM identified the lack of strong project management as a major contributor to project failure and client satisfaction issues across the corporation. These factors led IBM to decide to become a project-based business, applying and integrating project management disciplines across its core businesses, processes and systems.

In November 1996, IBM consolidated its efforts to become a project-based enterprise and established the Project Management Center of Excellence (PM/COE). The Project Management Center of Excellence charter is to develop a project management competency necessary to effectiveness and success within its matrixed enterprise. It drives IBM's transformation to a project based enterprise by developing and supporting project management professionals worldwide.

---

1. ©2013 by IBM Corporation. Reproduced by permission. All rights reserved.

2. Gerstner, Louis V., Who Says Elephants Can't Dance?, ©2002, Harper Business Publishers, an Imprint of Harper Collins Publishers, Appendix A, pages 313–314.

Since its inception, this charter is periodically reviewed to ensure its value and relevance. In 2010, program management was added to the charter in response to the increasing complexity of projects and the industry recognition of the need for qualified program managers. The addition of program management not only presented the need for additional education but also a path for career advancement.

The Project Management Center of Excellence was recognized in 2010 as the Project Management Office of the Year by PM Solutions and PMOSIG and in 2011 as a Best Practice by APQC.

**The Charter: Raising Project and Program Management to a Core Competence in IBM**

IBM identified project management as key to its ability to reliably deliver its commitments. As in many businesses in the 1990s, the lack of strong project management contributed to failure to deliver commitments on time, within budget, and with appropriate profit. Analysis of project management capabilities against industry best practices showed that IBM:

- Often did not recognize a work effort as a project and, therefore, did not apply the disciplines needed to deliver the work successfully;
- Did not have skilled project managers or a pipeline to cover project opportunities; and
- Had an organizational culture and systems (management and business) that were not project-based and, therefore, did not support, and sometimes actually impeded, project work.

IBM took definitive steps to change its culture and support systems to improve its business posture. The company took bold steps in how to organize, execute, and track work and began to organize work into projects that produced services, products, and solutions for our customers. This approach was applied to both external and internal development projects.

Since its inception, the IBM Project Management Center of Excellence, working hand in hand with all business units worldwide, worked to ensure that IBM became a project-based enterprise that applies and integrates project management disciplines into all core business processes and systems. (See Figure 15–1.) The initial charter drove more consistent and broader use of project management disciplines, important to IBM's ability to meet customer needs. However, project management cannot be leveraged without pervasive organizational competence in project management skills. Actions drove the formalization of the position of project manager throughout IBM as well as focus on improving the project management skills of all IBM employees.

The addition of program management in 2010 was in response to the business units' requirements to address increasing project complexity and the focus on programs. Actions to support program management are included in the discussions with the Executive Steering Team and the line organizations.

**Organizational Competence in Project and Program Management**

Core competence in project and program management requires an environment that focuses on projects and programs not functions. The environment must develop a cadre of skilled project and program managers worldwide. IBM encourages and expects all employees associated with projects to understand how their roles affect delivery and successful client engagements. It requires financial, human resource, and management systems to support project management activities. (See Figure 15–2 below)

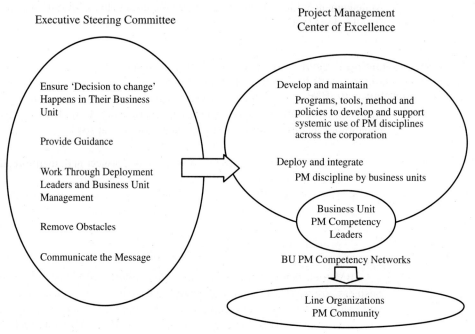

**Figure 15–1.** Involvement at all levels of the enterprise is critical to the systemic acceptance and use of project management disciplines.

| Build Skills | Deliver Key Enablers | Make PM Systemic |
|---|---|---|
| Skills<br>Education<br>Qualification<br>Knowledge<br>    Sharing | Methods<br>Tools<br>Assessments<br>Financial Systems | Executive Involvement<br>Measurements<br>Communications<br>Integration:<br>    Business, Cultural,<br>    Organization |

Individual ⟹ Enterprise

**Figure 15–2.** The roadmap to achieve project management competency reaches from the individual project management professional across the enterprise, ensuring organizational effectiveness and acceptance.

To achieve organizational competence in project management, the PM/COE drives the following action across its culture:

- Work is organized along project lines; efforts and initiatives are driven by cross-functional teams that focus on achieving specific goals in defined timeframes.
- Projects are consistently funded, staffed, and managed in all business units.
- Project management has a consistent role in overall business processes.
- Individuals involved in projects—including executives, project team members, and functional managers—understand their role in planning, carrying out, and evaluating individual and multiple projects.
- Project outcomes can be predicted with high degrees of certainty. When those predictions are wrong, the organization has the knowledge and ability to adjust decisions and take proper corrective action early.
- Timely, accurate, quantitative methods are available for monitoring progress, predicting results, and evaluating risk; systems are in place to support those techniques; and results are continually improved.
- Project managers are qualified and assigned according to stringent professional criteria.
- Project management is designated as a critical professional discipline with programs for continuous skill growth and professional development.

The Project Management Center of Excellence established the following actions to achieve such organizational competence and a project-based culture across IBM.

**Action 1:** IBM adopted one consistent approach for project and program management. This approach embodies core methodologies, practices, tools and techniques common to all IBM organizations, but flexible enough to accommodate individual business considerations.

Initially, there were multiple approaches to project management. Inclusive of different vocabularies, processes, and tools. Project managers were trained differently in different business units. Even the role of the project manager differed from one organization to the next. With the addition of program management, the PM/COE is providing a path for addressing the increasing complexity of the business and development environment.

An organizational competence requires a consistent approach. Consistency improves the timeliness and quality, and it reduces costs. Communication is simplified and timelier with standardized tools and formats. Project and program managers across IBM benefit from the experience of others as they no longer waste time reinventing techniques already learned once through experience. Clearly, an identical detailed project management process or program approach will not be applicable to all business units and could even interfere with individual business processes and needs. However, there are key characteristics common to all successful projects and programs regardless of their type or magnitude. We have developed a core process based on best practices within IBM and from our peers and competitors. Institutionalizing this core process ensures that the project and program management disciplines critical to success are now applied throughout the corporation.

**Action 2:** Selected, certified project and program managers manage significant projects and programs while the management systems enable this process. IBM project and program managers meet appropriate requirements for professional education and experience, commensurate with position requirements.

IBM project and program managers must meet educational standards. A progressive qualification structure, including the requirement to pass independent industry examinations, ensures that IBM project managers obtain the requisite knowledge and experience as they move through their careers. A detailed competency study was used to map the experience and education critical to developing the best project and program managers and ensuring their professional vitality.

The career path and qualification approach is supported by an integrated human resource framework and professional development program. These approaches are aligned and have converged into a single IBM-wide approach to education and professional vitality. The approach is flexible so that it can also embrace new Human Resource programs, which ensure skill levels attained are consistent not only across project management roles but against other IBM job roles.

Equally important to development and qualification is a refinement of the process by which they are assigned. Projects and programs are assessed based on size, revenue implications, risk, customer significance, time urgency, market necessity, and other characteristics; qualified professionals are assigned to them based on required education and experience factors.

**Action 3:** IBM monitors project and program performance and measures progress in terms of technical, financial, and schedule performance. Project and program managers and their sponsoring executives are accountable for specific results. To aid in these efforts, methods and tools necessary to analyze performance are also available.

Project management is too often thought of as simply managing tasks in a schedule. However, project success is also determined by the degree to which requirements are met and by the financial performance of the project. An effort that is completed on schedule but doesn't meet requirements or significantly overruns its budget projections may well end up costing the company clients, market share, or marketplace leadership as well as immediate profit and employee morale.

In evaluating performance, all aspects – technical, schedule, and financial performance – are consistently assessed. IBM processes and systems support this approach and provide timely and comprehensive project and program data. There are a number of systems available to support task and schedule management as well as to evaluate the financial performance of individual projects. The PM/COE has, since its inception, garnered historical databases that allow for the identification and evaluation of trends and historical comparisons. It continues to ensure that efforts underway to improve systems are supported and expanded to provide a project orientation for all work across the corporation.

**Action 4:** IBM established a project management Center of Competence, the mission of which is to support the practice of professional project management throughout the IBM Corporation. This mission was expanded to include program management in 2010.

IBM designated an executive sponsor and then established and staffed a PM Center of Excellence. The PM/COE is the virtual source and clearinghouse for project and program management expertise across the corporation. The PM/COE is staffed with certified project managers on rotational assignments from all business units. It is modeled after the highly successful approach of the IBM Academy and utilizes the governance model developed by Global Services for their IBM Solutions Institute.

Specific PM/COE responsibilities encompass:

- Developing an IBM-wide strategy and plans for the development and support of project and program management as an organizational competency as well as an individual profession
- Driving the development of processes, practices, tools, and curricula for the achievement of that strategy and coordinating related efforts in all business units
- Providing subject matter expertise and assistance on project and program management across the corporation
- Maintaining a professional project and program management community within IBM and coordinating that community with other related internal and external communities

**Action 5:** IBM project and program managers further the advancement of project and program management as a professional discipline through "giveback" activities during or in between assignments.

A professional community thrives and grows only through the contributions of its members as they help each other become more proficient in the practice of their discipline. These efforts include mentoring, teaching, project assessment and assurance activities as well as the sharing of experience through lessons learned exercises and published papers. Such giveback serves to refresh and renew project managers, and it is considered an honor to contribute to one's profession in this way. However, such professional growth and enhancement can occur only with conscious organizational support for these activities.

According to Debi Dell, PMP, Manager of IBM's Project Management Center of Excellence,

> "It is only through giveback and similar activities that IBM benefits from its learning, growing project management profession. We have set an expectation that our best and brightest project managers will contribute to the PM community—for they truly have the most to give. Participation in giveback activities is built into the qualification requirements for project managers across the corporation. Leadership must continue to endorse the profession and these activities as valuable and directly contributing to IBM's effectiveness."

In conclusion, the PM/COE develops and implements corporate-wide strategy and plans for achieving organizational competence in project and program management. It establishes and drives a consistency of approach, a network of knowledgeable practitioners, and supportive business processes and systems. See Figure 15–3. Finally, it establishes and maintains a professional project and program management community within IBM and acts as the interface between IBM's community and other internal and external professional communities.

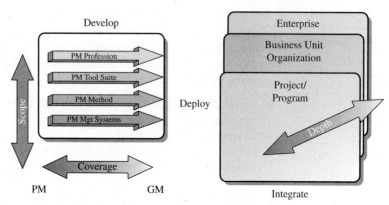

**Figure 15–3.** Enterprise Project Management.

Over time, the PM/COE has driven the transformation and integration of project management into the fabric of IBM. IBM professionals are more experienced and capable in their abilities to get their work done using project management disciplines. In summary, project management disciplines have become systemic across IBM, resulting in successful projects and higher client value. It is now focused on achieving the same with program management.

## Management Support

Strong executive backing began at the top of IBM when the Corporate Execute Committee, led by Lou Gerstner (CEO), declared in November 1996 that IBM would become a project based enterprise with common methods, tools, and a recognized project management profession. This transformation continues to receive strong executive support across the corporation.

> The shaping and deployment of IBM's Enterprise Project Management strategy is a glorious example of successful organizational transformation. Initiated by a CEO charter from Lou Gerstner in 1996 to make IBM into a project based business, IBM's Project Management Center of Excellence has directed the progressive phases of the program across these past thirteen years. Key to the PM/COE's success has been sponsorship by a senior executive committee, who acted as program champions for project management across a diverse set of global business units. The organizational benefits of this transformation that were realized over that time are readily evident in IBM's current business success.
>
> **Steve DelGrosso,**
> **Director, IBM Project Management Center of Excellence**

Many executives actively support project and program management within their organizations and send strong messages throughout their business unit in this regard. This is critical for getting others within the organization to support project management.

Management support must start at the top with a statement of strong executive support, which is communicated across and within the organization. This, in turn, is instrumental in getting middle management and line management to support the initiatives.

The PM/COE works with the executives within business units without this top down support to increase their understanding of the importance of project management to achieving their business objectives and to subsequently articulate their support and messages to their organization.

Following the attainment of general executive support in an organization, education is provided to middle managers, line managers, and practitioners about IBM's project and program initiatives and how they can be fully embraced. This education takes many different forms, such as courses, monthly articles, periodic meetings, conferences, etc.

Finally, a critical component to the integration of these disciplines within all business units is a Competency Network. A core team of community leaders exist in most of IBM's major geographies, lines of business, and business units (BU) who are responsible for their project and program management community. These leaders and teams see that activities, tools, methods and education are implemented within their geography or business unit.

## Project, Program, and Portfolio Management

IBM is structured around hardware, software, and services organizations globally. Implementing common processes, procedures, methods, tools, and skills that can cross over business units and geographies and be effective was initially difficult. However, by allowing organizations to tailor the approach to fit each organization's and geography's business model, there remains a consistency, commonality, and effectiveness across PM practices worldwide.

Within IBM's project-based enterprise, it is important for all levels of management to understand how projects, the basic construct for accomplishing work, related to programs and portfolios. This relationship is critical in order to apply the appropriate management approach, management authority, and management controls to ensure success. Definitions across the business for these constructs include:

**Projects** are temporary endeavors undertaken to produce a unique product or service. Projects are tactical endeavors as they are short term and have a defined scope or objectives.

**Programs** are long term endeavors undertaken to implement a strategy or mission to meet business or organizational goals. A program is implemented through multiple interrelated projects and ongoing activities. Many times Complex projects should more correctly be managed as Programs. In IBM typical programs include contracts, product development.

**Project/Program Portfolios** is a business view of projects and/or programs that share specific common characteristics and are viewed as a group for management purposes. In IBM typical portfolios are: IBM Global Services service offerings, projects under a principal, and programs under a sector.

The Project Management Center of Excellence developed corporate practices aimed at supporting the management of each of these areas. The practices are short, concise documents that are a good reference to use in verifying that a team is addressing all common areas that have been proven as success factors in a project based construct. The practices

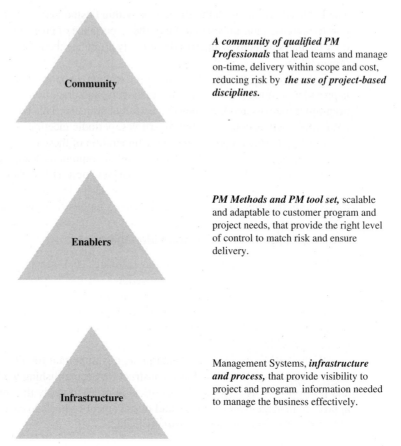

*A community of qualified PM Professionals* that lead teams and manage on-time, delivery within scope and cost, reducing risk by *the use of project-based disciplines.*

*PM Methods and PM tool set,* scalable and adaptable to customer program and project needs, that provide the right level of control to match risk and ensure delivery.

Management Systems, *infrastructure and process,* that provide visibility to project and program information needed to manage the business effectively.

**Figure 15–4.**   The relationship between projects, programs and portfolios and supporting materials.

specifically address the following areas of PM discipline and form a common base for management in all business units (See Figure 15–4):

- The relationship of projects, programs and portfolios;
- The responsibilities of the management and stakeholders of projects, programs and portfolios; and
- Project and Program Management Practices.

Compliance to a practice is recommended, except in cases where a business unit policy or process makes compliance mandatory. By using an IBM Corporate Practice, individual business units can make the appropriate decision as to compliance to the practice relevant to their business objectives.

*Project Management Offices (PMO)*

As part of its project-based mission, the IBM PM/COE focuses on the understanding and implementation of Project Management Offices (PMOs) to assist in the PM transformation process.

PMOs, defined as a coordinated group consisting of one or more individuals within an organization, are established within IBM to perform project management functions for a single project or for a portfolio of projects to make the organization more effective. In fact, **the PM/COE is structured as a PMO.**

Within IBM, two types of PMOs primarily exist:

- An internal project office that provides project management support to a single project or program and
- A centralized project office that provides project management expertise to multiple projects, programs and organizations or that has a charter beyond executing PM tasks.

The PM/COE developed and defined the Project Office program by providing a body of knowledge on the effective design and implementation of project office capabilities, both within IBM and on behalf of clients. The Project Office program provides assistance to project offices through a number of tools including the Project Office Guide, Setup Guide, and Staffing Roles Guidance.

Additionally, members of the PM/COE present the value of PMOs across the business, encouraging their use when appropriate. In today's world of highly complex solution delivery, a Project Management Office (PMO) increases efficiency and effectiveness. A PMO should be organized to support organizational goals and to provide benefits such as:

- Enabling project managers to take a consistent, systematic and repeatable approach to project management
- Continuously improving the skill set of project managers
- Providing management visibility and control of projects to reduce the occurrence of troubled projects
- Optimizing project resources, project selection and prioritization
- Performing project, program, and portfolio administrative functions to free up the project manager to focus on project/contract
- Supporting culture shift to project management

Executive focus and commitment is needed in establishing and enforcing the use of the project office. Executives need to sponsor and commit time, attention, and financial resources to making the project office successful. Management must then assign a respected project manager capable of working effectively across multiple projects as the project office manager and also ensure the PMO is staffed with experienced, capable, and full-time individuals. As always, management must effectively communicate and ensure acceptance of the project office's roles and responsibilities across the organization.

Finally, if the project office is well managed, additional benefits result, both at the project and organizational level. Projects see improved delivery times, reduced cost, improved collection of project intellectual materials and historical data for subsequent reuse, and improved estimates. Organizations realize improved communications, improved ability to anticipate rather than react to problems, enhanced "what if" analysis and corrective action planning, and improved business management skills throughout the organization.

*Professional Development - Qualification*

The project management profession is one of several IBM global professions established to ensure availability and quality of professional and technical skills within IBM. The PM Professional Development initiative includes worldwide leadership of IBM's project management profession, its qualification processes, IBM's relationship with Project Management Institute (PMI), and project management skills development through education and mentoring. These programs are targeted to cultivate project and program management expertise and to maintain standards of excellence within the profession. The bottom line is to develop practitioner competency.

What is the context of a profession within IBM? IBM professions are self-regulating communities of like minded and skilled IBM professionals and managers who do similar work. Their members perform similar roles wherever they are in the organizations of IBM and irrespective of their current job title. Each profession develops and supports its own community including providing assistance with professional development, career development and skills development. The IBM professions:

● Help IBM develop and maintain the critical skills needed for its business;
● Ensure IBM clients were receiving consistent best practices and skills in the area of project management; and
● Assist employees in taking control of their career and professional development.

All IBM jobs have been grouped into one of several different functional areas, called job families. A job family is a collection of jobs that share similar functions or skills. If data is not available for a specific job, the responsibilities of the position are compared to the definition of the job family to determine the appropriate job family assignment.

Project managers and, for the most part, Program Managers fall into the Project Management Job Family. Project Management positions ensure customer requirements are satisfied through the formulation, development, implementation and delivery of solutions. Project management professionals are responsible for the overall project plan, budget, work breakdown structure, schedule, deliverables, staffing requirements, managing project execution and risk and applying project management processes and tools. Individuals are required to manage the efforts of IBM and customer employees as well as third-party vendors to ensure that an integrated solution is provided to meet the customer needs. The job role demands significant knowledge and skills in communication, negotiation, problem solving, and leadership. Specifically, project management professionals need to demonstrate skill in:

● Relationship management skills with their teams, customers, and suppliers
● Technology, industry, or business expertise
● Expertise in methodologies
● Sound business judgment

Guidance is provided to management on classifying, developing and maintaining the vitality of IBM employees. In the context of the PM profession, vitality is defined as professionals meeting project management skill, knowledge, education, and experience requirements (qualification criteria) as defined by the profession, at or above their current level. Minimum qualification criteria are defined for each career milestone and used as individual's business commitments or development objectives, in addition to business unit and individual performance targets.

Skilled project and program management professionals are able to progress along their career paths to positions with more and more responsibility. For those with the right blend of skills and expertise, it is possible to move into program management, project executive, and executive management positions. Growth and progression in the profession are measured by several factors:

- General business and technical knowledge required to be effective in the job role;
- Project management education and skills to effectively apply this knowledge;
- Experience that leverages professional and business-related knowledge and skills "on the job"; and
- Contributions to the profession, known as "giveback," through activities that enhance the quality and value of the profession to its stakeholders.

IBM's project and program management profession has established an end-to-end process to "quality assure" progress through the project management career path. This process is called "qualification" and it achieves four goals:

- Provides a worldwide mechanism that establishes a standard for maintaining and enhancing IBM's excellence in project and program management. This standard is based on demonstrated skills, expertise, and success relative to criteria that are unique to the profession.
- Ensures that consistent criteria are applied worldwide when evaluating candidates for each profession milestone.
- Maximizes customer and marketplace confidence in the consistent quality of IBM project management professionals through the use of sound project management disciplines (i.e., a broad range of project and program management processes, methodologies, tools and techniques applied by project management professionals in IBM).
- Recognizes IBM professionals for their skills and experience.

The IBM project and program management profession career path allows employees to grow from an entry level to an executive management position. Professionals enter the profession at different levels depending upon their level of maturity in project management. Validation of a professional's skills and expertise is accomplished through the qualification process. The qualification process is composed of accreditation (at the lower, entry levels), certification (at the higher, experienced levels), recertification (to ensure profession currency), and/or level moves (moving to a higher certification milestone). See Figure 15–5.

**Accreditation** is the entry level into the qualification process. It occurs when the profession's qualification process evaluates a project management professional for Associate and Advisory milestones.

Career path allows growth from an entry level to an executive
management position.

**Figure 15–5.** IBM's Project and Program Management Career Growth Path.

**Certification** is the top tier of the qualification process and is intended for the more experienced project or program manager. It occurs when the profession's qualification process evaluates a project management professional for Senior and Executive Project Management milestones. These milestones require a more formal certification package to be completed by the project manager. The manager authorizes submission of the candidate's package to the Project Management Certification Board. The IBM Project Management Certification Board, composed of profession experts, administers the authentication step in the certification process. The Board verifies that the achievements documented and approved in the candidate's certification package are valid and authentic. Once the Board validates that the milestones were achieved, the candidate becomes certified as a Senior or Executive PM.

**Recertification** evaluates IBM certified project management professionals for currency at Senior and Executive Project Management milestones. Recertification occurs on a three-year cycle and requires preparing a milestone package in which a project manager documents what he/she has done in project management, continuing education, and giveback since the previous validation cycle.

IBM continues to be committed to improving its project management capabilities by growing and supporting a robust, qualified project management profession and by providing quality project management education and training to its practitioners.

The career path and qualification approach is supported by an integrated human resource framework and professional development program, the IBM Career Framework. IBM Career Framework is an enterprise-wide career model to guide employees' careers and develop capabilities which matter most to our clients. The Framework provides a formal process for **validating demonstrated capabilities and existing certification programs**. See Figure 15–6.

Validation across levels provides the milestones that ensure practitioners
are acquiring the knowledge, skills, capability and experience that is
required to provide client value at each level.

| Capabilities | Entry | Accreditation | Foundation | Experienced | Certification | Expert | Thought Leader |
|---|---|---|---|---|---|---|---|
| **Qualifications** | | | Associate PM | Advisory PM | | Senior PM | Executive PM |
| **Skills** | Technical Skills | | PM Intermediate Skills | PM Basic Skills | | PM Specialty Skills | Leadership Skills |
| **Experience** | Novice capability | | Acquired capability | Applied capability | | Mastery capability | Executive capability |

*Validation*

**Figure 15–6.** Career framework validation.

It gives practitioners integrated career-related processes and tools while providing greater clarity and guidance on the skills, abilities, and experience employees need to succeed in their current role. The Career Framework is integral to strengthening career and development discussions between managers and employees especially related to certification.

Equally important to project manager development and certification is a refinement of the process by which project managers are assigned. Projects are assessed based on size, revenue implications, risk, customer significance, time urgency, market necessity, and other characteristics; certified project managers are assigned to them based on required education and experience factors. See Figure 15–7.

Guidance is provided to management on classifying, developing and maintaining the vitality of IBM employees. In the context of the PM profession, vitality is defined as professionals meeting project management skill, knowledge, education, and experience requirements (qualification criteria) as defined by the profession, at or above their current level. Minimum qualification criteria are defined for each career milestone and used as individual's business commitments or development objectives, in addition to business unit and individual performance targets.

*PM Curriculum*

The IBM PM Curriculum is a global curriculum. The development and delivery of this **cohesive, world class PM curriculum** was a major element in the establishment of a consistent base of terminology and working knowledge for the thousands of project and program management professionals worldwide in IBM. A Curriculum Steering Committee (CSC) with global representation manages its development, delivery and deployment. Every classroom course is delivered the same way in every country using the same classroom materials and facilities with local instructors certified to teach the course. The e-learning courses are on servers that are accessible 24 hours a day, 365 days a year by

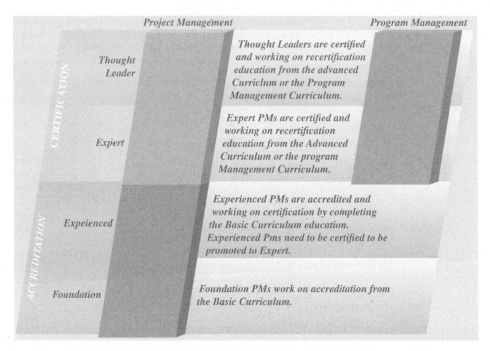

**Figure 15–7.** The refinement process.

students around the world. IBM's PM education is delivered via e-learning with over a hundred thousand student days of PM education delivered in 34 countries.

As the needs of the PM professionals and the skills at each job level were better understood and matured, the PM curriculum designed a multitier architecture (see Figure 15–8) that mapped course content against the skills requirements necessary to develop a project management competence within IBM. The architecture is composed of: PM Basic education that addresses the fundamentals of project management for employees with limited or no project management experience; PM Enabling education that deepens and broadens project management skills with deeper topical content; and Program Management Education that builds on PM skills by enhancing general business skills and providing project based tools and techniques to manage large programs with multiple projects and business objectives. Once the architecture and curriculum were set, we created roadmaps so students could easily determine what courses to take.

### Organization of the Project Management Curriculum

Most IBM project and program managers begin their IBM project management education by taking courses following the roadmap through the **PM Basic Education**. This education is designed for project managers with limited or no prior project management experience who are working toward IBM certification. Within the PM Basic Education curriculum, there is also a roadmap for those experienced project managers who are working towards certification. The Basic Curriculum includes PMI examination preparatory courses including PMI Examination Preparation and PMP Exam Prep Intensive Workshop.

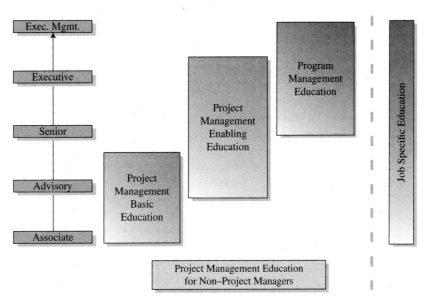

**Figure 15–8.** IBM's Project Management Curriculum Architecture.

To ensure that students exit the Basic Curriculum with the required IBM PM skills, exams are given at the end of most courses. The content and rigor of our curriculum led to George Washington University declaring that the IBM Basic Curriculum was equivalent to their master's certificate in project management. Several courses in the curriculum are also accredited by the American Council on Education for both graduate and undergraduate credit in North America and by the European National Academic Recognition Information Centre (NARIC) in Europe and Asia.

**Project Management Enabling Education** is designed to deepen and broaden project management skills. The audience for this education varies by course. In general, the just-in-time courses and short 2-to-4 hour e-learning courses require that students have a fundamental understanding of project management but are not required to have completed the Basic Education. However, the more in-depth classroom or e-learning courses do generally require completion of the Basic program.

**Program Management Education** is designed to build on project management skills by enhancing general business skills and providing tools and techniques to manage complex projects, programs and project portfolios. IBM certified project managers are the audience for this education as they will generally be tapped for the larger, more significant projects or programs.

**Project Management Education for Non–Project Managers** contains courses in the project management curriculum to help managers and project team members who support or work on project teams.

In addition to project and program management specific education, all project managers need education that prepares them for the specific environment in which they work.

This education may be unique to their specialty, business unit, industry, country, or other job elements. Job specific education is not specifically addressed by the project management curriculum but is recognized as a necessary part of a project manager's training. **Specialty education**, a part of job specific education, is required for IBM project management certification. After certification, credit is given for job specific education for recertification.

In addition to skill development, **PM University** provides curriculum information in an easy to use format. A web-based "location," this tool makes finding project and program management education and lessons learned sessions easier to find.

IBM continues to be committed to improving its project management capabilities by growing and supporting a **robust, qualified project management profession** and by providing quality project management education and training to its practitioners.

PM University allows for continuous improvement in all types of education. One example was the requirement for hands-on education to teach students how to get started using the PM methodology. This requirement drove the creation of several Advanced global courses addressing the application of the global PM methodology and tools. Another example was in the area of Contracting. Different countries have different laws. The global course was not modified, but several different country courses were created to address the specific country contracting requirements. After the students completed the global course to understand the contracting process, they took the country specific course to understand the special processes required for their country. PM University provides the road map and enrollment capability to address specific practitioner learning requirements.

### Project Management Methodologies
### Tony DeBellis/Lee Gay

To provide its teams with consistent methods for implementing project management worldwide, IBM developed the Worldwide Project Management Method (WWPMM), which establishes and provides guidance on the best project management practices for defining, planning, executing and controlling a wide variety of projects. IBM's **Worldwide Project Management Method (WWPMM)** is a response to the Corporate Executive Council (CEC) action to establish a single, common project management method for IBM projects worldwide. WWPMM is an implementation of the PMI *PMBOK® Guide* designed for the IBM environment. In 2012, WWPMM was enhanced to include program management extensions as the complexity of project and program delivery intensified.

Activities within the project management process can be classified into four basic groups: Defining, Planning, Executing and Controlling, and Closing. A project usually consists of a series of phases, known as the project life cycle, and these groups of process activities can be applied to each phase individually or to a set of multiple phases.

In the Defining group, agreement is reached on the objectives of the project, the scope of the project is established, the initial organization is defined, responsibilities are assigned, and the assessment of situational factors is documented.

In the Planning group, detailed work and risk plans are drawn up, the organization is confirmed, and staff assignments are made. No significant amount of resource may be expended on the project (that is, execution does not begin) until clear plans are in place and authorization to proceed has been received at the end of this phase.

In the Executing and Controlling group, the plans and controls are used to execute and manage the project as project development and delivery work is performed. As work proceeds, plans are expanded or refined as necessary.

In the Closing group, the sponsor agrees to close the project, the project is closed down, and the evaluation report (also known as lessons learned) is produced.

Recursively, during the life of a program and that of projects, multiple of these activities can be occurring and recurring concurrently. As programs are defined and planned, projects are identified which, in turn, are defined, planned, executed, controlled, and closed. As the programs are executed and controlled, periodic reassessment of goals can lead to identification of new projects to be defined as well as to the closure of projects that no longer provide value. As a project is executed, it is usual that plans are refined at the end of some phases, to prepare for the execution of the next phases.

In order to be generic and applicable across IBM, the project management method does not describe lifecycle phases, but rather PM activity groups that can be used repeatedly across any life cycle. This allows the flexibility for the method to be used with any number of technical approaches and life cycles. (See Figure 15–9.) WWPMM is how projects and programs are managed in IBM and is deployed worldwide through a variety of business unit specific methods and management systems such as Global Services Method and Integrated Product Development to name a few.

WWPMM contains reference documentation (domains), process steps (work patterns), and work products or templates for PM system components as well as support

Example: Within the IPD Concept Phase, the majority of the tasks fall within the WWPMM Defining and Planning process groups, though there are tasks which also fall within the Executing/Controlling and Closing WWPMM process groups

**Figure 15–9.**  IBM Worldwide Project Management Method.

material that explains WWPMM usage. WWPMM describes how to shape and plan a project and then manage its execution through three interrelated views (See Figure 15–10):

WWPMM helps define the Project Management System, a collection of plans, procedures and records that direct all PM activities. It describes the current state and history of the project, including the metrics necessary to track and measure the project. Generic templates are provided in downloadable form, from the WWPMM reference page and through various PM tools. When used with appropriate tools and integrated with business and technical management systems, this material provides a comprehensive PM environment. It also defines the documentation that will be created and delivered as well as how and where that documentation will be stored.

Keeping with the need to be flexible, the Project Management System templates and work products can be tailored to meet geography, business line, or client specific requirements while still maintaining our commitment for "one common project management method." Managers of project, programs, or portfolio managers can ensure their project managers are using a project management system appropriate for the risk and complexity of their specific project and business line needs.

The project management system is unique to each project. An organization's current policies, business and technical methods, and the PM methods of the Worldwide Project Management Methodology (WWPMM) provide a basis for a comprehensive project management system (See Figure 15–11). The project management system should then be tailored to the needs of each individual project to account for complexity of the project and the current business climate of the project environment. Project managers determine what procedures, processes, and tools are best for their projects based on the standards or guidance that has been defined by the business unit. Using a PM system helps ensure that each project teams is working in a structured and controlled environment, working towards the same project goals.

But, even with our own methodology, IBM stays in sync with what is happening in the industry and with responding to business unit requirements. **Agile** is another technical approach used to develop software and IT solutions within IBM. While the industry has employed it for years, IBM is now expanding its use as clients increasingly request IBM to utilize Agile techniques.

In IBM today, most Agile projects are pieces of larger development or services projects. Thus, a project manager is needed to ensure that the deliverables of the larger project are delivered and the requirements of the business management system are met. For projects where Agile development comprises the entire project, business management

**Figure 15–10.**   The interrelated views of IBM WWPMM.

**Figure 15–11.** The Worldwide Project Management Methodology (WWPMM) provide[s] a basis for a comprehensive project management system.

requirements and controls must still be followed. To address these business management system requirements, the IBM Worldwide PM Method (WWPMM) is designed to be both scalable and adaptable to meet the needs of all projects, including Agile projects. IBM provides guidance to its project managers on how WWPMM can be streamlined for Agile projects and how PM disciplines and behaviors can be applied to embrace agility while still maintaining the necessary business controls.

A method, however robust, is generally not as effective without a tool to enable its practical application. One effective way to implement WWPMM is through tools that provide the capability to deliver virtually all of a project management system in a single web accessible location as well as implementing and integrating a technical method. Through the use of its preapproved templates and routine monitoring, a project team can implement the appropriate project management system.

### Best practices

A web search of the keywords "ibm + best practice" yields literally millions of hits. A best practice can be defined as a method, process, or technique, to deliver an outcome that has been shown to provide a more reliable and predictable desired result than other techniques. IBM has throughout its history developed many best practices for many different aspects of information technology. These have been produced in many different levels of rigor and formality. In the technical world, these best practices may be published as white papers or, more formally, within IBM's Redbook series of publications. Within the project management discipline at IBM, it is no different. Since 1996, IBM has developed many project management best practices. The PM/COE has the job of communicating and educating the company's project management stakeholders on these best practices.

As discussed earlier in this chapter, a set of key project management initiatives was established as the core approach to deploying project management within the enterprise. Together in partnership, the PM/COE and business units throughout IBM worked to deploy eight key initiatives that formed the foundation for IBM's vision of its enterprise project management deployment. The eight key initiatives include (See Figure 15–12):

- Education
- Qualification
- Method
- Select Projects
- Tools
- Portfolio Management
- Maturity
- Policy

Within each initiative, best practices were developed to support deployment within the organization. Some were referenced previously but this snapshot provides the basis for the measurements against which IBM's progress to becoming a project based enterprise is measured.

*Education*

IBM's approach to PM education was outlined in a previous section. The PM curriculum, roadmap, and approach to education have been a very formal best practice within IBM for a number of years. All project managers in IBM are taught the same curriculum and the same messages, which have evolved over time as IBM has matured in project management. This best practice is maintained in a set of documents that are overseen by a small team of staff and a steering committee.

*Qualification*

IBM's approach to qualifying and certifying project managers through their career has proven to yield value to our clients, our project managers, and the company. This approach is documented as a best practice in the Project Management Profession Guide and maintained by approval of a professional development committee. A small team oversees the administration of this best practice. The material and updates are communicated to the

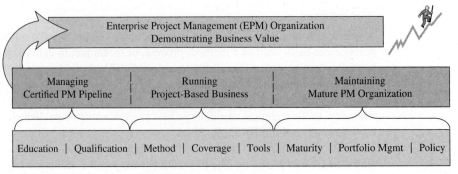

**Figure 15–12.** Within each of the initiatives, best practices were developed to support deployment within the organization.

project management community through the PM/COE's communications program to ensure that all project managers worldwide are adhering to the same qualification process.

### Method

IBM has been very method driven for many years. The challenge within project management was to bring the company together in a unified method of managing projects. While there are many dozens of technical methods maintained by IBM to define the different aspects of performing work, there is only one method for project management used by all parts of IBM in many different types of projects. WWPMM (Worldwide Project Management Method) was described in an earlier section, and is IBM's best practice for managing projects. It is maintained by the PM/COE and updated periodically to reflect changes in requirements. IBM business units have, in most cases, adapted WWPMM into their management processes.

### Selected Projects

While Education, Qualification, and Method are IBM best practices with very formal documentation and support mechanisms, Selected Projects is a best practice that is maintained much differently. One of the founding principles put in place when the PM/COE was formed was that "selected" key projects would be managed by the most qualified project managers. Because of the variety in how the different parts of IBM operate and staff their engagements, this best practice is defined in concept at a high level, and the PM/COE works with the different business units to develop a model that works for their business. It is still considered a best practice because the premise is very important to successful project management, yet you will not find detailed process documentation for the practice.

### Tools

When IBM started the journey to enterprise project management, many different tools were used to manage our projects. This was very inefficient, both in the cost to maintain many tools, as well as the resource inefficiency with staffing projects with individuals that knew the tools. The PM/COE, working together with the different parts of the business, selected a strategic tool to use for project management. This allowed for better investment in and integration of the tool with key elements such as labor accounting. A team working with the PM/COE has maintained best practices for the implementation of the tool into the business. As new units begin to deploy the tool for their organization, these best practices are followed to minimize problems and streamline the implementation.

### Portfolio Management

Project Portfolio Management groups, views, analyzes, and manages projects together to maximize positive business results within an organization's resource constraints. Projects are either included in, or excluded from, the portfolio based on their alignment with the portfolio strategy and their performance against the business objectives. IBM has a formal published policy which is our best practice for portfolio management. The PM/COE works with business units to adapt this policy into their business processes for managing their portfolio of projects.

### Maturity

IBM developed a standard tool and best practice, the Project Management Progress Maturity Guide (PMPMG), for measuring project management maturity. PMPMG enables an organization to assess the maturity of its project management capabilities at both the project and organizational levels. By understanding its strengths and weaknesses in these areas, an organization can identify actions for continuous improvement in managing projects/programs and achieving its business objectives. The PMPMG consists of a set of yes/ no questions separated into 5 levels that seek information on the PM processes in use. Upon completion, a weighted average of answers is used to compute a relative value for the organization's maturity. Within the guide, the questions apply either to the organization as a whole, or to projects within the organization. Successful execution of the tool will lead to a score of 1 to 5 on maturity, and a list of action items to address deficiencies. The PM/COE maintains the PMPMG as a Lotus Notes tool and database, and assists different business units with the performance of assessments.

### Policy

The PM/COE maintains formal corporate practices for project and program management that define the recommended set of management practices that should be used for all IBM projects and programs. These practices are formal corporate level best practice that the whole company is expected to adhere to. In order to deal with the variety of types of businesses in IBM, the PM/COE then works with each unit to develop their own policy statement that adopts the corporate practices and other initiatives and adapts them all to the specifics of their organization. These units' policies are sponsored by the project management executive sponsors for the organizations and published on the PM/COE intranet site so that all project management stakeholders can easily view them.

These eight initiatives comprise a large collection of best practices in project management that have been the pillars of IBM's move to enterprise project management. The PM/COE measures the progress of IBM as a company by monitoring these initiatives using a Red/Amber/Green scorecard that is updated quarterly for each part of the business and each geography around the world. As changes occur, such as a reorganization, market change, or executive leadership move, we see movement in the performance against our scorecard. The PM/COE works with the Executive Steering Team to help correct areas of concern to keep the company on track to maintaining our progress in maturity.

But, the best practices do not stop there. Support of our professionals is also at the core of the project-based transformation.

### Knowledge Sharing

While the eight initiatives are more formalized best practices, an opportunity exists to leverage lessons learned on engagements by IBM's project managers. The PM Knowledge Network (PMKN) supports IBM's becoming a project-based enterprise by leveraging knowledge through sharing and reusing assets (intellectual capital). The PMKN repository supports the PMKN Community with a wide range of assets that include templates, examples, case studies, forms, white papers, and presentations on all aspects of project management. Practitioners may browse, download, or reuse any of the more than 2,400 entries to aid their projects, their proposals, or their understanding.

### Community and Communications

According to Debi Dell, PMP, Manager of the Project Management Center of Excellence has driven a strong sense of community for its global project management professionals; this is a best practice among IBM's professions.

Within IBM, a community is defined as a collection of professionals who share a particular interest, work within a knowledge domain, and participate in activities that are mutually beneficial to building and sustaining performance capabilities. Our community focuses on its members and creating opportunities for members to find meaning in their work, increase their knowledge and mastery of a subject area, and feel a sense of membership—that they have resources for getting help, information, and support in their work lives. Knowledge sharing and intellectual capital reuse are an important part of what a community enables but not the only focus. Communities provide value to the business by reducing attrition, reducing the speed of closing sales, and by stimulating innovation.

Communities are part of the organizational fabric but not defined or constrained by organizational boundaries. In fact, communities create a channel for knowledge to cross boundaries created by workflow, geographies, and time and in so doing strengthen the social fabric of the organization. They provide the means to move local know-how to collective information, and to disperse collective information back to local know-how. Membership is totally based on interest in a subject matter and is voluntary. A community is NOT limited by a practice, a knowledge network, or any other organizational construct.

The PM Knowledge Network (PMKN) Community is sponsored by the PM/COE. Membership is open to all IBM employees with a professional career path or an interest in project management. The PMKN is a self-sustaining community of practice with almost 12,000 members who come together for the overall enhancement of the profession. Members share knowledge and create PM intellectual capital. The PMKN offers an environment to share experiences and network with fellow project managers.

Upon entering the PM community, professional hires into IBM are often asked the question: "What is the biggest cultural difference you have found in IBM compared to the other companies in which you have worked?" The most common answer is that their peers are extremely helpful and are willing to share information, resources and help with job assignments. The culture of IBM lends itself graciously to mentoring. As giveback is a requirement for certification, acting as a mentor to candidates pursuing certification not only meets a professional requirement but also contributes to the community.

Mentoring can be a fragile relationship. A corporate worldwide study revealed that the most common reason that mentoring relationships ended prematurely was due to the lack of a formal structure or goal to keep the momentum going. In addition, a worldwide mentoring tool was not available. The Project Management Center of Excellence responded to the study by creating a "Project Management Skills Mentoring Process." In line with the discipline of Project Management, a mentoring process was designed to follow the basic constructs of a project. A mentoring relationship had a definite beginning, a desired project outcome (raise the level of a skill or knowledge area), a plan, a schedule, milestones, tracking and completion. Sample mentoring plans, agreements, milestones, and closing documentation along with a PM Skills Mentoring Guide were provided on the web site.

To address the communities and all project management professionals. primary channels of communications include the PM/COE web site and focused newsletters and announcements Project management can even be added into a PM's corporate web profile. However, as the project management profession grows so do the requirements for projects targeted at specific communities. The Project Management Center of Excellence has developed information for the following communities with the PM profession:

- New to the Profession
- Managers of Project Managers
- Middle Management

- PM Knowledge Network
- Specific geographies' PM communities

But IBM's best practices are not just recognized within the company. Many have received recognition from industry sources.

According to Debi Dell, the Project Management Center of Excellence tracked its awards and achievements and shared these accomplishments across IBM, the PM community, and with clients in various proposals. This is a partial list of activities that highlight where IBM was recognized for its excellence in project management.

*Awards*

- 2012 Brandon Hall Best in Certification Programs (Gold)
- 2012 APQC Project Management Office of the Year
- 2012 PMI Continuing Education Provider for PM University
- 2011 Brandon Hall Excellence in Learning — Bronze
- 2011 PMI Continuing Education Product of the Year: Agile Learning Suite
- 2010 PM Solutions Project Management Office of the Year
- 2009 PMI Continuing Education Provider for PM Curriculum
- 2007 PMI Distinguished Project (IBM Stockholm Congestion Tax Project)
- 2006 PMI Professional Development Provider of the Year Award
- 2006 PMI Education Provider of the Year Award
- 2005 PMI Professional Development Provider of the Year Award
- 2005 PM/COE Director (C. Wright) PMI Distinguished Contribution Award
- 2004 PMI Professional Development Provider of the Year Award
- 2001 Patent award for "Learn How. . . Do It Now. . ." — PM Curriculum Team
- 2001 International Society for Performance Improvement (ISPI) Award of Excellence for Contracting for Project Management Lotus LearningSpace course
- 2000 American Society for Training and Development Excellence in Practice Award recognizing excellence in Training and career development processes.
- March 1998: Giga Information Group Gold Award: IBM's Intellectual Capital Management system

*Curriculum*

- PM curriculum continues to be recognized with our "Contracting for Project Managers" e-learning course by the International Society for Performance Improvement (ISPI) with an Award for Excellence.
- Recognized by American Society for Training & Development (ASTD) with an Excellence in Practice award for its project management curriculum initiative
- The American Council on Education (ACE) has recommended our Tier 1 courses for credit toward undergraduate / graduate degree programs.

*External*

- IBM PM Center of Excellence Director keynote speaker at IIL's International Project Management Day webinar (2011)
- IBM key participant in PMI Corporate Project Management Council—ongoing

*Publications*

- *Forbes Magazine* "Pulse of the Profession with PMI" (March 2012)
- *Keys to Building a Successful Project Management Office* (© 2011 IBM)
- *PMO of the Year* e-book (© 2010 PM Solutions)
- IBM PM history (Chapter 15), *Best Practices in Project Management*, Harold Kerzner (2009)
- IBM PM Center of Excellence recognized as one of the top 25 project-based organizations in 2007 by PMI
- IBM PM Practices cited in *Technical Business Review Report* in 2007

*Primary Submission Coordinator*

**Deborah (Debi) A. Dell, PMP®** is Manager of IBM's Project Management Center of Excellence. In her thirty-three years experience in IBM, Debi has held positions in project management, business planning, mobile and wireless technologies, and market and product planning. She co-authored *ThinkPad: A Different Shade of Blue* in 1999. Debi is PMI certified and has a master's of science in the management of technology and a master's and bachelor's in business administrations.

*Contributors*

**Mike Collins** is currently an Executive Project Manager working in IBM's Project Management Center of Excellence. Mike's current duties include supporting a global business unit's deployment of project management, and leading IBM's Value of Project Management and Worldwide PM Metrics projects. Formerly, Mike worked as a Delivery Manager and Project Executive in IBM Global Business Services managing a portfolio of development projects for NY State. Mike has been an IBM Certified Executive Project Manager and PMP for over 10 years, has 26 years industry experience, and is a member of IBM's PM Certification Board.

**Steve DelGrosso, PMP®** is Director of IBM's Project Management Center of Excellence. In his thirty-four years experience in different specialization areas of the IT industry Steve has served as IBM's North American PM Profession Leader, System Engineering manager, developer and instructor for IBM PM curriculum, and also teaches Project Management to graduate students currently at the University of Miami in Coral Gables, FL. Steve is a member of the PMI College of Performance Management, GovSIG, and *PM Journal* Editorial Review Board. Steve has published articles in *Industrial Engineer* magazine, *Personal Systems Journal*, and was featured on the cover of PMI's April 2004 *PM Network* magazine.

**John Marrine** is an executive PM in IBM's PM Center of Excellence (PM/COE). He has spent his last 20 years in Project Management and Project Executive roles. John has also held several management positions during his tenure at IBM. John holds IBM Executive Project Management, IBM Project Executive, and PMI PMP certifications and has achieved bachelor degrees in electrical engineering and computer science.

**Sandy Steinruck** is a mechanical engineer by degree but has been doing project management for more than 25 years in IBM. She started in product design, managing printer and copier development projects and then moved to FL where she spent 10 years managing software development and integration projects, in the telecommunications industry. For her project management work she was awarded the PM Excellence award in 1998. Sandy has been a member of IBM's PM Center of Excellence as the Program Manager for IBM's award winning PM curriculum for the last 12 years.

## 15.2 COMPUTER ASSOCIATES TECHNOLOGIES (CA): SUCCESSFUL PROJECT DELIVERY AND MANAGEMENT[3]

**Introduction**

CA Technologies is the leading independent software vendor of IT management solutions. Customers look to CA to provide solutions that manage and secure complex IT environments across the enterprise for greater business results. CA Technologies professional services organization, CA Services, has been helping customers deploy, upgrade and optimize solutions for more than 35 years. During this time, the application of lessons learned and continuous improvement have resulted in a mature framework for successful project delivery and management.

Today, most companies utilize industry standard project management techniques, and yet the success rates for projects remain low. Project success is typically dependent on the skills, and sometimes heroism, of the people assigned to the project, whereas world class companies are able to consistently repeat project success. CA Services recognized that to achieve repeatable project success, we must go beyond the customary project management best practices: project teams need an arsenal of proven tools, processes, and support.

When an organization completes a project successfully, the team likely had solid project management, high quality deliverables, a product that provided value to stakeholders, and a methodology for delivery that worked. Most organizations do not take advantage of these success factors to leverage them on other projects. The project documentation is archived and the team disbands, then the process starts from scratch all over again on the next project. Similar challenges apply to failed projects, where lessons learned, issues, and risks are not shared or used to improve the execution for later projects. CA Services is very focused on providing tools and techniques to our project teams that deliver higher quality solutions and repeatable success.

**CA Technologies Approach**

CA Technologies approach to delivering solutions starts with understanding the customer's vision: what is the full scope of the business challenge, and what is the best solution to address the challenge that will also provide flexibility and support future growth.

When viewed this way, the solution can require extensive requirements gathering, design, testing, process design, and training. While doable, this can require too much elapsed time before a solution is deployed. Customers demand an approach that delivers the solution and business value incrementally.

---

3. ©2013 by Computer Associates. Reproduced by permission. All rights reserved. Material on Computer Associates was provided by Robert J. Zuurdeeg, Sr. Consultant, Global Practices, and Mark Elkins, Director, Services PMO.

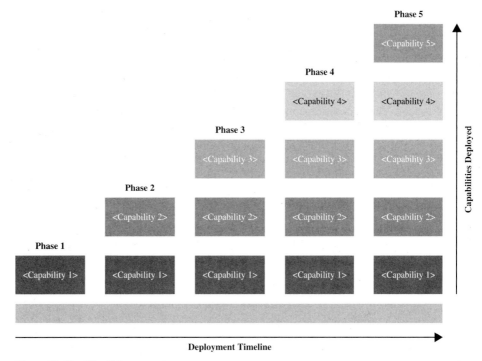

**Figure 15–13.** The CA approach.

Figure 15–13 depicts a realistic approach to achieving the vision: rather than attempting to define and implement many new business processes and supporting tools at once, one should take a phased approach to achieve the vision focused on delivering certain capabilities during each phase. This is how CA Services recommends customers plan for reaching the vision, in logical phases with each phase delivering business value.

"Each phase delivering business value" is critical to project success. But how does CA explain the business value and how does the customer understand and recognize the value delivered in each phase? Very simply: each phase delivers specific, agreed-upon capabilities, which are defined by (functional) use cases.

**Definitions**

**Capability:** An ability that an organization, person or system possesses. Capabilities are typically expressed in general and high-level terms and require a combination of organization, people, processes and technology to achieve.

**(Functional) Use Case:** A term used in the context of a solution. This is a specific use case that describes a functional goal that supports the overall functional requirements of the capability.

Capabilities and use cases names and descriptions use straightforward language to explain the functionality that is being delivered. An example: the capability "Flexible, On-Demand Provisioning" would include use cases such as "Deploy Application,"

"Manage Application," "Scale Application" and "Migrate Application," among others. The capability and use cases each have descriptions further explaining the functionality being provided.

This is important for CA Technologies, partners and customers — rather than relating or documenting that the project team will deploy product xxx with features yyy and zzz, they use capabilities and use cases to communicate the functionality that they will deliver. The relevant capabilities and their use cases are included in CA selling and scoping tools, statements of work and customer deliverables produced during the deployment engagement. This minimizes ambiguity about the functionality CA Technologies will deliver; that functionality is described from the onset of selling, and tracked through the delivery of the solution.

## Starting Out Right

Many projects are doomed to failure before they even begin. Simply following project management standard practices is not enough to ensure success. CA Technologies created several tools and implemented standard processes to help projects teams begin with a project that is set up to be successful from the start. The following sections outline some of these methods that are utilized prior to kicking off a project.

### Defining Project Success

One obvious question that is usually not asked is, "How do we define success?" The answer that most often comes to mind is whether the project is on budget, on schedule, and the scope statement completed; however every organization should examine this question more closely. Based on an evaluation of project goals, CA Services seeks project success that includes the following criteria:

- Customer Solution Adoption – Projects where new processes are defined, people are trained and IT solutions are deployed, only for those products to be shelved or used sparingly, are a key challenge in IT today. This happens when the customer's end users aren't trained effectively on the solution, management isn't aware of the solution's features and benefits, and/or the customer's goals evolve and it is unclear how the new solution will support their change in direction. Regardless, success should ultimately be defined to include the adoption of the solution by the customer because it addresses key business needs.
- Customer Satisfaction – Short-term and long-term customer satisfaction is critical to CA Technologies. This is measured informally at project closure by meeting with the customer and reviewing project accomplishments. It is also measured more formally through a survey process.
- Customer Time to Value – This includes the ability to move solutions into production and help the customer realize ROI as quickly as possible. Measures include Time to Contract, Days to Production Deployment, and Solution ROI.
- Solution Standardization and Best Practices – Over the long term, customer success can be negatively impacted if the solution deployed is highly customized and utilizes functions and processes that are outside the norm. Such customizations make it challenging for customers to take advantage of future product capabilities; therefore, CA Technologies believes that long term success includes defining

solution best practices and guiding customers to take advantage of those best practices. Measures include deployments of standardized solutions, use cases, and best practices.

- Project Performance – This includes bringing in the project on budget and on schedule. Measures include CPI, SPI, and overall project margin.

This definition of project success requires a different approach to managing projects. Typical IT project management activities start after the services agreement (a Statement of Work, or SOW) is signed, and finish with the solution being "rolled-out" into production. Achieving the above success factors requires full end-to-end management from the proposal activities, the deployment, and the transition to production with full adoption. Measuring and monitoring these success factors can be difficult, but critical for driving continuous improvement.

CA Services utilizes a dashboard within CA Clarity™ Project & Portfolio Management to help manage the portfolio of projects. These success measurements are captured and displayed within the Services Dashboard to make the data visible to various groups in CA Technologies. This open sharing of data on project success promotes cross-functional support of projects and feedback for continuous improvement.

### Project Focus

It is critical to recognize what an organization does well. The natural tendency is to think that we can apply management best practices to execute any project successfully. The reality is that most organizations have a sweet spot of projects they do very well; projects that include specific technologies, are within a particular scope or certain industry verticals, solve specific problems, are of a certain size, and so on. Organizations need to do an honest evaluation of the projects they do well, and the business needs they are able to solve. Failed projects jeopardize business relationships and future business.

For CA Services, the focus is on deploying, managing and optimizing the needed business solutions, utilizing CA software. The results of our evaluation indicated the need for:

- Standard solution offerings that deliver common capabilities (functionality) with optional capabilities.
- Standard offering tools to support the positioning, scoping and proposing of the right solution.
- A Service Offering Catalog; a repository from which the standard offering tools can be readily accessed.
- An independent solution review and validation for those solutions that include multiple CA Technologies products and/or which need to integrate with a number of non-CA products.

### Standard Solution Offerings

To achieve the defined success factors, it is important to understand why some projects did not meet success targets. In practice during engagements at CA Technologies, we discovered a common factor in projects that performed extensive requirements analysis, asked the customer what they wanted, and executed the scope statement perfectly. The capturing and documenting of excessively detailed requirements may seem like a good formula for project success, but in reality it created risk in prolonged analysis and delayed time to value. The customer did not receive optimized solutions and best practices because too much customization impacted long term value. A better approach is to define standard offerings, as seen in Figure 15–14, that address the majority of what most customers need, and allow customers to quickly achieve ROI. Most customers would prefer to receive guidance on what they should be implementing, what works best, and what solutions are being used within their industry. By defining packaged offerings, creating reference architectures, and utilizing standard use cases that can be used for projects with similar goals, CA Technologies is able to improve Time to Value for customers.

Ideally every offering would execute exactly the same scope, project after project. Although most situations are fairly similar in scope, there will often be some unique functionality required in each project. This challenge can be controlled by identifying options for scope within each offering. CA Technologies accomplishes this by defining Foundational Services and Acceleration Services for each offering.

Foundational Services are the minimum scope and functionality deployed for each solution, and establishes a common framework that can easily be expanded. Additional functionality aligned to customer business needs can then be provided using the Acceleration Services. Acceleration Services are a set of use cases that can be implemented in a common manner, taking advantage of best practices. Doing this maintains standardization and increases repeatable success, while at the same time providing capabilities aligned to customer needs.

### Standard Offering Tools

To ensure that sales teams can effectively position, scope and propose the right solution, including the standard offering(s), a number of focused tools, as shown in Table 15–1, have been produced to support them.

### Services Offerings Catalog

To help the CA Technologies organization easily access the Standard Offerings and their supporting tools, CA Services created the Services Offerings Catalog. This is a repository where tools are provided to support the positioning, scoping and selling of the Standard Offerings.

The Services Offerings Catalog undergoes continuous improvement and is a dynamic repository. Offerings are added, updated, and retired based on customer needs. Feedback

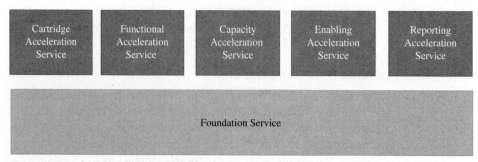

**Figure 15–14.** Standard Solution Offerings.

**TABLE 15–1.  STANDARD OFFERINGS TOOLS**

| Standard Offering Tool | Tool Description |
| --- | --- |
| Service Description | The Service Description is a one page front-and-back description of the offering, the value and benefits to the customer and specific deliverables. In addition to the Catalog, this document is available on ca.com. |
| Customer Presentation | The Customer Presentation is a set of slides about the specific offering to be incorporated in a larger solution presentation. It includes a description of the offering, the value and benefits to the customer and specific deliverables. |
| Solution Calculator | The Solution Calculator is a spreadsheet which estimates the effort (in terms of hours) for the project team delivering a Standard Offering; based on variables unique to the solution. |
| Statement of Work (SOW) | A SOW is the contract between the Customer and CA. It defines the solution CA will deploy in the form of capabilities and use cases. |
| Scoping Questionnaire | A tool used to allow the selling team to gather the functional and technical information needed by CA Services so the right solution is scoped and proposed. |
| Sizing Guidelines | A tool used to define the infrastructure resources needed to host the solution being proposed by CA Technologies |

is also used to improve the structure and ease of use. For example, focus groups recently provided the following feedback/input on the tools needed and the repository layout:

- Provide fewer but more comprehensive and focused tools
- Use as few keystrokes/mouse clicks possible to locate available documents
- Identify when in the Opportunity Life Cycle each tool is to be used

The Services Offerings Catalog allows a user to enter the catalog (which has been built on Microsoft SharePoint), quickly locate a Standard Offering, view all of the available tools and download the tools needed at that point in the sales cycle.

### Solution Review/Validation

Another way that CA Technologies improves the likelihood for project success is to utilize various programs and processes that verify the proposed solution's fit during the sales cycle. This is especially valuable for large and complex solutions where additional review may be appropriate. For example, solutions that include multiple CA Technologies products that need to integrate with a number of non-CA products in a large, complex environment. No one person can possibly be an expert in all enterprise software solutions, versions, platforms and integrations; for that reason, a team of technologists with various skills and experience reviews and approves the right architecture—before CA Technologies proposes the solution to a customer.

It may also be beneficial to spend additional time with the customer on complex engagements to validate the business needs and ensure that the proposed solution components will meet those needs and their expectations. In some cases this may slightly delay the sales cycle, but the extra focus helps ensure a successful project. The goal is to identify potential risks to meeting customer expectations and needs as early as possible, ideally before the project starts.

### Why the Focus on the Sales Cycle?

If the focus of this book is Best Practices in Project Management, why are we devoting so much time to the sales cycle?

The answer is very simple: a successful services project, which results in the deployment of the right solution and a satisfied customer, comes from:

1. Understanding the customer's business needs
2. Analyzing existing processes and supporting technologies
3. Customer agreement of the solution (Capabilities and Use Cases) to be deployed
4. Correctly scoping the project and defining it in the contractual documents
5. Properly setting customer expectations

*Customer benefits Include:*
- Significantly lower implementation risk, compared to not adhering to best practices
- Customers get into production quickly
- Increased satisfaction with defined outcomes delivering value
- Faster return on investment

## Project Execution

Once the services engagement is sold, CA Services prepares to deliver the needed solution. It's delivered by following a standard CA Services Solution Delivery Methodology, supported by deployment reference guides and deliverable templates and managed through CA Services' Project Management Methodology.

*Solution Delivery Methodology*

The CA Services Solution Delivery Methodology shown in Figure 15–15 is built on fundamental industry proven systems engineering principles that formalize a consistent, repeatable approach for a rapid implementation or multiphased solution deployments. The methodology focuses on ensuring that the solution aligns to customer business needs, and that the required deliverables to provide that solution.

*The methodology consists of:*

**Project Setup and Initiation**

Prior to CA Services arriving at the customer site, a series of joint meetings and/ or conference calls are held to prepare both CA Services and the customer for the

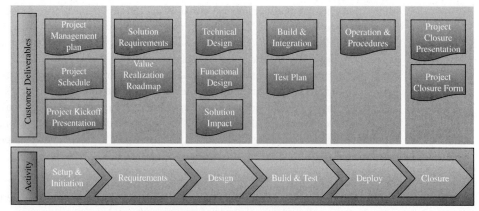

**Figure 15–15.** Solution Delivery Methodology.

beginning of the deployment. Key activities include creating and receiving approval of the Project Management Plan and the Project Schedule, conducting a Risk Assessment and validating the project scope and prerequisites.

### Solution Requirements Definition

The requirements should have already been gathered during an assessment or the sales process. In this stage, a series of requirements gathering sessions with key stakeholders, business managers, customer architects and operations team members are used to verify and further discover the customer's business drivers, functional and non-functional requirements as necessary. The outcomes are defined solution requirements that the customer and CA Technologies agree will produce the requested solution. This will be in the form of defined capabilities and use cases.

### Solution Architecture and Design

Through a series of collaborative workshops and discussions, CA Technologies analyzes the current state of the customer's environment. A planned end state is defined, key design decisions and configuration components are identified and the rationale for such decisions is documented. The outcome includes both a functional and technical design along with documented solution impacts.

### Solution Build and Test

1. The solution is installed and configured into the initial environment (usually development). Individual components are deployed to the appropriate environment and then integrated to the solution.
2. Upon completion of integration and unit testing, the testing staff (CA Services and customer) executes a test suite to validate functional requirements, quality attributes and solution constraints. These tests provide traceability from the working solution to the defined use cases to the project scope in the contract and align to customer business needs. Results of the testing are recorded in the Test Report. Any significant issues which require remediation will be addressed and then re-tested.

### *Production Solution Deployment*

To support the deployment of the solution into the customer's production environment, a predeployment planning template is utilized to define detailed plans for the deployment activities. This assists the project manager when determining the readiness to move the solution to production. The operational procedures are documented to assist the administration team that will be maintaining the solution.

### *Project Handoff and Closure*

The CA and customer project managers conduct a project closure meeting to validate the solution is in accordance with the requirements identified in the SOW and to discuss proposed next steps and/or deployment phases. This includes validation that all project documentation and deliverables have been provided along with the completion of knowledge transfer.

### *Deployment Templates*

Templates that align to each of the standard offerings support the Solution Delivery Methodology. These templates provide the framework for the standard deliverables produced within each stage. This promotes repeatable success by packaging what works, and helps others apply lessons learned. The deployment Playbook is shown in Figure 15–16.

- Methodical instructions that enable CA Services to deliver capabilities in a consistent and repeatable manner
- All of the documents and tools that you will need throughout the 9 stages of a project

*Deployment Playbook Guide*

*Solution Requirements Specification*

*Solution Design Specification*

*Solution Integration Specification*

*Solution Test Plan*

**Figure 15–16.** The deployment playbook.

The templates provide tremendous benefits during the execution of a project:

- Consistency – As team members move from project to project, they know the project execution will be very similar to other projects they have worked on. The time to bring the team up to speed and bring value to the project is greatly reduced.
- Standard Deliverables – This standardization promotes reuse and reduces time to create.
- Roadmap for the PM – The project manager doesn't need to create project plans from scratch; there is a predefined roadmap that makes it much easier to manage against.
- Continuous Improvement – Provides a repository of standard deliverables, tools, and techniques that can be continuously improved from lessons learned and project feedback.

Templates support every stage of the Deployment Methodology. Table 15–2 lists the templates provided to support the deployment of the solution.

*Project Management Templates*

Templates and tools for the project manager reinforce project management best practices. CA Technologies project managers utilize these templates to provide project governance throughout the project. As shown in Figure 15–17, governance is a continuous process that crosses borders from project stage to stage.

*The key benefits of project governance to CA Technologies projects include:*
- Minimize communication gaps
- Efficient use of skilled resources reduces project cost
- Effective financial management and consolidated budget reporting

**TABLE 15–2.  A LISTING OF TEMPLATES**

| Stage | Template | Description |
|---|---|---|
| Project Setup and Initiation | Kickoff Meeting Presentation | A presentation template to be used during a project kickoff meeting |
| Solution Requirements Definition | Requirements Gathering Workshop | Outline for a workshop to collect requirements |
| | Requirements Gathering Questionnaire | A questionnaire used for the collection of requirements for the solution |
| | Solution Requirements | Details requirements for the deployment of the solution |
| Solution Architecture & Design | Functional Design | Outlines the proposed solution from a functional perspective and how the solution is designed to mitigate the current challenges. Focus is on alignment between business needs and solution capabilities. Also outlines any process changes/improvements or personnel re-alignments required to support the new solution |
| | Technical Design | Outlines the proposed solution from a technical perspective. Details the components, the integrations and network configuration of the technology |
| Solution Build and Test | Solution Test Plan | Outlines required tests to validate the solution in each of the deployment environments |
| | Test Report | Details the results of the validation tests |
| | Testing Register | Summarizes the tests and test results |
| Production Deployment | Predeployment Planning | Requirements and detailed planned to proceed from a development or QA environment to the production environment |
| | Operations & Procedures | Procedures to maintain the solution after deployment |
| Project Handoff and Closure | Project Summary and Proposed Next Steps Presentation | Outlines "proposed next steps" and additional projects needed to increase the customer's maturity within the solution parameters. Overview of lessons learned that can be applied to future projects |

**Figure 15–17.**  Key elements of project governance.

- Single point of contact for all escalations, financials, resources, and planning
- Management of project success and standards
- Established project quality standards
- Increased customer satisfaction with project results

Table 15–3 highlights some of the key templates a project manager utilizes to support management activities.

*Packaged Work Products–Standard Components to Speed Implementation*

As solutions are developed for customers, it is not unusual to find some use cases and functional requirements that are not available in the reference architecture or out-of-the-box functionality. CA Services has a Global Delivery organization to support the delivery of such use cases. A customer may have a need for a specific report or function that is not part of the standard offering; in this situation, the development work is performed by Global Delivery in a specific code module. However, a big advantage of using a single group for any extended functionality work is the ability to track customer needs across all projects. It then becomes obvious which use cases are common to multiple customers, and where the solution can be expanded to provide the most value. Using this data, Global Delivery is able to package this work product into prebuilt accelerators and components. The ultimate goal is to minimize one-off customizations to a single customer, and lead customers to take advantage of the prebuilt components. This provides faster time to value, reduced project risk, and greater consistency.

## Project Closure

One of the stages that CA Technologies emphasizes is Project Closure. Many organizations view this stage as a formality that provides little value. However, CA has found this stage to be critical for long-term customer success, and sets the stage for additional

**TABLE 15–3. HIGHLIGHTS OF KEY TEMPLATES**

| Stage | Template | Description |
|---|---|---|
| All Stages | Meeting Minutes Template | A document containing a high level outline for meeting minutes |
| | Project Status Report | Reports project progress |
| | Project Issue Log | Tracks project issues including details, steps for remediation, and status |
| | Milestone Acceptance Form | Customer sign-off of deliverables |
| | Timesheet Report | Tracks team project time |
| Project Setup and Initiation | Project Management Plan | Plan for how the project will be managed |
| | Risk Assessment | Tool for assessing project risk |
| | Risk Inventory and Mitigation | Identifies and tracks project risks, mitigation plans, and status |
| | Financial Tracker | Tracks project financial plans, actual costs, and financial metrics |
| | Contact List | Identifies all individuals participating in the project and contact information |
| | Project Schedule | Defines the tasks, milestones and associated resources for the project |
| | Managed Products Register | Tracks the WBS deliverables and their status |
| Project Handoff and Closure | Implementation Feedback | Used internally to identify how the methodology, processes, templates, tools, or offerings could be improved to support future projects |

business. A formal project closure process is published to reinforce best practice. It outlines a project closure meeting with the customer and defines a specific agenda. Table 15–4 outlines the key agenda items during the project closure meeting and how they contribute to project and customer success.

## Continuous Improvement

One of the biggest challenges organizations face is their ability to apply lessons learned from a single project to improve planning and execution for future projects. CA Services has given this responsibility to two primary groups within CA Services—the Global Practices, and the PMO. At any given time a project manager is primarily focused on the successful execution of a single project for a customer, the Global Practices and PMO are responsible for the portfolio of projects and improving our ability to repeat project success.

As projects are completed, lessons learned are captured and deltas between the plan and actual execution identified; this feedback is used to update the offerings, calculators, tools, and templates as appropriate. Most organizations find making use of lessons learned a big challenge, but at CA Services this is supported by having documented offerings and a central repository to take action on lessons learned.

**TABLE 15–4. KEY AGENDA ITEMS**

| Project Closure Agenda Item | Contribution to Project and Customer Success |
| --- | --- |
| Recap of Business Requirements and Goals | Confirmation that the project met the customer's objectives and ROI has been positioned or realized. |
| Final Scorecard | Final measurements on how well the project was executed. Reviewing these metrics can help identify weaknesses for improvement or reinforce execution success. |
| Project Challenges and Lessons Learned | It is important to get the customer's perspective on any challenges the project had to overcome or lessons learned. Opening up this conversation with the customer brings in a different perspective and potentially fresh ideas for improvement. |
| Project Documentation | All project deliverables and key documentation is packaged and presented to the customer. This helps ensure the customer has copies of the final project work in a consolidated package for easy reference. |
| Training Plan | A review of education completed and recommendations for additional education can help ensure the customer is ready to take ownership of the deployed solution. This is critical for long-term customer success and ROI. |
| Next Steps | The completed project may be one of many phases or additional opportunities may have been identified for additional ROI. Reviewing next steps helps ensure the customer is aware of recommendations and sets expectations for future activities. |
| Transition from Services to Support | It's important to have a seamless transition from the team that just completed the project to the support team that will work with the customer long term. This activity reviews with the customer the process for working with Support and transitions key documentation to the Support organization. |
| Project Closure Sign-off | This confirms that the project is ready to close. |
| Customer Survey | CA Technologies surveys customers to confirm satisfaction and evaluate potential areas for improvement. Reinforcing this during project closure helps improve survey completion rates. |
| Customer Reference | This activity outlines the CA Technologies reference program and determines if the customer is a good fit. Including this activity in project closure helps improve the number and quality of references that can be used for future projects. |
| Meeting Closure | This activity opens up the discussion to any final comments from the customer and documents closure minutes for distribution. |

**TABLE 15–5.   USING LESSONS LEARNED FOR CONTINUOUS IMPROVEMENTS**

| Lesson Learned | Area for Improvement | Benefit to Project Success |
| --- | --- | --- |
| Customer architecture not 100% aligned to reference architecture | Updates to Reference Architectures | Ability to more quickly produce architecture and design docs; increased standardization to designs |
| Additional tasks or deltas in effort to execute project | Updates to Solution Calculator | More accurate WBS and task estimates |
| Use cases not included in standard offering | Packaging components by Global Delivery or creation of new Acceleration Services | Reduce time/cost to deploy; standardization of solutions; improved ROI to the customer |
| Improvements to content in deliverables | Template updates | Higher-quality deliverables |
| Methodology improvements | Solution Delivery and/or Project Management Methodology updates | Improve Time to Value and quality |
| Better alignment of scope to customer business needs | Standard Offerings / Services Offerings Catalog updates | Better solution alignment to customer business needs; Improve Time to Value |
| Challenges not identified early in life cycle | Updates to Risk Assessment Templates | Improved mitigation of risks and reinforcement of best practices; Improved Project Success |

Historically, this continuous improvement of project execution and project success has been very challenging for organizations. An organization may often capture lessons learned or document results from project reviews, but this information may be hidden away and never used for continuous improvement. One of the biggest advantages to having a repository of standard offerings and templates is the ability to improve over time. When a project does something very well or has challenges, this information is gathered and improvements made to the repository of tools thus allowing future projects to be even more successful. Table 15–5 outlines how some of these lessons learned are used for continuous improvement.

CA Technologies utilizes multiple methods for capturing lessons learned and data throughout the execution of the project. Figure 15–18 outlines these methods, and at what stage they are utilized for continuous improvement.

*Methods for process improvement:*

**Project Health Checks:** The PMO and Global Practices may conduct a project health check at anytime throughout a project. These are not audits, but rather a method for confirming the tools are providing value and gathering lessons learned. Specific challenges are addressed and additional assistance to the project team provided. The feedback is primarily used to improve the calculators, templates, and solution offerings.

**Solution Ratification:** For large projects with multiple technology integrations, a team of highly experienced specialists provide additional oversight to the project during the opportunity, requirements, and architecture design stages. The Global Practices provides experts to the project to review the customer's business needs, and to ensure the project design will meet these needs. The results of these reviews can then be used to make further improvements to the reference architectures and solution offerings.

**Global Escalation Management Team – Root Cause Analysis:** If a project encounters problems during the quality assurance and test or deployment stages, the Global Escalation Management Team may bring additional experts to the project that specialize

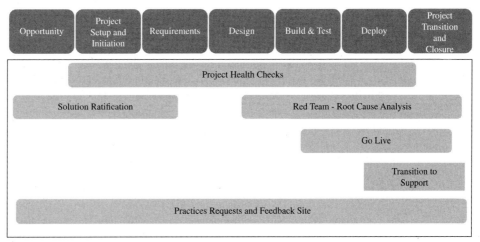

**Figure 15–18.** Methods for capturing lessons learned.

in troubleshooting and issue resolution. This team's immediate purpose is to resolve the problem, but a root cause analysis is also performed. The results of this analysis can provide insight for areas of improvement; especially for reference architectures.

**Go Live:** One of the more critical stages of a project is the deployment to production. CA has created a process where representatives from cross-functional groups come together to provide increased project support prior and during production deployment. This team is especially focused on any issues or obstacles that are impacting the project team's ability to successfully deploy the solution. Results of these checks can then provide feedback to continuously improve the methodology, tools, or templates.

**Transition to Support:** It is critical for long term customer success that a project seamlessly transition from the project team that deployed a solution to the support team that will work with the customer long term. During this transition period, the customer's architecture and design is shared with the support team along with knowledge transfer. Throughout the post-deployment phase customer success is monitored and feedback used to improve solution offerings.

**Global Practices Requests and Feedback Site:** This is a tool that allows anyone within CA Technologies to record feedback, make suggestions regarding any of the standard offerings, and request resources for assistance. Having this data in a centralized tool supports trend analysis, ensures that suggestions are not lost, and tracks the required action items to implement improvements.

## 15.3 MICROSOFT CORPORATION

There are training programs that discuss how to develop good methodologies. These programs focus on the use of "proven practices" in methodology development rather than on the use of a single methodology. Microsoft has developed a family of processes that embody the core principles of and proven practices in project management. These processes combined with tools and balancing people are called Microsoft Solutions Framework (MSF).[4] What appears in the remainder of this section is just a brief summary of MSF. For more information and a deeper explanation of the topic, please refer to Mike Turner's book in the reference at the bottom of the page.

MSF was created 16 years ago when Microsoft recognized that IT was a key enabler to help businesses work in new ways. Historically, IT had a heritage of problems in delivering solutions. Recognizing this, MSF was created based on Microsoft's experience in solution delivery.

MSF is more than just project management. MSF is about solutions delivery of which project management (aka governance) is a key component. Successful delivery is balancing solution construction with governance. According to Mike Turner:

> At its foundation, MSF is about increasing awareness of the various elements and influences on successful solutions delivery—no one has a methodological silver bullet; it is next to impossible to provide best practice recipes to follow to ensure success in all projects. . . MSF is about understanding your environment so you can create a methodology that enables a harmonious balance between managing projects and building solutions.
>
> Another key point with regard to MSF is that project management is seen as a discipline that all must practice, not just the project managers. Everyone needs to be accountable and responsible to manage their own work (i.e., project manager of themselves)—that builds trust among the team (something very needed in projects with Agile-oriented project management), not so much with formally run projects (still very top-down project management).
>
> The main point that MSF tries to get across is that customers and sponsors want solutions delivered—they frankly see project management as a necessary overhead. Everyone needs to understand how to govern themselves, their team and the work that the project does—not just the project managers.

Good frameworks focus on the understanding of the need for flexibility. Flexibility is essential because the business environment continuously changes, and this in turn provides new challenges and opportunities. As an example, Microsoft recognizes that today's business environment has the following characteristics:

- Accelerating rates of change in business and technology
  - Shorter product cycles
  - Diverse products and services
  - New business models
- Rapidly changing requirements
  - Legislation and corporate governance
  - Growing consumer demand
- New competitive pressures
- Globalization

---

4. M. S. V. Turner, *Microsoft Solutions Framework Essentials*, Microsoft Press. All rights reserved. The author is indebted to Mike Turner for providing the figures for this section in the book. The figures with numbers in the lower-right corner indicate page number sources in Mike Turner's book.

Typical challenges and opportunities include:

- Escalating business expectations
  - Technology is seen as a key enabler in all areas of modern business
- Increasing business impact of technology solutions
  - Risks are higher than ever before
- Maximizing the use of scarce resources
  - Deliver solutions with smaller budgets and less time
- Rapid technology solutions
  - Many new opportunities, but they require new skills and effective teams to take advantage of them

With an understanding of the business environment, challenges and opportunities, Microsoft created MSF.[5] MSF is an adaptable framework comprising:

- Models (see Figure 15–19)
- Disciplines (see Figure 15–19)
- Foundation principles
- Mindsets
- Proven practices MSF is used for successfully delivering solutions faster, requiring fewer people, and involving less risk, while enabling higher quality results. MSF offers guidance in how to organize people and projects to plan, build and deploy successful technology solutions.

MSF foundation principles guide how the team should work together to deliver the solution. This includes:

- Foster open communications
- Work toward a shared vision

**Models**

**Team**
Model

**Governance**
Model

**Disciplines**

**Project**
Management
Discipline

**Risk**
Management
Discipline

**Readiness**
Management
Discipline

**Figure 15–19.** MSF models and disciplines. *Source:* M. S. V. Turner, *Microsoft Solutions Framework Essentials*, Microsoft Press. All rights reserved.

---

5. MSF is part of a symbiotic relationship between the classic "build it" framework and the "run it"' framework. MSF is the "build it" and Microsoft Operations Framework (MOF) is the "run it."

- Empower team members
- Establish clear accountability, shared responsibility
- Deliver incremental value
- Stay agile, expect and adapt to change
- Invest in quality
- Learn from all experiences
- Partner with customers

MSF mindsets orient the team members on how they should approach solution delivery. Included are:

- Foster a team of peers
- Focus on business value
- Keep a solution perspective
- Take pride in workmanship
- Learn continuously
- Internalize qualities of service
- Practice good citizenship
- Deliver on your commitments

With regard to proven practices, Microsoft continuously updates MSF to include current proven practices in solution delivery. All of the MSF courses use two important project management best practices. First, the courses are represented as a framework rather than as a rigid methodology. Frameworks are based upon templates, checklists, forms, and guidelines rather than the more rigid policies and procedures. Inflexible processes are one of the root causes of project failure.

The second best practice is that MSF focuses heavily on a balance between people, process, and tools rather than only technology. Effective implementation of project management is a series of good processes with emphasis on people and their working relationships: namely, communication, cooperation, teamwork, and trust. Failure to communicate and work together is another root cause of project failure.

MSF focuses not only on capturing proven practices but also on capturing the right proven practices for the right people. Mike Turner states:

> The main thing that I think sets MSF apart is that it seeks to set in place a common-sense, balanced approach to solutions delivery, where effective solutions delivery is an ever changing balance of people, processes and tools. The processes and tools need to be "right sized" for the aptitude and capabilities of the people doing the work. So often "industry best practices" are espoused to people who have little chance to realize the claimed benefits.

MSF espouses the importance of people and teamwork. This includes:

- A team is developed whose members relate to each other as equals.
- Each team member is provided with specific roles and responsibilities.
- The individual members are empowered in their roles.
- All members are held accountable for the success of their roles.
- The project manager strives for consensus-based decision-making.
- The project manager gives all team members a stake in the success of the project.

The MSF team model is shown in Figure 15–20. The model defines the functional job categories or skill set required to complete project work as well as the roles and responsibilities of each team member. The team model focuses on a team of collaborating advocates rather than a strong reliance on the organizational structure.

On some projects, there may be the necessity for a team of teams. This is illustrated below in Figure 15–21.

Realistic milestones are established and serve as review and synchronization points. Milestones allow the team to evaluate progress and make midcourse corrections where the costs of the corrections are small. Milestones are used to plan and monitor progress as well as to schedule major deliverables. Using milestones benefits projects by:

- Helping to synchronize work elements
- Providing external visibility of progress
- Enabling midcourse corrections
- Focusing reviews on goals and deliverables
- Providing approval points for work being moved forward

There are two types of milestones on some programs: major and interim. Major milestones represent team and customer agreement to proceed from one phase to another. Interim milestones indicate progress within a phase and divide large efforts into workable segments.

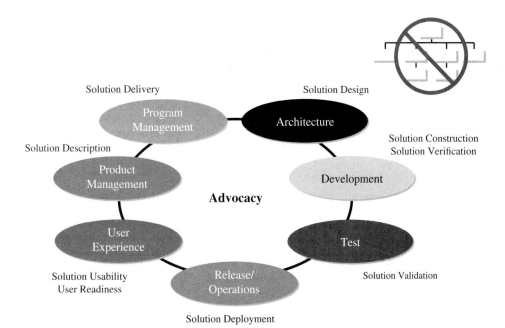

**Figure 15–20.** MSF team model. *Source:* M. S. V. Turner, Microsoft Solutions Framework Essentials, Microsoft Press. All rights reserved.

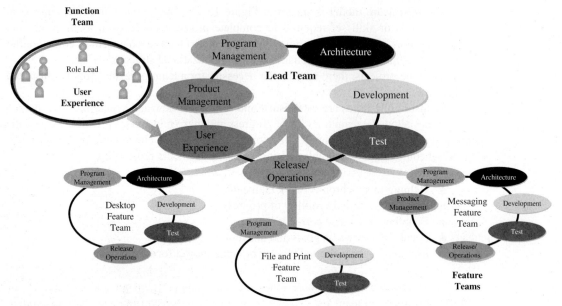

**Figure 15–21.** MSF team of teams. *Source:* M. S. V. Turner, *Microsoft Solutions Framework Essentials*, Microsoft Press. All rights reserved.

For each of the major milestones and phases, Microsoft defines a specific goal and team focus. For example, the goal of the envisioning phase of a program might be to create a high-level review of the project's goals, constraints, and solution. The team focus for this phase might be to:

- Identify the business problem or opportunity
- Identify the team skills required
- Gather the initial requirements
- Create the approach to solve the problem
- Define goals, assumptions, and constraints
- Establish a basis for review and change

MSF also establishes quality goals for each advocate. This is a necessity because there are natural "opposing" goals to help with quality checks and balances—that way realistic quality is built in the process and not as an afterthought.

This is shown in Table 15–6.

It is often said that many programs can go on forever. MSF encourages baselining documents as early as possible but freezing the documents as late as possible. As stated by Mike Turner:

The term "baselining" is a hard one to use without the background or definition. When a team, even if it is a team of one, is assigned work and they think they have successfully completed that work, the milestone/checkpoint status is called "Complete" (e.g., Test Plan Complete); whereas "Baseline" is used when the team that is assigned to verify the work

**TABLE 15–6. QUALITY GOALS AND MSF ADVOCATES**

| MSF Advocate | Key Quality Goals |
| --- | --- |
| Product management | Satisfied stakeholders |
| Program management | Deliver solution within project constraints Coordinate optimization of project constraints |
| Architecture | Design solution within project constraints |
| Development | Build solution to specifications |
| Test | Approve solution for release ensuring all issues are identified and addressed |
| User experience | Maximize solution usability Enhance user effectiveness and readiness |
| Release/operations | Smooth deployment and transition to operations |

*Source:* M. S. V. Turner, *Microsoft Solutions Framework Essentials,* Microsoft Press. All rights reserved.

**Figure 15–22.** MSF iterative approach. *Source:* M. S. V. Turner, *Microsoft Solutions Framework Essentials*, Microsoft Press. All rights reserved.

agrees that the work is complete (e.g., Test Plan Baselined). After the Baseline milestone/checkpoint, there is no more planned work. At the point when the work is either shipped or placed under tight change control is when you declare it

"Frozen"—meaning any changes must be made via the change control process. This is why you want to put off formal change management as late as possible because of the overhead involved.

This also requires a structured change control process combined with the use of versioned releases, as shown in Figure 15–22. What the arrows on the left mean is that as the solution is delivered, the solution completion increases, the knowledge of the solution space increases, and the overall risk to solution delivery goes down. The benefits of versioned releases include:

- Forcing closure on project issues
- Setting clear and motivational goals for all team members
- Effective management of uncertainty and change in project scope
- Encouraging continuous and incremental improvement
- Enabling shorter delivery time

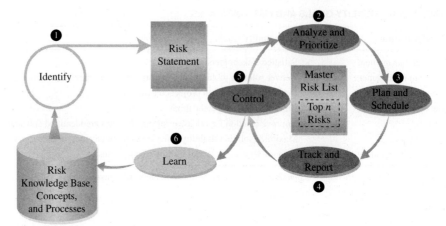

**Figure 15–23.** MSF risk management process. *Source:* M. S. V. Turner, *Microsoft Solutions Framework Essentials*, Microsoft Press. All rights reserved.

One of the strengths of MSF is the existence of templates to help create project deliverables in a timely manner. The templates provided by MSF can be custom-designed to fit the needs of a particular project or organization. Typical templates might include:

- Project schedule template
- Risk factor chart template
- Risk assessment matrix template
- Postmortem template

The MSF process for risk management is shown in Figure 15–23. Because of the importance of risk management today, it has become an important component of all project management training programs.

MSF encourages all team members to manage risk, not just the project managers. The process is administered by the project manager.

- *MSF Risk Management Discipline:* A systematic, comprehensive, and flexible approach to handling risk proactively on many levels.
- *MSF Risk Management Process:* This includes six logical steps, namely identify, analyze, plan, track, control, and learn.

Some of the key points in the MSF risk approach include:

- Assess risk continuously.
- Manage risk intentionally—establish a process.
- Address root causes, not just symptoms.
- Risk is inherent in every aspect and at all levels of an endeavor.

There are numerous ways to handle risk and MSF provides the team with various options. As an example, Figure 15–24 shows two approaches for risk prioritization.

Simple Prioritization Approach

| Probability/Impact | Low | Medium | High | Critical |
|---|---|---|---|---|
| High | M | M | H | C |
| Medium | L | M | M | H |
| Low | L | L | M | M |

Multiattribute Prioritization Approach

| Rating | Cost Overrun | Schedule | Technical |
|---|---|---|---|
| Low | Less than 1% | Slip 1 week | Slight effect on performance |
| Medium | Less than 5% | Slip 2 weeks | Moderate effect on performance |
| High | Less than 10% | Slip 1 month | Severe effect on performance |
| Critical | 10% or more | Slip more than 1 month | Mission cannot be accomplished |

**Figure 15–24.** Risk prioritization example. *Source:* M. S. V. Turner, *Microsoft Solutions Framework Essentials*, Microsoft Press. All rights reserved.

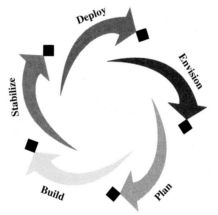

**Figure 15–25.** MSF governance model: enactment tracks. *Source:* M. S. V. Turner, *Microsoft Solutions Framework Essentials*, Microsoft Press. All rights reserved.

In Figure 15–19, we showed that MSF is structured around a team model and a governance model. The governance model is shown in Figure 15–25. This model appears on all of the MSF figures, illustrating that governance is continuously in place.

There are two components to the MSF governance model: project governance and process enactment:

- Project governance
  - Solution delivery process optimization
  - Efficient and effective use of project resources
  - Ensuring that the project team is and remains aligned with:
    - External (strategic) objectives

- Project constraints
- Demand for oversight and regulation
- Process enactment
- Defining, building, and deploying a solution that meets stakeholders' needs and expectations

The MSF governance model, as shown in Figure 15–25, is represented by five enactment tracks. Figures 15–26 through 15–30 provide a description of each of the enactment tracks. MSF has established success criteria for each of the tracks:

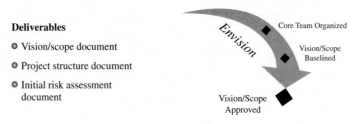

**Deliverables**

- Vision/scope document
- Project structure document
- Initial risk assessment document

**Goals**

- Develop a clear understanding of what is needed and all project constraints
- Assemble the necessary team to envisage possible solution(s) with options and approaches that best meet those needs
- Establish a basis for change for the remainder of the project life cycle

**Figure 15–26.** MSF envision track. *Source:* M. S. V. Turner, *Microsoft Solutions Framework Essentials*, Microsoft Press. All rights reserved.

**Deliverables**

- Functional specifications
- Master project plan
- Master project schedule

**Goals**

- Evolve solution concept into tangible designs and plans so it can be built in the developing track
- Find out as much information as possible, as early as possible
- Know when you have enough information to move forward

**Figure 15–27.** MSF plan track. *Source:* M. S. V. Turner, *Microsoft Solutions Framework Essentials*, Microsoft Press. All rights reserved.

**Deliverables**

- Completed solution
- Training materials
- Documentation
  - Deployment processes
  - Operational procedures
  - Support and troubleshooting
- Marketing materials
- Updated master plan, schedule, and risk document

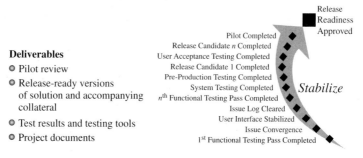

**Goals**

- Build all aspects of the solution in accordance with deliverables from the plan track (e.g., designs, plans, requirements)
  - Develop solution features and components (code and infra), complete all documentation deliverables and other elements of the solution (training material, etc.)
  - Test all aspects of the solution to assess the state of quality of the solution

**Figure 15–28.** MSF build track. *Source:* M. S. V. Turner, *Microsoft Solutions Framework Essentials*, Microsoft Press.

**Deliverables**

- Pilot review
- Release-ready versions of solution and accompanying collateral
- Test results and testing tools
- Project documents

**Goals**

- Improve solution quality to meet release criteria for deployment to production
- Validate solution meets stakeholder needs and expectations
- Validate solution usability from a user perspective
- Maximize success and minimize risks associated with solution deployment and operations in its target environment(s)

**Figure 15–29.** MSF stabilize track. *Source:* M. S. V. Turner, *Microsoft Solutions Framework Essentials*, Microsoft Press.

*Envision Track*

- Agreement by the stakeholders and team has been obtained on:
  - Motivation for the project
  - Vision of the solution
  - Scope of the solution
  - Solution concept
  - Project team and structure

**Deliverables**

- Operations and support information systems
- Revised processes and procedures
- Repository of all solution collateral

**Goals**

- Place solution into production at designated environment(s)
- Facilitate smooth transfer of solution from project team to operations team as soon as possible

**Figure 15–30.** MSF deploy track. *Source:* M. S. V. Turner, *Microsoft Solutions Framework Essentials*, Microsoft Press. All rights reserved.

- Constraints and goals have been identified.
- Initial risk assessment has been done.
- Change control and configuration management processes have been established.
- Formal approval has been given by the sponsors/and or key stakeholders.

*Plan Track*

- Agreement with stakeholders and team has been obtained on:
  - Solution components to be delivered
  - Key project checkpoint dates
  - How the solution will be built
- Supporting environments have been constructed.
- Change control and configuration management processes are working smoothly.
- Risk assessments have been updated.
- All designs, plans, and schedules can be tracked back to their origins in the functional specifications and the functional specification can be tracked back to envisioning track deliverables.
- Sponsor(s) and/or key stakeholders have signed off.

*Build Track*

- All solutions are built and complete, meaning:
  - There are no additional development of features or capabilities.
  - Solution operates as specified.
  - All that remains is to stabilize what has been built.
  - All documentation is drafted.

*Stabilize Track*

- All elements are ready for release.
- Operations approval for release has been obtained.
- Business sign-off has been obtained.

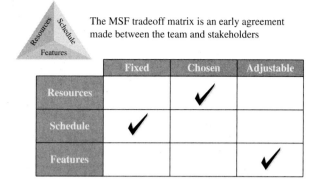

The MSF tradeoff matrix is an early agreement made between the team and stakeholders

|  | Fixed | Chosen | Adjustable |
|---|---|---|---|
| Resources |  | ✓ |  |
| Schedule | ✓ |  |  |
| Features |  |  | ✓ |

Given a **fixed** <u>schedule</u>, we will **choose** <u>resources</u>, and **adjust** <u>features</u> as necessary.

**Figure 15–31.** Project trade-off matrix. *Source:* M. S. V. Turner, *Microsoft Solutions Framework Essentials*, Microsoft Press. All rights reserved.

### Deploy Track
- Solution is completely deployed and operationally stable.
- All site owners signed off that their deployments were successful.
- Operations and support teams have assumed full responsibility and are fully capable of performing their duties.
- Operational and support processes and procedures as well as supporting systems are operationally stable.

MSF focuses on proactive planning rather than reactive planning. Agreements between the team and the various stakeholder groups early on in the project can make trade-offs easier, reduce schedule delays, and eliminate the need for a reduction in functionality to meet the project's constraints. This is shown in Figure 15–31.

## 15.4 DELOITTE: ENTERPRISE PROGRAM MANAGEMENT[6]

**Introduction**

Many organizations initiate more projects than they have the capacity to deliver. As a result, they typically have too much to do and not enough time or resources to do it. The intended benefits of many projects are frequently not realized, and the desired results are seldom fully achieved.

6. Material on Deloitte was provided by Daniel Martyniuk, Christine Lyman, PMP, and Rusty Greer Copyright © 2013 Deloitte Development LLC. All rights reserved. As used in this document, "Deloitte" means Deloitte Consulting LLP, a subsidiary of Deloitte LLP. Please see www.deloitte.com/us/about for a detailed description of the legal structure of Deloitte LLP and its subsidiaries. Certain services may not be available to attest clients under the rules and regulations of public accounting. This publication contains general information only and Deloitte is not, by means of this publication, rendering accounting, business, financial, investment, legal, tax, or other professional advice or services. This publication is not a substitute for such professional advice or services, nor should it be used as a basis for any decision or action that may affect your business. Before making any decision or taking any action that may affect your business, you should consult a qualified professional advisor. Deloitte shall not be responsible for any loss sustained by any person who relies on this publication.

*There are several factors that can make delivering predictable project results that much more difficult:*

- Added complexity of the transformational nature of many projects
- Constant drive for improved efficiency and effectiveness
- Renewed pressures to demonstrate accountability and transparency
- Accelerating pace of change and constantly shifting priorities

Traditional methods of coordinating and managing projects are becoming ineffective and can lead to duplication of effort, omission of specific activities or poor alignment and prioritization with business strategy. Making the right investment decisions, maximizing the use of available resources and realizing the expected benefits to drive organizational value have never been more important.

This submission will explore Deloitte's project portfolio management methods, techniques, approaches and tools ranging from translating organizational strategy into an aligned set of programs and projects, to tracking the attainment of the expected value and results of undertaken transformational initiatives.

**Enterprise Program Management**

Organizations are facing increased pressures to "do more with less." They need to balance rising expectations for improved quality, ease of access and speed of delivery with renewed pressures to demonstrate effectiveness and cost-efficiency. The traditional balance between managing the business, i.e., day-to-day operations, and transforming the business, i.e., projects and change initiatives, is shifting. For many organizations, the proportion of resources deployed on projects and programs has increased enormously in recent years. However, the development of organizational capabilities, structures and processes to manage and control these investments continues to be a struggle.

Furthermore, there has been a significant increase in project interdependency and complexity. While many projects and programs may likely have been confined to a specific function or business area in the past, increasingly, we see that there are strong systemic relationships between specific initiatives. Most issues do not exist in isolation and resolutions have links and knock-on impacts beyond the scope of one problem. Not only do projects increasingly span people, process and technology they also cross functional, geographical and often organizational boundaries. Without a structured approach to their deployment, projects and programs can fail to deliver the expected value. The need for strategic approach to project, program and portfolio management is great.

Deloitte's approach to project portfolio management is represented by the guiding Enterprise Program Management (EPM) framework that provides a model within which projects, programs and portfolios fit into a hierarchy where project execution and program delivery is aligned with enterprise strategy and can lead to improved realization of desired benefits. This approach aims to strike a balance between management of results (effectiveness) and management of resources (efficiency) to deliver enterprise value.

In Figure 15–32, *Strategy* includes the definition of the organization's vision and mission, as well as the development of strategic goals, objectives and performance measures. The *Portfolio Management* capability translates the organization's enterprise strategy into

**Figure 15–32.** Deloitte Enterprise Program Management (EPM) Framework. © Deloitte & Touche LLP and affiliated entities.

reality and manages the portfolio to determine effective program alignment, resource allocation and benefits realization. *Program Management* focuses on structuring and coordinating individual projects into related sets to determine realization of value that may likely not have been attained by delivering each project independently in isolation. Disciplined *Project Management* helps enable that defined scope of work packages are delivered to the desired quality standards.

**Strategy and Enterprise Value**    Today's business leaders live in a world of perpetual motion, running and improving their enterprises at the same time. Tough decisions need to be made every day—setting directions, allocating budgets and launching new initiatives—all to improve organizational performance and, ultimately, create and provide value for stakeholders. It is easy to say stakeholder value is important, though much more difficult to make it influence the decisions that are made every day: where to spend time and resources, how best to get things done, and, ultimately, how to win in the competitive marketplace or in the public sector, effectively deliver a given mandate.

Supporting the *Strategy* component of the Enterprise Program Management framework, Deloitte's Enterprise Value Map™ is designed to accelerate the connection between taking action and generating enterprise value. It facilitates the process of focusing on important areas, identifying practical ways to get things done, and determining if chosen initiatives provide their intended business value. The Enterprise Value Map™ can make this process easier by accelerating the identification of potential improvement initiatives and depicting how each can contribute to greater stakeholder value.

The Enterprise Value Map™, as illustrated at a summary level in Figure 15–33, is powerful and appealing because it strikes a very useful and practical balance between:

● Strategy and tactics
● What can be done and how it can be done

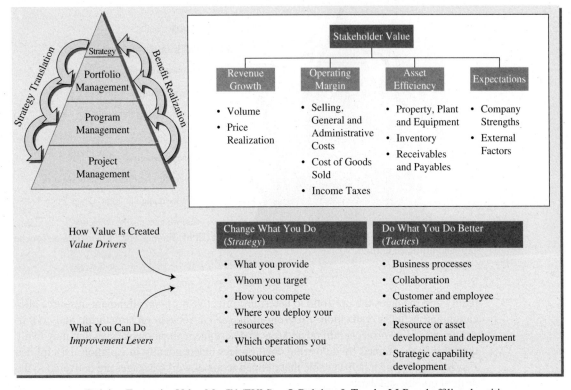

**Figure 15–33.** Deloitte Enterprise Value Map™ (EVM).   © Deloitte & Touche LLP and affiliated entities.

- The income statement and the balance sheet
- Organizational capability and operational execution
- Current performance and future performance

Overall, the Enterprise Value Map™ helps organizations focus on the applicable things and serves as a graphic reminder of what they are doing and why. From an executive perspective, the Enterprise Value Map™ is a framework depicting the relationship between the metrics by which companies are evaluated and the means by which companies can improve those metrics. From a functional perspective, the Enterprise Value Map™ is a one-page summary of what companies do, why they do it, and how they can do it better. It serves as a powerful discussion framework because it can help companies focus on the issues that matter most to them.

*The Enterprise Value Map™ is leveraged by Deloitte to help clients:*

- Identify things that can be done to improve stakeholder value
- Add structure to the prioritization of potential improvement initiatives
- Evaluate and communicate the context and value of specific initiatives
- Provide insights regarding the organization's current business performance

- Depict how portfolio of project and programs aligns with the drivers of value
- Identify pain points and potential improvement areas
- Depict past, current and future initiatives

*Stakeholder value is driven by four basic "Value Drivers": Revenue Growth, Operating Margin, Asset Efficiency, and Expectations:*

1. *Revenue Growth*: Growth in the company's "top line," or payments received from customers in exchange for the company's products and services.
2. *Operating Margin*: The portion of revenues that is left over after the costs of providing goods and services are subtracted. An important measure of operational efficiency.
3. *Asset Efficiency*: The value of assets used in running the business relative to its current level of revenues. An important measure of investment efficiency.
4. *Expectations*: The confidence stakeholders and analysts have in the company's ability to perform well in the future. An important measure of investor confidence.

There are literally thousands of actions companies can take to improve their stakeholder value performance, and the Enterprise Value Map™, in its full version, depicts several hundred of them. While the actions are quite diverse, the vast majority of them revolve around one of three objectives:

- Improve the effectiveness or efficiency of a business process
- Increase the productivity of a capital asset
- Develop or strengthen a company capability

The individual actions in the Value Map start to identify how a company can make those improvements. Broadly speaking, there are two basic approaches to improvement:

1. Change what you do (*change your strategy*): these actions address strategic changes—altering competitive strategies, changing the products and services you provide and to whom, and changing the assignment of operational processes to internal and external teams.
2. Do the things you do better (*improve your tactics*): these actions address tactical changes – assigning processes to different internal or external groups (or channels), redesigning core business processes and improving the efficiency and effectiveness of the resources executing those processes.

## Portfolio Management

Portfolio Management is a structured and disciplined approach to achieving strategic goals and objectives by choosing the most effective investments for the organization, and determining the realization of their combined benefits and value while requiring the use of available resources.

The Portfolio Management function provides the centralized oversight of one or more portfolios and involves identifying, selecting, prioritizing, assessing, authorizing, managing, and controlling projects, programs, and other related work, to achieve specific strategic goals and objectives. Adoption of a strategic approach to Portfolio Management enables organizations to improve the link between strategy and execution. It helps them to set priorities, gauge their capacity to provide and monitor achievement of project outcomes to drive the creation and delivery of enterprise value.

Deloitte's approach to Portfolio Management can enable an organization to link its strategic vision with its portfolio of initiatives and manage initiatives as they progress. It provides the critical link that translates strategy into operational achievements. As illustrated in Figure 15–34, the Portfolio Management Framework helps to answer the questions of "Are we doing the 'right' things?," "Are we doing enough of the 'right' things?," and "How well are we doing these things?"

Once implemented, the framework helps to transform the business strategy into a coordinated portfolio of initiatives that work together to increase stakeholder value. Additionally, it can provide the tools and techniques to keep projects on track, greatly improving an organization's chances of achieving the desired results. It focuses an organization on initiatives that offer high value-creation opportunities, and can also provide a structure and discipline to drive performance improvement initiatives and aid in the identification of continuous improvement opportunities. Lastly, it confirms that the appropriate resources and budget are made available for critical assignments and provides the tools and techniques to effectively manage an organization's portfolio of initiatives.

The first crucial step in the process of developing an effective project portfolio is the establishment of a method for determining which projects will be within and which will fall outside of the scope of the portfolio. A clear definition of what constitutes a "project" in needed, as well as identification of the criteria that will be applied to place a particular initiative inside or outside the boundaries of the portfolio. Daniel Martyniuk, a manager

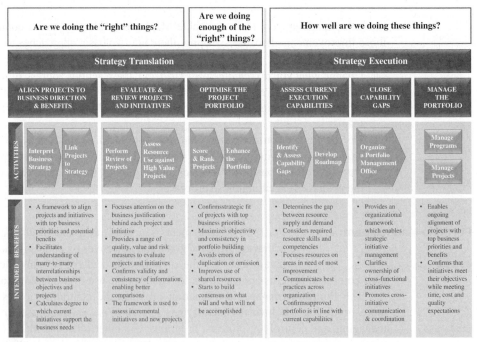

**Figure 15–34.** Deloitte Portfolio Management Framework.  © Deloitte & Touche LLP and affiliated entities.

in Deloitte Consulting LLP's Strategy & Operations practice that specializes in project portfolio management, highlights:

> While this first step may seem basic insofar as its aim is to provide a basic framework in which to define, sort and categorize projects, the critical component is carefully capturing all projects that are currently undertaken or proposed for approval. Many hard-to-define projects are often missed, as they may take the form of day-to-day activities or may take place 'out-of-sight'. As such, it is essential to define clear boundaries between day-to-day operations and project work – failure to do so may lead to ambiguity and inaccurate representation of the true count of projects in the organization.

A consistent categorization method, such as the Deloitte Investment Framework illustrated in Figure 15–35, helps to answer the question of "why are we allocating resources

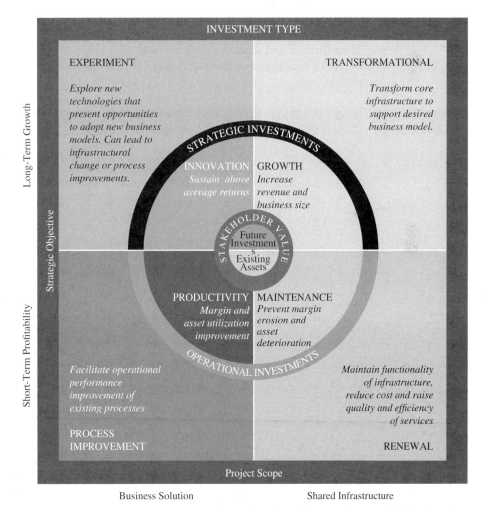

**Figure 15–35.** Deloitte Investment Framework. © Deloitte & Touche LLP and affiliated entities.

to this project?" It aims to define the differences between initiatives allowing for immediate recognition and categorization of projects, and it provides the context for comparing projects that are different in nature or scope. It also facilitates allocation of resources by type first, followed by prioritization of projects within a type. Most importantly, it provides common ground to facilitate dialogue and prioritization discussions.

Once the scope of the portfolio has been set, the organization may require a disciplined process to enable continuous alignment of projects to strategic objectives, evaluation, prioritization and authorization as well as the on-going management of progress, changes and realization of benefits.

The Deloitte Portfolio Management Process, as illustrated in Figure 15–36, can serve as a basis for the definition of common portfolio management sequence. It allows for coordination across projects to capitalize on synergies and reduce redundancies. It also helps to outline and identify projects in a comparable format when there are a multiple project opportunities and/or organizational pain points to increase the value created by the organizations initiatives, while balancing risk and reward.

When the approved list of projects making up the project portfolio is established, project registration and sequencing becomes the next critical step. Just because a project is now part of the "approved" project registry, it does not mean that it should or will be started right away.

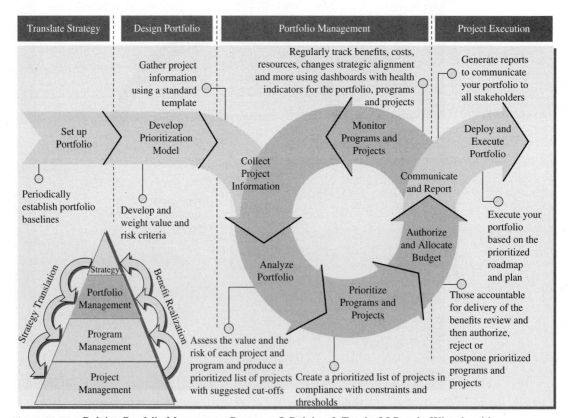

**Figure 15–36.** Deloitte Portfolio Management Process. © Deloitte & Touche LLP and affiliated entities.

There are a number of factors that should be considered when determining the appropriate sequence for project execution. Some of the important decision criteria for project sequencing include:

- *Strategic priority* – the level of importance placed on this project by stakeholders or organizational leadership; fast track the start of those initiatives that directly contribute to the realization of the stated business objectives.
- *Window of opportunity* – some initiatives need to be completed within a certain period of time in order to gain the desired benefits; give those initiatives the required consideration to help ensure that the opportunity to provide value is not missed.
- *Project interdependencies* – confirm that all dependencies between related projects have been identified and considered when making project sequencing and initiation decisions; also, consider other dependencies that could impact the start or the completion of projects, such as timing of important decisions, budget cycle, etc.
- *Resource availability* – a project cannot be started until the applicable resources become available to begin working on that particular project; however, remember that "availability" is not a skill, and in addition to getting resources assigned to your projects, make sure that they are the "right" resources in terms of their knowledge, ability and experience.
- *Risk* – consider the level of risk being taken on as a result of undertaking a given set of projects; it is a good idea not to initiate high risk projects all at once, all within the same delivery period; high-risk projects should be closely monitored and you should strive to find a applicable mix of high-risk and low-risk projects; whenever possible, you should consider staggering the execution of high-risk projects and conduct a full risk analysis to determine and agree on appropriate risk mitigation strategies.
- *Change* – consider the novelty of the undertaking, and the amount of change to be introduced into your organization as a result of implementing the proposed set of projects; confirm that your organization is ready to accept the amount and level of changes being created – there is only so much change that an organization can handle; stagger those projects that introduce significant changes and sequence them accordingly to limit change fatigue in your organization.
- *Cost/Benefit* – different initiatives have varying cost/benefits associated with it; similar to risk, it is imperative to understand which projects will provide the biggest benefits for the lowest cost; you don't want to start all of your higher cost projects at the same time especially if you are not going to reap all of the benefits upfront.

A factor to proper sequencing, and as a result appropriate portfolio balancing, is having a sound understanding of the organization's capacity to deliver as well as the capabilities of its resources. Organizations should know who is available to work on projects and what type of skills they have. It is often easy to determine how many people there are—so creating a resource inventory is typically not a problem. The challenge comes when trying to determine what the resources are currently working on and how much availability they have for project work or for additional projects, if they are already working on a project. One of the methods available to get that correct picture is to do time tracking of project resources.

The expected long-term outcomes and the benefits of adopting a consistent Portfolio Management Framework and Process can include, but are not limited to, having the ability and capability to:

- Make conscious choices in selecting projects for implementation based on correct and up-to-date information such as strategic alignment to business priorities, expected benefits, estimated costs and identified risks.
- Determine capacity, i.e., the number of concurrent projects, for managing small, medium and large-size projects to enable prioritization.
- Proactively manage risks associated with small to medium, as well as large and complex transformation projects.
- Increase core competencies in project management across the organization and adopt a portfolio management approach to executive decision making.
- Streamline and standardize processes related to the management of single, as well as the management of related, multiple projects and project portfolios.
- Maintain a current list of all projects, active and inactive, phase the initiation of projects to match capacity, and improve the delivery in accordance to requirements, of approved projects.
- Maximize use of internal resources and rationalize the use of external resources to supplement internal staff, with greater ability to facilitate value and efficient completion of the approved projects portfolio.
- Measure actual, real-time performance and track the realization of project and/or program benefits; with the ability to identify actual progress made in the achievement of tangible outcomes and real results.

**Program Management**      In accordance with Project Management Institute's practice standards, a program is a group of related projects managed in a coordinated way to obtain benefits and controls not available from managing them individually. Programs may include elements of related work (e.g., ongoing operations) outside the scope of the discrete projects in a program. Some organizations refer to large projects as programs. If a large project is split into multiple related projects with explicit management of the benefits, then the effort becomes a program. Managing multiple projects via a program may enhance schedules across the program, provide incremental benefits, as well as enable staffing to be optimized in the context of the overall program's circumstances.

As depicted in Figure 15–37, Deloitte's approach to Program Management highlights four core responsibilities for the program management function: program integration, dependency awareness, standards adherence, and program reporting. The figure further illustrates additional primary and secondary activities that fall within the scope of work for this function.

While time, cost and scope/quality are important performance measures at the individual project-level; coordination, communication and sequencing are the factors at the program-level that help enable the desired results. This is because Program Management involves grouping and managing a series of projects in an integrated manner, and not just completing individual projects. In the end, good project management can help to deliver the program according to the planned scope. Good program management will also provide a better understanding of the linkages and dependencies between projects and programs across the overall projects portfolio.

**Figure 15–37.** Deloitte Program Management Framework. © Deloitte & Touche LLP and affiliated entities.

## Project Management

At the program level, consistency can breed desirable results. This operating rhythm allows for regular reporting and monitoring across multiple projects. At the project level that consistency does not always make sense. Internal project variance can be a product of a number of factors:

- Type of project (e.g., strategy development, technology implementation, organizational change deployment, etc.)
- Geographical/organizational scope (e.g., single site, single country, global)
- Project implementation model (e.g., Agile, waterfall, iterative, etc.)
- Project team size

These variances lead to different needs and constraints that impact some project management processes. The implication is that enterprise and program guidelines need to be standardized in some areas while retaining flexibility in others. This balance, when appropriately struck, can enable project managers to tailor processes in certain areas (e.g., status reporting, work planning), while still helping to meet minimum standards of performance.

Furthermore, there are other factors external to the project that also drive variability and therefore should be considered. Some of these additional factors include:

- Industry
- Environment (e.g., public sector, regulated, commercial)
- Technology being implemented (e.g., cloud solution, ERP, etc.)

Even with so many differentiating elements, there are a number of processes and guidelines that will stay fixed regardless of which project management model they select. This includes laws or regulations, organizational policies, company standards, project controls and financial management processes/policies. See Figure 15–38.

It is important that the organization understand where variability is required versus where standardization is required and effective. The objective is to make it easier for project teams to deliver solutions well, not to be dogmatic or overly theoretical. It also strives to help enable the placement of acceptable safeguards to identify and manage "out of control" situations proactively.

As each project is initiated, the particulars of the project are considered. The result is a tailored set of project management processes that align with enterprise and program standards while reflecting the specific nature of the project itself.

## The Methodology

A holistic Project Management solution is concerned with the definition and delivery of specific work streams within an overall Enterprise Program Management framework. It includes standards, processes, templates, training, job aids, and tools. The more that these

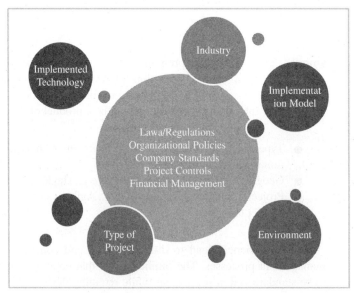

**Figure 15–38.** Comparison of fixed organizational factors versus variable factors. © Deloitte & Touche LLP and affiliated entities.

components can be standardized, the easier it can be to deploy them; teams understand the expectations, know the tools, and have lived the processes.

Enterprise Value Delivery (EVD) for Project Management is Deloitte's method for delivering consistent Project Management solutions to our consulting clients. The method systematically addresses specific components of project management and is a common, standards-based approach supported by leading enabling tools, experienced coaching, and training. This method is scalable and flexible; it can be integrated into other Deloitte methods in whole or in part to address relevant project management issues. It incorporates standards, while also allowing for individual projects to tailor the parts that make sense for their individual situation.

*Designed to help Deloitte practitioners manage their projects, EVD for Project Management is:*

- *Scalable* – uses a modular design to increase its flexibility and can fit the majority of projects accounting for a variety of project sizes or scope
- *Deliverables-based* – allows for the iterative nature of project management processes.
- *Prescriptive* – includes tools, detailed procedures, templates, and sample deliverables specific to the management of the project that help practitioners initiate, plan, execute, control, and close the project.
- *Rich in information* – houses extensive information about method processes, work distribution, and deliverable creation.
- *Based on experience* – allows practitioners to leverage reusable material developed through the vast industry experience and knowledge of our practice.
- *Based on leading practices* – reflects Deloitte leading practices and industry research and experience, allowing Deloitte practitioners to share a common language worldwide.
- *Practical* – provides realistic and useful information, focusing on what truly works.

Deloitte's Project Management method content is aligned with the Project Management Institute's (PMI's) Project Management Body of Knowledge (PMBOK Guide) and the Software Engineering Institute's (SEI's) Capability Maturity Model Integration (CMMI). Divided into two disciplines of work, Project Management and Quality Management, the method includes detailed task descriptions, step-by-step instructions, and considerations for task completion essential in delivering a PM solution. Multiple development aids including guidelines, procedures, as well as tools supplement each task.

*A number of benefits can result from consistently applying the defined project management tasks and deliverables:*

- Helps project managers see the "big" picture and accelerates work
- Provides a consistent approach and a common language
- Includes deliverable templates and tools
- Incorporates Quality and Risk Management, making it easier to improve quality and reduce risk of project deliverables
- Can be used to manage programs as well as projects

## The Tools

Deloitte has found that leveraging tools focused on enabling project management processes can help drive the adoption of sound project management processes within an organization. There are numerous tools available for organizations to use and they all have their own sets of advantages and disadvantages. See Figure 15–39. Selecting the appropriate tool can help facilitate management consistency throughout the project, but it is important to differentiate between processes and tools. It is less important which tool is deployed, so long as a rigorous process discipline is retained. Having a ready-to-use and leading-edge tool available is extremely advantageous but, the balance is between flexibility (use the appropriate tool for the job) and standardization (regardless of the tool being used, you should do a prescribed set of tasks).

If an organization does not have a tool available to utilize, Deloitte has a solution that may benefit clients. The customized tool provides sophisticated features while being simplistic enough to deploy to a project team quickly. Security is provided in a multitenant environment and practitioners are trained on the tool prior to engaging on a project.

### "Agile" project management

Deloitte's EVD solution establishes a foundation for performing common project management tasks, while providing the flexibility to tailor processes to address known variability, by:

- Defining typical usage models for frequent solution scenarios that incorporate the full spectrum of solution components (documentation, training, samples, etc.)
- Providing guidelines to enable projects to leverage the appropriate processes for the specific project circumstances
- Allowing flexibility to define an appropriate governance structure to support the specific risk/cost profile for the project

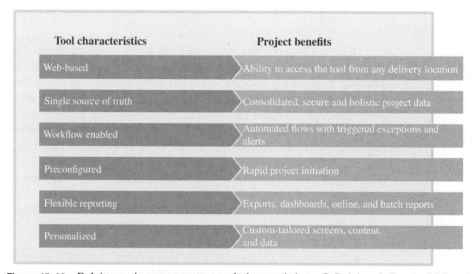

| Tool characteristics | Project benefits |
| --- | --- |
| Web-based | Ability to access the tool from any delivery location |
| Single source of truth | Consolidated, secure and holistic project data |
| Workflow enabled | Automated flows with triggered exceptions and alerts |
| Preconfigured | Rapid project initiation |
| Flexible reporting | Exports, dashboards, online, and batch reports |
| Personalized | Custom-tailored screens, content, and data |

**Figure 15–39.** Deloitte project management tool characteristics. © Deloitte & Touche LLP and affiliated entities.

Emerging market conditions have also driven the need to evaluate standard project management processes especially in an "Agile" environment. Agile approaches are typically used in projects where requirements are unclear or subject to frequent change, and/ or where frequent delivery of solution components with the greater value is required. Projects leveraging an Agile methodology perform certain project management processes in a similar manner to projects using waterfall or iterative approaches (e.g., risk and issue management, financial management, and some aspects of status reporting), while others are performed in a very different manner. EVD provides guidance for Project Managers from both perspectives. Furthermore, for those aspects that are different, it prescribes how they can be approached using techniques that are specific to Agile.

Specifically, scope management and work planning and tracking in Agile is significantly different than in projects using a waterfall or iterative methodology. EVD for Agile describes how projects develop and manage the Product and Sprint Backlogs, and includes guidelines for documentation of user stories that are sufficiently granular to be addressed within a single sprint. It describes the metrics analysis (using velocity, capacity, and burn up/down) that project teams use to size sprints and user stories, forecast delivery of components, and includes the productivity reports that are used to monitor progress. The important factor is to retain sufficient project management discipline, even if the techniques for managing the work are dramatically different. To summarize, although Agile may be vastly different in how a project is executed, basic project management functions still exist and Deloitte has identified a way to accomplish this.

## The Project Team

Even with the most intricate method and set of tools, without a dedicated and trained project team, the project can still falter. Resources with significant project or program management experience typically function at the Program Management level while resources with limited to no project experience can be found at the project level. With that said, having a project management approach that takes that into account is essential.

To make resources effective throughout the project, there are a few things to consider in tailoring your project management deployment as shown in Figure 15–40.

- **Understanding the internal change management implications.** Change Management processes can be more rigid in some places than others. The project has to be fully aware of what circumstances to go through the change management process and what information is required at what times. This is critical in facilitating changes to get routed correctly and efficiently. Having to review every change request for accuracy slows down the project and can have an unanticipated adverse effect on project timelines.
- **Start basic activities early**. Recognize that while the enterprise and program management layers are typically staffed by dedicated, full-time PM professionals, project teams typically include operational resources who have little to no actual project experience. Make sure the project team knows what is expected from them and knows the "basics." Examples include:
  - How often am I reporting status?
  - What is the overall project timeline?
  - Where can I find scope documentation for this project?

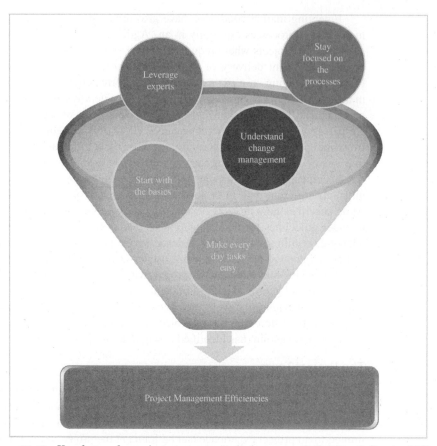

**Figure 15–40.** Key factors for project management deployment. © Deloitte & Touche LLP and affiliated entities.

Performing a project kick-off and implementing appropriate process rigor early are instrumental in introducing the team to the specific dynamics of that project. This is when the overall timeline, project management protocols, roles and responsibilities, among other topics are presented. At this time, the project team can also start other planning activities including template identification or development.

- **Make it easy to do the simple things well.** For everyday tasks like storing updated deliverables, recording time or status, or updating the work plan, the effort should be minimal regardless of the level of experience. These activities should not cause additional overhead to the project. The more time taken to perform simple things, the less productive time there is for the substantial design or build work on a project.

- **Leverage specialists when more advanced project management activities are required.** Certain activities of project management do require significantly more skill and should not be handled by a project novice. These areas center frequently on work planning efforts. Work planning activities occur throughout the project

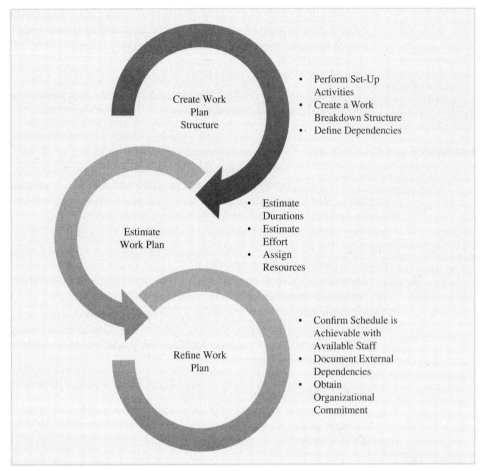

**Figure 15–41.** Activities required to develop a work plan. © Deloitte & Touche LLP and affiliated entities.

and require a deep understanding of managing dependencies, identifying critical path items and performing meticulous resource allocation. The activities shown in Figure 15–41 are focused only on the initial development of the work plan. Resources also need to understand what adjustments are required to the work plan as changes occur either to scope or resources.

- **Maintain focus on defined project management processes when things get precarious**. As changes occur throughout the project, sometimes it feels like the project can get out of control. One of the ways to avoid this is to uphold the project management processes that were approved at project initiation. A solid project management discipline enables the project manager to pull the project out of whatever uncertain situations may arise. The process can determine the overall effect to a project's scope, timeline, resources or budget if managed diligently. It is critical to maintain the discipline when things are starting to go awry—just when the average person says "I don't have time for that" is when it becomes more critical.

Excellent project management is a result of a clear understanding of the project context. Deployment of standardized processes and tools can make things easier, but only if balanced to reflect the variations of each particular project. Once this balance is determined, it's primarily a matter of blocking and tackling. Communicate expectations and establish disciplines early, so that it becomes second nature. This allows the project team to focus creative energies on building the best solution, which is, after all, why there is a project in the first place.

**Leadership and Governance**    There are additional factors that influence an organization's ability to generate value and deliver transformation results that go beyond having the right project, program and portfolio management processes or templates. "The importance of proper governance, leadership and accountability cannot be underestimated", adds Martyniuk. "In my experience implementing project portfolio management, having the right framework to guide project stakeholders through the myriad of decisions that needs to be made on a constant basis is a critical differentiating factor between a project's success or failure."

The main purpose of governance is to specify decision rights, clarify accountabilities, and encourage desirable behaviors. Governance is about bringing the appropriate individuals to the table to have the desired conversation under the relevant process to make the preferred decisions given available information. Governance frameworks depict the structures and processes by which decisions are made, and they define sets of principles and practices for managing:

- *What* decisions need to be made
- *Who* has the authority and accountability for making decisions, and with whose input, and
- *How* decisions get implemented, monitored, measured, and controlled

As illustrated in Figure 15–42, effective governance requires strong executive sponsorship, clear "business" ownership, and sound technical advisory to facilitate compliance with established regulations, standards, and guidelines. It also requires some type of benefit and value oversight. This can be done through a committee of select stakeholders who understand the qualitative aspects of a project's value. Most importantly, each function of the chosen governance framework should be empowered with the required decision making authority within their area of responsibility.

**People and Organizational**    Lastly, though most importantly, it is *people* that are the critical ele-
**Change Management**    ment to achieving project and transformation objectives. They are also a leading cause of transformation results falling short. Integrated people and organizational change management, HR and learning services should be delivered across the portfolio at the program and project levels to drive consistency, alignment, and effective delivery across the overall transformation effort.

Illustrated in Figure 15–43, the Deloitte People Dimension of Transformation is a broad framework that aligns with the business strategy and can address everything from risk assessment and leadership alignment to behavioral change, communication, training, organizational design, and more.

**Figure 15–42.** Deloitte Project Portfolio Governance Framework © Deloitte & Touche LLP and affiliated entities.

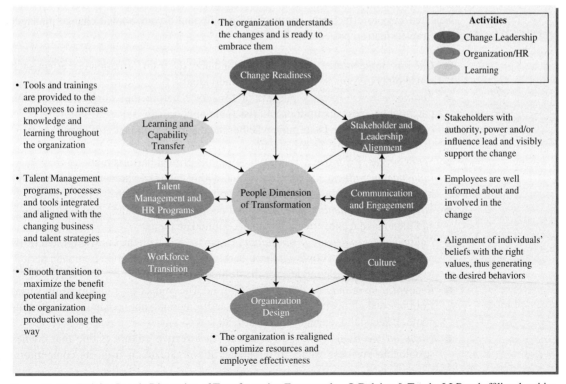

**Figure 15–43.** Deloitte People Dimension of Transformation Framework. © Deloitte & Touche LLP and affiliated entities.

One of the major causes of a transformation not achieving its desired objectives is the stakeholders' inability to see and feel the compelling reason for the change. As a result, fear, anger, or complacency can take root and cause resistance. In cases where change is more effective, individuals have a sense of passion. Create compelling, eye-catching and dramatic situations to help people see and visualize problems, solutions or progress in addressing complacency, lack of empowerment, or other important issues.

Sustained transformation also requires deep, personal commitment at every level of the organization. Some stakeholders will be co-creators who help shape the transformation vision and plans. Some will be interpreters. Other stakeholders will be consumers of the transformation. Effective transformation requires contributions and involvement from all types of players. Alignment and internal commitment start at the top—leaders should be aligned, willing to seek out resistance, and committed to leading the transformation by example.

Transformation projects usually alter structures, work processes, systems, relationships, leadership styles, and behaviors that together create what we know as organizational culture. Creating the culture the organization wants—or preserving the one it already has—may require a deliberate program that aligns with other transformation activities. Without a conscious effort, it is easy to end up with an organization stuck in between new ways of working and old modes of behavior.

To enhance the investment in new business models, technologies, and processes, a formal and deliberate program of education and skills development for the people affected by the transformation is essential. Yet education and training are usually near the bottom of the transformation to-do list.

Selected keys to effectively approaching people and organizational change management on transformation projects include:

- *Get your stakeholders correct*. Understand how the transformation can affect each stakeholders group as well as specific individuals.
- *Anticipate risks*. Identify pockets of resistance before they surface, along with potential business disruptions and risks that might arise.
- *Assess the situation*. Determine whether the magnitude and pace of change is energizing or paralyzing the organization.
- *Set priorities*. Prioritize activities, tackling the critical barriers first.
- *Influence the influencers*: Identify people within each stakeholder group who command respect, and then get them involved as champions for the transformation.
- *Strive for real commitment*: Understand people's circumstances and aspirations— and then make a concerted effort to accommodate them.
- *Equip leaders to drive transformation*: Equip leaders with the knowledge and skills needed to help their people get through this challenging and often traumatic period. Make leaders the role models for the desired behavior.
- *Recognize there may be winners and losers*: The impact of transformation varies from one stakeholder group to the next, and some may not be happy with the outcome. Understand, engage, and inform all stakeholders.
- *Focus on the things that really matter*. An effective culture is one that creates sustainable business value, differentiates the organization from its competitors,

supports the specific requirements of the industry, and helps customers get what they really want.

- *Be consistent.* Things that drive behavior and culture should align with one another. Misalignment simply confuses people.
- *Reinforce.* Align people-related initiatives—particularly rewards and incentives—to help foster the new culture. Establish effective leadership models and introduce new words and vocabulary that highlight the desired behavior.
- *Retain select staff.* Identify top performers and other select staff who are critical to the organization's future results. Let them know they are not at risk.
- *Capture knowledge.* Establish formal processes and systems to transfer and capture organizational knowledge—particularly for sourcing transformations.
- *Be kind but confident.* Decision makers should be gentle, but not show doubts that decisions were required, appropriate, and final.

**Conclusion**

The adoption and consistent application of a standard project, program and portfolio management frameworks, as well as implementation of the relevant governance along with effective people and organizational change management techniques, can lead the organization to the realization of a number of benefits, including:

- *Improved executive decision making* – enhanced ability to determine which projects to continue/stop, based on correct, up-to-date project status/progress information.
- *Financial transparency and accountability* – improved ability to manage budget under and over runs; and, shift funds within the portfolio to better manage and respond to unforeseen circumstances and changes in priorities.
- *Enhanced resource capacity management* – availability of needed information and data to make effective use of available resources; and, ability to shift resources within the portfolio to enhance resource utilization across projects.
- *Proactive issues and risk management* – ability to foresee and respond to challenges before they escalate into major problems; a mechanism for bringing select issue resolution or risk mitigation decision/action requirements to the attention of the executives.
- *Standardization and consistency* – apples-to-apples comparison between projects; improved and more timely internal and external communications with staff, clients, executives and other stakeholders.
- *Increased collaboration and better results* – enhanced realization of benefits through the joint management of initiatives as an integrated portfolio; co-operation and improved removal of roadblocks to results.

Although not exhaustive, the topics addressed in this submission outline selected critical factors to achieve the desired results that, based on our practical experience, can guide an organization in the "right" direction as it embarks on the road to implement sustained project portfolio management capability to deliver real, tangible enterprise value.

## 15.5 COMAU[7]

In the second edition of my text, *Strategic Planning for Project Management Using a Project Management Maturity Model*, I stated that the path to maturity can be accelerated with (1) the implementation of a PMO early on in the process; (2) having the PMO report directly to the executive levels of the company; and (3) having visible executive-level support for project management. Companies that accomplish all three of these items seem to outperform their competitors with regard to project management. Such was the case with COMAU.

**COMAU's Background**

COMAU is a worldwide leader in manufacturing flexible, automatic systems and integrating products, processes and services that increase efficiency while lowering overall costs. With an international network that spans 13 countries, COMAU uses the latest technology and processes to deliver advanced turnkey systems and consistently exceed the expectations of its customers. COMAU specializes in Body Welding, Powertrain Machining & Assembly, Robotics & Maintenance as well as Environmental Services for a wide range of industrial sectors. The continuous expansion and improvement of its product range enables COMAU to guarantee customized assistance at all phases of a project—from design, implementation and installation, to production start-up and maintenance services. COMAU core competences: spot, laser & arc-welding, sealing, drilling & riveting, machining, assembly & test, monitoring & control, handling & logistics, maintenance services, energy efficiency consultancy, and project management services.

**Description of the Problem**

During the 1980s, COMAU was enjoying significant success and as such, recognized the opportunity that could come from acquisitions. In the 1990s, COMAU started pursuing a strategy of global acquisitions. The problems with managing the acquired companies soon became apparent because each had a different level of maturity in terms of project management, and there were no corporate standards for project management. Until a few years ago, project management in COMAU was typically executed in a very fragmented way. There was a general lack of culture, methodologies, processes, and guidelines surrounding the project management process. By 2007 there was an urgent need to implement an efficient and global approach. The goal was simple: to integrate project management knowledge across the entire global company in order to give COMAU a powerful competitive advantage.

**The Solution—"From a Cluster to an Effective Network . . ."**

In 2007, COMAU decided to reinforce the project management culture by establishing the Contract & Project Management corporate function. As with most companies that understand the importance of project management and recognize the need for executive leadership in project management,

---

7. ©2013 by COMAU. Reproduced by permission. All rights reserved. Material on COMAU has been provided by Roberto Guida, COMAU Contract & Project Management Vice President, Angelo Putiri, COMAU PMO Manager, Claudio Samarotto, COMAU Risk Manager, Paolo Vasciminno, COMAU PM Academy Manager.

the new organization was headed by a vice-president for contract and project management. The main guidelines for the organization's mission included:

1. Company organizational development and the implementation of global organizational policies related to project management
2. Reinforcement of the corporate project management policy and Project Management Office structure
3. Creation of Comau Project Management Academy—the continuous development of knowledge of both hard and soft skills

COMAU correctly recognized the importance of the PMO in achieving this mission. Unlike less mature companies, COMAU viewed itself as a solution provider whose goal was satisfying the business needs of its global customers. The Contract & Project Management Office was therefore considered as an internal business solutions provider.

*COMAU approached the three main guidelines as follows:*
- *Company Development and Implementation of the Organizational Policies.* During 2007, a global project management policy was developed together with an intensive training program based on the *PMBOK® Guide*. Project management benchmarks were established to measure the company's maturity level and an action plan was created to continuously improve the maturity process. The global policy, which would be applied across the entire COMAU global enterprise, was a project management policy describing the tasks that all the project teams must perform. The policy was directly connected to the PMI *PMBOK® Guide* best practices. It should be noted that the company decided to integrate contract management and project management into a single unit. COMAU was convinced this was an important innovation and that it would produce positive results for the company at large.
- *Reinforcement of the Corporate Project Management Policy and Project Management Office Structure.* Today the COMAU PMO portfolio consists of a multimillion-euro revenue group covering global automotive commercial vehicle, renewable energy, and aerospace projects carried out in more than 30 countries. COMAU Project Management Office, as part of the Contract and Project Management Office, coordinates the efforts of each of the regional PMOs: PMO North America, PMO South America, PMO Europe, and PMO Asia. The COMAU global organization is shown in Figure 15–44.
- The Project Management team, international experts in project, program, and portfolio management, is included as part of the Project Management Family and is composed of project managers, program managers, planners, and team members. The missions of the COMAU PMO are shown in Figure 15–43.
- *Creation of the COMAU Project Management Academy.* In 2007 we established a specific organizational entity aimed at developing culture and skills in project management: the COMAU PM Family, which is the community of COMAU Project Management Professionals. We also understood that it would be effective to have a structure with the specific responsibility to enhance this community. Thus, the Project Management Academy was launched.

Italy

France

Germany

U.K.

Poland

Romania

Russia

U.S.A.

Mexico

Brazil

Argentina

China

India

INDUSTRIAL OPERATIONS     ENGINEERING

INNOVATION CENTER

**Figure 15–44.** COMAU worldwide. Reproduced by Courtesy of Comau.

*Since the beginning, the PM Academy objectives have been:*
- To spread project management knowledge and culture throughout the company
- To analyze the training needs of COMAU Business Units,
- To design specific training activities for our professionals in project management
- To organize initiatives (conferences, workshops, etc.) to foster the dissemination of Project Management know-how

To support COMAU's mission, the PMO first prepared a high-level roadmap for 2007–2009 which included the following:

- 2007:
  - Perform benchmarking and maturity base lining
  - Define the concept of the PMO
  - Develop operational policies for project management
  - Develop project management training programs
- 2008:
  - Establish a corporate PMO
  - Establish regional PMOs
  - Implement innovation actions according to the maturity assessment
  - Establish the Project Management Academy as a PMI REP
- 2009:
  - Manage ongoing activities for projects, programs, and projects portfolio
  - Perform external benchmarking on project management maturity
  - Expand the activities of the Project Management Academy
  - Manage selected strategic and special projects

Managing COMAU projects portfolio globally; closing the gap between **strategy and actions** and allowing experience flow between different countries and businesses

**Harmonizing** project management procedures (adoption of PMI guidelines) and tools worldwide

COMAU PMO Mission

Providing **contractual & risk specialists** during project execution phase

Managing the **COMAU PM Academy** (which is a PMI Registered Education Provider)

**PM Family**

Managing multinational projects with the right empowering and visibility

**Figure 15–45.** COMAU PMO missions. Reproduced by Courtesy of Comau.

As stated previously, COMAU viewed the PMO as the primary mechanism for providing internal business solutions. Some of the benefits achieved by COMAU included:

1. Customer recognition as the number one integrator of complex work, thus adding value to the value/supply chain.
2. Development of a high-standard, international class-compliant project management culture and approach.
3. Better support for the sales team resulting in increased project success through proactive project planning and risk reduction strategies.
4. Development of a culture capable of synchronizing language with COMAU customers, reducing misunderstandings within projects definitions and executions, and supporting trust-oriented communications throughout the projects.
5. Development of one of the best workload optimization techniques capable of reducing costs for its customers.
6. Development of a shared technical language when working to standardize a global approach, e.g., WBS Powertrain Italy and France. This enables COMAU to exchange parts between products and project teams in different countries, thereby achieving better planning, execution, cost control, workload planning as well as more balanced risk management, communications, and quality.
7. Identification and the management of "out-of-project-scope" situations, which results in better benefits for customers, COMAU, and providers.
8. Optimization of processes and reporting systems, which reduces wasted time and creates more time for managing critical issues.
9. Contribution to the company-wide integration of work and processes, sharing information, visions, and strategies, which includes the start-up of strategic projects.
10. Creation a strong team of high-skilled managers and technicians capable of supporting difficult projects and high-pressure situations.

COMAU PMO is actively managing the Global Project Portfolio, reporting to the CEO and working to achieve the highest degree of alignment between project management and business strategy as shown in Figure 15–46.

The PMO team has become the change agents inside each of COMAU's organizational business units. The result has been several "quick win" solutions. COMAU has been able to get better control of its indirect costs while providing value-added opportunities for both the company and its global customer base. All managers are now delegating authority to a greater extent than in the past, and the Vice Presidents and Regional Managers are functioning as strong sponsors.

A second high-level road map has been developed for 2010–2013 including the following objectives:

- 2010:
  - Develop at a worldwide level the concept of the Contract Management
  - Increase Portfolio Management
  - Develop project management training programs for external customers
- 2011:
  - Develop the concept of Risk Management at worldwide level
  - Develop Global Portfolio Management rules & methodologies
- 2012:
  - Develop Global Project Management process
  - Develop risk management training programs

Following the road map, the Contract & Project Management modified its structure in order to reach a configuration based on the coordinated sharing of activities between the Project Management Office, Contract Management, the Risk Management Office and the PM Academy, as well as on a geographical scale by local PMOs and Contract Management. The regional PMOs report to the Corporate PMO and also to the regional Managers. These Region Managers have demonstrated a sincere desire to function as executive-level sponsors and make the PMO evolution happen.

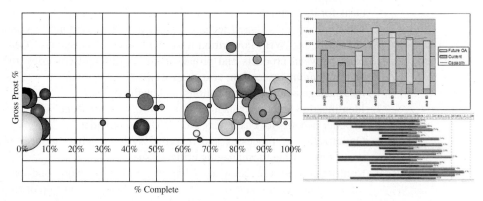

**Figure 15–46.**   Comau Portfolio Management.   Reproduced by Courtesy of Comau.

From the geographical point of view, the present global structure of the Contract & Project Management Office is reported in Figure 15–47.

**The First COMAU Global Project Management Process**

A high degree of fragmentation at international level along with business sustainability problems led COMAU to think about Project Management as an integration tool. As a result, common language and tools have been developed and spread across the organization.

Today, the high occurrence of multicountry projects (e.g., a global client issues an order to COMAU Inc. for an automated line to be placed in a plant in India. COMAU Inc. develops the engineering of the plant while COMAU China performs manufacturing and COMAU India manages the installation) and the creation of Global Business Units has led to a much greater degree of integration, forcing COMAU to take a step forward.

It therefore became necessary to review the Project Management Processes with the aim of harmonizing our way of working and managing projects throughout the entire organization.

To this extent, the Quality Department and Contract and Project Management Offices jointly sponsored an initiative to produce a global revision of the Project Management Processes (see Figure 15–48), which was accomplished by cross-functional and cross-country teams, coordinated by the corporate PMO.

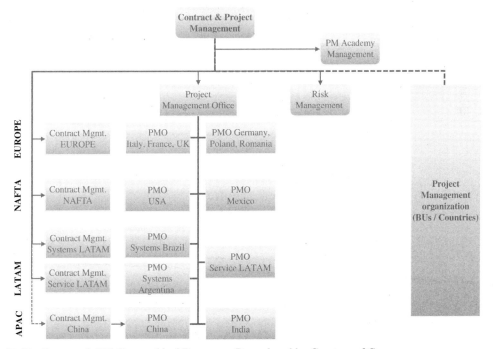

**Figure 15–47.** Contract & PM Geographical Presence. Reproduced by Courtesy of Comau.

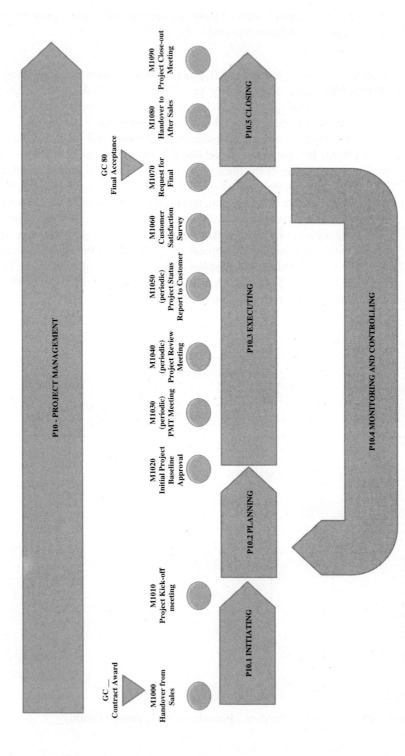

**Figure 15–48.** COMAU global project management process   Reproduced by Courtesy of Comau.

We based the creation and development of the Global Project Management upon the followings sources (see Figure 15–49):

● Local Project Management Processes in use in the different COMAU branches
● COMAU lessons learned
● PMI® model
● COMAU Global Policy of Execution

In order to ensure the creation of a real global process we used the paradigm shown in the Figure 15–50.

The Paradigm Pyramid has "Policy" on the top (at the Sector level), which inspires the development of Global Procedures/Processes, "Operating Instructions" (at the Global Level) and the application of "Local Procedures / Operating Instructions" at the Country Level.

**Comau Approach to Risk Management**

In 2006, COMAU started addressing Risk Management with a more focused, strategic approach, recognizing it as an essential part of the successful completion of projects. A Risk Management aspect was included in the Project Management Policy, thereby introducing the general rules for planning, assessing, handling and monitoring variable factors to ensure favorable outcomes.

The increased organizational complexity and internationality of order management made it necessary to find more structured and refined tools for managing uncertainties. Consequently, in 2010, we created a Risk Management Office as part of the Contract & Project Management. The purpose of which is to better cope with the worldwide business environment and the increased project size and complexity, and provide internal support and governance to all ongoing projects/programs in the different Business Units and countries. Its task is to improve risk management capabilities and to collect the experiences learned during execution of COMAU's projects.

Assuming that effective risk management should be proactive, it is fundamental to identify issues that can potentially impact a project and work to diminish their consequences rather than simply reacting as issues emerge. As such, specific responsibilities of the Risk Management Office may include:

● Define corporate Risk Management rules;
● Improve Risk Management approach throughout the project life cycle;
● Perform Project Risk Assessment (Project Health Check);
● Provide support to BU/Profit Center for Risk Management in bidding phase
● Provide support to BU/Profit Center for Risk Management at Portfolio level

In 2011, following the need to harmonize the risk management approach throughout the worldwide organization, a common Risk Register tool (refer Figure 15–51) was created and disseminated within the complete COMAU organization. After a period of training and tuning, the tools were included in the COMAU Project Management tools set. and included within the book, "Project and People Management – An Operational Guide" published in March 2013.

Starting from mid-2012, as part of the opportunity to define the new Global Project Management Process, the Risk Management Processes & Rules have been established

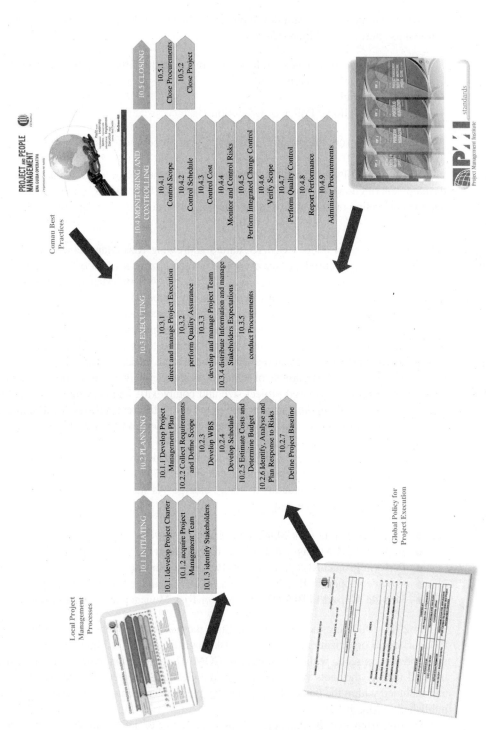

Local Project Management Processes

Coman Best Practices

Global Policy for Project Execution

**10.1 INITIATING**

10.1.1 develop Project Charter

10.1.2 acquire Project Management Team

10.1.3 identify Stakeholders

**10.2 PLANNING**

10.1.1 Develop Project Management Plan

10.2.2 Collect Requirements and Define Scope

10.2.3 Develop WBS

10.2.4 Develop Schedule

10.2.5 Estimate Costs and Determine Budget

10.2.6 Identify, Analyse and Plan Response to Risks

10.2.7 Define Project Baseline

**10.3 EXECUTING**

10.3.1 direct and manage Project Execution

10.3.2 perform Quality Assurance

10.3.3 develop and manage Project Team

10.3.4 distribute Information and manage Stakeholders Expectations

10.3.5 conduct Procurements

**10.4 MONITORING AND CONTROLLING**

10.4.1 Control Scope

10.4.2 Control Schedule

10.4.3 Control Cost

10.4.4 Monitor and Control Risks

10.4.5 Perform Integrated Change Control

10.4.6 Verify Scope

10.4.7 Perform Quality Control

10.4.8 Report Performance

10.4.9 Administer Procurements

**10.5 CLOSING**

10.5.1 Close Procurements

10.5.2 Close Project

**Figure 15—49.** Project management process sources.

**Figure 15–50.** Project management process paradigm

and included in the standard COMAU processes as well. In this environment, Risk Management Offices act as providers of training and tips to support the teams during the start-up of the Risk Management activities on the relevant projects.

**COMAU Approach to Contract Management**

COMAU established the requirement to have the support of a Contract Manager for the program/project team clearly emerged when the PMO was first launched in order to provide the team with an additional "point of view" not only focused on technical or managerial aspects.

Project Management is also assisted by Contract Management, which increases both customer satisfaction and supplier performance through the design of clear and well-balanced contract agreements. The goal is to guarantee efficient project execution by focusing on common expected project goals and utilizing competent, fair negotiation.

Contract management specialists, located in our main geographical business areas, combine legal and project management skills and support Sales and Project managers, as well as legal and finance staff. Competent in specific local laws and regulations, they work with all COMAU Business Units worldwide during the bidding, execution, and closing phases.

The purpose of the central Contract Management activity is to ensure the project is carried out as requested by the customer, and supported by the suppliers as per the joint contract agreements.

*Main Contract Management activities are:*
- Contract understanding
- Contract completeness

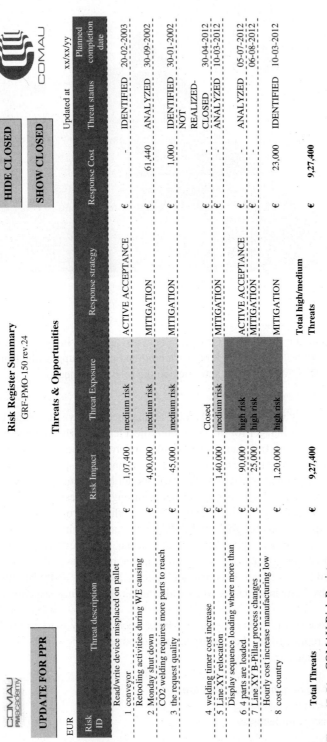

**Figure 15–51.** COMAU Risk Register

- Contract matching
- Contract modification
- Disputes and claims
- Contract closure
- Lessons learned

**COMAU Project Management Academy**

The first challenge the PM Academy had to deal with was its international nature: it had to provide a useful service for COMAU companies in all the countries. To reach colleagues all over the world, we decided to create a knowledge exchange network that would be enriched by their contributions. In fact, the Academy promotes the direct involvement of COMAU PM Experts to share knowledge and expertise. People performing day-to-day project management tasks have the best perspective regarding how to develop effective project management. This was the second challenge: the PM Academy should be a cooperation center. The teaching staff is made by qualified project managers who deliver training and support to project management professionals. In other words the COMAU PM Academy is "open": everyone in the company can ideally participate in it (it is just a question of skills and interest). The goal of this approach is to achieve mutual growth: the involved individual increases his skills and knowledge; the PM Academy receives a constant supply of new "vital blood."

For the people involved in it, the PM Academy offers the opportunity to "reinterpret" their own professional experience, allowing them to mature a stronger knowledge of the key points to be considered in managing complex projects. The Academy is also a "gym" for some of the competencies that are usually called "soft skills," such as managing interpersonal relationships, meetings, and presentations, or the ability to give and receive feedback and empathy Today the COMAU Project Management Academy offers training on several topics:

- Courses on the essentials of Project Management (for professionals who do not have direct responsibility for project management)
- Courses on the fundamentals of Project Management (for team leaders and junior Project Managers)
- Specialized courses (for experienced Project Managers)
- Direct cooperation activities to support projects and PM Family (seminars, workshops, coaching, mentoring, insights)

## The PM Academy and the PMP Certification

Since COMAU chose to adapt its own project management processes to PMI®'s standards and recommendations, the PM Academy's activities are PMI® compliant as well. So it was a natural progression for us to apply to become a PMI® Registered Education Provider. The PM Academy has been a R.E.P. since 2008.

Thanks to this recognition, our Academy can provide valuable training to support project managers to obtain the PMP certification. As a consequence, the Academy has had to face a new challenge: to deliver training not only to COMAU employees but also to customers. Today our educational activities are appreciated and requested because they are the result of a mix of qualified content, good methodology, and the concrete professional experience of our trainers (validated Project Management Professionals).

**Some Lessons Learned**　　The true success of a project is not only based on profitability, but also on knowledge that can benefit the entire corporation. COMAU has written down a list of lessons learned collected during its international projects.

- The ability to share best practices and organizational standards is even more important when working globally. It allows the exchange of information, workload sharing, resource leveling, etc.
- People's participation is critical. It is difficult to get people to understand the future benefits of managing a project and the importance of effective participation because it is related to personal knowledge, experiences, and empowerment. The ability to get people to participate is critical for change, speed, and buy-in.
- The development of "quick wins" is an important driver.
- Make people feel confident about the future is a buy-in driver. Today the importance of project management as a business solution for a global company is a strong driver for the PM Family.
- More effective communication, transparency and use of the "open door" philosophy are strong drivers.
- Developing the Project and Program Management approach is important but not enough; the whole company must understand the business model of management by projects.
- Education in effective leadership is a key issue on a daily basis.

The project management approach has demonstrated that other functional groups must be equally receptive to the acceptance of project management, thereby reducing barriers and creating better products.

In the second edition of my text *Strategic Planning for Project Management Using a Project Management Maturity Model*, I stated that the path to maturity can be accelerated with (1) the implementation of a PMO early on in the process; (2) having the PMO report directly to the executive levels of the company; and (3) having visible executive-level support for project management. Companies that accomplish all three of these items seem to outperform their competitors with regard to project management and are able to compete effectively in a global market. Such was the case with COMAU.

## 15.6 FLUOR CORPORATION: KNOWLEDGE MANAGEMENT FOR PROJECT EXECUTION[8]

**Knowledge Management for Project Execution at Fluor**

**Fluor Corporation Background**

Fluor Corporation is one of the world's largest publicly traded engineering, procurement, construction, maintenance, and project management companies. Over the past century, Fluor, through its operating subsidiaries, has become a trusted global leader by providing exceptional services and technical knowledge.

---

Consistently rated as one of the world's safest contractors, Fluor's primary objective is to develop, execute, and maintain projects on schedule, within budget, and with excellence. Fluor is a FORTUNE 500 company and is ranked No. 1 in the "Engineering, Construction" category of America's largest corporations. In addition, Fluor ranks No. 1 on ENR (*Engineering News-Record*) magazine's list of Top 100 Design-Build Firms and No. 2 on its Top 400 Contractors list. For Fluor's clients, this recognition emphasizes Fluor's ability to successfully execute large, financially complex projects around the globe.

Fluor's revenue in 2012 totaled a record $27.6 billion, up 18 percent from the previous year. New awards were $27.1 billion, and the company's business backlog at year end was $38.2 billion. Through their individual and collective expertise, Fluor's global workforce of more than 43,000 employees provides cost-effective, intelligent solutions in a timely manner.

Fluor maintains a network of offices in more than 25 countries across 6 continents. This workforce provides Fluor with the capability to execute diverse scopes of work on projects, both large and small, and the flexibility to staff projects in accordance with project needs.

Fluor serves clients in a wide variety of industries, including oil and gas, chemicals and petrochemicals, commercial and institutional, government services, life sciences, manufacturing, microelectronics, mining, power, telecommunications, and transportation.

Fluor is aligned into five principal operating segments: Energy & Chemicals, Industrial & Infrastructure, Government, Global Services and Global Supply Chain. Fluor projects include designing and building manufacturing facilities, refineries, pharmaceutical facilities, power plants, and telecommunications and transportation infrastructures. Many of Fluor's projects are the largest, most remote, complex and challenging capital projects in the world.

## Knowledge Communities in Support of Global Project Management

Fluor's business environment is global, mobile, cyclical, and collaborative. Employees are located across the globe and must work together closely for distributed project execution, when multiple offices around the globe work on the same project concurrently. Work force scarcity and mobility is a reality at Fluor and access to experts from anywhere around the world is critical to its business environment. In addition, due to the organization's aging work force, knowledge retention is an increasingly important issue.

Through our knowledge management capability, Fluor integrates and leverages the collective intellectual capital of our employees for project execution excellence. Our vast knowledge base, called Knowledge OnLine$^{SM}$, enables employees throughout our worldwide offices to access our corporate knowledge and contribute with their own knowledge and expertise. This system promotes collaboration and provides a systematic way to capture, share, and reuse the company's knowledge. To support global project execution, Fluor organizes around knowledge communities and its Knowledge OnLine technology platform allows distributing work without moving people around. Fluor's knowledge management technology platform is positioned as the delivery mechanism for all practices and procedures, training material, collaboration and expertise location. The overall strategic direction for knowledge retention, sharing, and collaboration tightly links with Fluor's business environment.

At Fluor, we define knowledge management as the way the organization identifies, creates, captures, acquires, shares, and leverages knowledge- it is a culture and management process, not a product or simply an IT installation. Knowledge communities are designated networks of people that share information and knowledge. Community members share, collaborate, and learn from one another face to face or virtually. Communities are held together by a common goal and desire to share experiences, insights, and best practices within a topic or discipline using shared norms and processes. Communities are also accountable for capturing best practices and stewarding a body of knowledge on behalf of the organization. Communities at Fluor are formally launched and have community leaders, managers and designated experts.

Fluor's knowledge management efforts have received several third-party awards for excellence, including the North American and Global Most Admired Knowledge Enterprise (MAKE) Awards, achieving "Hall of Fame" status for creating an environment for collaborative knowledge sharing. These awards recognize solutions that systematically leverage an organization's knowledge and people to measurably improve organizational responsiveness, innovation, competency and efficiency. In addition, Fluor's approach has been featured in the *Harvard Business Review* and the *Wall Street Journal*, and other books and articles.

Most of the targeted workforce is involved in one or more knowledge communities, sharing knowledge globally, enabling work processes, and bringing new people up to speed quickly. Fluor encourages knowledge-sharing behaviors across the organization. Any employee can join one or more knowledge communities, and can post a question or answer a question. Answers to forum questions are usually received within 24 hours of posting—in keeping with Fluor's promise of dependable and responsive communities— ensuring that all employees can have a high degree of confidence in the system.

Today, there are 49 established knowledge communities in the organization. Fluor has 30,000 active community members dispersed globally, and over 4,000 subject matter experts (SMEs) in over 1,200 subject areas within those communities. There is a high volume of community activity; there are more than 10,000 searches daily, 2,600 attachment views or downloads daily, and 10,000 forum reads on a weekly basis.

## Knowledge Management Strategies to Support Project Execution

Fluor has a strong history of knowledge sharing, much of which is attributed to a culture that consistently supports such behaviors. The passion to build careers and knowledge is at the heart of Fluor's business. The organization promotes a culture that is based on sharing knowledge across a network of employees. Leaders play an important role in sustaining this culture.

Fluor's formal knowledge management process began in 1999, when it launched its knowledge management strategy. Fluor used a two-pronged approach by implementing both a codification strategy and a personalization strategy:

- The *codification strategy* involves documenting the knowledge and storing it in a database. The emphasis here is on knowledge re-use.
- The *personalization strategy* focuses on connecting people to people, creating people networks, and emphasizing customized solutions to unique problems.

Fluor's aim is that new content should be moved quickly from a personalization focus into a codification focus in order to enable re-using the knowledge that is gathered.

One way that Fluor keeps communities relevant is by updating content on a regular basis. The organization implemented processes to ensure that communities frequently review and revise content. Each community is expected to take responsibility for its intellectual assets.

Why use this two-pronged approach? The knowledge management team realized that emphasizing explicit knowledge over tacit knowledge would not have worked because it would have been looked upon as "just another tool." On the other hand, focusing solely on the tacit knowledge would have resulted in a lack of documented knowledge and a lack of global, cross-functional impact.

### *Enterprise Projects, Enterprise Thinking*

Fluor's approach to knowledge management is highly unique, requiring enterprise thinking and a global mindset—an approach rarely seen in other organizations. This approach is taken to support their global projects. At Fluor, every employee has access to every community, a rigorous community deployment process is followed, community performance measurement and auditing programs are in place, and knowledge sharing behaviors are integrated into all aspects of company operations.

Fluor's enterprise wide knowledge management vision was to have one technology solution that included communities with integrated content, discussions, and people profiles to promote a global mindset. Leveraging the collective intellectual capital of all employees to support business strategic direction was a key objective. Communities were designed to provide optimal solutions to customers by sharing knowledge across geographic and business line boundaries, using a robust search capability, and allowing global access. Enterprise-wide KM was also targeted to enhance the skill sets of employees through ready access to knowledge, training material and by connecting them to expertise. Finally, enterprise-wide KM protects intellectual property in the organization. The system monitors activity to safeguard intellectual property.

### *Knowledge Communities: Where People Gather*

Knowledge communities are supported by leaders, knowledge managers, a global core KM team, and subject matter experts. A centralized KM team oversees community activities and works closely with leaders, knowledge managers, and subject matter experts across the enterprise.

Fluor utilizes a strong governance model, as shown in Figure 15–52, for community creation, defining community objectives and measuring performance. Fluor has observed other organizations which offer an ad hoc community creation model or offer a core KM Team developed set of templates to any group interested in creating a community. Experience has shown these open community creation models result in ineffective and redundant communities.

Fluor's enterprise-wide approach to knowledge management requires an expanded mindset for deploying and maintaining performing communities beyond what is required when the KM approach is targeted to a segment of the company, is regional, or is not open to all employees. Fluor applies four enterprise thinking concepts across its KM program:

1. ***Stewardship*** – Taking responsibility for the intellectual assets entrusted to the community (people and explicit content), Don't manage content that should be

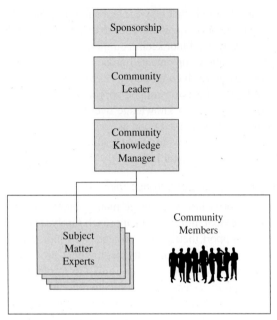

**Figure 15–52.**  Fluor governance model.

managed by another community, Report gaps – don't create it unless you really own it, If you own it – keep it up to date;

2. *Member Eye View* – Creating an environment that allows all employees to "get it" without expending excess effort. Something may make sense while self-contained in one community, but from a member's eye view (and across the enterprise), it can be confusing, so we help employees think about how to solve engineering and business problems, not how to decipher custom community (website) navigation;

3. *Client Credibility Test* – Expect that a client will see it, live demonstrations to potential customers are frequent;

4. *Community Franchise Expectations* – Support the Fluor KM brand, be a leader among community peers, leverage KM best practices across communities, create pride across all communities.

Knowledge communities do have flexibility: each community establishes its own objectives that are aligned with the overall organization strategy, but specific enough to define what the community of people needs to achieve in order to support the strategic direction. Community leaders are the highest authority within the function. In this leadership role, they are responsible for improving overall performance within their respective functions, selection, implementation and updates of best practices, selection and support of software tools unique to their function, and functional people development. A network of functional community leaders (Global Excellence Leaders) that create consistency across functions.

When launching a new knowledge community, each goes through a management process that includes the following concepts over a period of time determined by the urgency of the community leadership:

- Readiness assessment
- Community leadership strategy
- Community kickoff
- Community structure
- Content collection concepts
- Priority content identification
- Content collection update
- Launch strategy
- Community launch
- Deployment to performance transition
- Community performance plan
- Ongoing performance meetings
- Initial Deployment
- Performance

Within each community, a structure of categories is established based on core competencies, as stewardship of the explicit assets is critical. This includes creating a process to have an expert review and approve new content, expert review and update of content as it reaches a review date, and dependable and responsive expert answers to questions in discussion forums.

Community performance is measured by how well a community meets its objectives. Within the technology implementation, numerous statistics are collected. These statistics, while not a real measure of performance, provide an indication of activity levels and usage and track some levels of compliance.

## Expertise Forms the Backbone of the Knowledge Community

Real-world experience is no substitute for an application or algorithm. Fluor's network of subject matter experts (SMEs) forms the backbone of the company's knowledge efforts. Engaging these experts is critical to community, project and company success. These SMEs form the nucleus of every community. The frequency with which an SME participates in a community impacts the quality and quantity of community content. Experts are chosen by the leadership of each community—one can't just declare themselves and expert- they have to be recognized by others as such. SMEs receive KM process training that communicates critical expectations so they can function effectively. They subscribe for automatic discussion notifications and receive updates when documents require review or feedback. Aligning content and forums with areas of expertise makes it easier to automatically notify SMEs when their expertise is needed.

## Success Story: Access to Design Alternatives Saves Project 1 Million Euro

A process engineer in a start up office in the Middle East was looking for alternatives to a costly piece of equipment that was in the design basis of the project. He wanted to challenge the need for the piece of equipment, because it came at a very high total installed cost to the project.

Although he was in an office without all typical functional resources, through access to Fluor's global knowledge base, he was able to access design manuals, gain expert responses to forum questions, and past project references. This resulted in him being able to recommend the elimination of the expensive equipment, which saved the client 1,000,000 euro plus additional operational and maintenance costs.

The client was so pleased by the short response time, and the quick access to Fluor's worldwide knowledge and expertise, that a new 700,000 Euro work order was awarded for a similar study at another facility.

The SME Protégé Program is another platform designed to promote early engagement of future experts. The program pairs entry- and mid-level employees with senior-level experts and promotes accelerated learning. SMEs work with employees to create relationships, discuss project technical areas, develop new knowledge, review existing knowledge, and help answer forum questions.

### Leadership Is the Key to Success

Effective leadership is the best predictor of success for Fluor's knowledge communities. Knowledge leaders serve and communicate openly and frequently with communities and are visible role models of desired behaviors. Community leaders recognize that they are helping to create a knowledge sharing culture and facilitate people-to-people connections. Leaders also make sure it is "safe" for employees to ask questions, share openly, and trust the answers they receive. In addition, knowledge leaders recognize that a community is a human community, not a technology implementation. Leaders understand they are responsible for the human capital component of their communities and encourage lifelong learning.

It is also important that knowledge leaders acknowledge the talent and intelligence within the organization. Leaders inspire community members to combine experience and innovative thinking to create new knowledge, and they encourage the community to work smarter through collaboration.

Finally, leaders connect the community with strategic direction and drive for business results. They know that greater energy is available when people are connected through a sense of purpose.

### Work Processes for Knowledge Management Communications

An integral part of Fluor's knowledge management strategy has been effective communication. Fluor utilizes several communication principles to ensure effective and timely communications. Fluor believes it is important to leverage all the communication channels within the organization and proactively look for opportunities to communicate about knowledge management and how it enables project execution

Fluor has numerous communication strategies in place in support of knowledge retention and transfer. A few of these include:

New hire orientation. Employees are exposed to Fluor's knowledge sharing culture during recruitment and new-hire orientation. Specifically, the new-hire orientation materials challenge employees to share knowledge across Fluor's global networks in order to succeed in the organization. Since career mobility is enhanced by community participation, many employees quickly engage in community activities.

- Knowledge Manager Training
- Global teleconferences.
- Department meetings.
- Lunch-and-learn sessions.
- Online community front-page success stories which change every two to three days.

Another work process improvement is linked to the organizational communication strategy. The old practice was to disseminate information via the organizational hierarchy. However, not everyone saw those communications. Now, communities send messages directly to the entire community membership. As a result, messages have a broader readership (greater penetration), which keeps people "in the know" and helps attract new members. Each message is sent as an e-mail with a link to supporting information in the knowledge community. The link approach is for tracking and sometimes, for compliance purposes. Employees frequently reply to messages and are encouraged to join a particular community as part of a routine follow-up.

## Recognizing KM Pioneers . . . and Beyond

Since Fluor began on its knowledge sharing in support of global projects path, it recognized that it had to recognize those so instrumental to its success and adoption. Early on, it recognized "KM Pioneers" those that looked outside of normal geographical or business line boundaries to see how knowledge sharing and collaboration could be institutionalized globally. Now, Fluor has two successful recognition programs in place. The first program is for peer recognition of knowledge sharing behaviors, and the second program recognizes the value generated through knowledge sharing and collaboration—this campaign helps Fluor to tell the stories of why knowledge sharing and collaboration is important not just for Fluor, but for individuals.

Fluor developed an annual knowledge management peer recognition award known as the KM Pacesetter. Peers nominate one another year round for this prestigious award. A KM Pacesetter is someone who excels in knowledge-sharing behaviors. In 2012 alone there were over 1,000 nominations from Fluor employees all over the world and to date hundreds of people have been recognized as KM Pacesetters. Fluor feels this campaign is successful because it is employee driven—employees get to recognize the good work their peers are doing, and in turn the company is able to highlight them as KM Pacesetters.

The second campaign is an annual KM success story contest as part of an annual "Knowvember" celebration. The KM Team helps gather KM success stories that help to articulate the value that is being brought to people and projects around the world. These stories are varied, and have different value propositions. This is important, as there is no "one size fits all" value proposition when it comes to a workforce of 30,000. The secret to the value of the success stories is that usually, there is something that resonates with everyone and they can share the success with their project team or replicate it themselves on their project.

"Knowledge sharing is now clearly an institutionalized part of our culture, whereby value is created for our Clients through an open, transparent and collaborative structure, which is interestingly enough very reflective of today's modern and successful societies," Peter Oosterveer, Group President, Energy & Chemicals.

### Knowledge Management in the Context of Project Management

When knowledge management is considered in the context of project management, there are two high level areas that must be addressed:

- Knowledge sharing and collaboration across projects
- Knowledge sharing and collaboration within a project

Each area has unique aspects and opportunities to deliver value to the project and the organization. Figure 15–53 provides a view of these two types of knowledge sharing and collaboration.

#### Knowledge Sharing across Projects

Project execution organizations develop a considerable base of knowledge gained from the number of past projects executed. Some of this knowledge is explicit and is leveraged through a codification strategy. Other knowledge is tacit and requires a personalization strategy to tap the potential value.

Documented knowledge takes many forms, including project completion reports; practices, guidelines, lessons learned and other content that provide the framework for project execution processes and can be stewarded in a readily accessible sharing platform. Other examples of documented knowledge include risk mitigation reports, unique

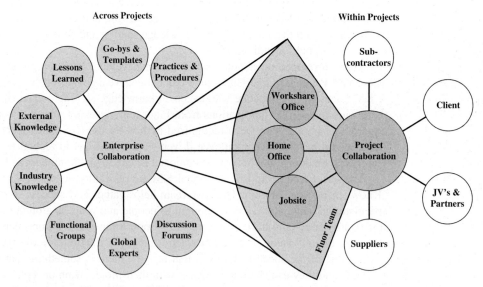

**Figure 15–53.** Knowledge collaboration.

solutions developed on the project, analysis reports of alternatives and recommendations; and value awareness suggestions.

A significant amount of accumulated knowledge however is tacit and based on personal experience that employees gain over many years on many projects in many locations for many clients. This knowledge resides inside the heads of the organization's most experienced people. It is not written down, and even if it were, the context and interpretation would make it very difficult to use properly. Unless this knowledge is brought to bear on new projects it is, in effect, a company asset that goes untapped.

To be successful with knowledge sharing across projects, content and expertise must also be accessible in ways that meet the needs of the project participants seeking the knowledge.

The goal of knowledge sharing knowledge across projects is to continually improve project performance through the application of knowledge and experience acquired during execution of prior projects. For employees, knowledge sharing across projects creates opportunities to learn from peers around the world, new employees can be brought up to speed quicker, and a sense of confidence is developed knowing that the global community is available and willing to assist.

As an example of knowledge sharing across projects, a recent engineering graduate was hired by Fluor and was sent to a jobsite. One day he was challenged by the client and site manager. At a recently installed pipe rack column, one of the mounting studs did not stick up high enough to properly tighten the nut and washer. This was a never before seen situation for the new employee, but his reply was, "don't worry; I'll have an answer by the end of the day." He went back to the construction trailer, typed his problem into a discussion forum and subject matter experts provided the necessary procedures within one hour. Instilling this type of confidence in all employees, not just new grads, is a huge benefit when knowledge is readily shared across projects.

As the organization has recognized the value of global collaboration, many examples of work process innovations have emerged. For instance, knowledge management is integrated into Fluor's operating system requirements, which is the overall quality program for the company. All of Fluor's key operating documents are delivered through, and maintained in, Knowledge Online. These work process documents are updated frequently, based on project experience and changes to industry codes. During the update process, the proposed changes are posted in the appropriate knowledge community and any employee can make suggestions.

### Knowledge Sharing within Projects

Collaboration and knowledge sharing within a project is as important as sharing across projects. This type of collaboration is different in that the parties involved can include the contractor, the client, joint venture partners, suppliers, fabricators and other organizations. Requests for information (RFI's), interface management, vendor document reviews, design and progress reviews are examples of interactions that benefit from knowledge sharing and collaboration within a project.

Challenges that may be encountered when trying to share knowledge and collaboration within a project include situations where each organization may have their own KM technology platform, a culture for knowledge sharing may not be present, and concerns regarding intellectual property. The overall goal however should be that each organization that makes up the extended project team not only assigns individual members, but also has the capabilities to bring the combined intellectual assets of their organization to the project. As new knowledge is developed through execution of the project, as applicable, it should be available for reuse by each organization.

Communication and knowledge sharing approaches should be part of an alignment session when a project kicks off and when fabricators or suppliers are brought on board.

### Key Success Factors

Research into knowledge management initiatives indicate a very high number fail to deliver on expectations—as high as 80%. With over a decade of successful knowledge sharing and collaboration, Fluor has applied the following key success factors to guide their knowledge management activities.

*Align KM Objectives with Business Strategy*   Knowledge sharing and collaboration should be applied to a business challenge or strategic opportunity with a business mindset. Simply standing up a knowledge management or collaboration platform without solid strategic reasoning is a leading cause for knowledge management failures. The organization is more likely to adopt and demonstrate desired knowledge sharing behaviors when clear business needs and expectations are defined and communicated.

*Employees Form the Core of Fluor's Knowledge-Based Service Strategy*   Enthusiastic people, active, supportive and involved leadership, and strong people networks are critical to success. Knowledge retention and transfer behaviors must span the employment lifecycle from recruitment through retirement. It is equally important to define and communicate knowledge sharing expectations and to ingrain desired knowledge sharing behaviors into the organization culture.

*Use Technology to Support the Global Community*   While successful knowledge management programs place a strong emphasis on people, connecting a globally distributed workforce requires a robust technology platform. To minimize the learning curve and user confusion, creating an enterprise-wide, "one stop shop" to connect employees regardless of location or time differences eases adoption.

*Emphasis on Communications*   To sustain communities, it is important to make sure that early success stories become daily occurrences. Strong community sponsorship must communicate value on a regular basis and promote community engagement.

*Share Knowledge across Boundaries*   Incorporate enterprise-wide thinking into your knowledge management strategy. Make sure your knowledge management processes engage the global community at the outset.

*Leadership Has a Direct Influence on Knowledge Management Success*   Knowledge management success is directly tied to leadership support. Project managers are in a position where their visible support directly influences the strength of knowledge sharing and collaboration both on their specific projects and across the whole organization. Project managers in particular need to leverage innovations, encourage project team members to seek out expert advice, encourage members to submit new knowledge, and should recognize the talent and intelligence within the organization.

Knowledge management allows project managers to leverage the best knowledge and expertise of their organization, across the boundaries of time and space. KM informs project team members, helping them make better decisions and improving the final results of the project.

## Future Directions for Knowledge Management on Projects

When you have successfully applied KM and achieved measurable results—where do you go next? Consider the following areas of opportunity:

### Anticipatory Delivery of Knowledge

When project teams become comfortable sharing and leveraging knowledge, the project benefits. The responsibility for knowledge sharing today still falls largely on the shoulders of each project team member, accessing knowledge or expertise when a need arises. How can you manage the process, anticipating the need for knowledge and expertise? How do you ensure that the project leverages the best available knowledge, regardless of the initiative of the individual project team member?

The future lies in the *anticipatory delivery of knowledge*. This starts with a proactive approach to delivery using a facilitated knowledge-assist session—sometimes called a *peer assist*.

### The Knowledge Assist

The knowledge assist is planned, facilitated session to bring together all of the potential project resources for the purpose of sharing experiences and knowledge with the project team prior to project execution kickoff. Results of the session are documented with identifiable actions or suggestions to be used and followed. The goal is to learn before doing.

A successful knowledge assist requires planning and preparation. You need to have the right people available, and hold the assist at the proper time, for the value of the session diminishes over time. The earlier you can hold a knowledge assist session, the greater the potential positive impact on the project.

---

### Anticipatory Delivery of Knowledge—The Project Manager's Role

In order to learn before doing, the project manager needs to reach out and get the right people involved in the knowledge assist process from the start. The greatest opportunity for impact is at the beginning, and leveraging the experience from past projects gives your project a head start. The project manager must be open to outside ideas, and be willing to apply them to the specific project requirements.

---

### Connections beyond the Project Team

Most knowledge is *tacit*. It is the hidden know-how that sits in the heads of each project team member and other outside resources. Most of this know-how will never be exposed in a document or online source. It surfaces at the intersection of need and experience. At that intersection, a transaction occurs—an exchange of knowledge from one person to another. People connecting with each other remain the most important conduit for knowledge transfer on projects.

When managing a project, there are several emerging technologies that can help facilitate those people-to-people connections. Purposeful social computing, using social collaboration tools allows people to connect across boundaries of time and space, leveraging the best available know-how. And enabling these connections to occur where people are, using whatever device they are on. In many cases, this will mean via a wireless mobile or tablet device.

---

**Connections beyond the Project Team—The Project Manager's Role**

When you allow project team members to engage their extended networks, participating and collaborating through social business systems, your project benefits. Your team members are able to leverage that network for better informed decision making on the project.

---

*Knowledge in Context*

Starting the project well-informed by a knowledge assist session is essential, but not the only point where knowledge and expertise are needed. As the project team grows comfortable with sharing and leveraging knowledge, they will take the initiative, reaching out or sharing when needed. The challenge is to get full participation. To do this, you need to have context about the task at hand. As we know the task at hand for a given project team member, we can begin to deliver the knowledge and expertise related to that task. This *contextual delivery of knowledge* informs all project team members and helps ensure that are fully informed.

Contextual delivery of knowledge relies on an understanding of a project team member's current activity, and the ability to connect related knowledge and expertise in the stream of their work process. Leveraging and sharing knowledge does not involve a conscious decision or an extra step; it is embedded within the activities that they are already doing.

For example, an engineer working on a stair tower design could quickly connect to past examples; connect with other experts or reference industry codes and practices to inform their design decisions.

---

**Knowledge in Context—The Project Manager's Role**

Set expectations with your project team, encouraging them to identify weaknesses in current processes that could be improved by an injection of knowledge or expertise. Challenge them to look for ways to improve the process, especially when that makes leveraging knowledge and expertise more transparent.

---

*Accelerated Expertise Development*

On January 1, 2011 the first Baby Boomers reached retirement age. Since that day, over 10,000 Boomers reach retirement age every single day. This pace will continue through the year 2030. The risk of knowledge loss due simply to a retiring workforce is tremendous, but can be mediated.

In the book *Outliers*, Malcolm Gladwell suggests that ". . . the idea that excellence at performing a complex task requires a critical minimum level of practice surfaces again and again in studies of expertise. In fact, researchers have settled on what they believe is the magic number for true expertise: ten thousand hours." There are no shortcuts to expertise, but there are ways to apply attention to the issue by encouraging person to person transfer of knowledge.

The first step is to identify a career path that acknowledges and encourages expertise growth. Recognizing that your project team may work together on future projects, consider the opportunities to stretch and develop team members.

For the current experts, pair them up with mentees. The aspiring protégé can assist the expert, taking some of the workload from a highly valued, in-demand team member. And it provides a clear opportunity for knowledge transfer while developing the next generation of experts.

---

### Accelerated Expertise Development—The Project Manager's Role

Project managers need to consider the development of project team members, both for their career satisfaction and growth, but also for the benefit of your project and future projects. Knowledge transfer still largely occurs through the development of expertise.

---

Knowledge management gives project managers the ability to leverage the best available expertise, often including resources outside of the immediate project team. By learning from other projects, your project gets a head start towards success. The key is to start early, and provide clear leadership to your team. Define the expectations, and encourage them to participate in sharing knowledge and expertise within the project, as well as across project boundaries.

Your understanding and support of knowledge management provides the necessary foundation for smarter project execution.

## 15.7 SIEMENS PLM SOFTWARE: DEVELOPING A GLOBAL PROJECT MANAGEMENT METHODOLOGY

For decades, large companies have allowed their multinational divisions tremendous autonomy in the way they do business. This works well as long as the various units do not have to interact and work together on projects. But when interaction is required, and each division has a different approach to project management—using different tools and processes—unfavorable results can occur. Today, the trend is toward the development of an enterprise-wide methodology. Siemens PLM Software is an example of a company that has successfully developed such a methodology.[9]

---

9. The remainder of this section has been provided by Jan Hornwall, Global Services PMO, Siemens PLM Software. For more information on Siemens PLM Software products and services, visit www.siemens.com/plm. Note: Siemens and the Siemens logo are registered trademarks of Siemens AG.

**About Siemens PLM Software** Siemens PLM Software, a business unit of the Siemens Industry Automation Division, is a leading global provider of product life-cycle management (PLM) software and services with 5.9 million licensed seats and 56,000 customers worldwide. Headquartered in Plano, Texas, Siemens PLM Software works collaboratively with companies to deliver open solutions that help them turn more ideas into successful products.

**About Siemens Industry Automation Division** The Siemens Industry Automation Division (Nuremberg, Germany) is a worldwide leader in the fields of automation systems, low-voltage switchgear and industrial software. Its portfolio ranges from standard products for the manufacturing and process industries to solutions for whole industrial sectors that encompass the automation of entire automobile production facilities and chemical plants. As a leading software supplier, Industry Automation optimizes the entire value added chain of manufacturers—from product design and development to production, sales, and a wide range of maintenance services. With around 42,900 employees worldwide Siemens Industry Automation achieved in fiscal year 2008 total sales of EUR 8.7 billion.

**Abstract** A methodology has been developed by Siemens PLM Software, a business unit of the Siemens Industry Automation Division and a leading global provider of product lifecycle management (PLM) software and services, which includes project and program management, technical activities and project governance. It is based on the use of an internal web site allowing quick access for all employees and has been successfully deployed and taught globally. The remainder of this section describes the background of the project methodology and identifies best practices for similar efforts in the future.

**Project Background** Siemens PLM Software develops and deploys enterprise software for Product Lifecycle Management which includes solutions for computer aided design, manufacturing and engineering analysis (CAD/CAM/CAE) as well as for data management, collaboration, and digital factory simulation. The company's global sales and services organization is responsible for configuring and deploying the solutions at customer sites. Engagements can range from small projects over a few man-months to large multiyear global programs with hundreds of people. These projects have been executed against several different methodologies at locations worldwide. Due to the increasingly global nature of manufacturing companies and the increased demand for a variety of subject matter experts from around the world, the initiative to create a single global methodology for Siemens PLM Software was started.

**Business Benefits** The following business benefits drove the development of the new methodology:

- Sharing best practices and good examples across projects and geographies
- Accelerating project deployment through quick access to tools, guides, templates and best practices

- Establishing a common methodology "language" that is used across all geographies and projects
- Sharing resources around the world and quickly developing new hires and external employees
- Enabling increased repeatability and predictability resulting in reduced risk and faster delivery times
- Providing a structured project/program governance framework
- Increasing reuse of information in projects; laying the foundation for knowledge management
- Presenting a unified and consistent project management experience for global customers

**Methodology Development**     Once the initiative was started, the first decision was to either develop our own methodology or procure an off-the-shelf PM methodology.

Management, together with the project team, decided on developing our own methodology based on existing in-house experience. The key decision criteria was that the methodology must cover not only the project management activities but also the technical activities specific to our business. It was crucial to work on both the project management and the technical tracks together in each phase since quite a few projects are done in small teams and sometimes even the project manager has the dual role of also being the chief technical solution architect. Another key criteria was that we wanted to leverage what we already had in terms of processes and templates. We also anticipated a much higher adoption rate if the people in the field recognized parts of the methodology.

The project was planned and a project management plan was written and signed off by the project sponsor. The project team consisted of key persons from all zones; Americas, Europe, Asia-Pacific and our in-house offshore team in India who implemented the methodology web site. In total the core team consisted of approximately ten people.

The scope of the project was developed and included the following:

- The methodology must cover the entire lifecycle of a service project, from it's [*sic*] inception to close down. In addition, it must contain program management and governance.
- General PM activities will be included in a section which is the same for all phases; "Manage Project."
- Technical activities will cover methods on how to identify out-of-the-box solutions while keeping customer customizations to a minimum; as well as modern techniques such as rapid prototyping and iterative development.
- The activities in each phase are structured and had to be described; responsibilities of each task had to be defined; and template guidelines and supporting tools had to be available
- The methodology had to be aligned with surrounding processes such as sales process and post project support process.
- Consolidate various existing service delivery methods, leveraging recognized best practices already existing within the company.

- The project management track must be aligned with PMI (Project Management Institute) process and terminology,
- Start with a comprehensive methodology, then a "small projects" version has to be developed.
- Technology would include a web site with content management and a feedback tool with tracking functionality
- A downloadable version must be available for all employees working outside the internal network, such as in the defense industry.
- Training and deployment globally

The team worked virtually through conference calls but also had three face-to-face meetings in various locations around the globe, each four to five days long. The project to develop the methodology lasted ten months. If we had been all in the same location the time would have been significantly shorter.

At the start of the project, Siemens PLM Software was known as UGS, an independent and privately owned company. By the end of the project UGS was acquired by Siemens and renamed Siemens PLM Software. It was kept as an intact business unit. Since the methodology is tailored to our PLM business, management decided to continue the project and deploy the methodology. Alignment with the mandatory Siemens project management aspects would take place in consecutive releases.

**Resulting Methodology—PLM VDM—Description**   The product lifecycle management value delivery methodology (PLM VDM) provides a structured process for delivering a PLM solution. (See Figure 15–54.) PLM VDM emphasizes the unique aspects of delivering an enterprise wide solution using Siemens PLM Software products and has been adopted across the Siemens PLM Software services organization.

PLM VDM encompasses both project management and technical delivery work streams. It is structured to allow iterative and flexible project delivery while maintaining "quality gates" and milestones between phases.

The seven methodology phases are:

**Pre-Align**   The purpose of the Pre-Align phase is to gain a sufficient understanding of customer requirements and the scope of the project, to be able to define the high level solution outline and statement of work.

The pursuit team works with sales and the customer to establish the overall project scope, determine a preliminary project schedule, define the services strategy, conduct an infrastructure assessment and develop the initial project budget.

**Align**   In the Align phase, the project team works with the customer to transform the solution concepts that were defined during the Pre-Align activities into a well defined overall solution architecture.

- The objectives of the Align phase are to establish a common understanding between the customer and the implementation team on all aspects of the project by

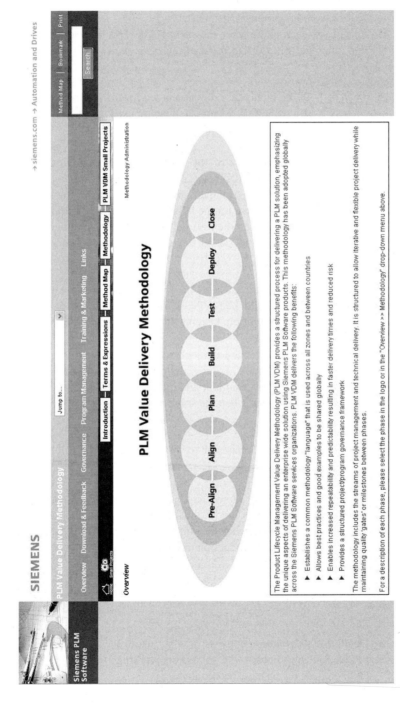

**Figure 15–54.** The PLM VDM internal web site.

capturing a complete and accurate project definition through technical workshops, use case definition, rapid prototyping and aligning the solution requirements to the "out of the box" product capabilities.

● This phase is complete when the customer accepts the use cases and requirements and authorizes the work to proceed.

**Plan**

In the Plan phase the project team works with the customer to develop the remaining documents that are used to execute and control the project and to develop the technical design.

● Depending on complexity of the solution, the team defines detailed plans for scope, schedule, cost, skills, resources, risks, quality and communication.
● In addition to the test environment, the team "baselines" the system infrastructure to create a stable platform for the development, test and training environments.
● This phase is complete when all necessary project management plans and the required functional and design specifications have been reviewed and base lined.

**Build**

In the Build phase the team works with the customer to create the defined solution, keeping strict adherence to the requirements.

● During the Build phase the technical team configures and tests the solution, implements the data migration strategy and develops the training materials. The Build phase also includes internal unit and integration testing.
● This phase is complete when the solution is ready for customer testing.

**Test**

In the Test phase the team validates that the solution is ready for production use.

● During the Test phase, representatives from the user community perform functional and system tests to verify that the system fulfills the requirements.
● This phase is complete when the solution is accepted by the customer and is ready for deployment into a production environment.

**Deploy**

In the Deploy phase the team delivers the production-ready solution to the end users.

● Deploying the solution consists of ensuring all the data has been migrated to the production environment, the solution is working with all interfaces and the users and helpdesk teams have been trained.
● The Deploy phase completes when the solution has been turned over to the customer for production use.

**Close**

In the Close phase the team assures that all administrative aspects of the project are complete.

- During the Close phase the project team completes and archives project documents and conducts project retrospective to capture and document lessons learned. The project team is released

The additional sections of the methodology are:

**Program Management**

Following the PMI standard, the five phases of program management are described including supporting templates:

- Preprogram Setup
- Program Setup
- Establishing Program Management & Technical Infrastructure Setup
- Delivering Incremental Benefits
- Closing the Program

**Small Projects**

Projects smaller than US$100,000 in total revenue can select a simplified methodology, with shorter activity descriptions and simplified templates.

**Governance**

Project Governance is about the line organization making sure that the project is governed correctly and ensuring efficient and effective decision-making, steering the project to success. This is done by ensuring that the following is in place:

- Project Charter
- Assigning Project Manager Authority
- Project Steering Board
- Management Review Board (MRB)
- Technical Review Board (TRB)
- Project Healthchecks and Project Retrospectives
- Approval process for new projects

Best practices are described in these sections together with supporting templates. This is an area where the alignment with the mandatory Siemens processes is done.

**Training and Marketing**

This important section covers material for training, release updates, links to PLM VDM training on the internal training site, presentations of the methodology internally and for customers.

**Launch and Deployment**     The launch and deployment of the methodology globally included the following activities:

- Announcement at conferences in each of the geographies. Management reserved time for the methodology to be presented which sent a positive and powerful message to all employees.
- Developing an one hour overview of the methodology as a voice-over presentation. This was made available on the in-house training infrastructure and participants could view this globally over intranet.
- Developing training material for a two day classroom training. This training was then given in many classes on four continents to approximately 600 persons over six months. This training covered all roles and included exercises.
- Additional live presentations in conference calls for several hundred additional people
- Developing marketing collateral; factsheet for the methodology—same format as for our products, developing a logotype and customer presentations
- Follow-up activities; monitoring adoption through KPIs and acting on users' feedback

**Lessons Learned and Best Practices**     Developing a methodology in a remote/virtual team is possible but plan for several face to face meetings. These meetings were crucial to the success of the project; to bring all together in one room for longer workshops, divide work and make break-out sessions, collect all again and take final decisions. It is also important to get to know each other and have fun in off-hours—this makes working remotely afterwards much more efficient. These meetings also served as a recognition of the contribution made by all.

Involve key persons in the various geographies in the development early on. It takes much longer than a pure top-down approach by a central PMO but these persons became champions in each geographic location during the deployment and significantly increases long-term adoption.

If you need a pure project management methodology, consider procuring an off-the-shelf product.

If you need a business specific project methodology, consider developing your own. Most often there is tremendous knowledge in the company, it is just a matter of getting hold of it, put it in writing and making it available globally within the company

To get management buy-in is crucial; to secure the active participation of key persons, to get slots at conferences for the announcement and to get the logotype of the methodology visible in top management presentations which is essential for adoption.

Don't underestimate the time it takes to develop training material and giving the training globally. The success of this methodology is thanks to all the hard work done by the dedicated people successfully completing these tasks!

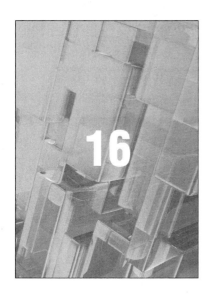

# Value-Driven Project Management

## 16.0 INTRODUCTION

Over the years, we have come to accept the traditional definition of project success, namely meeting the triple constraint. More recently, we modified our definition of success by stating that there must be a valid business purpose for working on the project. Success was then recognized as having both a business component and a technical component.

Today, we are modifying the definition of success further by adding in a "value" component, as seen in Figure 16–1. In others words, the ultimate purpose of working on the project should be to provide some form of value to both the client and the parent organization. If the project's value cannot be identified, then perhaps we should not be working on the project at all.

Value can be defined as what the stakeholders perceive the project's deliverables as being worth. Each stakeholder can have a different definition of value. Furthermore, the actual value may be expressed in qualitative terms rather than purely quantitative terms. It simply may not be possible to quantify the actual value.

The importance of the value component cannot be overstated. Consider the following statements:

Completing a project on time and within budget does not guarantee success if you were working on the wrong project. Having the greatest enterprise project management methodology in the world cannot guarantee that value will be there at the end of the project.

- Completing a project on time and within budget does not guarantee that project will provide value at completion.

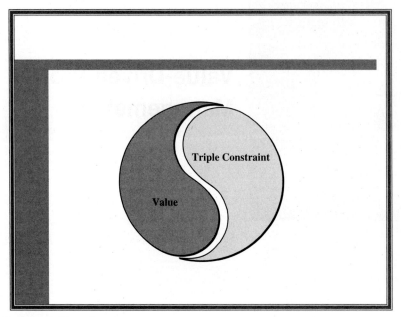

**Figure 16–1.** Definition of success.

These three statements lead us to believe that perhaps value is now the dominating factor in the selection of a project portfolio. Project requestors must now clearly articulate the value component in the project's business case or run the risk that the project will not be considered.

## 16.1 VALUE OVER THE YEARS

Surprisingly enough, numerous research on value has taken place over the past 15 years. Some of the items covered in the research include:

- Value dynamics
- Value gap analysis
- Intellectual capital valuation
- Human capital valuation
- Economic value-based analysis
- Intangible value streams
- Customer value management/mapping
- Competitive value matrix
- Value chain analysis
- Valuation of IT projects
- Balanced scorecard

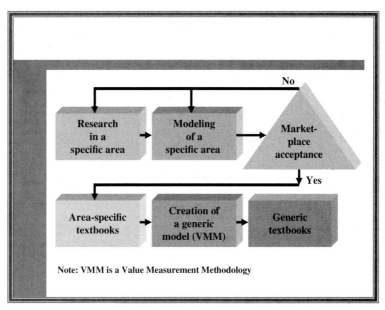

Note: VMM is a Value Measurement Methodology

**Figure 16–2.** Evolution of value-based knowledge.

The evolution of value-based knowledge seems to follow the flowchart in Figure 16–2. Research seems to take place in a specific research area such as calculating the value of software development projects or calculating shareholder value. The output of such research is usually a model that is presented to the marketplace for acceptance, rejection, and/or criticism. Soon others follow with similar models but in the same research area such as software development. Once marketplace acceptance concurs on the validity of these models, textbooks appear discussing the pros and cons of one or more of the models.

With the acceptance of modeling in one specific area, modeling then spreads to other areas. The flowchart process continues until several areas have undergone modeling. Once this is completed, textbooks appear on generic value modeling for a variety of applications. The following list contains some of the models that have occurred over the past 15 years:

● Intellectual capital valuation
● Intellectual property scoring
● Balanced scorecard
● Future Value Management™
● Intellectual Capital Rating™
● Intangible value stream modeling
● Inclusive Value Measurement™
● Value performance framework
● Value measurement methodology (VMM)

There is some commonality among many of these models such that they can be applied to project management. For example, Jack Alexander created a model called "value performance framework" (VPF). The model is shown in Figure 16–3. The model focuses on building shareholder value rather than creating shareholder value.[1] The model is heavily biased toward financial key performance indicators. However, the key elements VPF can be applied to project management as shown in Table 16–1. The first column

**Figure 16–3.** The VPF model. *Source:* J. Alexander, *Performance Dashboards and Analysis for Value Creation*, Hoboken, NJ: Wiley, 2007, p. 5. Reproduced by permission of John Wiley & Sons.

**TABLE 16–1.   APPLICATION OF VPF TO PROJECT MANAGEMENT**

| VPF Element | Project Management Application |
|---|---|
| Understanding key principles of valuation | Working with the project's stakeholders to define value |
| Identifying key value drivers for the company | Identifying key value drivers for the project |
| Assessing performance on critical business processes and measures through evaluation and external benchmarking | Assessing performance of the enterprise project management methodology and continuous improvement using the PMO |
| Creating a link between shareholder value and critical business processes and employee activities | Creating a link between project values, stakeholder values, and team member values |
| Aligning employee and corporate goals | Aligning employee, project, and corporate goals |
| Identifying key "pressure points" (high leverage improvement opportunities) and estimating potential impact on value | Capturing lessons learned and best practices that can be used for continuous improvement activities |
| Implementing a performance management system to improve visibility and accountability in critical activities | Establishing and implementing a series or project-based dashboards for customer and stakeholder visibility of key performance indicators |
| Developing performance dashboards with high-level visual impact | Developing performance dashboards for stakeholder, team and senior management visibility |

*Source:* J. Alexander, *Performance Dashboards and Analysis for Value Creation*, Hoboken, NJ: Wiley, 2007, p. 6. Reproduced by permission of John Wiley & Sons.

1. J. Alexander, *Performance Dashboards and Analysis for Value Creation*, Hoboken, NJ: Wiley, 2007, p. 5. Reproduced by permission of John Wiley & Sons.

contains the key elements of VPF from Jack Alexander's book and the second column illustrates the application to project management.[2]

## 16.2 VALUES AND LEADERSHIP

The importance of value can have a significant impact on the leadership style of project managers. Historically, project management leadership was perceived as the inevitable conflict between individual values and organizational values. Today, companies are looking for ways to get employees to align their personal values with the organization's values.

Several books have been written on this subject and the best one, in this author's opinion, is *Balancing Individual and Organizational Values* by Ken Hultman and Bill Gellerman.[3] Table 16–2 shows how our concept of value has changed over the years.[4] If you look closely at the items in Table 16–2, you can see that the changing values affect more than just individual versus organization values. Instead, it is more likely to be a conflict of four groups, as shown in Figure 16–4. The needs of each group might be:

- Project manager
  - Accomplishment of objectives
  - Demonstration of creativity
  - Demonstration of innovation
- Team members
  - Achievement
  - Advancement
  - Ambition
  - Credentials
  - Recognition
- Organization
  - Continuous improvement
  - Learning
  - Quality
  - Strategic focus
  - Morality and ethics
  - Profitability
  - Recognition and image
- Stakeholders
  - Organizational stakeholders: job security
  - Product/market stakeholders: high-quality performance and product usefulness
  - Capital markets: financial growth

---

2. Ibid., p. 6.

3. K. Hultman and B. Gellerman, *Balancing Individual and Organizational Values*, San Francisco: Jossey-Bass/ Pfeiffer, 2002.

4. Ibid., pp. 105–106.

**TABLE 16–2.   CHANGING VALUES**

| Moving Away From: Ineffective Values | Moving Toward: Effective Values |
| --- | --- |
| Mistrust | Trust |
| Job descriptions | Competency models |
| Power and authority | Teamwork |
| Internal focus | Stakeholder focus |
| Security | Taking risks |
| Conformity | Innovation |
| Predictability | Flexibility |
| Internal competition | Internal collaboration |
| Reactive management | Proactive management |
| Bureaucracy | Boundaryless |
| Traditional education | Lifelong education |
| Hierarchical leadership | Multidirectional leadership |
| Tactical thinking | Strategic thinking |
| Compliance | Commitment |
| Meeting standards | Continuous improvements |

*Source:* Adapted from K. Hultman and B. Gellerman, *Balancing Individual Organizational Values*, San Francisco: Jossey-Bass/Pfeiffer, © 2002.

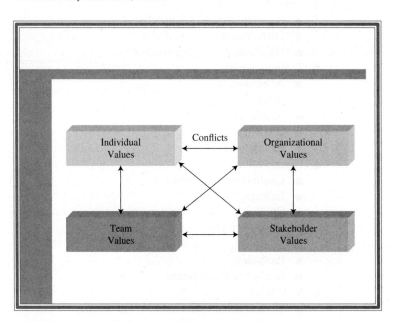

**Figure 16–4.**   Project management value conflicts.

There are several reasons why the role of the project manager and the accompanying leadership style have changed. Some reasons include:

● We are now managing our business as though it is a series of projects.
● Project management is now viewed as a full-time profession.

- Projects manager are now viewed as both business managers and project managers and are expected to make decisions in both areas.
- The value of a project is measured in business terms rather than solely technical terms.
- Project management is now being applied to parts of the business that traditionally have not used project management.

The last item requires further comment. Project management works well for the "traditional" type of project, which includes:

- Time duration of 6–18 months.
- The assumptions are not expected to change over the duration of the project.
- Technology is known and will not change over the duration of the project.
- People that start on the project will remain through to completion.
- The statement of work is reasonably well defined.

Unfortunately, the newer types of projects are more nontraditional and have the following characteristics:

- Time duration over several years.
- The assumptions can and will change over the duration of the project.
- Technology will change over the duration of the project.
- People that approved the project may not be there at completion.
- The statement of work is ill defined and subject to numerous changes.

The nontraditional types of projects have made it clear why traditional project management must change. There are three areas that necessitate changes:

- New projects have become:
  - Highly complex and with greater acceptance of risks that may not be fully understood during project approval
  - More uncertain in the outcomes of the projects and with no guarantee of value at the end
  - Pressed for speed-to-market irrespective of the risks
- The statement of work (SOW) is:
  - Not always well-defined, especially on long-term projects
  - Based upon possibly flawed, irrational, or unrealistic assumptions
  - Inconsiderate of unknown and rapidly changing economic and environmental conditions
  - Based upon a stationary rather than moving target for final value
- The management cost and control systems [enterprise project management methodologies (EPM)] focus on:
  - An ideal situation (as in the *PMBOK® Guide*)
  - Theories rather than the understanding of the work flow
  - Inflexible processes

- Periodically reporting time at completion and cost at completion but not value (or benefits) at completion
- Project continuation rather than canceling projects with limited or no value

Over the years, we have taken several small steps to plan for the use of project management on nontraditional projects. This included:

- Project managers are provided with more business knowledge and are allowed to provide an input during the project selection process.
- Because of the preceding item, project managers are brought on board the project at the beginning of the initiation phase rather than the end of the initiation phase.
- Projects managers now seem to have more of an understanding of technology rather than a command of technology.

The new types of projects combined with a heavy focus on business alignment and value brought with it a classification system, as shown in Figure 16–5.

- *Operational Projects:* These projects, for the most part, are repetitive projects such as payroll and taxes.

They are called "projects," but they are managed by functional managers without the use of an enterprise project management methodology.

- *Enhancement or Internal Projects:* These are projects designed to update processes, improve efficiency and effectiveness, and possibly improve morale.

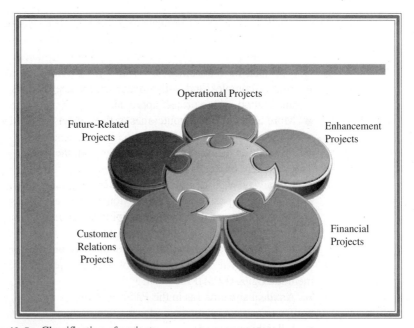

**Figure 16–5.** Classification of projects.

● *Financial Projects:* Companies require some form of cash flow for survival. These are projects for clients external to the firm and have an assigned profit margin.
● *Future-Related Projects:* These are long-term projects to produce a future stream of products or services capable of generating a future cash flow. These projects may be an enormous cash drain for years with no guarantee of success.
● *Customer-Related Projects:* Some projects may be performed, even at a financial loss, to maintain or build a customer relationship. However, performing too many of these projects can lead to financial disaster.

These new types of projects focus more on value than on the triple constraint. Figure 16–6 shows the traditional triple constraint, whereas Figure 16–7 shows the value-driven triple constraint. With the value-driven triple constraint, we emphasize stakeholder satisfaction and decisions are made around the four types of projects (excluding operational projects) and the value that is expected on the project. In others words, success is when the value is obtained, hopefully within the triple constraint. As a result, we can define the four cornerstones of success using Figure 16–8. Very few projects are completed without some trade-offs. This holds true for both the tradition projects and the value-driven projects. As shown in Figure 16–9, traditional trade-offs result in an elongation of the schedule and an increase in the budget. The same holds true for the value-driven projects shown in Figure 16–10. The major difference is performance. With traditional trade-offs, we tend to reduce performance to satisfy other requirements. With value-driven projects, we tend to increase performance in hopes of providing added value, and this tends to cause much larger cost overruns and schedule slippages than with traditional trade-offs. Projects managers generally do not have the sole authority for scope/performance increases or decreases. For

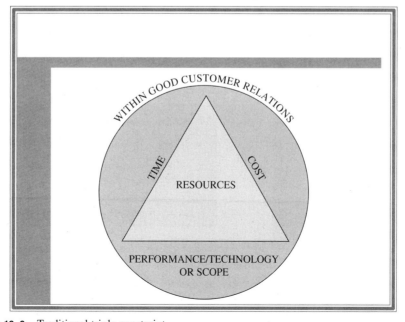

**Figure 16–6.** Traditional triple constraint.

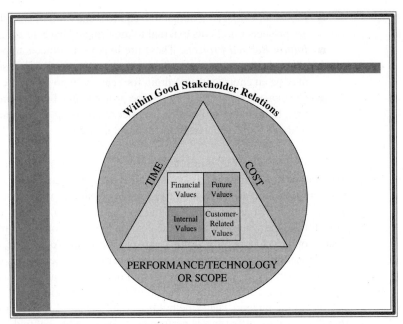

**Figure 16–7.** Value-driven triple constraint.

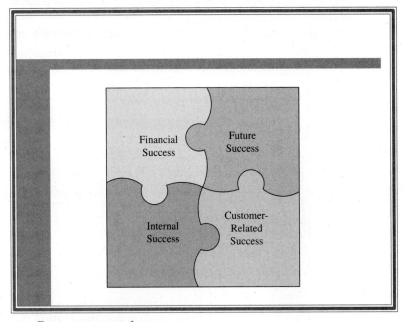

**Figure 16–8.** Four cornerstones of success.

**Figure 16–9.** Traditional trade-offs.

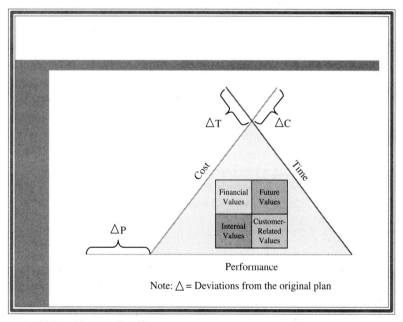

**Figure 16–10.** Value-driven trade-offs.

traditional trade-offs, the project manager and the project sponsor, working together, may have the authority to make trade-off decisions.

However, for value-driven projects all or most of the stakeholders may need to be involved. This can create additional issues, such as:

- It may not be possible to get all of the stakeholders to agree on a value target during project initiation.
- Getting agreement on scope changes, extra costs, and schedule elongations is significantly more difficult the further along you are in the project.
- Stakeholders must be informed of this at project initiation and continuously briefed as the project progresses; that is, no surprises!

Conflicts among the stakeholders may occur. For example:

- During project initiation, conflicts among stakeholder are usually resolved in favor of the largest financial contributors.
- During execution, conflicts over future value are more complex, especially if major contributors threaten to pull out of the project.

For projects that have a large number of stakeholders, project sponsorship may not be effective with a single-person sponsor. Therefore, committee sponsorship may be necessary. Membership in the committee may include:

- Perhaps a representative from all stakeholder groups
- Influential executives
- Critical strategic partners and contractors
- Others based upon the type of value

Responsibilities for the sponsorship committee may include:

- Taking a lead role in the definition of the targeted value
- Taking a lead role in the acceptance of the actual value
- Ability to provide additional funding
- Ability to assess changes in the enterprise environment factors
- Ability to validate and revalidate the assumptions

Sponsorship committees may have significantly more expertise than the project manager in defining and evaluating the value in a project.

Value-driven projects require that we stop focusing on budgets and schedules and instead focus on how value will be captured, quantified, and reported. Value must be measured in terms of what the project contributes to the company's objectives. To do this, an understanding of four terms is essential.

- *Benefits:* An advantage.
- *Value:* What the benefit is worth.

- *Business drivers:* Target goals or objectives defined through benefits or value and expressed more in business terms than technical terms.
- *Key performance indicators (KPIs):* Value metrics that can be assessed either quantitatively or qualitatively.

Traditionally, business plans identified the benefits expected from the project. Today, portfolio management techniques require identification of the value as well as the benefits. However, conversion from benefits to value is not easy.[5] Table 16–3 illustrates the benefit-to-value conversion. Also, as shown in Figure 16–11, there are shortcomings in the conversion process that can make the conversion difficult.

We must identify the business drivers, and they must have measurable performance indicators using KPI. Failure to do so may make true assessment of value impossible. Table 16–4, illustrates typical business drivers and the accompanying KPIs.

KPIs are metrics for assessing value. With traditional project management, metrics are established by the enterprise project management methodology and fixed for the duration

**TABLE 16–3. MEASURE VALUE FROM BENEFITS**

| Expected Benefits | Value Conversion |
| --- | --- |
| Profitability | Easy |
| Customer satisfaction | Hard |
| Goodwill | Hard |
| Penetrate new markets | Easy |
| Develop new technology | Medium |
| Technology transfer | Medium |
| Reputation | Hard |
| Stabilize work force | Easy |
| Utilize unused capacity | Easy |

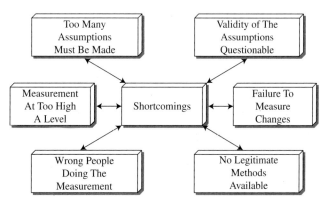

**Figure 16–11.** Shortcomings.

5. For additional information on the complexities of conversion, see J. J. Phillips, T. W. Bothell and G. L. Snead, *The Project Management Scorecard*, Oxford, UK: Butterworth Heinemann, 2002, Chapter 13.

**TABLE 16–4.   BUSINESS DRIVERS AND KPI**

| Business Drivers | Key Performance Indicators |
| --- | --- |
| Sales growth | Monthly sales or market share |
| Customer satisfaction | Monthly surveys |
| Cost savings | Earned-value measurement system |
| Process improvement | Time cards |

**TABLE 16–5.   MEASURING VALUES**

| Easy (Soft/Tangible) Values | Hard (Intangible) Values |
| --- | --- |
| Return on investment (ROI) calculators | Stockholder satisfaction |
| Net present value (NPV) | Stakeholder satisfaction |
| Internal rate of return (IRR) | Customer satisfaction |
| Cash flow | Employee retention |
| Payback period | Brand loyalty |
| Profitability | Time-to-market |
| Market share | Business relationships |
| | Safety |
| | Reliability |
| | Reputation |
| | Goodwill |
| | Image |

of the projects life cycle. But with value-driven project management, metrics can change from project to project, during a life-cycle phase and over time because of:

- The way the company defines value internally
- The way the customer and contractor jointly define success and value at project initiation
- The way the customer and contractor come to an agreement at project initiation as to what metrics should be used on a given project
- New or updated versions of tracking software
- Improvements to the enterprise project management methodology and accompanying project management information system
- Changes in the enterprise environmental factors

Even with the best possible metrics, measuring value can be difficult. Some values are easy to measure while others are more difficult. The easy values to measure are often called soft or tangible values, whereas the hard values are often considered intangible values. Table 16–5 illustrates some of the easy and hard values to measure. Table 16–6 shows some of the problems associated with measuring both hard and soft values.

The intangible elements are now considered by some to be more important than tangible elements. This appears to be happening on IT projects where executives are giving significantly more attention to intangible values. The critical issue with intangible values is not necessarily in the end result, but in the way that the intangibles were calculated.[6]

---

6. For additional information on the complexities of measuring intangibles, see note 5, Chapter 10. The authors emphasize that the true impact on a business must be measured in business units.

**TABLE 16–6.  PROBLEMS WITH MEASURING VALUES**

| Easy (Soft/Tangible) Values | Hard (Intangible) Values |
| --- | --- |
| Assumptions are often not disclosed and can affect decision making. | Value is almost always based upon subjective-type attributes of the person doing the measurement. |
| Measurement is very generic. | It is more of an art than a science. |
| Measurement never meaningfully captures the correct data. | Limited models are available to perform the measurement. |

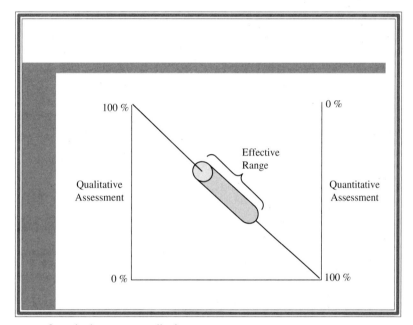

**Figure 16–12.**  Quantitative versus qualitative assessment.

Tangible values are usually expressed quantitatively, whereas intangible values are expressed through a qualitative assessment. There are three schools of thought for value measurement:

- *School 1:* The only thing that is important is ROI.
- *School 2:* ROI can never be calculated effectively; only the intangibles are what are important.
- *School 3:* If you cannot measure it, then it does not matter.

The three schools of thought appear to be an all-or-nothing approach where value is either 100 percent quantitative or 100 percent qualitative. The best approach is most likely a compromise between a quantitative and qualitative assessment of value. It may be necessary to establish an effective range, as shown in Figure 16–12, which is a compromise among the three schools of thought. The effective range can expand or contract.

The timing of value measurement is absolutely critical. During the life cycle of a project, it may be necessary to switch back and forth from qualitative to quantitative assessment and, as stated previously, the actual metrics or KPIs can change as well. Certain critical questions must be addressed:

- When or how far along the project life cycle can we establish concrete metrics, assuming it can be done at all?
- Can value be simply perceived and therefore no value metrics are required?
- Even if we have value metrics, are they concrete enough to reasonably predict actual value?
- Will we be forced to use value-driven project management on all projects or are there some projects where this approach is not necessary?
  - Well defined versus ill defined
  - Strategic versus tactical
  - Internal versus external
- Can we develop a criterion for when to use value-driven project management or should we use it on all projects but at a lower intensity level?

For some projects, assessing value at closure may be difficult. We must establish a time frame for how long we are willing to wait to measure the value or benefits from a project. This is particularly important if the actual value cannot be identified until some time after the project has been completed. Therefore, it may not be possible to appraise the success of a project at closure if the true economic values cannot be realized until some time in the future.

Some practitioners of value measurement question whether value measurement is better using boundary boxes instead of life-cycle phases. For value-driven projects, the potential problems with life-cycle phases include:

- Metrics can change between phases and even during a phase.
- Inability to account for changes in the enterprise environmental factors.
- Focus may be on the value at the end of the phase rather than the value at the end of the project.
- Team members may get frustrated not being able to quantitatively calculate value.

Boundary boxes, as shown in Figure 16–13, have some degree of similarity to statistic process control charts. Upper and lower strategic value targets are established. As long as the KPIs indicate that the project is still within the upper and lower value targets, the project's objectives and deliverables will not undergo any scope changes or trade-offs.

Value-driven projects must undergo value health checks to confirm that the project will make a contribution of value to the company. Value metrics, such as KPIs, indicate the current value. What is also needed is an extrapolation of the present into the future. Using tradition project management combined with the traditional enterprise project management methodology, we can calculate the time at completion and the cost at completion. These are common terms that are part of earned-value measurement systems. But as stated previously, being on time and within budget is no guarantee that the perceived value will be there at project completion.

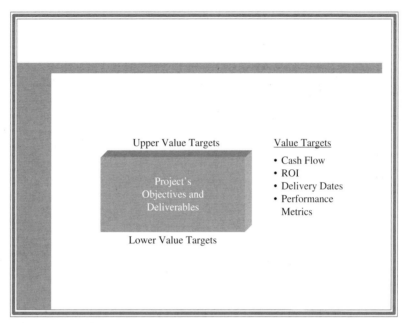

**Figure 16–13.** The boundary box.

Therefore, instead of using an enterprise project management methodology, which focuses on earned-value measurement, we may need to create a VMM which stresses the value variables. With VMM, time to complete and cost to complete are still used, but we introduce a new term entitled value (or benefits) at completion. Determination of value at completion must be done periodically throughout the project. However, periodic reevaluation of benefits and value at completion may be difficult because:

● There may be no reexamination process.
● Management is not committed and believes that the reexamination process is unreal.
● Management is overoptimistic and complacent with existing performance.
● Management is blinded by unusually high profits on other projects (misinterpretation).
● Management believes that the past is an indication of the future.

An assessment of value at completion can tell us if value trade-offs are necessary. Reasons for value trade-offs include:

● Changes in the enterprise environmental factors
● Changes in the assumptions
● Better approaches found, possibly with less risk
● Availability of highly skilled labor
● Breakthrough in technology

As stated previously, most value trade-offs are accompanied by an elongation of the schedule. Two critical factors that must be considered before schedule elongation takes place are:

- Elongating a project for the desired or added value may incur risks.
- Elongating a project consumes resources which may have already been committed to other projects in the portfolio.

Traditional tools and techniques may not work well on value-driven projects. The creation of a VMM may be necessary to achieve the desired results. A VMM can include the features of earned-value measurement systems (EVMSs) and enterprise project management systems (EPMs), as shown in Table 16–7. But additional variables must be included for the capturing, measurement, and reporting of value.

**TABLE 16–7.   COMPARISON OF EVMS, EPM, AND VMM**

| Variable | EVMS | EPM | VMM |
|----------|------|-----|-----|
| Time | ✓ | ✓ | ✓ |
| Cost | ✓ | ✓ | ✓ |
| Quality | | ✓ | ✓ |
| Scope | | ✓ | ✓ |
| Risks | | ✓ | ✓ |
| Tangibles | | | ✓ |
| Intangibles | | | ✓ |
| Benefits | | | ✓ |
| Value | | | ✓ |
| Trade-offs | | | ✓ |

# 17 Effect of Mergers and Acquisitions on Project Management

## 17.0 INTRODUCTION

All companies strive for growth. Strategic plans are prepared identifying new products and services to be developed and new markets to be penetrated. Many of these plans require mergers and acquisitions to obtain the strategic goals and objectives. Yet even the best-prepared strategic plans often fail. Too many executives view strategic planning as planning only, often with little consideration given to implementation. Implementation success is vital during the merger and acquisition process.

## 17.1 PLANNING FOR GROWTH

Companies can grow in two ways: internally and externally. With internal growth, companies cultivate their resources from within and may spend years attaining their strategic targets and marketplace positioning. Since time may be an unavailable luxury, meticulous care must be given to make sure that all new developments fit the corporate project management methodology and culture.

External growth is significantly more complex. External growth can be obtained through mergers, acquisitions, and joint ventures. Companies can purchase the expertise they need very quickly through mergers and acquisitions. Some companies execute occasional acquisitions, whereas other companies have sufficient access to capital such that they can perform continuous acquisitions. However, once again, companies often neglect to consider the impact on project management. Best practices in project management may not be transferable from one company to another. The impact on project management systems resulting from mergers and acquisitions is often irreversible, whereas joint ventures can be terminated.

**Effect of Mergers and Acquisitions on Project Management**

This chapter focuses on the impact on project management resulting from mergers and acquisitions. Mergers and acquisitions allow companies to achieve strategic targets at a speed not easily achievable through internal growth, provided that the sharing of assets and capabilities can be done quickly and effectively. This synergistic effect can produce opportunities that a firm might be hard-pressed to develop themselves.

Mergers and acquisitions focus on two components: preacquisition decision making and postacquisition integration of processes. Wall Street and financial institutions appear to be interested more in the near-term financial impact of the acquisition rather than the long-term value that can be achieved through better project management and integrated processes. During the mid-1990s, companies rushed into acquisitions in less time than the company required for a capital expenditure approval. Virtually no consideration was given to the impact on project management and whether or not the expected best practices would be transferable. The result appears to have been more failures than successes.

When a firm rushes into an acquisition, very little time and effort appear to be spent on postacquisition integration. Yet, this is where the real impact of best practices is felt. Immediately after an acquisition, each firm markets and sells products to each other's customers. This may appease the stockholders, but only in the short term. In the long term, new products and services will need to be developed to satisfy both markets. Without an integrated project management system where both parties can share the same best practices, this may be difficult to achieve.

When sufficient time is spent on preacquisition decision making, both firms look at combining processes, sharing resources, transferring intellectual property, and the overall management of combined operations. If these issues are not addressed in the preacquisition phase, unrealistic expectations may occur during the postacquisition integration phase.

## 17.2 PROJECT MANAGEMENT VALUE-ADDED CHAIN

Mergers and acquisitions are expected to add value to the firm and increase its overall competitiveness. Some people define value as the ability to maintain a certain revenue stream. A better definition of value might be defined as the competitive advantages that a firm possesses as a result of customer satisfaction, lower cost, efficiencies, improved quality, effective utilization of personnel, or the implementation of best practices. True value occurs *only* in the postacquisition integration phase, well after the actual acquisition itself.

Value can be analyzed by looking at the value chain: the stream of activities from upstream suppliers to downstream customers. Each component in the value chain can provide a competitive advantage and enhance the final deliverable or service. Every company has a value chain, as illustrated in Figure 17–1. When a firm acquires a supplier, the value chains are combined and expected to create a superior competitive position. Similarly, the same result is expected when a firm acquires a downstream company. But it may not be possible to integrate the best practices.

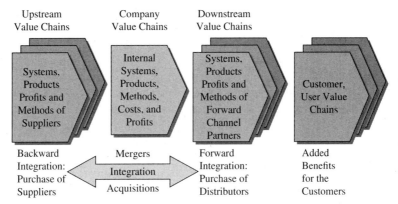

**Figure 17–1.** Generic value-added chain.

**Figure 17–2.** Project management value-added chain.

Historically, value chain analysis was used to look at a business as a whole.[1] However, for the remainder of this chapter, the sole focus will be the project management value-added chain and the impact of mergers and acquisitions on the performance of the chain.

Figure 17–2 shows the project management value-added chain. The primary activities are those efforts needed for the physical creation of a product or service. The primary activities can be considered to be the five major process areas of project management: project initiation, planning, execution, control, and closure.

---

1. M. E. Porter, *Competitive Advantage*, New York: Free Press, 1985, Chap. 2.

The support activities are those company-required efforts needed for the primary activities to take place. At an absolute minimum, the support activities must include:

- *Procurement management:* The quality of the suppliers and the products and services they provide to the firm.
- *Effect of mergers and acquisitions on project management:* The ability to combine multiple project management approaches, each at a different level of maturity.
- *Technology development:* The quality of the intellectual property controlled by the firm and the ability to apply it to products and services both offensively (new product development) or defensively (product enhancements).
- *Human resource management:* The ability to recruit, hire, train, develop, and retain project managers. This includes the retention of project management intellectual property.
- *Supportive infrastructure:* The quality of the project management systems necessary to integrate, collate, and respond to queries on project performance. Included within the supportive infrastructure are the project management methodology, project management information systems, total quality management system, and any other supportive systems. Since customer interfacing is essential, the supportive infrastructure can also include processes for effective supplier–customer interfacing.

These four support activities can be further subdivided into nine of the ten knowledge areas of the *PMBOK® Guide*. The arrows connecting the nine *PMBOK® Guide* areas indicate their interrelatedness. The exact interrelationships may vary for each project, deliverable, and customer.

Each of these primary and support activities, together with the nine process areas, is required to convert material received from your suppliers into deliverables for your customers. In theory, Figure 17–2 represents a work breakdown structure for a project management value-added chain:

- *Level 1:* Value chain
- *Level 2:* Primary activities
- *Level 3:* Support activities(which can include the Stakeholder Management knowledge area)
- *Level 4:* Nine of the ten *PMBOK® Guide* knowledge areas

The project management value-added chain allows a firm to identify critical weaknesses where improvements must take place. This could include better control of scope changes, the need for improved quality, more timely status reporting, better customer relations, or better project execution. The value-added chain can also be useful for supply chain management. The project management value-added chain is a vital tool for continuous improvement efforts and can easily lead to the identification of best practices.

Executives regard project costing as a critical, if not the most critical, component of project management. The project management value chain is a tool for understanding a project's cost structure and the cost control portion of the project management methodology. In most firms, this is regarded as a best practice. Actions to eliminate or reduce a cost or schedule disadvantage need to be linked to the location in the value chain where the cost or schedule differences originated.

The "glue" that ties together elements within the project management chain is the project management methodology. A project management methodology is a grouping of forms, guidelines, checklists, policies, and procedures necessary to integrate the elements within the project management value-added chain. A methodology can exist for an individual process such as project execution or for a combination of processes. A firm can also design its project management methodology for better interfacing with upstream or downstream organizations that interface with the value-added chain. Ineffective integration at supplier–customer interface points can have a serious impact on supply chain management and future business.

## 17.3 PREACQUISITION DECISION MAKING

The reason for most acquisitions is to satisfy a strategic and/or financial objective. Table 17–1 shows the six most common reasons for an acquisition and the most likely strategic and financial objectives. The strategic objectives are somewhat longer term than the financial objectives that are under pressure from stockholders and creditors for quick returns.

The long-term benefits of mergers and acquisitions include:

- Economies of combined operations
- Assured supply or demand for products and services
- Additional intellectual property, which may have been impossible to obtain otherwise
- Direct control over cost, quality, and schedule rather than being at the mercy of a supplier or distributor
- Creation of new products and services
- Pressure on competitors through the creation of synergies
- Cost cutting by eliminating duplicated steps

Each of these can generate a multitude of best practices.

The essential purpose of any merger or acquisition is to create lasting value that becomes possible when two firms are combined and value that would not exist separately.

**TABLE 17–1.  TYPES OF OBJECTIVES**

| Reason for Acquisition | Strategic Objective | Financial Objective |
| --- | --- | --- |
| Increase customer base | Bigger market share | Bigger cash plow |
| Increase capabilities | Provide solutions | Wider profit margins |
| Increase competitiveness | Eliminate costly steps | Stable earnings |
| Decrease time-to-market (new products) | Market leadership | Earnings growth |
| Decrease time-to-market (enhancements) | Broad product lines | Stable earnings |
| Closer to customers | Better price–quality–service mix | Sole-source procurement |

The achievement of these benefits, as well as attaining the strategic and financial objectives, could rest on how well the project management value-added chains of both firms integrate, especially the methodologies within their chains. Unless the methodologies and cultures of both firms can be integrated, and reasonably quickly, the objectives may not be achieved as planned.

Project management integration failures occur after the acquisition happens. Typical failures are shown in Figure 17–3. These common failures result because mergers and acquisitions simply cannot occur without organizational and cultural changes that are often disruptive in nature. Best practices can be lost. It is unfortunate that companies often rush into mergers and acquisitions with lightning speed but with little regard for how the project management value-added chains will be combined. Planning for better project management should be of paramount importance, but unfortunately is often lacking.

The first common problem area in Figure 17–3 is the inability to combine project management methodologies within the project management value-added chains. This occurs because of:

- A poor understanding of each other's project management practices prior to the acquisition
- No clear direction during the preacquisition phase on how the integration will take place
- Unproven project management leadership in one or both of the firms
- A persistent attitude of "we–them"

Some methodologies may be so complex that a great amount of time is needed for integration to occur, especially if each organization has a different set of clients and different types of projects. As an example, a company developed a project management methodology to provide products and services for large publicly held companies. The company

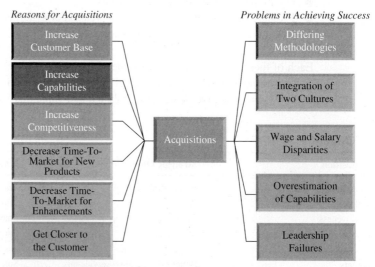

**Figure 17–3.** Project management problem areas after an acquisition.

then acquired a small firm that sold exclusively to government agencies. The company realized too late that integration of the methodologies would be almost impossible because of requirements imposed by the government agencies for doing business with the government. The methodologies were never integrated and the firm servicing government clients was allowed to function as a subsidiary, with its own specialized products and services. The expected synergy never took place.

Some methodologies simply cannot be integrated. It may be more prudent to allow the organizations to function separately than to miss windows of opportunity in the marketplace. In such cases, "pockets" of project management may exist as separate entities throughout a large corporation.

The second major problem area in Figure 17–3 is the existence of differing cultures. Although project management can be viewed as a series of related processes, it is the working culture of the organization that must eventually execute these processes. Resistance by the corporate culture to support project management effectively can cause the best plans to fail. Sources for the problems with differing cultures include:

- A culture in one or both firms that has limited project management expertise (i.e., missing competencies)
- A culture that is resistant to change
- A culture that is resistant to technology transfer
- A culture that is resistant to transfer of any type of intellectual property
- A culture that will not allow for a reduction in cycle time
- A culture that will not allow for the elimination of costly steps
- A culture that must "reinvent the wheel"
- A culture in which project criticism is viewed as personal criticism

Integrating two cultures can be equally difficult during both favorable and unfavorable economic times. People may resist any changes in their work habits or comfort zones, even though they recognize that the company will benefit by the changes.

Multinational mergers and acquisitions are equally difficult to integrate because of cultural differences. Several years ago, a U.S. automotive supplier acquired a European firm. The American company supported project management vigorously and encouraged its employees to become certified in project management. The European firm provided very little support for project management and discouraged its workers from becoming certified, using the argument that their European clients do not regard project management in such high esteem as do General Motors, Ford, and Chrysler. The European subsidiary saw no need for project management. Unable to combine the methodologies, the U.S. parent company slowly replaced the European executives with American executives to drive home the need for a singular project management approach across all divisions. It took almost five years for the complete transformation to take place. The parent company believed that the resistance in the European division was more of a fear of change in its comfort zone than a lack of interest by its European customers.

During the past 40 years, Philip Morris has systematically pursued a diversification strategy to reduce its dependence on cigarettes. In addition to its cigarette business, it owns Clark Chewing Gum, Kraft General Foods, Miller, Miller Lite, Lowenbrau, Jell-O, Oscar Meyer, Maxwell House, Sealtest, Orowheat Baked Goods, and Louis Kemp Seafood. The strategy of Philip Morris was to acquire well-established industry leaders. Each organization acquired had a distinctive corporate culture. Yet even though all of the organizations were successful, some of the expected synergies were not realized because of cultural dissimilarities. Each acquisition was then treated as a stand-alone organization. Although it could be argued that there was no reason to merge these firms into one culture, it does show that even highly successful firms are often resistant to change.

Sometimes there are clear indications that the merging of two cultures will be difficult. When Federal Express acquired Flying Tiger, the strategy was to merge the two into one smoothly operating organization. At the time of the merger, Federal Express employed a younger workforce, many of whom were part time. Flying Tiger had full-time, older, longer-tenured employees. Federal Express focused on formalized policies and procedures and a strict dress code. Flying Tiger had no dress code and management conducted business according to the chain of command, where someone with authority could bend the rules. Federal Express focused on a quality goal of 100 percent on-time delivery, whereas Flying Tiger seemed complacent with a 95–96 percent target. Combining these two cultures had to be a monumental task for Federal Express. In this case, even with these potential integration problems, Federal Express could not allow Flying Tiger to function as a separate subsidiary. Integration was mandatory. Federal Express had to address quickly those tasks that involved organizational or cultural differences.

Planning for cultural integration can also produce favorable results. Most banks grow through mergers and acquisitions. The general belief in the banking industry is to grow or be acquired. During the 1990s, National City Corporation of Cleveland, Ohio, recognized this and developed project management systems that allowed National City to acquire other banks and integrate the acquired banks into National City's culture in less time than other banks allowed for mergers and acquisitions. National City viewed project management as an asset that has a very positive effect on the corporate bottom line. Many banks today have manuals for managing merger and acquisition projects.

The third problem area in Figure 17–3 is the impact on the wage and salary administration program. The common causes of the problems with wage and salary administration include:

- Fear of downsizing
- Disparity in salaries
- Disparity in responsibilities
- Disparity in career path opportunities
- Differing policies and procedures
- Differing evaluation mechanisms

When a company is acquired and integration of methodologies is necessary, the impact on the wage and salary administration program can be profound. When an acquisition takes place, people want to know how they will benefit individually, even though they know that the acquisition is in the best interest of the company.

The company being acquired often has the greatest apprehension about being lured into a false sense of security. Acquired organizations can become resentful to the point of physically trying to subvert the acquirer. This will result in value destruction, where self-preservation becomes of paramount importance to the workers, often at the expense of the project management systems.

Consider the following situation. Company A decides to acquire company B. Company A has a relatively poor project management system in which project management is a part-time activity and not regarded as a profession. Company B, on the other hand, promotes project management certification and recognizes the project manager as a full-time, dedicated position. The salary structure for the project managers in company B was significantly higher than for their counterparts in company A. The workers in company B expressed concern that "We don't want to be like them," and self-preservation led to value destruction.

Because of the wage and salary problems, company A tried to treat company B as a separate subsidiary. But when the differences became apparent, project managers in company A tried to migrate to company B for better recognition and higher pay. Eventually, the pay scale for project managers in company B became the norm for the integrated organization.

When people are concerned with self-preservation, the short-term impact on the combined value-added project management chain could be severe. Project management employees must have at least the same, if not better, opportunities after acquisition integration as they did prior to the acquisition.

The fourth problem area in Figure 17–3 is the overestimation of capabilities after acquisition integration. Included in this category are:

- Missing technical competencies
- Inability to innovate
- Speed of innovation
- Lack of synergy
- Existence of excessive capability
- Inability to integrate best practices

Project managers and those individuals actively involved in the project management value-added chain rarely participate in preacquisition decision-making. As a result, decisions are made by managers who may be far removed from the project management value-added chain and whose estimates of postacquisition synergy are overly optimistic.

The president of a relatively large company held a news conference announcing that his company was about to acquire another firm. To appease the financial analysts attending the news conference, he meticulously identified the synergies expected from the combined operations and provided a timeline for new products to appear on the marketplace. This announcement did not sit well with the workforce, who knew that the capabilities were overestimated and that the dates were unrealistic. When the product launch dates were missed, the stock price plunged and blame was placed erroneously on the failure of the integrated project management value-added chain.

The fifth problem area in Figure 17–3 is leadership failure during postacquisition integration. Included in this category are:

- Leadership failure in managing change
- Leadership failure in combining methodologies
- Leadership failure in project sponsorship
- Overall leadership failure
- Invisible leadership

● Micromanagement leadership
● Believing that mergers and acquisitions must be accompanied by major restructuring

Managed change works significantly better than unmanaged change. Managed change requires strong leadership, especially with personnel experienced in managing change during acquisitions.

Company A acquires company B. Company B has a reasonably good project management system but with significant differences from company A. Company A then decides that "We should manage them like us," and nothing should change. Company A then replaces several company B managers with experienced company A managers. This was put into place with little regard for the project management value-added chain in company B. Employees within the chain in company B were receiving calls from different people, most of whom were unknown to them and were not provided with guidance on who to contact when problems arose.

As the leadership problem grew, company A kept transferring managers back and forth. This resulted in smothering the project management value-added chain with bureaucracy. As expected, performance was diminished rather than enhanced.

Transferring managers back and forth to enhance vertical interactions is an acceptable practice after an acquisition. However, it should be restricted to the vertical chain of command. In the project management value-added chain, the main communication flow is lateral, not vertical. Adding layers of bureaucracy and replacing experienced chain managers with personnel inexperienced in lateral communications can create severe roadblocks in the performance of the chain.

Any of the problem areas, either individually or in combination with other problem areas, can cause the chain to have diminished performance, such as:

● Poor deliverables
● Inability to maintain schedules
● Lack of faith in the chain
● Poor morale
● Trial by fire for all new personnel
● High employee turnover
● No transfer of project management intellectual property

## 17.4 LANDLORDS AND TENANTS

Previously, it was shown how important it is to assess the value chain, specifically the project management methodology, during the preacquisition phase. No two companies have the same value chain for project management as well as the same best practices. Some chains function well; others perform poorly.

For simplicity sake, the "landlord" will be the acquirer and the "tenant" will be the firm being acquired. Table 17–2 identifies potential high-level problems with the landlord–tenant relationship as identified in the preacquisition phase. Table 17–3 shows possible postacquisition integration outcomes.

**TABLE 17–2. POTENTIAL PROBLEMS WITH COMBINING METHODOLOGIES BEFORE ACQUISITIONS**

| Landlord | Tenant |
|---|---|
| Good methodology | Good methodology |
| Good methodology | Poor methodology |
| Poor methodology | Good methodology |
| Poor methodology | Poor methodology |

**TABLE 17–3. POSSIBLE INTEGRATION OUTCOMES**

| Methodology | | |
|---|---|---|
| Landlord | Tenant | Possible Results |
| Good | Good | Based upon flexibility, good synergy achievable; market leadership possible at a low cost |
| Good | Poor | Tenant must recognize weaknesses and be willing to change; possible cultural shock |
| Poor | Good | Landlord must see present and future benefits; strong leadership essential for quick response |
| Poor | Poor | Chances of success limited; good methodology may take years to get |

The best scenario occurs when both parties have good methodologies and, most important, are flexible enough to recognize that the other party's methodology may have desirable features. Good integration here can produce a market leadership position.

If the landlord's approach is good and the tenant's approach is poor, the landlord may have to force a solution upon the tenant. The tenant must be willing to accept criticism, see the light at the end of the tunnel, and make the necessary changes. The changes, and the reasons for the changes, must be articulated carefully to the tenant to avoid cultural shock.

Quite often a company with a poor project management methodology will acquire an organization with a good approach. In such cases, the transfer of project management intellectual property must occur quickly. Unless the landlord recognizes the achievements of the tenant, the tenant's value-added chain can diminish in performance and there may be a loss of key employees.

The worst-case scenario occurs when neither the landlord nor the tenant have good project management systems. In this case, all systems must be developed anew. This could be a blessing in disguise because there may be no hidden bias by either party.

## 17.5 BEST PRACTICES: CASE STUDY ON JOHNSON CONTROLS, INC.[2]

The Automotive Systems Group (ASG) of Johnson Controls, Inc. (JCI) is one of the best-managed organizations in the world, with outstanding expertise in the management of the project value-added chain. ASG is a global supplier of automotive interior systems and batteries.

---

2. D. Kandt, formerly Group Vice President, Quality, Program Management and Continuous Improvement, and Alok Kumar of the Automotive Systems Group of Johnson Controls, Inc., provided much of the information in the remainder of this section. Originally published in Harold Kerzner, *Advanced Project Management: Best Practices on Implementation*, Hoboken, NJ: Wiley, 2004, pp. 520–524.

During the 1990s, JCI automotive business expanded more than 10-fold, with some of the growth coming from strategic acquisitions. ASG was increasingly working on development projects that required multiple teams in multiple locations developing products for a single customer. It was clear that ASG needed to have one common global project management methodology or process that would allow *all* ASG employees and teams to communicate better and improve the efficiency of the development process. Without this, the results could have been devastating. ASG was no longer supplying simply products. It was now providing complete solutions for its customers: namely, the interior of the vehicle.

**Johnson Controls, Inc. Automotive Systems Group Integration with Prince and Becker Corporation**

JCI purchased both Prince and Becker approximately 10 years ago, with the intent of becoming an integrated interior supplier to the automotive industry. The addition of Prince's overhead systems, door panel, and instrument panel capabilities to Becker's interior plastic trim capability and ASG seating products established ASG immediately as an interior supplier. Organizational changes were necessary to truly integrate the companies and their capabilities. The result was a new business model from which the company would be reorganized into vehicle platforms so as to provide complete solutions for its customers rather than merely components.

All three companies had different project management systems, project management position descriptions, and organizational structures. For example, Becker in Europe had project accounters, which were combined project managers and sales personnel. Prince, on the other hand, in addition to project managers, had project coordinators who handled the administrative functions for the project managers. JCI followed a traditional approach, with a well-defined matrix organization and project managers who were fully responsible for all aspects of project execution and success. JCI decided quickly that the three project management systems had to be integrated and commonized. This proved to be a difficult but productive task.

**Integrating the Methodologies** Johnson Controls' Automotive Systems Group had a 15-year history of project management with a project management methodology that was well integrated with total quality control. Between 1995 and 2000, ASG had received more than 164 quality awards from its customers and other organizations, and much of the credit was given to the way in which projects were managed. This system had nine life-cycle phases. Prince's system was referred to as a new product development/product development process and had seven life-cycle phases. Becker's system had six life-cycle phases and was called the Projekt Management Handbuch (Project Management Handbook). Not only were the systems different, the languages were different. Each company thought that its system was superior and should be adopted by the entire company. Reaching agreement just on position titles required extensive discussion.

**Corporate Cultures** In addition to different organizations and systems, the three companies had different values and all were understandably proud of their own ways of operating. Prince Corporation had a very well integrated culture, with a strong focus on the leadership that had been provided by the founder, Ed Prince. Prince Corp. (located in Holland, Michigan) had never had any real need to consider European interests, especially those of Becker, which had traditionally been a competitor or supplier. JCI had a very strong presence in Europe, with its central office in Burscheid, Germany. The Burscheid office followed North America's lead for project management systems but had strong opinions on how these systems should function in the European culture and followed a much more laissez-faire approach to managing projects. In keeping with the European culture and traditional separation of European countries, the various ASG development centers located in different countries in Europe operated pretty much on their own, with little centralized influence. The end result was a lot of opinions on how to integrate the project management systems, with some common principles and some widely varying values.

**Product Focus** In addition to the differences listed above, the purchase of Prince and Becker was intended to bring JCI into an automotive total interior systems position with the vehicle companies. This meant that not only differences in culture, organization, and systems had to be overcome, but also differences in product, equipment, and core manufacturing processes. ASG had to find a way to commonize. In addition, the ASG had to position the new systems to allow the development and launch of total vehicle interiors, a fundamentally different scope for all three of the companies involved. The newly developed integrated project management value-added chain would be a completely new approach for all three companies.

**The Integration** A team was created to integrate the various systems, organization, and values of the three companies. Project managers from all three companies, including North America and Europe, were appointed. Also, representatives from all functional departments were represented. Representatives from Quality, Engineering, Manufacturing, and Finance were all part of the team and were able to influence the direction.

The team was challenged to create a project management methodology (and multinational project management value-added chain) that would achieve the following goals:

- Combine best practices from all existing project management methodologies and project management value-added chains
- Create a methodology that encompasses the entire project management value-added chain from suppliers to customers
- Meet the industry standards established by the Automotive Industry Action Group (AIAG), Project Management Institute (PMI), and International Organization for Standardization (ISO)
- Share best practices among all ASG global locations
- Achieve the corporate launch goals of timing, cost, quality, and efficiency
- Accommodate all ASG products
- Optimize procedures, deliverables, roles, and responsibilities
- Provide clear and useful documentation

The system that evolved, called PLUS (Product Launch System), incorporated all three companies' ideas and best practices.[3] PLUS had five phases (plus an initial

---

3. The complexities in the development and roll-out of PLUS and the mapping of PLUS against PMBOK® can be found in R. Spigarelli and C. Allen, "Implementation of an Automotive Product Launch System," *Proceedings of the Project Management Institute Annual Seminars and Symposiums*, November 1–10. 2001, Nashville, Tennessee.

phase called phase 0 for ideation, which was Prince's product creation phase). All three companies were able to identify with this new system and provided the basic project management structure for JCI to pursue its strategic target of providing vehicle interiors. PLUS repackaged and improved the three existing systems into a new process with optimized procedures, roles, responsibilities, and deliverables. The new system was also painstakingly mapped to the Quality System Requirements QS9000 (AIAG, March 1998), Advanced Product Quality Planning (APQP) and Control Plan (AIAG, June 1994), the Automotive Project Management Guide (AIAG, 1997), and *PMBOK® Guide* (2000).

There were still some problems that had to be resolved. This became evident when teams with more than one year of development time were required to convert to PLUS. PLUS was introduced to the workers through an online PLUS Overview Course that was not sufficient to prepare the teams for using PLUS. Additional training programs were then put in place with customization for particular audiences of workers. The training program provided guidance on why PLUS was developed, how to apply it, and procedures for suggested improvements.

**The Outcome** PLUS was well received by all three companies. More than 350 programs were using PLUS. The team that created the new system was given the Chairman's Award for Customer Satisfaction. ASG now had a basic system that allowed everyone in the company to utilize a common language for product development and launch. In fact, this development paved the way for an organizational change that came one year later, referred to as the New Business Model, which reorganized the company into vehicle platform teams, emphasizing JCI's strategic direction to become a total vehicle interior supplier.

This change allowed ASG to function as an integrated company. The platform teams (project offices) had become true teams and felt a sense of identity based on the entire vehicle rather than on doors, seats, and cockpits. The major metrics, such as timing, profitability, and customer relations, were now viewed from the perspective of the entire vehicle. The customers were happier because they now had a single point of contact for their vehicle instead of multiple points of contact.

**Follow-on Work** The benefits of PLUS were clear: integrated system, common roles and responsibilities, and the ability to manage projects as integrated platforms. The downside was that PLUS was designed by a committee. Because everyone had a vote, the system was somewhat more bulky and less elegant than it could be. The next step was obvious—PLUS needed to be simplified and streamlined. Out of respect for the strong opinions and cultures of the three companies, PLUS was allowed to remain as it was for approximately 1 to 1.5 years. After that, a central group of project management directors took a minimalist approach to PLUS, removed some unnecessary content, focused on the critical deliverables, and made it much simpler and user friendlier. The revised system, launched in October 2001, focused on the critical deliverables and on a one-deliverable, one-owner principle. The results were well accepted in Europe and North America.

The system is revised twice a year at an interval of six months. The focus continues on making it intuitive and keeping it flexible, driven by critical deliverables and having clear accountability. Proper integration of methodologies, along with possibly the entire project management value-added chain, can provide excellent benefits. At JCI, the following benefits were found:

- Common terminology across the entire organization
- Unification of all companies
- Common forms and reports
- Guidelines for less experienced project managers and team members
- Clearer definition of roles and responsibilities
- Reduction in the number of procedures and forms
- No duplication in reporting
- Reduction in the number of timeline items from 184 to 110

**Recommendations for Other Companies**     The following recommendations can be made:

- *Use a common written system for managing programs.* If new-companies are acquired, bring them into the basic system as quickly as is reasonable.
- *Respect all parties.* You cannot force one company to accept another company's systems. There has to be selling, consensus, and modifications.
- *It takes time to allow different corporate cultures to come together.* Pushing too hard will simply alienate people. Steady emphasis and pushing by management are ultimately the best way to achieve integration of systems and cultures.
- *Sharing management personnel among the merging companies helps to bring the systems and people together quickly.*
- *There must be a common "process owner" for the project management system.* A person on the vice-presidential level would be appropriate.

## 17.6 INTEGRATION RESULTS

The best prepared plans do not necessarily guarantee success. Reevaluation is always necessary. Evaluating the integrated project management value added after acquisition and integration is completed can be done using the modified Boston Consulting Group Model (BCG), shown in Figure 17–4. The two critical parameters are the perceived value to the company and the perceived value to customers.

If the final chain has a low perceived value to both the company and the customers, it can be regarded as a "dog." The characteristics of a dog include:

- There is a lack of internal cooperation, possibly throughout the entire value-added chain.
- The value chain does not interface well with the customers.
- The customer has no faith in the company's ability to provide the required deliverables.
- The value-added chain processes are overburdened with excessive conflicts.
- Preacquisition expectations were not achieved, and the business may be shrinking.

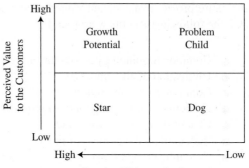

**Figure 17–4.** Project management system after acquisition.

*Possible strategies to use with a dog include:*
- Downsize, descope, or abandon the project management value-added chain.
- Restructure the company to either a projectized or departmentalized project management organization.
- Allow the business to shrink and focus on selected projects and clients.
- Accept the position of a market follower rather than a market leader.

The *problem child* quadrant in Figure 17–4 represents a value-added chain which has a high perceived value to the company but is held in low esteem by customers.

*The characteristics of a problem child include:*
- The customer has some faith in the company's ability to deliver but no faith in the project management value-added chain.
- Incompatible systems may exist within the value-added chain.
- Employees are still skeptical about the capability of the integrated project management value-added chain.
- Projects are completed more on a trial-by-fire basis rather than on a structured approach.
- Fragmented pockets of project management may still exist in both the landlord and the tenant.

*Possible strategies for a problem child value chain include:*
- Invest heavily in training and education to obtain a cooperative culture.
- Carefully monitor cross-functional interfacing across the entire chain.
- Seek out visible project management allies in both the landlord and the tenant.
- Use of small breakthrough projects may be appropriate.

The *growth-potential* quadrant in Figure 17–4 has the potential to achieve preacquisition decision-making expectations. This value-added chain is perceived highly by both the company and its clients.

*The characteristics of a growth potential value-added chain include:*
- Limited, successful projects are using the chain.
- The culture within the chain is based upon trust.
- Visible and effective sponsorship exists.
- Both the landlord and the tenant regard project management as a profession.

*Possible strategies for a growth potential project management value-added chain include:*
- Maintain slow growth leading to larger and more complex projects.
- Invest in methodology enhancements.
- Begin selling complete solutions to customers rather than simply products or services.
- Focus on improved customer relations using the project management value-added chain.

In the final quadrant in Figure 17–4 the value chain is viewed as a *star*. This has a high perceived value to the company but a low perceived value to the customer. The reason for the customer's low perceived value is because you have already convinced the customer of the ability of your chain to deliver, and your customers now focus on the deliverables rather than the methodology.

*The characteristics of a star project management value-added chain include:*
- A highly cooperative culture exists.
- The triple constraint is satisfied.
- Your customers treat you as a partner rather than as a contractor.

Potential strategies for a star value-added chain include:

- Invest heavily in state-of-the-art supportive subsystems for the chain.
- Integrate your PMIS into the customer's information systems.
- Allow for customer input into enhancements for your chain.

## 17.7 VALUE CHAIN STRATEGIES

At the beginning of this chapter the focus was on the strategic and financial objectives established during preacquisition decision making. However, to achieve these objectives, the company must understand its competitive advantage and competitive market after acquisition integration. Four generic strategies for a project management value-added chain are shown in Figure 17–5. The company must address two fundamental questions concerning postacquisition integration: Will the organization now compete on cost or uniqueness of products and services?

- Will the postacquisition marketplace be broad or narrow?

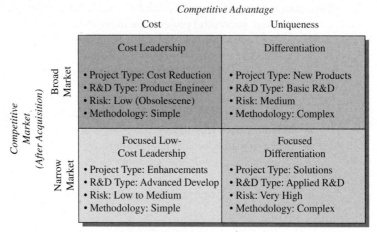

**Figure 17–5.** Four generic strategies for project management.

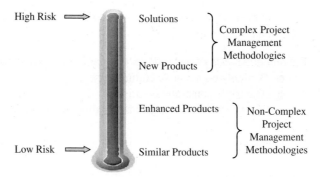

**Figure 17–6.** Risk spectrum for type of project.

The answer to these two questions often dictates the types of projects that are ideal for the value-added chain project management methodology. This is shown in Figure 17–6. Low-risk projects require noncomplex methodologies, whereas high-risk projects require complex methodologies. The complexity of the methodology can have an impact on the time needed for postacquisition integration. The longest integration time occurs when a company wants a project management value-added chain to provide complete solution project management, which includes product and service development, installation, and follow-up. It can also include platform project management, as was the case with Johnson Controls, Inc. Emphasis is on customer satisfaction, trust, and follow-on work.

Project management methodologies are often a reflection of a company's tolerance for risk. As shown in Figure 17–7, companies with a high tolerance for risk develop project management value-added chains capable of handling complex R&D projects and become market leaders. At the other end of the spectrum are enhancement projects that focus on maintaining market share and becoming a follower rather than a market leader.

**Figure 17–7.** Risk spectrum for the types of R&D projects.

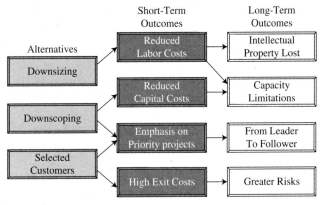

**Figure 17–8.** Restructuring outcomes.

## 17.8 FAILURE AND RESTRUCTURING

Great expectations often lead to great failures. When integrated project management value-added chains fail, the company has three viable but undesirable alternatives:

- Downsize the company.
- Downsize the number of projects and compress the value-added chain.
- Focus on a selected customer business base.

The short- and long-term outcomes for these alternatives are shown in Figure 17–8.
Failure often occurs because the preacquisition decision-making phase was based on illusions rather than fact. Typical illusions include:

- Integrating project management methodologies will automatically reduce or eliminate duplicated steps in the value-added chain.
- Expertise in one part of the project management value-added chain could be directly applicable to upstream or downstream activities in the chain.

- A landlord with a strong methodology in part of its value-added chain could effectively force a change on a tenant with a weaker methodology.
- The synergy of combined operations can be achieved overnight.
- Postacquisition integration is a guarantee that technology and intellectual property will be transferred.
- Postacquisition integration is a guarantee that all project managers will be equal in authority and decision making.

Mergers and acquisitions will continue to take place regardless of whether the economy is weak or strong. Hopefully, companies will now pay more attention to postacquisition integration and recognize the potential benefits.

# Index

# If you enjoyed this book, you may also like these:

**Project Management**
**by Harold Kerzner**
ISBN: 9781118022276

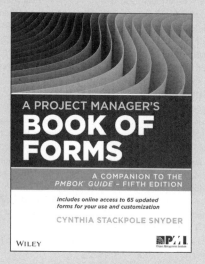

**A Project Manager's**
**Book of Forms**
**by Cynthia Stackpole Snyder**
ISBN: 9781118430781

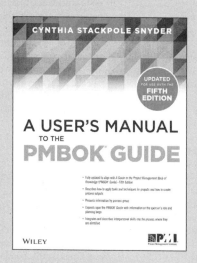

**A User's Manual to**
**the PMBOK® Guide**
**by Cynthia Stackpole Snyder**
ISBN: 9781118431078

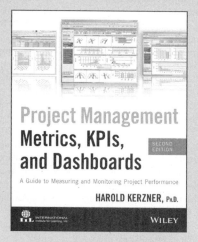

**Project Management**
**Metrics, KPIs, and**
**Dashboards**
**by Harold Kerzner**
ISBN: 9781118524664